SOL-GEL TECHNOLOGY
FOR THIN FILMS, FIBERS, PREFORMS, ELECTRONICS, AND SPECIALTY SHAPES

MATERIALS SCIENCE AND PROCESS TECHNOLOGY SERIES

Editors

Rointan F. Bunshah, University of California, Los Angeles *(Materials Science and Technology)*

Gary E. McGuire, Microelectronics Center of North Carolina *(Electronic Materials and Processing)*

DEPOSITION TECHNOLOGIES FOR FILMS AND COATINGS; **Developments and Applications:** by *Rointan F. Bunshah et al*

CHEMICAL VAPOR DEPOSITION FOR MICROELECTRONICS; **Principles, Technology, and Applications:** by *Arthur Sherman*

SEMICONDUCTOR MATERIALS AND PROCESS TECHNOLOGY HANDBOOK; **For Very Large Scale Integration (VLSI) and Ultra Large Scale Integration (ULSI):** edited by *Gary E. McGuire*

SOL-GEL TECHNOLOGY FOR THIN FILMS, FIBERS, PREFORMS, ELECTRONICS, AND SPECIALTY SHAPES: edited by *Lisa C. Klein*

HYBRID MICROCIRCUIT TECHNOLOGY HANDBOOK; **Materials, Processes, Design, Testing and Production:** by *James J. Licari* and *Leonard R. Enlow*

HANDBOOK OF THIN FILM DEPOSITION PROCESSES AND TECHNIQUES; **Principles, Methods, Equipment and Applications:** edited by *Klaus K. Schuegraf*

Related Titles

ADHESIVES TECHNOLOGY HANDBOOK: by *Arthur H. Landrock*

HANDBOOK OF THERMOSET PLASTICS: edited by *Sidney H. Goodman*

HANDBOOK OF CONTAMINATION CONTROL IN MICROELECTRONICS; **Principles, Applications and Technology:** edited by *Donald L. Tolliver*

SOL-GEL TECHNOLOGY FOR THIN FILMS, FIBERS, PREFORMS, ELECTRONICS, AND SPECIALTY SHAPES

1988

Edited by

Lisa C. Klein

Center for Ceramics Research
College of Engineering
Rutgers—The State University of New Jersey
Piscataway, New Jersey

np NOYES PUBLICATIONS
Park Ridge, New Jersey, U.S.A.

Library of Congress Catalog Card Number: 87-34780
ISBN: 0-8155-1154-X
Printed in the United States

Published in the United States of America by
Noyes Publications
Mill Road, Park Ridge, New Jersey 07656

10 9 8 7 6 5 4 3 2 1

Library of Congress Cataloging-in-Publication Data

Sol-Gel technology for thin films, fibers, preforms, electronics, and speciality shapes.

Bibliography: p.
Includes index.
1. Ceramic materials. 2. Glass fibers.
3. Thin films. 4. Colloids. I. Klein, Lisa C.
TP862.S65 1988 666'.15 87-34780
ISBN 0-8155-1154-X

To my daughter, Martha Ann Kinsella, who was born while the book was in progress.

and

To Dennis Ravaine, my scientific collaborator and friend, who passed away suddenly in 1986.

Preface

This book covers the principles, developments, techniques, and applications of sol-gel processing. The sol-gel process is not new, however, a few commercial successes in the recent past have revived interest. The commercial successes are largely in the area of thin films. These films have been developed for optical, mechanical and electrical applications. About one-third of this book covers thin films.

The second area where there has been commercial success is fibers. These fibers whether spun or drawn may be continuous or woven. The applications realized and projected are refractories, composite reinforcement and thermal insulation. About one-third of this book covers fibers.

The third area encompasses the special applications such as preforms, microballoons and electronics. Discussion of the chemistry, polymerization, drying and characterization are all necessary parts of a treatment of sol-gel processing.

The anticipated product of this effort is a book that covers the background and fundamentals. Also, it evaluates the present technology and projects new directions short range and long range.

The graduate students at Rutgers University, P. Anderson, H. deLambilly, T. Gallo, T. Lombardi and J. Ryan, are thanked for their editorial assistance. Visiting scientists Jean-Yves Chane-Ching and Henry Wautier served as reviewers.

Lisa C. Klein

Center for Ceramics Research
Rutgers–The State University of New Jersey
Piscataway, New Jersey
December 1987

Contributors

Carol S. Ashley
Sandia National Laboratories
Albuquerque, New Mexico

John B. Blum
Norton Company
Northboro, Massachusetts

Jean Pierre Boilot
Groupe de Chimie du Solide
Laboratoire de Physique de la Matiere Condensee
Ecole Polytechnique
Palaiseau, France

C. Jeffrey Brinker
Sandia National Laboratories
Albuquerque, New Mexico

Richard K. Brow
Department of Materials Science and Engineering
The Pennsylvania State University
University Park, Pennsylvania

Lee A. Carman
Department of Materials Science and Engineering
The Pennsylvania State University
University Park, Pennsylvania

Philippe Colomban
Groupe de Chimie du Solide
Laboratoire de Physique de la Matiere Condensee
Ecole Polytechnique
Palaiseau, France

Helmut Dislich
Schott Glaswerke
Mainz, Federal Republic of Germany

Raymond L. Downs
KMS Fusion, Inc.
Ann Arbor, Michigan

Matthias A. Ebner
KMS Fusion, Inc.
Ann Arbor, Michigan

Jochen Fricke
Physikalisches Institut der Universitat Am Habland
Wurzburg, West Germany

Stephen H. Garofalini
Ceramics Department
Rutgers—The State University of New Jersey
Piscataway, New Jersey

Lisa C. Klein
Ceramics Department
Rutgers—The State University of New Jersey
Piscataway, New Jersey

William C. LaCourse
Alfred University
Alfred, New York

Wayne J. Miller
KMS Fusion, Inc.
Ann Arbor, Michigan

Shyama P. Mukherjee
IBM Corporation
Endicott, New York

George F. Neilson
Microgravity Science and Applications Group
Jet Propulsion Laboratory
California Institute of Technology
Pasadena, California

Richard B. Pettit
Sandia National Laboratories
Albuquerque, New Mexico

Carlo G. Pantano
Department of Materials Science and Engineering
The Pennsylvania State University
University Park, Pennsylvania

Eliezer M. Rabinovich
AT&T Bell Laboratories
Murray Hill, New Jersey

Scott T. Reed
Sandia National Laboratories
Albuquerque, New Mexico

Sumio Sakka
Institute for Chemical Research
Kyoto University
Uji, Kyoto-Fu, Japan

Harold G. Sowman
3M
St. Paul, Minnesota

Ian M. Thomas
Lawrence Livermore National Laboratory
University of California
Livermore, California

Michael C. Weinberg
Microgravity Science and Applications Group
Jet Propulsion Laboratory
California Institute of Technology
Pasadena, California

Masayuki Yamane
Department of Inorganic Materials
Tokyo Institute of Technology
Tokyo, Japan

NOTICE

To the best of the Publisher's knowledge the information contained in this publication is accurate; however, the Publisher assumes no liability for errors or any consequences arising from the use of the information contained herein. Final determination of the suitability of any information, procedure, or product for use contemplated by any user, and the manner of that use, is the sole responsibility of the user.

Mention of trade names or commercial products does not consititute endorsement or recommendation for use by the Publisher.

The book is intended for informational purposes only. The reader is warned that caution must always be exercised when dealing with hazardous materials, and expert advice should be obtained at all times when implementation is being considered.

Contents

Part I

Chemistry and Phase Transformations

1

Multicomponent Glasses from the Sol-Gel Process

Ian M. Thomas

Lawrence Livermore National Laboratory
University of California
Livermore, California

HISTORICAL INTRODUCTION

The preparation of multicomponent glasses by the sol-gel process dates from 1950; at that time Della and Rustum Roy[1] prepared a number of compositions from silicon tetraethoxide and metal nitrate salts primarily for phase equilibrium studies where homogeneous samples are essential. This new method gave products which, when melted just once, gave a glass more homogeneous than the best glasses obtained after three successive melting and crushing operations of samples prepared in the conventional manner from individual oxides. Hamilton and Mackenzie in 1960,[2] Luth and Ingamells in 1965,[3] Hamilton and Henderson in 1968[4] and Biggar and O'Hara in 1969[5] all described variations of the same process primarily directed towards the preparation of silicate mixtures for phase equilibrium studies.

An investigation of bulk glass systems by the sol-gel process was started at Owens-Illinois, Inc. in 1967 by Levene and Thomas.[6] This work resulted in the commercialization of several four component sputtering glass target discs in 1969 and six component planar dopant discs a few years later. The process involved a combination of metal alkoxides and metal acetates as oxide sources and was used because of the high purity requirements not specifically for homogeneity. Dried gel products were melted and fabricated by conventional means.

Sol-gel coating systems were extensively studied at Schott Glass starting in the 1950's. Although the initial emphasis appeared to have been on the preparation of single oxide optical coatings, mainly TiO_2 and SiO_2, mixed oxide materials were also investigated and this led to commercial products. The work is described by Schroeder first in 1962[7] and in more detail in 1969;[8] a summary is also given by Dislich.[9] Later both Schroeder and Dislich investigated bulk gel preparation by an all alkoxide route described by Dislich in 1971[10] and resulting in a number of patents.[11,12]

The increasing interest of many investigators in the sol-gel process became apparent in the mid-1970's and the amount of published work has blossomed since that time. In the following sections various aspects of multicomponent glass preparation, properties and uses will be described. The review is restricted to multicomponent glass systems and so the emphasis will naturally be on silicate materials. It is of note, however, that a lot of work has been reported on silica alone and on both single oxide and multicomponent oxide ceramic systems. The ceramic systems have shown some impressive advantages, especially with respect to processing, over their conventional equivalents.

PREPARATION

General

The prime objective in all preparations of multicomponent oxide compositions is to obtain initially a solution of all components in the form of soluble precursor compounds; mixing can then be considered to be at the molecular level and if this level can be retained in the subsequent conversion to oxides a very homogeneous product should result. In most cases it does appear that oxide products prepared in this manner at low temperature are indistinguishable from those obtained in the conventional manner by fusing the relevant oxide mixtures at high temperature. This will be discussed in more detail later.

There are a number of different types of precursor materials that can be used. All should be soluble in organic solvents and easily converted to the relevant oxide preferably by hydrolysis but alternatively by chemical reaction or thermal or oxidative decomposition. Several preparative methods are available dependent on the nature of the starting materials and these are described below.

All-Alkoxide Method

Probably the best starting materials for sol-gel preparations are the class of materials known as metal alkoxides. All metals form alkoxides and they have the following general formula:

$$M(OR)_x$$

where M is the metal, R is an alkyl group, and x is the valence state of the metal.

All metal alkoxides, with two notable exceptions, are rapidly hydrolyzed to the corresponding hydroxide or oxide. The method of hydrolysis can be varied and many times depends on the final use of the product. This will become apparent later but the overall reaction can be represented as follows:

$$M(OR)_x + xH_2O \rightarrow M(OH)_x + xROH$$

$$2M(OH)_x \rightarrow M_2O_x + xH_2O$$

The by-product, ROH, is an aliphatic alcohol and readily removed by volatilization.

The two notable exceptions are the alkoxides of silicon and phosphorus. Silicon alkoxides require an acid or basic catalyst for hydrolysis and even with these the reaction rate is slow. Trialkylphosphates are very difficult to hydrolyze and this precludes their use as a source of phosphorus in sol-gel preparations.

There is also a limited class of compounds known as double alkoxides. These contain two different metals in the same compound and have the general formula:

$$M'_xM''_y(OR)_z$$

where M′ and M″ are metals, R is an alkyl group and x, y and z are integers.

The physical properties of metal alkoxides can be varied by changing the alkyl group and for most metals soluble, and in some cases even liquid, products can be obtained. In addition, many alkoxides are volatile and can readily be purified by distillation; this can lead to very pure oxide products. Double alkoxides have the added advantage of not only being volatile but retaining exact molecular stoichiometry between the metals. An excellent source of information on metal alkoxides is the book by Bradley et al[13] which is highly recommended reading for anyone starting in the sol-gel field.

The simplest method of preparation of multicomponent systems involves making a solution of all the components as alkoxide precursors in a suitable organic solvent and then reacting the solution with water to form the oxide mix. The reaction can be represented for a three component system as follows:

$$M(OR)_a + M'(OR)_b + M''(OR)_c + xH_2O \rightarrow MO_{a/2} + M'O_{b/2} + M''O_{c/2} + yROH$$

This was the method first used by Schroeder[11] and Dislich[10] and has since been used by many other investigators. The reaction is far more complex than the simple hydrolysis shown in the equation above. It involves first hydrolysis of metal alkoxide groups to metal hydroxide groups and subsequent condensation of these groups with each other or with unhydrolyzed alkoxide groups to give products containing M-O-M linkages (metallometalloxane polymers).

$$M{-}OR + H_2O \rightarrow M{-}OH + ROH.$$

$$M{-}OR + M{-}OH \rightarrow M{-}O{-}M + ROH.$$

$$M{-}OH + M{-}OH \rightarrow M{-}O{-}M + H_2O$$

The products can contain one or more metal atoms in the same molecule depending on the relative hydrolysis and condensation reaction rates of the component metal alkoxides. The more alkoxides present in the original mixture the more complex can the polymerization become. Ultimately, the polymeric products become insoluble due to cross-linking and gellation or precipitation results. The complexity of a multicomponent system makes investigation of the reaction mechanisms extremely difficult. Brinker and Scherer[14] give an excellent summary of research in this field and describe their own extensive investigations on polymer growth and gel formation.

The variation in reaction rates especially in the initial hydrolysis can give rise to inhomogeneities in the final product. This is particularly the case for those materials containing silicon which of course includes all silicate glasses. When hydrolysis is rapid, as is the case with the addition of an excess of liquid water, the hydrolysis rate of silicon alkoxides is so slow that they can remain substantially unreacted when all other components in the mixture have already precipitated as oxides. Gross inhomogeneities then result.

One way of avoiding this problem is to carry out the hydrolysis very slowly. The alkoxide mixture can be made up in a suitable solvent then exposed in bulk to atmospheric moisture, or wet alcohol, sometimes containing acidic or basic catalysts, can be slowly added. In both cases, soluble polymerized products are initially formed and this is followed by a viscosity increase and eventual gelation; this process combined with concurrent solvent evaporation can take from days to weeks. The exact method of hydrolysis depends on the final use of the product; at the viscous, but still soluble, stage fibers have been spun from certain compositions; monolithic dried gels have been obtained when the process is taken to the gel stage. Both of these topics are covered in much greater detail in other sections of this book.

Another method for avoiding the variable hydrolysis rate problem was devised by Levene and Thomas[6] in 1972. This involved a partial hydrolysis of the silicon alkoxide (usually the ethoxide) with an equimolar quantity of water using an acid catalyst to give a trialkoxysilanol which remained in solution:

$$(RO)_4Si + H_2O \rightarrow (RO)_3SiOH + ROH.$$

The addition of other alkoxides then followed and this resulted in reaction between them and the silanol derivative to form soluble metallosiloxane derivatives:

$$(RO)_3SiOH + M(OR')_x \rightarrow (RO)_3Si\text{-}O\text{-}M(OR')_{x-1} + R'OH.$$

This reaction is well known and has been used in the preparation of many siloxy-metal monomers[15] and polymers[16] in non sol-gel applications, e.g.:

$$Ti(OC_3H_7)_4 + 4(CH_3)_3SiOH \rightarrow Ti[OSi(CH_3)_3]_4 + 4C_3H_7OH.$$

All metal alkoxides so far investigated react with silanol derivatives to give metallosiloxane products, so all alkoxides subsequently added to the partially hydrolyzed silicate mixture will react provided that there are sufficient silanol groups available for reaction. This is usually no problem in silicate glass compositions which normally contain a major portion of silica. While the formation of these metallosiloxanes is a complex reaction, especially if more than one additional alkoxide is added, it was found that these derivatives were susceptible to further hydrolysis and can be considered to be similar to single alkoxides except that they contain two or more metals in the same compound. Addition of more water ultimately gave homogeneous oxide products.

A similar method for titanium silicate systems was later used by Yoldas[17] in which he prehydrolyzed a titanium alkoxide and then reacted this with silicon tetraethoxide monomer to give a soluble titanosiloxane polymer:

$$Ti(OR)_4 + H_2O \rightarrow (RO)_3TiOH + ROH.$$

$$(RO)_3TiOH + Si(OC_2H_5)_4 \rightarrow (RO)_3Ti\text{-}O\text{-}Si(OC_2H_5)_3 + C_2H_5OH.$$

Further hydrolysis of the product ultimately gave a homogeneous oxide mixture.

It is fairly obvious that this method can be extended to cover the use of any metal alkoxide that can give soluble partially hydrolyzed materials containing reactive hydroxyl groups. The latter can then be reacted with the silicate to form a metallosiloxane polymer.

Alkoxide-Salt Method

For some metals it is inconvenient to use alkoxides because of preparation problems or unavailability and alternative starting materials must be found. This is particularly the case with Group I and Group II elements whose alkoxides are solid, non-volatile and in many cases of low solubility; consequently, they are sometimes difficult to obtain pure. Metal salts provide a viable alternative provided that they are readily converted to the oxide by thermal or oxidative decomposition and are preferably soluble in organic solvents. They also usually can be obtained as a high purity analytical grade. Salts of organic acids, in particular acetates but also formates, citrates, tartrates, etc. are potential candidates. Nitrates are really the only suitable inorganic salts because others, such as chlorides or sulfates, are more thermally stable and the anion may be difficult to remove.

The normal method of sol-gel preparation using salts is first to form a solution of all components which are to be added as alkoxides, as described in the preceding section, and then add one or more salts as solutions in alcohol or, if this is not possible, in the water that is to be used for further hydrolysis. All components are then uniformly dispersed and subsequent gelation should then freeze all elements in a gel network.

The first sol-gel preparations carried out by the Roys[18] used silicon tetraethoxide and solutions of the nitrates of Group I and Group II elements as well as those of aluminum, lead, iron, lanthanum, titanium, zirconium, thorium, nickel and gallium. A major drawback to the preparation of ultrahomogeneous glass using nitrates was pointed out by Roy and McCarthy[19] and this was the tendency for one or more nitrates to crystallize during dehydration thereby destroying homogeneity; this effect was especially bad with sodium, lead and barium nitrates. The solution to the problem was to evaporate a large portion of the water before the solution gelled which was possible if the pH was kept low.

A word of warning should be given on the use of nitrates in preparations involving more than a few hundred grams of material. Nitrates are strong oxidizing agents and there are initially large quantities of oxidizable material present in the system, this can lead to uncontrollable exotherms and even explosions during drying.

The use of acetate salts was first described by Levene and Thomas[6] and these were used in much the same way as nitrates. In some cases acetates are more soluble than the corresponding nitrates for example, those of sodium, potassium and barium, and their use therefore reduces the crystallization tendency described above. Also the explosion hazard is eliminated in large scale preparations. Disadvantages are that they do not thermally degrade as cleanly as nitrates and can be a source of carbonaceous residues; also solutions of many acetates are basic and therefore their use leads to rapid gelation in silicate systems due to the high pH but this can be partly negated by buffering with acetic acid.

Thomas[20] has found that the use of sodium acid tartrate as a sodium source in silicate systems led to stable solutions especially useful for coating applications where gelation must be avoided. This salt is acid and also contains alcohol groups which are potential reactants for alkoxides in the system. No doubt other acid tartrates could be used in like manner.

Nitrates and acetates continue to be the most commonly used salts in preparations described in the recent literature, their use, however, is much less frequent than the all alkoxide method. Some recent examples are Brinker and Mukherjee[21] who used sodium and barium acetates with aluminum, boron and silicon alkoxides; Holand, Plumat and Duvigneaud[22] used magnesium acetate with aluminum and silicon alkoxides; Phalippou, Prassas and Zarzycki[23] prepared lithium aluminosilicates from lithium and aluminum nitrates and ethyl silicate; and some calcium silicates were prepared from calcium nitrate and ethyl silicate by Hayashi and Saito.[24]

Sol-gel preparations involving salts are usually more complex than those with only alkoxides because the hydrolysis of the latter is more readily accomplished than the thermal or oxidative degradation required for the former. A novel use of certain acetates was developed by Thomas[20] mainly to reduce the amount of acetate groups that had to be removed by thermal degradation during oxide conversion. It was found that certain acetates react with some alkoxides to form metallometalloxane derivatives with the liberation of alkyl acetate, the first step in the reaction is as follows:

$$M(OR)_n + M'(OAc)_{n'} \rightarrow (RO)_{n-1}\ M{-}O{-}M'(OAc)_{n'-1} + ROAc$$

where Ac represents the group $-COCH_3$.

The reaction then continues with further reaction of acetate and alkoxide groups to increase molecular weight. The reaction is carried out by heating the reagents together in the absence of solvent and in most cases about 60-80% of the theoretical ester can be distilled out. It is preferable for the mol ratio of alkoxide to acetate to be greater than one and this leads to a soluble and hydrolyzable product that can subsequently be used in sol-gel preparations in a similar manner to a double alkoxide.

Some acetate-alkoxide pairs that can be reacted in this manner are as follows:

Calcium acetate–aluminum alkoxide

Magnesium acetate–aluminum alkoxide

Zinc acetate–aluminum alkoxide

Lead acetate–silicon alkoxide

Lead acetate–titanium alkoxide

Thomas[20] prepared $CaO\text{-}MgO\text{-}Al_2O_3\text{-}SiO_2$ and $PbO\text{-}B_2O_3\text{-}SiO_2$ glass compositions using this method and the reaction was also used by Gurkovich and Blum[25] to prepare monolithic lead titanate.

The acetate-alkoxide reaction is particularly useful for lead silicate containing compositions because lead acetate is particularly prone to leaving carbonaceous residues in the oxide product.

Other Methods

As an alternative to alkoxides or salts there are a few other materials that can be used for certain specific elements. Some metal oxides and hydroxides are soluble in alcohols because of reaction to form partial alkoxides; these solutions can then be used in sol-gel preparations in the same manner as alkoxides. The reactions with alcohols are reversible and solubility can sometimes be increased if the water is removed:

$$MO_x + 2nROH \rightleftharpoons MO_{x-n}(OR)_{2n} + nH_2O$$

$$M(OH)_x + nROH \rightleftharpoons M(OH)_{x-n}(OR)_n + n/2H_2O$$

Ultimately, of course, the fully substituted alkoxide will be formed when all the water is removed.

Examples of compounds that can be used in this manner are the oxides and hydroxides of all Group I metals, boric acid and oxide, phosphoric acid and oxide, lead monoxide and vanadium pentoxide. Perhaps the most commonly used material is boric acid which can be obtained in a high state of purity and is quite soluble in methanol (10%) because of partial alkoxide formation. Many workers have incorporated boron into sol-gel preparations in this manner though it is difficult to avoid loss of boron from the system during the drying process because of the high volatility of trimethylborate. It is better to dissolve the acid in a higher alcohol and remove the water if necessary to increase solubility. Dislich[26] reports that even using ethanol with boric acid there is considerable boron loss because of the volatility of triethylborate.

Phosphoric acid is the only good source of phosphorus as the phosphorus alkoxides are precluded because of stability; it reacts readily with silicon alkoxides in particular to give useful phosphorosiloxane intermediates.[27]

In cases where suitable oxide precursors are not available preparations can be carried out in which all components except one are made up in solution by any of the standard methods and the odd one then added as a solid oxide. The particle size of the latter oxide should be as small as possible and mixing then carried out vigorously enough to coat each particle with the solution of the other components. While the homogeneity is obviously not as good as a true sol-gel material it can be considerably better than a conventional oxide mix. Thomas[28] prepared lead and zinc aluminoborosilicates and borosilicates by suspending the relevant lead or zinc oxide in a solution of the other components then gelling and drying in the normal manner. It was then possible to obtain these low melting glasses free of carbonaceous residues an objective not possible when lead or zinc acetates were used as precursor materials.

Perhaps the ultimate in hybrid systems is the use of colloidal suspensions as precursor materials. These are the ultimate in particle size and dispersion and silica sols in particular have been investigated. These are commercially available in aqueous suspension, acid or base stabilized, with particle size down to 8 nm. The early investigators of sol-gel systems mentioned in the introduction such as Luth and Ingamells[3] and Hamilton and Henderson[4] used aqueous colloidal silica as an alternative to silicon tetraethoxide in aqueous systems with nitrate salts. More recently Rabinovich et al[29] prepared SiO_2-B_2O_3 compositions from aqueous colloidal silica and an aqueous solution of boric acid. It was possible to cast this mixture into molds and prepare large solid sinterable bodies on drying. The system obviously is limited to water soluble precursor reagents for the other components but these mixtures can be gelled by pH control and processed in much the same way as alkoxide derived products.

There is now considerable interest in aqueous metal oxide sols and mixtures in ceramic processing but this is beyond the scope of this article.

PROPERTIES

Homogeneity

The methods used for sol-gel preparations should in theory give very homogeneous products. If one assumes that a typical sol-gel solution prior to conversion contains all the components mixed at or near the molecular level and if this level can be maintained during the subsequent conversion, then the product should be very good. With silicate systems for example one might expect the product after drying at relatively low temperatures to be equivalent to or better than glass "cullet" which is normally prepared from an oxide melt. The first sol-gel preparations were carried out with the sole objective of obtaining homogeneous products and the results were very successful; later work has usually indicated that the homogeneity of the product was excellent although this was rarely the prime objective of the particular study.

A simple analysis of dried gel particles from a sodium aluminosilicate glass batch was carried out by Levene and Thomas.[6] Three samples approximately 10 mg each were removed at random from a 200 g batch which had been heated to 400°C and which was prepared from sodium acetate and aluminum and silicon alkoxides; these were analyzed for SiO_2 and Al_2O_3 with the following results:

SiO_2	63.9%	64.0%	64.5%
Al_2O_3	24.1%	24.3%	24.0%

Later Thomas[19] did the same thing with samples taken from a SiO_2–Al_2O_3–B_2O_3–MgO batch prepared from all alkoxide reagents with the following results:

SiO_2	56.2%	54.3%	55.0%	58.6%	56.1%	57.0%
Al_2O_3	21.8%	21.2%	21.4%	22.6%	21.8%	22.4%
MgO	7.3%	7.5%	7.4%	7.4%	7.2%	7.7%

Even allowing for analytical error these results show good but certainly not excellent homogeneity.

Mukherjee and Mohr[30] measured homogeneity using light scattering. They found that samples of a $Na_2O-B_2O_3-SiO_2$ composition prepared by two different sol-gel methods, all alkoxide and nitrate-alkoxide, when melted without stirring had a scattering intensity of only 1% of that of the best regions in a similar composition prepared in the conventional manner but also without stirring. The gel glasses were also striae-free and of optical quality.

Another study was carried out by Yamane et al.[31] They compared the homogeneity of a TiO_2-SiO_2 gel prepared by various methods with physical mixtures of the individual oxide gels. The method used involved glass formation by the addition of sodium nitrate and measurement of refractive index deviation and transmission of melted crushed samples. No samples were found to be of optical quality but the gel glasses were found to be superior to the physical gel mixture. It was also claimed that base hydrolysis of a physical alkoxide mixture gave better homogeneity than either acid hydrolysis of a physical alkoxide mixture or acid hydrolysis of a titanosiloxane polymer; the latter material is erroneously referred to in the paper as being prepared by the Yoldas method rather than the Levene and Thomas method.[6]

Comparison with Conventional Glass

Perhaps a more significant factor than homogeneity is whether there is any basic difference between sol-gel products and the same products prepared conventionally each under the best conditions. There have been several studies of this type. Vergano[32] concluded that there was little or no difference in the physical properties of $CaO-MgO-Al_2O_3-SiO_2$ and $CaO-Li_2O-Al_2O_3-SiO_2$ compositions prepared from conventional or sol-gel starting materials and melted in the usual manner. Slight differences in viscosity were attributed to the high hydroxyl content of the sol-gel glass due to its method of preparation. Minimal differences were also found in the physical properties of melted TiO_2-SiO_2 glasses by Kamiya and Sakka[33] and in $Na_2O-K_2O-Al_2O_3-B_2O_3-SiO_2$ compositions by Dislich.[10]

However, slight differences in properties such as liquidus temperature, crystallization behavior and phase separation were found by Weinberg and Neilson[34,35] who did an extensive study of the sodium silicate system and by Mukherjee et al[36] with $La_2O_3-SiO_2$, $La_2O_3-Al_2O_3-SiO_2$ and $La_2O_3-ZrO_2-SiO_2$ compositions. Both investigations attributed the differences to sharply different structural factors and high hydroxyl content.

Purity

The purity of sol-gel compositions obviously depends on the purity of the relevant starting materials and the degree of care taken in the conversion to oxide. Metal alkoxides are particularly good reagents for high purity products because many of them are volatile and hence readily purified by distillation. This is especially the case for the alkoxides of boron and silicon which can be fractionally distilled under nitrogen at atmospheric pressure to give extremely high purity materials. Other alkoxides normally require vacuum distillation which, while not as efficient as atmospheric pressure distillation, can give excellent material. In the absence of a suitable alkoxide it is now possible to obtain commercially many metal salts or metal oxides, which can be converted to salts, in super-pure grade. The use of these materials in clean apparatus in a clean room can give excellent high purity material with most common compositions.

Some purity studies have been carried out. Gossink et al[37] prepared SiO_2 and Al_2O_3 from distilled samples of silicon tetraethoxide and aluminum isopropoxide respectively and obtained products in which the major impurity was iron at the 30-50 ppb range. These oxides were to be used for the preparation of fiber optic waveguides where transition metal impurities were particularly detrimental. For a similar use Thomas[38] prepared a $Al_2O_3-SiO_2$ sol-gel glass from the alkoxides which had an iron content of 90 ppb.

FABRICATION AND USE

General

The fabrication and use of materials prepared by the sol-gel process can be divided into three main categories, bulk glass, coatings and fibers. The latter two are well covered in other chapters of this book and will not be considered here. The bulk glass category can be further divided into products that are prepared from dried gel by conventional melting at high temperatures and products that are fabricated at lower temperatures without melting. Each of these will be considered.

Bulk Glass by Melting

One of the first uses of the sol-gel process was in the fabrication of bulk glass by the conventional melting of dried gel. This was described in the introduction for the phase equilibrium studies by Roy and others and for the first commercial products from Owens-Illinois, Inc. While this method gives glass of excellent quality there are disadvantages as well as the obvious advantages of the gel being homogeneous and amorphous leading to lower melting temperatures and no stirring requirement.

The disadvantages are that the gel must be heated very slowly to the melting temperature to ensure that all carbonaceous residues, water and hydroxyl groups are removed. A very seedy and foamy melt can be obtained if this process is not carried out correctly. Another disadvantage is that the raw material cost is very high compared to the cost of the individual oxides and the end result must therefore justify this charge. In Roy's case cost was of little consequence in view of the quality of the product and the quantity prepared. With the sputtering target discs manufactured at Owens-Illinois the improved purity justified the cost, however such would not be the case nowadays because of the ready availability of high purity oxides not available at that time.

Much of the justification of fabrication in this manner will depend on the composition under consideration. High melting, high viscosity glasses are very difficult to melt conventionally but quite readily fabricated using sol-gel starting materials. A good example is the zero expansion composition 92% SiO_2-8% TiO_2. When this was prepared by the sol-gel process and melted at 1590°C a seedy but otherwise homogeneous glass was obtained within 6 hours. A standard batch from the individual oxides was only partially melted after 89 hours at 1590°C.[6]

An interesting potential use of unmelted gel was proposed by Neilson and Weinberg[39] who suggested that gel precursors would probably be the preferred starting materials for preparing glasses in space because of their microhomogeneity and amorphous properties.

Bulk Glass Without Melting

Many attempts have been made to prepare bulk materials directly from gels without going through the melting process. There are a number of potential methods and some have been quite successful. The simplest is the preparation of monolithic pieces by the slow drying of alcogels which have been molded directly from sol-gel solutions. This subject is covered in detail in another chapter of this book but briefly some difficulty arises from the extremely large surface tension forces in effect during drying due to the very small pore sizes in the gel. This makes the preparation of a large crack-free body difficult; more success has been obtained with gels prepared from colloidal solutions in which pore sizes are much larger and surface tension effects much reduced. The difficulties of the system are well summarized by Zarzycki et al[40] who concluded that the method which gives the most consistent results in regard to crack-free samples is that of supercritical drying of the alcogel. This eliminates surface tension effects and gives a product known as an aerogel.[41] Large monolithic dense pieces have been obtained by heating aerogels to high temperatures.[40]

An alternative use for aerogels was described by Thomas.[42] Aerogels of several glass-ceramic compositions were prepared in the normal manner and then crushed to give a very fine powder. This powder was then cold pressed in a steel mold and the green body sintered by heating in the conventional manner. Dense transparent glass samples were obtained at temperatures as low as 800°C, further heating allowed crystallization to take place to give a final molded glass ceramic body. This process avoided the necessity of obtaining completely crack-free aerogels, very much reduced the shrinkage of going from gel to dense product and allowed a number of different shapes to be fabricated by ceramic techniques.

Ceramic techniques were also used by Decottignies et al[43] who hot pressed gels to obtain dense bodies. The compositions investigated were SiO_2, La_2O_3–SiO_2 and B_2O_3–SiO_2 and it was found that careful drying and calcining of the gels prior to pressing was required to ensure removal of carbonaceous residues and water. These were high melting compositions and pressing temperatures of 1400°C were used to obtain dense samples.

Bulk glass can also be of use in a porous granular form rather than as a dense shaped body especially in the catalyst field. The high surface area of aerogels is particularly advantageous and Teichner et al[44] describe a NiO–Al_2O_3–SiO_2 aerogel prepared from nickel acetate and alkoxides which was found to be a very selective catalyst for the partial oxidation of isobutylene.

Commercial Products

Commercialization of multicomponent sol-gel products has been slow. The bulk glass products produced at Owens-Illinois, Inc. in 1970 were discontinued because of cost and it may be that no large quantities of sol-gel glass will ever be produced simply because the raw material and processing costs are very high.

Coatings are much more promising because on a square foot basis raw material and processing costs are low. Schott Glass continues to produce coated products for architectural and other uses and these are presumably profitable. More recently Owens-Illinois and several other suppliers have produced a series of metal silicate solutions used to dope silicon wafers with arsenic, antimony, boron or phosphorus for semi-conductor use.[45]

There may also be commercial internal uses in which sol-gel processing precedes the final manufacturing step for products whose association with the sol-gel process is not immediately apparent.

One particularly promising aspect of sol-gel coatings that appears to be close to commercialization is in the field of optical coatings. The ease with which high purity oxide materials can be laid down from solution is particularly useful. One specific example is the development of a cheap SiO_2-TiO_2 antireflective coating for silicon solar cells by Yoldas and O'Keeffe.[46]

One assumes in view of the wide variety of potential uses that other commercial products will be forthcoming in the near future.

CONCLUSIONS

The variety of preparative methods and ready availability of a large number of starting materials means that almost any glass composition can now be prepared by the sol-gel method. Bulk glass, both porous and dense, coatings and fibers have all been obtained and some have been shown to have superior performance to their conventional equivalents in a number of applications. As is often the case however, theory lags behind application. The chemistry of the polymerization of even silicon tetraethoxide alone has not yet been fully determined; the tetrafunctionality of this material results in early branching and crosslinking to three dimensional structures during polymerization which is also affected by the type of catalyst, acidic or basic, used. Whether the far more complex multicomponent systems will ever be understood remains to be seen.

The potential applications have obviously not yet been fully realized as indicated for example by the disappointing lack of commercialization. However, there appears to be no reason to expect interest in this field to wane. This is a new method for preparing a very old product, a product having a broad variety of uses; research interest should therefore remain high and the field should be fruitful for many years to come.

REFERENCES

1. Roy, D.M. and Roy, R., *Am. Mineralogist* 40, 147 (1955).
2. Hamilton, D.L. and Mackenzie, W.S., *Journ. Petrology* 1, 56 (1960).
3. Luth, W.C. and Ingamells, C.O., *Am. Mineralogist* 50, 255 (1965).
4. Hamilton, D.L. and Henderson, C.M.B., *Min. Mag.* 36, 832 (1968).
5. Biggar, G.M. and O'Hara, M.J., *Min. Mag.* 37, 198 (1969).
6. Levene, L. and Thomas, I.M., U.S. Patent 3,640,093; February 8, 1972; assigned to Owens-Illinois, Inc.
7. Schroeder, H., *Opt. Acta.* 9, 249 (1962).

8. Schroeder, H., *Physics of Thin Films* 5, 87 (1969).
9. Dislich, H. and Hinz, P., *J. Non. Cryst. Solids* 48, 11 (1982).
10. Dislich, H., *Angew. Chem.* 10, 363 (1971).
11. Schroeder, H. and Gliemeroth, G., U.S. Patent 3,597,252; August 3, 1971; assigned to Jenaer Glaserk Schott and Gen.
12. Dislich, H., Hinz, P. and Kaufmann, R., U.S. Patent 3,759,683; Sept. 18, 1973; assigned to Jenaer Glaswerk Schott and Gen.
13. Bradley, D.C., Mehrotra, R.C. and Gaur, D.P., *Metal Alkoxides,* Academic Press (1978).
14. Brinker, C.J. and Scherer, G.W., *J. Non. Cryst. Solids* 70, 301 (1985).
15. Bradley, D.C. and Thomas, I.M., *J. Chem. Soc.* 3404 (1959).
16. Bradley, D.C., Polymeric metal alkoxides, organometalloxanes and organometalloxanosiloxanes, *Inorganic Polymers,* F.G.A. Stone and W.A.G. Graham editors, Academic Press (1962).
17. Yoldas, B.E., *Appl. Opt.* 21, 2960 (1982).
18. Roy, R., *J. Amer. Cer. Soc.* 344 (1969).
19. McCarthy, G.J. and Roy, R., *J. Amer. Cer. Soc.* 639 (1971).
20. Thomas, I.M., unpublished results.
21. Brinker, C.J. and Mukherjee, S.P., *J. Mater. Sci.* 16, 1980 (1981).
22. Holand, W., Plumat, E.R. and Duvigneaud, P.H., *J. Non. Cryst. Solids* 48, 205 (1982).
23. Phalippou, J., Prassas, M. and Zarzycki, J., *J. Non. Cryst. Solids* 48, 17 (1982).
24. Hayashi, A. and Saito, H., *J. Mater. Sci.* 15, 1971 (1980).
25. Gurkovich, S.R. and Blum, J.B., Preparation of monolithic lead titanate by a sol-gel process, *Ultrastructure Processing of Ceramics, Glasses and Composites,* L.L. Hench and D.R. Ulrich editors, John Wiley & Sons (1984).
26. Dislich, H., *Glastech Ber.* 44, 1 (1971).
27. Thomas, I., U.S. Patent 3,767,432; October 23, 1973; assigned to Owens-Illinois, Inc.
28. Thomas, I., U.S. Patent 3,799,754; March 26, 1974; assigned to Owens-Illinois, Inc.
29. Rabinovich, E.M., Johnson, D.W., MacChesney, J.B. and Vogel, E.M., *J. Amer. Ceram. Soc.* 66, 683 (1983).
30. Mukherjee, S.P. and Mohr, R.K., *J. Non. Cryst. Solids* 66, 523 (1984).
31. Yamane, M., Inoue, S. and Nakazawa, K., *J. Non. Cryst. Solids* 48, 153 (1982).
32. Vergano, P., private communication.
33. Kamiya, K. and Sakka, S., *J. Mater. Sci.* 15, 2937 (1980).
34. Weinberg, M.C. and Neilson, G.F., *J. Amer. Ceram. Soc.* 66, 132 (1983).
35. Neilson, G.F. and Weinberg, M.C., *J. Non. Cryst. Solids* 63, 365 (1984).
36. Mukherjee, S.P., Zarzycki, J. and Traverse, J., *J. Mat. Sci.* 11, 341 (1976).
37. Gossink, R.G., et al, *Mat. Res. Bull.* 10, 35 (1975).
38. Thomas, I.M., U.S. Patent 4,028,085; June 7, 1977; assigned to Owens-Illinois, Inc.
39. Neilson, G.F. and Weinberg, M.C., Glass research in space, *Advances in Ceramics* 5, 110 (1983).
40. Zarzycki, J., Prassas, M. and Phalippou, J., *J. Mater. Sci.* 17, 3371 (1982).
41. Kistler, S.S., *J. Phys. Chem.* 36, 52 (1932).

42. Thomas, I.M., U.S. Patent 3,791,808; February 12, 1974; assigned to Owens-Illinois, Inc.
43. Decottignies, M., Phalippou, J. and Zarzycki, J., *J. Mater. Sci.* 13, 2605 (1978).
44. Teichner, S.J., Nicolaon, G.A., Vicarini, M.A. and Gardes, G.E.E., Inorganic oxide aerogels, *Advances in Coll. and Interface. Sci.* 5, 245 (1976).
45. Thomas, I.M. and Tillman, J.J., Ger. Offen. DE 3,247,173, August 4, 1983; assigned to Owens-Illinois, Inc.
46. Yoldas, B.E. and O'Keeffe, T.W., *Appl. Opt.* 18, 3133 (1979).

2

Simulation of the Sol-Gel Process

Stephen H. Garofalini

Ceramics Department
Rutgers–The State University of New Jersey
Piscataway, New Jersey

INTRODUCTION

Sol-gel processing involves hydrolization and polymerization steps. Although polymerization products, reaction rates, and the effects of pH or different additives on these rates may be experimentally obtainable, the actual mechanisms of polymerization are not fully understood and can only be conjectured from experimental results. For instance, reactant species (i.e., monomer) and products (i.e., dimer, n-mer) can be observed in certain experiments,[1,2] but intermediate species and reaction paths are not observed or reported and reaction mechanisms must be inferred. In order to understand the dynamics of the early stages of the polymerization process in better detail, the molecular dynamics (MD) computer simulation technique is being used to study the interactions and kinetic behavior of $Si(OH)_4$ molecules, the mechanisms of reaction between these molecules, and the growth of these molecules into larger clusters. Silicic acid molecules are being used for this study for several reasons. First, although polymerization of SiO_2 in sol-gel processing probably involves molecules which are not fully hydrated at the onset of polymerization, the incorporation of the alkyl groups (R) in the simulations would require potential functions which describe not only Si–O, O–O, Si–Si, Si–H, O–H, and H–H pairs but also potential functions which describe all such combinations with the alkyl groups. This introduction of alkyl groups presents a significant increase in complexity which was not deemed necessary at this stage. Second, the silicic acid monomer, H_4SiO_4, the pyrosilicic acid 'dimer,' $H_6Si_2O_7$, and other Si–O–H containing molecules have been studied using molecular orbital, MO, calculations,[3-5] Such calculations have been compared to the experimental data of structures inferred from gas phase molecule studies and from a large number of studies of silicate minerals. The MD simulations of the monomer and dimer can be compared to the molecular orbital data. More importantly, the potentials used in these simulations can also be used in simulations of bulk silicate glasses with the results being compared to the wealth of experimental data on bulk silicates.

Both molecular dynamics simulations and computational quantum chemistry (molecular orbital calculations) are providing new insights concerning ceramic and glassy materials. Each technique provides a clearer understanding of the properties of these materials at an atomic or molecular level. The computational techniques provide structural and/or dynamic information which not only match experimentally observed or expected features, but also provide additional details which are currently not experimentally obtainable. Also, the gap between physical experiments and computer experiments is closing with the application of picosecond spectroscopies in time resolved relaxation studies[6] and high resolution ESR studies of defect structures at low concentrations (10^{12} defects per cm^3).

MO calculations using pyrosilicic acid molecules have been used to mimic bulk silicates.[3-5,7,8] Additions of cations to these molecules in the MO calculations have been used to provide insight into the structural changes caused by the addition of ions into crystalline and glassy silicates.[7,8] The effect of cations placed near the bridging oxygen in the siloxane (Si–O–Si) bridge on the siloxane bond angle and Si–O bond length has been evaluated.[8] Such results match trends observed in the crystalline analogs of the molecular systems and can be used in new interpretations of the effect of impurity or added species on the local structure of silicate glasses and the resultant implications of those effects on properties such as diffusion, viscosity, and chemical durability.

Molecular dynamics calculations involve a different approach to study atomic or molecular behavior than the MO calculations and can act as a bridge between static or nonthermal studies (as in the MO calculations) and physical experiments. MD simulations have been used for 20 years in studies of motion and structures at an atomic level.[9] However, only in the last 9 years has the technique been used to study oxide glasses,[10] and only in the last 3 years has this technique been applied to studies of the surfaces of oxide glasses.[11-14]

With respect to bulk silicate glasses, MD simulations reproduce many experimentally observed structural and dynamic features. In particular, the tetrahedral coordination of O around Si is found in the simulations, with an Si–BO (BO = bridging oxygen) bond length of 1.62 Å, and an appropriate RDF;[10] the Si–O–Si bond angle distribution correctly varies from 120° to 180°;[15] the frequency spectrum generated in the simulations through the Fourier transform of the velocity autocorrelation function shows the three major features at ~400 cm^{-1}, ~800 cm^{-1}, and ~1200 cm^{-1},[16] as seen in neutron scattering studies. Analysis of the structures of a large number of crystalline silicates indicate an increase in Si–BO bond length with decreasing Si–O–Si bond angle.[17] MD simulations of bulk v–SiO_2 also show this increase in Si–BO bond length with decreasing bond angle.[18] Additional studies indicate specific defects present in bulk tetrahedrally coordinated fluoride and silicate glasses.[19,20]

More recent studies of glass surfaces using the MD technique indicate the validity of this method for reproducing many of the known structural features at the glass surface[11-14] as well as predicting previously unknown structures.[18] In particular, the simulations reproduce the known[21] predominance of O ions at the outer surface of pristine v–SiO_2, with strained siloxane bonds (Si–O–Si bonds) and nonbridging oxygens (NBO).[11] Analysis of a large number of these simulations performed in our lab indicates that the surface concentrations of NBO and strained siloxane bonds observed in the simulated surfaces create a concentration of silanol (Si–OH) groups similar to that experimentally observed on v–SiO_2[22] under the assumption that the surface defect sites observed in the simulations would hydrolyze to silanols in the presence of H_2O. The Si–NBO bond length observed in the simulations is shorter than the normal Si–BO bond length and correlates with a number of experimental and theoretical studies,[23-25] although there is still some question concerning actual bond lengths at the hydroxyl site.[3-5]

Simulations of $R_2O{\cdot}3SiO_2$(R = Li, Na, or K) glass surfaces indicate the preferential rearrangement of the larger alkali ions, K and Na, to the outermost surface, but no such arrangement outward for Li.[12-14] These results correlate with ISS (ion scattering spectroscopy) studies of the outermost monolayers of such glass surfaces.[26,27] The simulations also indicate the effect of even a short time thermal pulse on the distribution of alkali in the surface. Significant temperature increases are believed to occur at the tip of a crack during fracture.[28]

Additional detailed analysis of simulated v–SiO_2 glass surfaces indicate a slight increase in Si–BO bond length in the top several angstroms of the surface[18,29,30] in comparison to the normal Si–BO bond length.

ESR studies of the v–SiO_2 surface fractured under ultrahigh vacuum conditions indicate the presence of E' centers (3 coordinated Si) in the surface[31] which are not observed in surfaces fractured at atmospheric conditions. MD simulations of the v–SiO_2 surface which have a perfect vacuum above the surface also have 3-fold Si forming in the surface,[32] although not to the extent seen in the formation of nonbridging oxygens in the surface.

It should be apparent from the above mentioned comments that molecular dynamics simulations correlate very well with many experimentally observed properties of silicate glasses and glass surfaces. Simulations can indicate trends and are capable of providing information which is difficult to obtain experimentally. Some simulations predict properties which are currently being corroborated experimentally; in other cases, new experimental procedures must be designed. In any case, the simulations can provide new ideas and insight concerning glasses at an atomic level. Application of these computational techniques to more complex systems will be at the forefront of new approaches being developed for studying glasses and ceramics and can provide a strong base for accurate and detailed atomic level studies which can complement physical experiments.

COMPUTATIONAL PROCEDURE

The molecular dynamics technique involves solving Newton's equation of motion for a system of atoms interacting via an assumed interatomic potential function, ϕ. The force, $\vec{F}$, used in the calculations is given as $\vec{F} = -\partial\phi/\partial\vec{r}$.

Atoms are initially given X, Y, and Z coordinates within some specified volume or box. When simulating a few atoms or molecules or a cluster, the box can essentially be infinite; when simulating condensed phases, the box size must be such that appropriate densities are achieved. Time derivatives of the coordinates are used in the algorithms in the solutions of the equations of motion.

From the atomic coordinates, forces, and time derivatives of the positions at a time t, a subsequent configurational and dynamic state of the system at time $t + \Delta t$ can be determined. A series of configurations are generated which present a time evolution of the system of atoms. The actual numerical procedure must be one in which the solutions to the equations of motion does not diverge with time from the true solution. The stability of the integration procedure can be monitored using a constant of motion (such as total energy). The time step which indicates the time separation between consecutive configurations can be used to control the stability of the integration scheme. Time steps are on the order of 10^{-15}-10^{-16} sec. for systems containing light elements (O, H, etc.) and 10^{-12} sec. for systems containing only heavy elements (Ar, Kr, Pt, etc.). That is, the time step should be 0.01 to 0.1 times the vibrational period of the lightest element in the system.

When studying several molecules, system size is not important. However, when studying a bulk-like system, system size is normally restricted to several thousand atoms due to limitations in computer time or storage. In order to remove size effects for large systems, periodic boundary conditions (PBC) are used. With PBC, a central cell of N atoms is constructed and repetitions or images of this cell are present in the directions of periodicity. Only atoms in the central cell are followed and the 'minimum distance' technique is used to calculate force and energy between a central cell atom i and its neighbors in the central cell. In this technique, the separation distance between atom i and its neighbor is taken as the minimum distance between atom i and the neighbor's central cell or image position (whichever is closer). Also, as an atom leaves the central cell by moving across the central cell boundary, its image enters the central cell from the opposite side. PBC would not be necessary in simulations of just a few molecules or isolated clusters.

The temperature, T, of a system of atoms is obtained using the classical relations of temperature to kinetic energy:

$$T = \frac{1}{3Nk} \sum_{i}^{N} m_i \vec{v}_i^{\,2}$$

where m_i = mass of particle i,

N = number of particles in system,

k = Boltzmann's constant,

v_i = velocity of particle i.

The velocity autocorrelation function, (t), is calculated from

$$\gamma(t) = \frac{\langle \sum_i \vec{v}_i\,(t_o).\vec{v}_i(t_o + t)\rangle}{\langle \sum_i \vec{v}_i\,(t_o).\vec{v}_i(t_o)\,\rangle},$$

where $v_i\ (t_o)$ is the velocity of atom i at some initial time, t_o, and $v_i(t_o + t)$ is the velocity of atom i at some later time, $t_o + t$.

The frequency spectrum, $D(\omega)$, is calculated from the Fourier transform of the velocity autocorrelation function, and is given as

$$D(\omega) = \int_0^\infty \gamma(t)\ \cos \omega t\ dt.$$

The pair distribution function, n(r), is calculated from

$$n_i(r) = \frac{1}{N_i} \langle \sum_{i \neq j}^{N} \sum \delta\ (r - r_{ij})\rangle$$

where $n_i(r)$ is the number of neighbors, j, at a distance r from a central atom, i. r_{ij} is the distance between atoms i and j. In the actual calculations, $n_i(r)$ is the number of neighbors between r and $r + \Delta\ r$. N_i is the number of central atoms and the brackets indicate a time average.

Because of the ability to label each atom in the simulations, the time evolution of individual atoms or specific groups or types of atoms can be monitored and evaluated in detail. Thus, the above-mentioned properties can be evaluated for a system as a whole, or for just specific species, such as Si atoms only. For example, the sum over i in the pair distribution function, $n_i(r)$, can be over all atoms as i or over just Si atoms as the i atoms (central atoms), or even over just one particular atom of interest. In addition, all atoms can be looked at as neighbors, j, or specific species can be selected as the neighbors, j. Thus, very specific structural data can be obtained–specific Si–O distances, O–O distances, Si–H distances, etc. In this way the simulations can be used to look at whole system pair distribution functions, similar to experimentally obtained radial distribution functions, as well as much more detail which might otherwise be lost in these distribution functions over a whole system of atoms.

In the work presented here, the modified Born-Mayer-Huggins (BMH) equation[10-16] and the revised Rahman-Stillinger-Lemberg (RSL2) equations[33,34] are being used to simulate the interactions in the silicic acid and pyrosilicic acid molecules.

The BMH equation has been used to simulate ionic systems and has a short range repulsive term, a coulomb term, and dispersion terms (the latter are often ignored). It has previously been used as an effective potential to simulate vitreous silica ($v{-}SiO_2$) and silicates with considerable accuracy.[10-16] It gives the interactions between all atom pairs in the $v{-}SiO_2$ system. With respect to the calculations done here, it is used for the Si–Si, O–Si, O–O, and Si–H interactions. In the Si–H interactions, only repulsive forces are considered for Si–H since the Si–H bond (at 1.48 Å) is not expected to disrupt the Si–O bond and remain stable in the presence of the O. In this work, Si–H forces are meant to keep the Si–O–H bond angle at reasonable values. The BMH equation gives the potential between atoms i and j, separated by a distance r_{ij}, as

$$\phi_{ij} = A_{ij} \exp\left(-\frac{r_{ij}}{\rho_{ij}}\right) + \frac{Z_i Z_j e^2}{r_{ij}} \quad \xi\left(\frac{r_{ij}}{\beta_{ij}}\right),$$

where Z is the ion valence, ρ_{ij} and β_{ij} are constants, and A_{ij} is an adjustable short range repulsive parameter based on ion sizes.

The RSL2 equations are used for the O–H and H–H interactions. These equations are given (in kcal/mol and Å units) as:

$$\phi_{HH}(r) = \frac{36.1345}{r} + \frac{18}{1 + \exp[40r - 2.05]}$$

$$-17\exp[-7.62177(r-45251)^2],$$

$$\phi_{OH}(r) = -\frac{72.269}{r} + \frac{6.23403}{r^{9.19912}} - \frac{10}{1+\exp[40(r-1.05)]}$$

$$-\frac{4}{1+\exp[5.49305(r-2.2)]}$$

The RSL2 potential was developed as an effective central force potential for simulations of water. The advantage of using this potential as opposed to other water potentials in our simulations is because dissociation of a water molecule is allowed by the RSL2 potential. Thus, specific O–H and H–H pair interactions can exist which are necessary in our simulation of the H_4SiO_4 and $H_6Si_2O_7$ molecules. Although an O–H interaction may be different in an Si–O–H combination than in an H–O–H combination, the RSL2 water potential is nonetheless an excellent starting point for these simulations.

The RSL2 potential was not altered in any of the simulations. Changes in potentials were only made to the BMH equation. Any changes made to the BMH equation for simulations of the H_4SiO_4 and $H_6Si_2O_7$ molecules were evaluated in simulations of bulk v–SiO_2 and compared with previous simulations of v–SiO_2 as well as experimental data. In this way, no changes in potentials were made which would generate properties of v–SiO_2 which would be inconsistent with existing data.

The H_4SiO_4 and $H_6Si_2O_7$ molecules are drawn schematically in Figure 1. The largest circles are oxygen, next largest are silicon, and smallest are hydrogen. In the MD simulations of H_4SiO_4, the coordinates of a Si, four O and four H were introduced into the computer in a distorted $Si(OH)_4$ configuration, with the 4 oxygens surrounding the Si and the hydrogens on the outside, away from the Si, one H near each O. Only O and H were allowed to move in this molecule (Si held immobile). In the simulations of the $H_6Si_2O_7$ molecule, the atoms were set up in a configuration similar to the 'dimer' in Figure 1. In the $H_6Si_2O_7$ molecule, one Si atom was not allowed to move, while all other atoms were allowed to move.

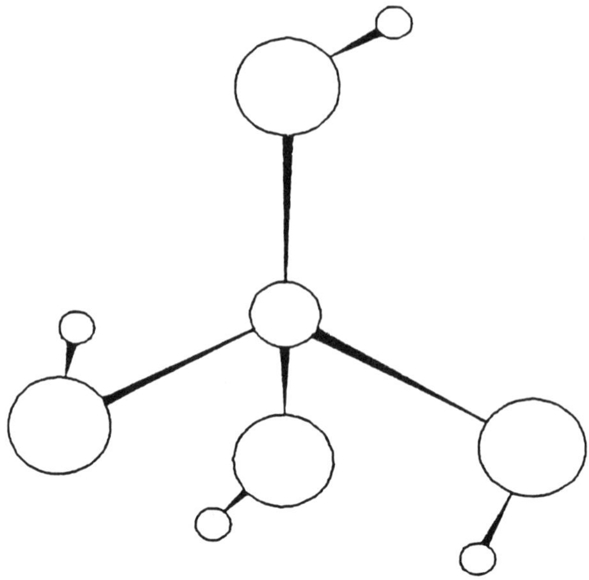

SILICIC ACID (MONOMER)
H_4SiO_4

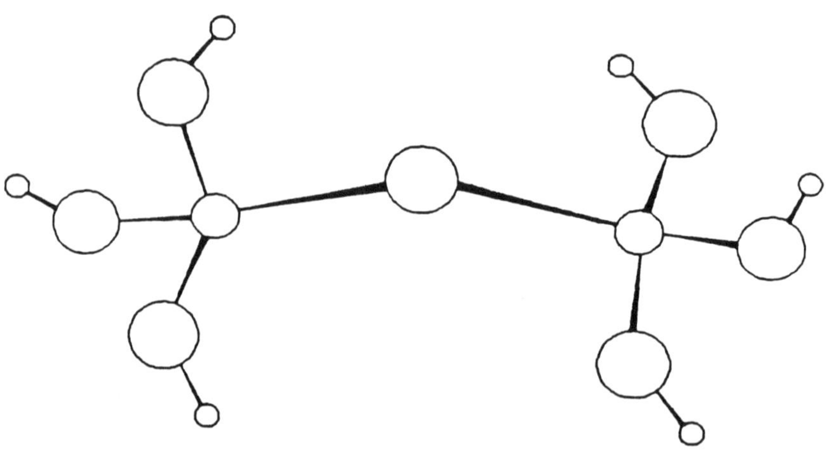

PYROSILICIC ACID (DIMER)
$H_6Si_2O_7$

Figure 1.

The forces were calculated on each moving atom due to *all* of the other atoms and moved an incremental distance in response to the resultant force. Each calculation of forces and movement of all moving atoms was considered a "move" or "time step." The time of each time step was 1.0×10^{-16} sec. Runs of at least 20,000 moves were performed.

Since these simulations use central force potentials, the H ion can be located anywhere around the O ion at a distance of approximately 0.96 Å. Attraction of the H ion to a nonbonding, or second neighbor, O may bring the H ion too close to the Si. In order to prevent this and keep the Si–O–H bond angle at reasonable values, repulsive Si–H interactions were used. The Si–O–H angle is believed to be 113° at the hydroxylated vitreous silica surface[35] and 129° in the H_4SiO_4 molecule.[5] In some simulations, only the short range repulsive portion of the BMH equation was used for the Si–H interactions; in other cases, both the short range term and the coulomb term were used. In the latter case, to keep the potential repulsive, the Si and H charges were +4 and +1 respectively.

Only the A_{ij} and β_{ij} parameters were altered in the BMH equations. Other parameters were kept constant. The ρ_{ij} parameters were constant at 0.29×10^{-8} cm for all pairs. $Z_{Si} = +4$, $Z_O = -2$, and, when the coulomb term for Si–H interaction was used, $Z_H = +1$.

RESULTS AND DISCUSSION

Table 1 shows bond length changes in the H_4SiO_4 monomer with changes in the A_{ij} parameter for Si–O; the coulomb portion of the BMH equation was not used for Si–H interactions in this case. A_{Si-H} was 9.95×10^{-9} ergs, and all β_{ij} values were 2.50×10^{-8} cm. The A_{Si-O} parameter equal to 3.25×10^{-9} ergs gave Si–O distances similar to those reported in MO studies of the monomer[5] and Si–O–H bond angles near 125° to 130° at 50°K to 100°K. However, the A_{Si-O} value of 2.96×10^{-9} ergs is that which is used in simulations of bulk silicates in order to obtain the 1.62 Å Si–O bond distance in those systems. Subsequent studies using the larger A_{Si-O} value in simulations of bulk v–SiO_2 gave results which were not as good as those using the 2.96×10^{-9} erg value.

Table 1: Effect of Changes in Repulsive Parameter on the Bond Lengths in the $Si(OH)_4$ Molecule

A_{Si-O} ($\times 10^{-9}$ ergs)	A_{Si-H} ($\times 10^{-9}$ ergs)	T°K	r_{Si-O} (Å)
4.00	9.95	100	>1.70
3.50	9.95	100	1.65
3.25	9.95	50	1.62
		100	1.62
3.00	9.95	100	1.60
		300	1.55
2.96	9.95	100	1.57

r_{O-H} = 0.96 (Å) throughout

In simulations of the $H_6Si_2O_7$ dimer, the use of the β_{ij} parameter equal to 2.50×10^{-8} cm for Si–Si, Si–O, and O–O interactions resulted in collapse of the $H_6Si_2O_7$ dimer to lower Si–O–Si bond angles (<120°). This collapse is very dependent upon the presence of the hydrogen ions, which, through hydrogen bonding, attract the oxygen from the adjacent tetrahedron. This effect must be counteracted in the potential functions without deleteriously altering the ability of these potentials to generate accurate bulk properties. In light of that, the BMH equation should be considered. This equation is commonly used for ionic systems. Because of the long range of the coulomb term in a full BMH equation, the Ewald method of calculating this potential is often used for computational speed. This method involves a real space summation over interacting pairs out to some relatively short cut-off distance, based on a box size and periodic boundary conditions, plus a reciprocal space summation. Rapid convergence of each summation simultaneously enhances calculation speed. The ξ function in the BMH equation shown above makes the coulomb term the first term of the Ewald sum, where $\xi(x) = \mathrm{erfc}(x)$; erfc is the complementary error function. β_{ij} is a convergence term which depends upon the size of the direct summation portion of the Ewald sum (size of the central box). Since these silicate systems are not fully ionic, use of an ionic potential would obviously be inaccurate. Even in simulations of bulk SiO_2, the full Ewald sum is seldom employed. Therefore, application of the ξ function and the β_{ij} parameter take on a less exact meaning in the BMH equation shown above and may be considered as more a part of an effective potential than as a truly and theoretically accurate potential. In addition, increasing the number of atoms in the system, and the central box size, increases the β_{ij} values. Without the compensating contribution from the reciprocal space summation, this creates a new potential for each system size and is probably not appropriate for these simulations. Therefore, the BMH equation shown above should be considered as an *effective* potential. In this sense, the ξ function can be considered as altering the formal charges on the ions in the coulomb term. As such, the effective charge on O ions at an O–O distance of 2.6 Å is about -0.7e, which is similar to that found from MO calculations[36] and that used in simulations of H_2O.[33,34] Increasing β_{ij} values increases the effective charges on the ions and increases the strength of the coulomb term.

The collapse of the $H_6Si_2O_7$ molecule with the 2.50×10^{-8} cm β_{ij} values indicated that perhaps the role of the nonbonded forces should be considered in the calculations. Ab-initio calculations showed that overlap populations between nonbonded Si–Si increased much more rapidly with decreasing bond angle than did nonbonded O–O interactions.[37] Using constant A_{ij} parameters, a series of β_{ij} values were used in simulations of the dimer in which Si–Si repulsions were enhanced with respect to the Si–O and O–O interactions. This was accomplished by increasing the β_{Si-Si} parameters in comparison to the β_{Si-O} and β_{O-O} values. Results of different β_{ij} combinations on bond lengths and bond angles are shown in Table 2.

Dimer structure does not appear to be overly sensitive to the selected β_{ij} values, except at the largest values. In addition, incorporation of some of these sets of terms in simulations of bulk v–SiO_2 and silica surfaces showed that structural results were good using the $\beta_{Si-Si} = 2.6x$ series. Even the vibrational spectra were generally good for these series.

Table 2: Effect of β_{ij} Parameters in BMH Potential on Dimer Structure

β_{Si-Si} ($\times 10^{-8}$ cm)	β_{Si-O} ($\times 10^{-8}$ cm)	β_{O-O} ($\times 10^{-8}$ cm)	Average Bond Lengths (Å)		Si-O-Si Bond Angles
			Si-O	Si-NBO	
3.0	2.7	2.7	1.67	1.55	149°
3.0	2.7	2.6	1.65	1.54	139°
2.9	2.7	2.7	1.66	1.55	143°
2.9	2.7	2.6	1.64	1.54	137°
2.7	2.7	2.7	1.63	1.55	153°
2.7	2.6	2.5	1.64	1.54	151°
2.65	2.55	2.50	1.63	1.55	138°
2.65	2.56	2.53	1.63	1.55	135°
2.63	2.55	2.53	1.63	1.55	147°
2.61	2.55	2.55	1.63	1.55	139°
2.61	2.55	2.50	1.63	1.55	137°
2.61	2.55	2.49	1.63	1.55	136°
2.60	2.55	2.60	1.63	1.55	136°
2.60	2.55	2.53	1.63	1.55	138°
2.60	2.55	2.50	1.63	1.55	140°
2.50	2.50	2.50	1.63	1.55	<120°

The deprotonation energy of the H_4SiO_4 was calculated as the difference between the H_4SiO_4 molecule and the $H_3SiO_4^-$ ion and compared with a similar energy difference obtained from MO calculations. The MD simulations were done at 70°K. The MO calculations give a deprotonation energy of 2.62×10^{-11} ergs/molecule.[5] MD simulations consistently produced deprotonation energies around 1.5×10^{-11} to 1.6×10^{-11} ergs/molecule using various sets of β_{ij} values. That is, even though the total energy of an individual molecule calculated by MO differs by two orders of magnitude from that calculated by MD, the energy differences between the H_4SiO_4 and $H_3SiO_4^-$ molecules calculated by each technique are within a factor of 2. This is a remarkably good result and indicates the potential for estimating energy differences using the MD simulation technique.

Based on the reasonable results presented above, simulations of the polymerization of silicic acid molecules were initiated. Simulations of the interaction between two silicic acid molecules, $2Si(OH)_4$, using potentials which give good dimer structure, $H_6Si_2O_7$, led nonetheless to collapse of the product molecule $H_8Si_2O_8$ with an edge shared structure. The difference between the dimer, which does not collapse, and the product molecule, is the presence of an extra HOH in the latter. (Note also that additional simulations showed that without the presence of the hydrogens on the product molecule, no collapse of the molecule occurs.) Each Si becomes pentacoordinate in the product molecule. Dissociation of the HOH from the product molecule would result in a normal dimer and an H_2O molecule ($H_8Si_2O_8 \rightarrow H_6Si_2O_7 + H_2O$). However, since these simulations were performed in a vacuum, there is probably a very large activation barrier for removal of the HOH from the product molecule. This large barrier and the short time of the simulations did not allow for this dissociative reaction. Elevated temperatures in the simulation only went into rotational energy and did not enhance the dissociation process. Introducing additional H_2O molecules around the reacting silicic acid molecules ($2Si(OH)_4 + xH_2O$) resulted in formation of hexacoordinate silicon,[38] which has been considered as a possible intermediate in silicic acid reactions.[3,22] The product molecule is corner shared with only one bridging oxygen between the silicons, although each silicon still remains hexacoordinate throughout the time of the simulation (~10 psec). Further simulations of these reactions are being performed.

CONCLUSIONS

The molecular dynamics computer simulation technique has been used to simulate the structure and early stages of polymerization of silicic acid molecules, H_4SiO_4. The effective potentials used in these simulations are similar to those used in simulations of bulk v–SiO_2 and water. The central force water potential was not altered in the simulations; changes in the potentials describing all Si and O pair combinations in the molecules were validated in simulations of bulk silicate glasses and compared to experimental data available for bulk systems.

Results showed that a fairly wide range in the combination of relevant parameters could reproduce appropriate molecular structures (as compared to molecular orbital calculations) as well as bulk v–SiO_2 structure and frequency spectra.

The presence of hydrogen on the silicic acid molecules enhances reactivity between molecules. The addition of water molecules creates the formation of hexacoordinate silicon.

REFERENCES

1. Harris, R.K., Knight, C.T., Hull, W.E., *Soluble Silicates,* ACS Symposium Series 194, Ed. J.S. Falcone, p. 79, ACS Washington, DC, 1982.
2. Cary, L.W., de Jong, B.H.W.S. and Dibble, W.E., *Geochim. Cosmochim. Acta,* 46, 1317 (1982).
3. Gibbs, G.V., *Amer. Mineral.* 67, 421 (1982).
4. Gibbs, G.V., Meagher, EP., Newton, M.D. and Swanson, D.K., *Structure and Bonding in Crystals,* 1 (1981) 195 M. O'Keefe and A. Navrotsky, Eds., Academic Press, NY.
5. O'Keeffe, M., Domenges, B. and Gibbs, G.V., *J. Phys. Chem.* 89, 2304 (1985).

6. Heilweil, E.J., Casassa, M.P., Cavanagh, R.R. and Stephenson, J.C., *J. Chem. Phys.* 81, 2856 (1984).
7. de Jong, B.M.W.S., Keefer, K.D., Brown, G.E. and Taylor, C.M., *Geochim. Cosmochim. Acta,* 45, 1291 (1981).
8. Geisinger, K.L., Gibbs, G.V. and Navrotsky, A., *Phys. Chem. Minerals,* 11, 266 (1985).
9. Rahman, A., *Phys. Rev.* 136, A405 (1964).
10. Woodcock, L.V., Angell, C.A. and Cheeseman, P., *J. Chem. Phys.* 65, 1565 (1976).
11. Garofalini, S.H., *J. Chem. Phys.* 78, 2069 (1983).
12. Garofalini, S.H., *J. Am. Ceram. Soc.,* 67, 133 (1984).
13. Garofalini, S.H., *Phys. Chem. Glasses,* 26, 166 (1985).
14. Garofalini, S.H. and Levine, S.M., *J. Am. Ceram. Soc.* 68, 376 (1985).
15. Soules, T.F., *J. Chem. Phys.* 71, 4570 (1979).
16. Garofalini, S.H., *J. Chem. Phys.* 76, 3189 (1982).
17. Gibbs, G.V., Hamil, M.M., Louisnathan, S.J., Bartell, L.S. and Hsiukang Yow, *Amer. Mineral.* 57, 1578 (1972).
18. Garofalini, S.H., presented at the Pacific Coast Meeting of the American Ceramic Society, San Francisco, 1984.
19. Brawer, S.A., *Phys. Rev. Lett.* 46, 778 (1981).
20. Garofalini, S.H., *J. Non-Cryst. Solids,* 63, 337 (1984).
21. McCune, R.C., *Anal. Chem.* 51, 1249 (1980).
22. Iler, R.K., *The Chemistry of Silica,* J. Wiley and Sons, 1979.
23. Lucovsky, G., *Phil. Mag.* B 39, 513 (1979).
24. Brown, G.E., Gibbs, G.V. and Ribbe, P.H., *Am. Mineral.* 54, 1044 (1969).
25. Brown, G.E., Gibbs, G.V., *Am. Mineral.* 55, 1587 (1970).
26. Kelso, J.F., Pantano, C.G. and Garofalini, S.H., *Surface Science,* 134, L543 (1983).
27. (a) Pantano, C.G., Kelso, J.F. and Suscavage, M.J., *Advances in Materials Characterization,* Ed. D.R. Rossington, R.A. Condrate and R.L. Snyder, Plenum 1983. (b) Pantano, C.G., personal communication.
28. Weichert, R. and Schonert, K., *J. Mech. Phys. Solids,* 26, 151 (1978), and 22, 127 (1974).
29. Levine, S.M. and Garofalini, S.H., Proceeding MRS Fall Meeting, 1985.
30. Garofalini, S.H., *Structure and Bonding in Noncrystalline Solids,* Ed. A.G. Revesz and G.E. Walrafen, in print.
31. Weeks, R., personal communication.
32. Levine, S.M. and Garofalini, S.H., submitted to *J. Chem. Phys.*
33. Stillinger, F.H. and Rahman, A., *J. Chem. Phys.* 68, 666 (1978).
34. Rahman, A., Stillinger, F.H. and Lemberg, H.L., *J. Chem. Phys.* 63, 5223 (1975).
35. Peri, J.B., *J. Phys. Chem.* 70, 2937 (1966).
36. Newton, M.D., Gibbs, G.V., *Phys. Chem. Minerals,* 6, 221 (1980).
37. Newton, M.D., O'Keeffe, M. and Gibbs, G.V., *Phys. Chem. Minerals* 6, 305 (1980).
38. Garofalini, S.H. and Melman, H., to be published.

3

Phase Transformation in Gels: A Comparison of the Phase Transformation Behavior of Gel-Derived and Ordinary Na_2O-SiO_2 Glasses

Michael C. Weinberg and George F. Neilson

Microgravity Science and Applications Group
Jet Propulsion Laboratory
California Institute of Technology
Pasadena, California

INTRODUCTION

A unique feature of the gel method of glass preparation is the ability to form glasses at relatively low temperatures. In the ordinary (or traditional) method of glass formation the ingredients are heated above the liquidus temperature and then cooled below the glass transition temperature, T_g. This procedure may be termed "glass formation from above" since the glass is formed by cooling from a high temperature to a temperature below T_g. The gel method, however, affords the option of "glass formation from below" since the gel precursor may be densified to form a glass by heating the former at temperatures below T_g. Since glass is a non-equilibrium state, the question arises as to whether glasses prepared from above and below are equivalent.

On the other hand, gel precursors may be heated in excess of T_g and cooled to form glasses from above. Here, one may inquire as to how high a temperature and for how long a time must the gel be heated in order to produce a glass which is equivalent to the glass prepared by standard methods.

The above questions are intimately related and neither has been satisfactorily resolved to date. Therefore, one does not know whether the properties and behavior of ordinary and gel glasses are identical and how the gel preparation procedure employed affects the latter.

One aspect of glass conduct which is of intrinsic importance, and which could be a quite sensitive probe of dissimilarities between gel and ordinary glasses, is its phase transformation behavior. Early experimental studies revealed qualitative differences between the phase separation and crystallization characteristics of gel and ordinary glasses. For example, Mukherjee, Zarzycki, and Traverse[1] noted differences in the phase separation of ordinary and gel produced $La_2O_3-SiO_2$ glasses. They observed that phase separation occurred with a uniform microstructure and with the production of small droplets in the gel glass, while the oxide glass exhibited a phase separation characterized by a non-uniform distribution of larger and more irregularly shaped particles. In the same study the crystallization behavior of gel and ordinary glasses in the $La_2O_3-SiO_2$, $La_2O_3-Al_2O_3-SiO_2$ and $La_2O_3-ZrO_2-SiO_2$ systems were examined. It was observed that, in general, there was a greater degree of crystallization in the gel glasses than in the ordinary glasses for any given heat treatment. Furthermore, it was noted that the crystallization morphology in the gel $La_2O_3-SiO_2$ glass was more fine-grained than that appearing in the corresponding ordinary glass.

These studies, while of great interest, were limited to qualitative observations due to the experimental techniques used. A focused heating source was employed which produced a certain degree of sample vaporization. Hence, it was not certain whether the gel and ordinary glasses were of precisely the same composition. Another complicating feature was the occurrence of phase separation during the quenching process due to the rapid unmixing in these systems. Thus, if the thermal histories experienced by these glasses were not identical, then one would expect the degree and morphology of the phase separation in the quenched glass to be dissimilar.

However, the observations made in Reference 1 led us to consider the hypothesis that the phase transformation behavior of gel and ordinary glasses may differ. In order to investigate this possibility, a series of studies were undertaken to contrast the phase separation and crystallization of gel glass with ordinary glass in a system in which these processes could be more conveniently studied. Hence, we anticipated that in such a system any such differences in behavior could be described more reliably and quantitatively.

The compositional system chosen for investigation was that of Na_2O-SiO_2. This binary has a region of metastable immiscibility which approximately spans the 0-33 mol percent Na_2O regime. Phase separation processes may be easily studied in this region[2-4] due to the relatively sluggish phase separation kinetics (especially on the higher side of the above Na_2O region). Also, the crystallization of this glass may be measured at easily accessible temperatures and in a temperature regime where the glass is still quite viscous. The latter features simplify the experimental techniques required for crystallization measurements. In addition, this system has the added virtue that data from previous studies are available concerning the crystallization and phase separation behavior of glasses prepared by standard methods.

Herein, our completed and ongoing work on the phase transformation behavior of Na_2O-SiO_2 glasses is reviewed. It will be seen that a resolution has been reached regarding the previously observed differences between gel and ordinary glasses processed at high temperatures. However, no definite conclusions may be drawn concerning the equivalence of gel glasses prepared by low temperature processing and conventionally melted glasses.

METASTABLE LIQUID-LIQUID IMMISCIBILITY IN Na_2O-SiO_2 GLASS

Immiscibility Temperatures

Many glass forming systems containing two or more components have been observed to undergo liquid-liquid phase separation in particular temperature regimes. The systems and compositions where phase separation has been noted and the thermodynamics and kinetics of phase separation have been reviewed by James[5] and Uhlmann and Kolbeck.[6]

Immiscibility regions in glass forming systems are usually categorized as being stable or metastable depending upon whether the temperature regime where the unmixing occurs lies above or beneath the liquidus temperature. The entire immiscibility region in the Na_2O-SiO_2 system is subliquidus, and hence metastable.

A schematic characteristic curve which delineates the two phase and single phase regions for given conditions of temperature and composition is shown in Figure 1. The region under the curve corresponds to the two phase region, while the region exterior to the curve is the single phase region. The bounding curve which indicates the temperature at which a given composition will transform from a single phase to a two phase region is termed the binodal (or immiscibility curve), and the transformation temperature is termed the immiscibility temperature, T_m.

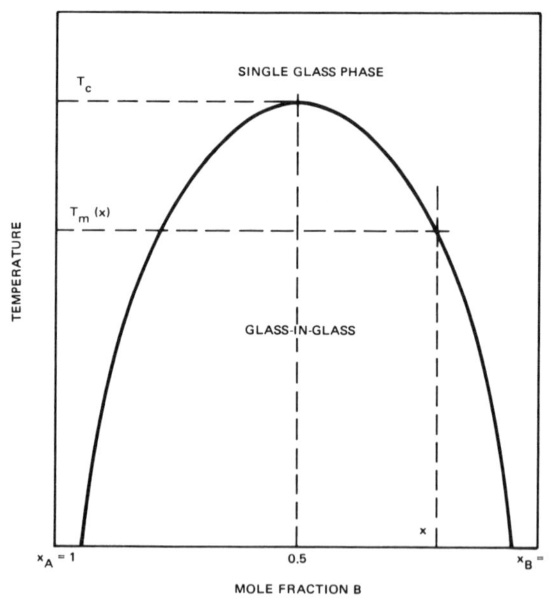

Figure 1: Schematic representation of an immiscibility curve for a regular solution of components A and B which has a critical immiscibility temperature, T_c.

The prediction of the binodal requires the use of a solution model which accurately describes the thermodynamics of mixing. The simplest one which has been used is the regular solution model,[7] which predicts a symmetrical binodal as shown in Figure 1. However, the experimental binodal curve for Na_2O-SiO_2 appears to be asymmetrical, and hence the regular solution model is inappropriate for the description of the thermodynamics of phase separation in this system. Haller and coworkers,[8] and others[9,10] have found that a modified regular solution model, though, provides an excellent fit to the immiscibility data for several binary systems, including Na_2O-SiO_2.

Experimental measurements of the binodal have been made, too. The miscibility gap in the Na_2O-SiO_2 glass system has been reported in Reference 8. Here, the immiscibility data gathered for this system by others[11,12] was also reviewed. In Figure 2, the T_m values given in References 8, 11 and 12 are displayed for part of the high soda portion of the immiscibility region. The line marked H is the binodal curve constructed from the data given by Haller et al.[8] The line designated as J is that constructed from Hammel's data.[11] The results of Andreev et al[12] are shown as the dashed line. It should be noted that for most compositions the values of T_m reported by these investigators disagree by amounts larger than that which could be attributed to temperature uncertainty, and they appear to be systematically different. Possible causes for these differences will be discussed subsequently.

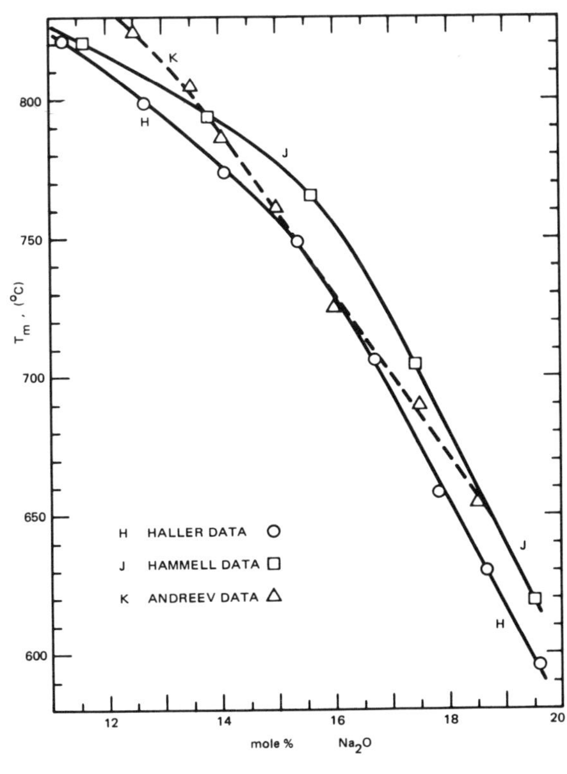

Figure 2: Experimental immiscibility curves for the soda-silica system in the high soda region of immiscibility as determined by Haller,[8] Hammel,[11] and Andreev.[12]

Initial Study of Phase Separation of Gel-Derived Glass

For the reasons discussed previously, an investigation was begun with the objective of contrasting the phase separation behavior of ordinary and gel derived glass. The initial study employed a Na_2O-SiO_2 glass composition consisting of 18.56 mol percent Na_2O.[13] The gel glass was prepared using tetraethyl silicate, and sodium nitrate was the soda source. The preparation procedure followed the one outlined by Levene and Thomas.[14] The dried gel was melted at 1565°C for 3 hours and then cast into stainless steel molds. The ordinary glass was made from Na_2CO_3 and Brazilian quartz. The batch was melted for 74 hours, at the same temperature, in a platinum crucible. The melt was stirred continuously. Chemical analyses of both gel and ordinary glass were performed.

Two aspects of the phase separation behavior of these glasses were studied; the morphologies of the developing second phase upon heat treatment and the location of the immiscibility temperature. These measurements are described in turn. The developing minor phases in both glasses were investigated using replication electron microscopy (REM) and small angle x-ray scattering (SAXS). Immiscibility temperature measurements were made by noting the clearing temperature of a phase separated sample and by the use of REM.

Morphology of Phase Separation. Samples of ordinary and gel glass were heated for 30 and 120 minutes at 630°C. SAXS data and REM were obtained for all samples. These measurements indicated that the glasses phase separated at this temperature, a result of which is consistent with all of the previous immiscibility measurements illustrated in Figure 2. However, the morphologies of the phase separation in the gel and ordinary glasses were strinkingly different. Figures 3(a) and 3(b) are the replication micrographs of the ordinary and gel glass, respectively, for samples which were heated 30 minutes at 630°C. The former glass phase separated only to a small degree with the formation of spherical particles with an average diameter of roughly 400 Å. However, as can be seen from Figure 3(b), the gel glass was extensively phase separated. Furthermore, the morphology exhibited in this figure is characteristic of phase separation via a spinodal or non-classical mechanism. The SAXS data shown in Figure 4 tends to corroborate this interpretation. This figure contains Guinier plots of the scattering intensities from gel (MOD) and ordinary (batch) glasses. The gel sample heated for 30 minutes at 630°C exhibits a high SAXS intensity having a large range of linearity at intermediate and large angles with downward deviations from linearity at small angles. These characteristics are indicative of a high concentration of closely-spaced particles of similar size and are suggestive of a nonclassical mode of phase separation.[3] After 120 minutes of heat treatment both gel and batch glasses show characteristics of coarsening. Hence, their differences in morphology are less pronounced at longer times.

Immiscibility Temperature. Clearing temperature measurements performed on the batch glass indicated that the immiscibility temperature for this composition is in excess of 630°C, but less than 640°C. Figure 5 shows a micrograph of a batch glass taken after a four hour heat treatment at 640°C. There are no signs of phase separation in this sample, and this observation confirms the findings of the clearing temperature measurements. These results are compatible

a

b

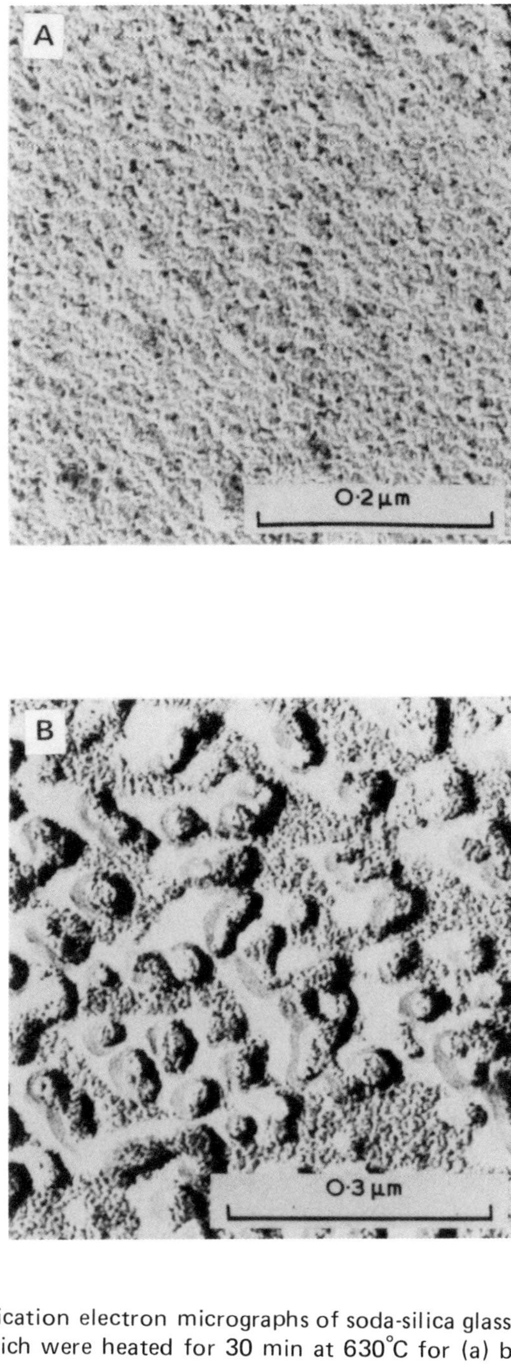

Figure 3: Replication electron micrographs of soda-silica glasses containing 18.6 mol % soda which were heated for 30 min at 630°C for (a) batch prepared and (b) gel prepared glasses. Reproduced from Reference 13.

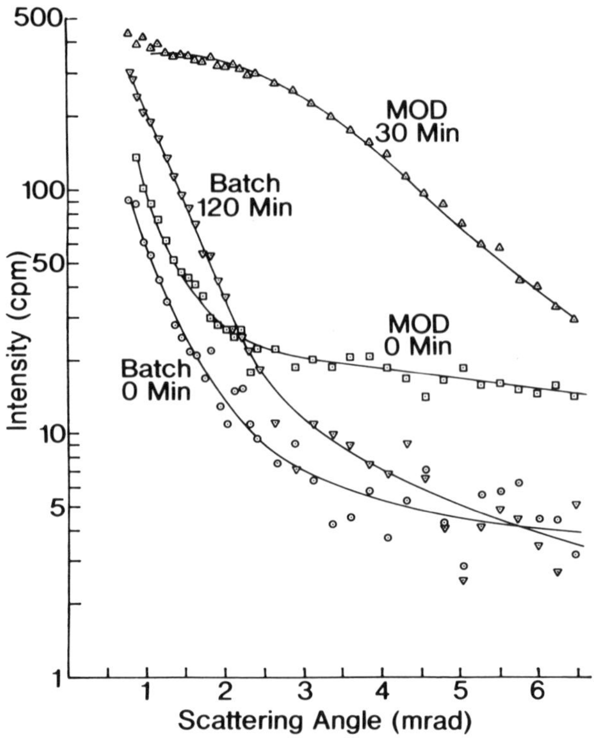

Figure 4: Guinier plots (log intensity vs scattering angle squared) of SAXS data of annealed gel (MOD) soda-silica glass containing 18.6 mol % soda, and of the gel (MOD) and batch glasses heated 30 min at 630°C shown in Figure 3. Reproduced from Reference 13.

with those of Haller, Blackburn, and Simmons (see Figure 2). However, it was observed that T_m for the gel prepared glass was well in excess of 640°C. Phase separated gel glass samples which were heated at 690°C, 710°C, and 755°C did not clear. Furthermore, REM taken from samples heated at these temperatures exhibited clear signs of phase separation. Sections of these micrographs are shown in Figures 6 through 8. Hence, it appears that T_m for the gel glass is at least 100°C higher than for the corresponding batch glass. This is an astounding result.

The elevation of the immiscibility temperature of the gel glass was explained in terms of its enhanced water content.[13] IR spectral measurements of the batch and gel glass revealed that the gel glass had a water concentration roughly 3½ times that of the ordinary glass. Water is contained in glass primarily in the form of bound hydroxyl groups. Molecular water is transformed into the network structure via the reaction.[15]

$$\equiv Si{-}O{-}Si\equiv \; + \; H_2O \rightarrow 2 \; \equiv SiOH \qquad (1)$$

Thus, glass with an enhanced water content would possess a smaller degree of structural coherency since as a result of Equation 1 such glass would be expected to have fewer bridging oxygens. This reduction in the connectivity of

Figure 5: Replication electron micrograph of 18.6 mol % soda batch glass after heat treatment at 640°C for 4 hrs. Reproduced from Reference 13.

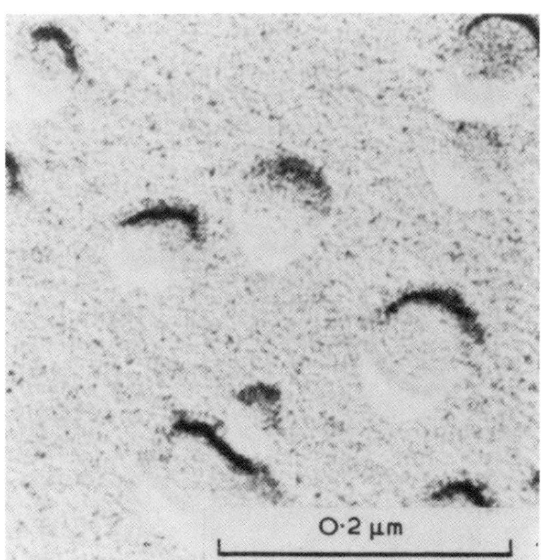

Figure 6: Replication electron micrograph of 18.6 mol % soda gel glass after heat treatment at 690°C for 1 hr. Reproduced from Reference 13.

the gel glass framework could possibly account for its elevated immiscibility temperature.[13]

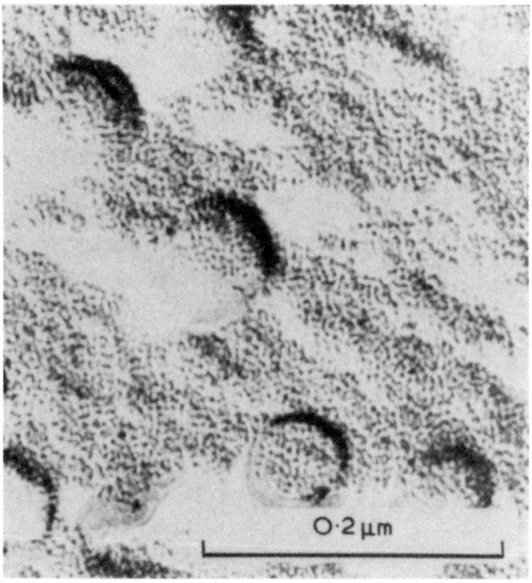

Figure 7: Replication electron micrograph of 18.6 mol % soda gel glass after heat treatment at 710°C for 1 hr. Reproduced from Reference 13.

Figure 8: Replication electron micrograph of 18.6 mol % soda gel glass after heat treatment at 755°C for 10 min. Reproduced from Reference 13.

Compositional Effects

If an enhanced OH^- content is present in gels of all compositions in this system and the relatively large hydroxyl ion concentration is responsible for elevating T_m, then one may anticipate that the entire binodal curve could be significantly raised for the gel glass. This question was investigated by measuring the immiscibility temperature and water concentration of a number of gel glasses with soda content varying from approximately 10 to 19 mol percent.[16]

The results of this study were quite surprising. They are summarized in Table 1. The first column of the table gives the analyzed (AA) glass composition. The second column contains the measured immiscibility temperatures of the gel glasses. Values of ΔT_m are shown in column 3. ΔT_m is the elevation in immiscibility temperature relative to the data reported for T_m by Haller et al.[8] In the fourth column the IR absorbances at 3600 cm^{-1} for these glasses are listed. This wave number corresponds to the location of the major water absorption. Another water absorption band is centered about 2700 cm^{-1}. The last column lists the ratio of the maximum peak intensities of the 2700 cm^{-1} to the 3600 cm^{-1} absorption bands. It should be mentioned that the gel glasses were fabricated using somewhat different melting histories, and the gels were subjected to different drying procedures. Hence, the water concentration in the gel glasses for different compositions varied.

Table 1: Immiscibility and IR Absorbance Results from Previous Studies

Glass Composition (mol %)	Tm (°C)	ΔTm (°C)	Absorbance (A/t at 3600 cm^{-1})	R
19.1	692	78	0.19	0.88
19.0	748	132	0.33	1.23
18.4	693	54	0.32	1.38
17.0	745	52	0.44	1.22
17.0	735	42	0.38	0.81
16.7	768	63	0.56	0.81
16.7	768	63	0.47	1.07
16.7	773	68	0.32	0.67
15.0	785	28	0.36	0.77
12.8	820	23	0.55	0.56
10.5	848	20	1.20	0.32
10.5	848	20	1.86	0.36
10.5	848	20	0.68	0.28
10.4	845	16	0.87	0.37

Although the data presented in this table are difficult to interpret in detail, certain general trends are apparent. Since ΔT_m is positive in all cases, the immiscibility temperature of the gel glass seems to be greater than that of the corresponding ordinary glass for all compositions. However, it is apparent, also, that ΔT_m generally decreases as the soda composition decreases and the top of the immiscibility dome is approached. In addition, the data in this table provide evidence that the water concentration of the glass cannot be the sole cause of the observed anomalous behavior of T_m. For example, the three glasses of composition 10.5 mol percent Na_2O have quite different water contents, but they exhibit identical immiscibility temperatures.

It was suggested[16] that perhaps the arrangement of hydroxyl groups in the gel glass was responsible for the immiscibility temperature elevation. One may observe from Table 1 that those compositions which exhibit larger ΔT_m generally show greater R values. However, close scrutiny of the data show there are exceptions to this trend.

Thus, the examination of the dependence of the immiscibility temperature of the gel glass upon composition raised doubts concerning water as the origin of the anomalous behavior. Also, a bothersome feature of the results was the somewhat sporadic nature of the reported values of ΔT_m. In view of these uncertainties, considerations were given to other factors which could be responsible for the observed differences in gel and ordinary glass phase separation behavior.

Factors Affecting Phase Separation Behavior

Trace Impurities. It is well known that minor amounts of certain oxide impurities can drastically affect the immiscibility temperature.[8,17-20] Usually these impurity oxides tend to lower the immiscibility temperature.[20] However, this is not always the case. Minor amounts of P_2O_5 have been observed to elevate the immiscibility temperature,[18] while Al_2O_3 has been reported to raise[17] or lower[17,19] the immiscibility temperature in the Na_2O–SiO_2 system depending upon the base glass composition and Al_2O_3 concentration.

Trace impurity inclusions could be partly responsible for the lack of agreement of the T_m data for batch glasses compiled by various investigators. Herein, however, the question is whether it can account for the anomalous behavior of the gel glasses.

In References 17 and 19 it is indicated that 1 weight percent Al_2O_3 may *lower* the immiscibility temperature by as much as 120°C. Thus, although the presence of Al_2O_3 drastically affects T_m, it shifts it in the wrong direction as far as the present question is concerned. One might postulate that T_m is not really raised for the gel glasses, but that all previous batch glass measurements of T_m have been too low. However, this argument is untenable since the cation impurity concentration differences between the batch and gel glass are relatively small. Also, the magnitude of impurity ion concentration in either glass is quite small (see below).

Atomic absorption analysis was used to measure the impurity levels of several metal oxides in three gel glasses, and one ordinary undoped and four doped glasses, all of which were of similar base glass composition. The results are displayed in Table 2. It is clear from an inspection of this table that oxide impurities play at most a minor role in explaining differences in gel and ordinary glass immiscibility temperatures. The most influential impurity cations alter the immiscibility temperature by at most 10°C/0.1 weight percent. Therefore, differences in metal ion concentrations can account for differentials in T_m of only a few degrees at most.

Table 2: Trace Cation Content and Immiscibility Temperatures of 14.2 Mol % Na_2O Gel Glasses and 15.5 Mol % Na_2O Batch Glasses

Glass	Tm (°C)	Impurity Ion Concentration (wt %)				
		Al	Fe	Cr	Mo	Zr
Gel-1	776	0.016	<0.003	<0.003	<0.003	<0.08
Gel-2	775	0.028	<0.006	<0.006	<0.006	<0.14
Gel-3	775	0.009	0.003	<0.003	<0.003	<0.06
Batch-1	761	0.062	0.012	<0.002	<0.002	<0.06
Batch-2	755	0.18	0.017			
Batch-3	751	0.25	0.014			
Batch-4	754	0.085	0.026			
Batch-5	754	0.11	0.042			

Water. It is now well known that water in glass may affect the kinetics of liquid-liquid immiscibility.[21,22] However, it does not appear that much work has been devoted to the study of the influence of water upon the immiscibility temperature.

Free water (i.e., water existing as molecular H_2O), may be readily removed from glass by heating at relatively low temperatures. However, bound hydroxyls, incorporated in the glass via Eq. (1) are difficult to eliminate. Adams[23] and Scholze[15] have attributed the bands in the 3 to 4 μm region formed in the spectra of silica and alkali silicates to bound hydroxyls. Adams has postulated that for alkali silicates the hydroxyl groups may be hydrogen bonded to either bridging or non-bridging oxygens (see Figure 9). He has associated the band near 3660 cm^{-1} with OH groups hydrogen bonded to bridging oxygens. The hydrogen bonding produces a slight shift from the fundamental Si–OH stretch and broadens the band somewhat. However, the OH groups which are hydrogen bonded to non-bridging oxygens exhibit bands which peak at about 2700 cm^{-1}. Typical IR spectra taken from Na_2O–SiO_2 glasses are shown in Figure 10. These spectra show the two bands discussed above.

The possible effect of water incorporation upon the immiscibility temperature was discussed previously. It was concluded that differences in T_m between the gel and ordinary glass could not be attributed to the disparity in water concentrations of these glasses as was postulated earlier. However, it was noted that in many cases where ΔT_m was large, R, the ratio of the maximum peak intensity of the 2700 cm^{-1} to the 3600 cm^{-1} band appeared to be large. Previously, it was

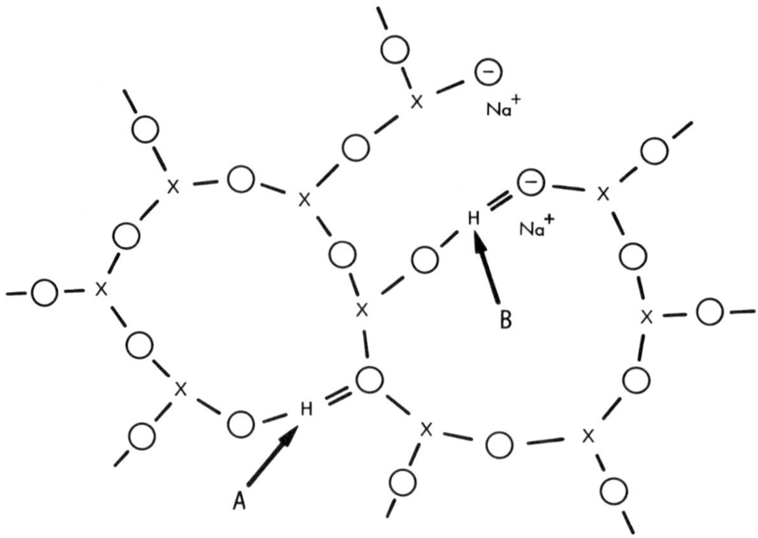

Figure 9: Schematic representation of a glass structure comprised of network forming atoms (X) and oxygen atoms (O). Also shown are two bound hydroxyl groups (--O--H), one of which (A) is hydrogen bonded to a bridging oxygen atom, and the other (B) to a non-bridging oxygen anion.

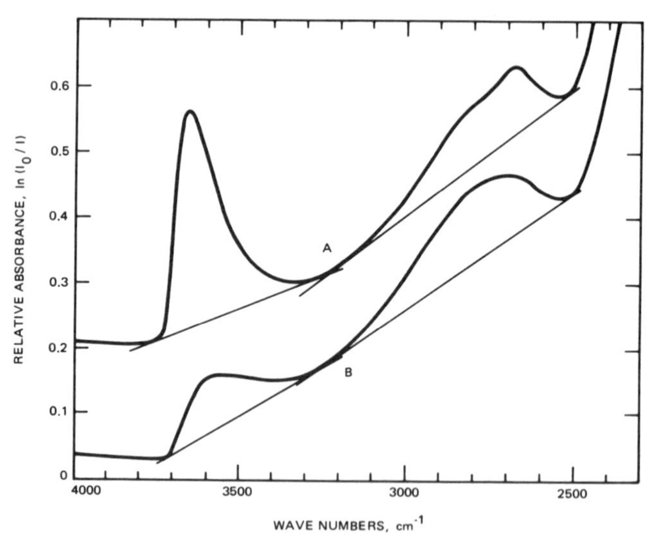

Figure 10: IR spectra in the near infrared region of soda-silicate glass slabs of nominal 2 mm thicknesses. Spectrum (A) is of a 12.6 mol % and (B) of a 18.0 mol % soda glass.

speculated that the value of R could be intimately related to the elevation of T_m in the gel glass.[13] However, herein it was observed that this pattern was not generally followed for gel glasses.

In order to investigate the above hypothesis further, two ordinary glasses of identical compositions (16.9 mol percent) were prepared using disparate melting procedures. IR analyses indicated that both the water concentrations and the R values of these glasses differed. Yet, they were found to have precisely the same immiscibility temperature. Thus, neither the water concentration nor the value of R appears to significantly affect T_m.

Structure. If gel and ordinary glasses of identical composition possess structural differences, then this could account for the exceptional behavior of gel glasses. However, structural determinations of glasses are fraught with difficulties. It has been indicated recently[24] that there is a lack of agreement concerning the structural details for even glasses as simple as the alkali silicates. MacKenzie[25] has reviewed the large number of possible structural units and theories which may be invoked to describe their structure. Therefore, detailed structural comparisons between gel and ordinary glasses would be difficult and, as will be demonstrated, probably unnecessary.

It is well known that the structure of the gel is not the same as that of the glass.[26] IR spectra of gels heated at successively higher temperatures appear increasingly similar to those of the glass (produced by melting).[27] However, IR spectroscopy provides a limited amount of structural information, and the similarity of two IR spectra does not establish the structural identity of the glasses from which the spectra were produced. However, as alluded to previously, glasses made by heating gels for a long enough time period, and at a high enough temperature, should be equivalent to glasses prepared by standard methods.

Although in general it is difficult to specify the precise set of initial conditions required to equilibrate gel and ordinary glass structures, in the present case an order of magnitude calculation should suffice to assess the possible importance of structural variations. One may employ the structural relaxation time, τ, as a guide to the timescale needed for the structure to come to equilibrium.[27] Rekhson[28] has found the following relationship between τ and η, the shear viscosity,

$$\tau = \eta/K_p \qquad (2)$$

where $K_p = 2.5 \times 10^{10}$ dynes/cm^2. Furthermore, Mazurin[29] found that K_p ranges from 10^{10} to 10^{11} dynes/cm^2 and is roughly temperature independent. Thus, it is simple to make an estimate of the time required for structural equilibration at the melting temperature. Taking η as roughly 10^3 poise at 1565°C and K_p as 10^{10} dynes/cm^2, then $\tau \sim 10^{-7}$ sec. Thus, it is quite hard to believe that structural features or "memory" of the gel persist after high temperature melting.

Phase Separation Kinetics. In References 1, 21 and 22 it was observed that the kinetics of phase separation and structural morphology of the developing phases appeared to differ in the gel and ordinary glass. Also, as previously mentioned, water can accelerate glass immiscibility kinetics. Thus, an investigation was initiated in our laboratory to disentangle the effects of OH content and glass preparation procedure upon phase separation by studying the rates of phase separation of glasses of similar composition and comparable water content prepared by gel and ordinary methods. The results of this study were presented at the 3rd International Workshop on Glasses and Glass-Ceramics from Gels,[30] and they are summarized below.

Ordinary and gel glasses were prepared by melting at 1565°C and casting bars. "Wet" and "dry" glasses were produced by bubbling during melting with nitrogen gas saturated with water and dried nitrogen gas, respectively. The composition and relative water concentration of the glasses are shown in Table 3. Here BW and BD signify the wet and dry ordinary (or batch) glasses, respectively, and G denotes a gel glass composition.

Table 3: Compositions and Relative OH Concentrations of Batch and Gel Glasses

Glass	Analyzed Composition*		Absorbance
	Wt % Na_2O	mol % Na_2O	(A/t at 3600 cm^{-1})
BW	17.4	16.9	0.26
BD	17.4	16.9	0.17
G	17.6	17.1	0.24

*Trace impurity analyses by AA showed that glass G contained 0.02 wt% Al and 0.01% Fe, while glasses BW and BD contained 0.06 wt% Al and 0.02 wt% Fe.

Samples of each glass were heated at 590°C for several time periods in order to promote phase separation. The development of the phase separation process was studied with the aid of SAXS and REM.

It should be noted (from Table 3) that the gel glass has very similar water content to that of the ordinary wet glass, BW. Hence, after differences in phase separation behavior exhibited by these two glasses could be considered a manifestation of the anomalous behavior which has been discussed. Samples of these glasses were heated at 590°C and subjected to SAXS analysis. Guinier plots were constructed from the scattering data and D (particle diameter) and N (particle number density) were computed. These results are shown in Table 4. It is apparent from an inspection of this table that the phase separation kinetics of the gel and ordinary glass are quite similar. Thus, this investigation showed no evidence for differences in kinetic behavior between gel and batch glasses of identical composition and OH content.

Table 4: SAXS Parameters of Gel and Wet Batch Glasses

Glass	Heating Time Hours at 590°	D (Å)	I(0) (Relative)	N (Relative)
BW	0.5	154	195	1460
BW	1.0	187	391	914
BW	5.0	325	2660	226
G	0.5	138	140	2030
G	1.0	180	347	1020
G	5.0	329	2720	214
G	9.6	420	5860	107

Recent Studies. In view of the above negative results, it was decided to reexamine the question of the elevation of the immiscibility temperature of the gel glasses. The immiscibility temperature in the Na_2O-SiO_2 system is a very sensitive function of composition, especially for the high Na_2O portion of the immiscibility region. During the performance of the kinetic studies it was ascertained that index of refraction measurements of the glasses could be used to obtain highly accurate and reliable compositional analyses. Also, it was discovered that previously obtained compositional analyses via atomic absorption were at times erroneous.

Consequently, measurements were made of T_m, once again, for ordinary and gel glasses of several compositions, and the results are shown in Table 5. In the first column of this table the glass type and designation are given along with the melting conditions used for glass preparation. The glass compositions, as determined by index of refraction measurements, are shown in columns 2 and 3 in weight percent and mol percent, respectively. In the fourth column the immiscibility temperatures are shown. These T_m values were obtained by clearing temperature measurements. The absorbancies due to the water bands at 3600 cm^{-1} and 2700 cm^{-1} are given in the last two columns of Table 5.

Table 5: Recent Glass Composition, Immiscibility Temperature, and IR Absorbance Results

Glass Description (melting conditions)	Na_2O (wt%)	Na_2O (mol%)	Tm (°C)	A_1 (3600)	A_2 (2700)
GEL JPL2-18.5 (1565° - 24 HR dry)	14.9	14.5	779	0.22	0.16
GEL OI-15.0 (1500° - 3 HR)	15.1	14.7	776	0.48	0.48
GEL OI-15.0 (1500° - 24 HR)	15.1	14.7	776	0.39	0.35
GEL OI-15.0 (1500° - 45 HR)	15.1	14.7	775	0.39	0.35
GEL OI-15.0 (1500° - 72 HR)	15.0	14.6	773	0.31	0.29
BATCH JPL-15 (1565° - 24 HR)	15.9	15.5	761	0.32	0.27
GEL JPL1-18.5 (1565° - 3 HR)	16.8	16.4	735	0.36	0.30
BATCH JPL-BW (1565° - 24 HR wet)	17.4	16.9	711	0.26	0.33
BATCH JPL-BD (1565° - 24 HR dry)	17.4	16.9	711	0.17	0.16
GEL JPL-G (1565° - 4 HR)	17.6	17.1	710	0.24	0.18
GEL JPL-CD (1565° - 24 HR dry)	17.6	17.1	704	0.19	0.15
GEL OI-18.5 (1565° - 3 HR)	18.5	18.0	693	0.49	0.73
BATCH OI-18.5 (1565° - 74 HR)	19.0	18.5	667	0.18	0.19

Several important features regarding these results should be noted. With the exception of the two compositions containing the highest soda content, all measured T_m fall within the bounds set by the immiscibility curves determined by Haller et al and Hammel (see Figure 2). The above applies *both to ordinary and gel compositions.* The high soda compositions exhibit T_m values slightly in excess of the values reported by Hammel. Furthermore, the differences between the T_m values which we found for ordinary and gel glasses of comparable composition were smaller than the differences between the immiscibility temperatures of the ordinary glasses reported by various investigators. Thus, it appears as though there is no anomalous phase separation behavior for the gel glass. In addition, from an inspection of the data in the table, it may be concluded that hydroxyl content has little (if any) effect upon T_m for the concentrations of OH typically found in these glasses.

A number of unresolved points do exist, however, which could have a bearing upon the discrepancies in T_m reported by various investigators. The high soda content glasses tend to surface devitrify when heated for long periods of time. Opalescence experiments for these compositions require long time heat treatments since the phase separation kinetics are rather sluggish (especially at temperatures near T_m). However, even a small surface layer of crystals of pure SiO_2 can cause a composition change which, although slight, can produce a significant effect upon the measured T_m. Also, it was found that for large Na_2O containing compositions, opalescence measurements are difficult. Similar conclusions were drawn by Hammel.[11]

CRYSTALLIZATION OF Na_2O-SiO_2 GEL AND GLASS

Past studies[1,31,32] appeared to indicate that both the gel and gel glasses exhibit crystallization tendencies which differ from that of the ordinary glass. Mukherjee et al.[1] noted that the degree of crystallization in gel glasses of the $La_2O_3-SiO_2$, $La_2O_3-Al_2O_3-SiO_2$, and $La_2O_3-ZrO_2-SiO_2$ systems was greater than that in the ordinary glasses for a given heating. Also, it was observed that the crystallization of the gel $La_2O_3-SiO_2$ glass exhibited a more fine-grained morphology than that in the corresponding ordinary glass. Mukherjee, Zarzycki, Badie and Traverse[31] studied the crystallization behavior of ground gel and batch $La_2O_3-SiO_2$ glass employing x-ray diffraction measurements. Both glasses were heated at 1100°C, and it was observed that the gel glass showed a greater degree of crystallization. This result was explained in terms of the enhanced hydroxyl content of the gel glass.

Phalippou et al[32] studied the effects of variation of several processing parameters upon the crystallization of $B_2O_3-SiO_2$, $Na_2O-B_2O_3-SiO_2$ and Na_2O-SiO_2 gels. Generally, gels were prepared either by the destabilization of colloidal silica by adjustment of the pH of solution (method 1), or by hydrolysis and polycondensation of metal-organics (method 2). The crystallization of the gel was studied primarily with the aid of x-ray diffraction analysis. Of central interest here is the crystallization behavior of the Na_2O-SiO_2 system. Two compositions were investigated; 10 Na_2O - 90 SiO_2, which was prepared by both methods and 0.4 Na_2O - 99.6 SiO_2, which was made only via method 2. Several factors were considered in assessing the crystallization; the pH of gel preparation (method 1), the Na_2O concentration, the method of gel preparation, and the crystallization temperature. When gels produced by method 1 were heated at 700°C, those gelled at pH 5.6 showed less crystallinity than those made at pH 7.2. However, all of the gels produced by method 1 showed the same amount of crystalline species when heated at 800°C. The crystalline phase was identified as α-cristobalite. The gels of the same composition which were produced by methods 1 and 2 exhibited identical overall crystalline behavior, with detectable crystallinity appearing in the 660°-680°C temperature region. However, the soda content of the gel had a marked influence on its crystallization characteristics. The 0.4 Na_2O - 99.6 SiO_2 gel, when heated at 850°C, showed three diffraction peaks (in addition to those of cristobalite) which were tentatively assigned to tridymite and a sodium silicate phase. Also, for this composition the minimum temperature for detectable crystallization was elevated by at least 50°C with respect to the higher soda composition.

The investigation of Phalippou et al detailed the effects of gel processing upon crystallization, but did not compare the crystallization characteristics of gels with the corresponding glasses. Neilson and Weinberg[33] performed such a study using gel and glass of a particular Na_2O–SiO_2 composition. The crystallization of (1) powdered materials; gel precursor, ground gel glass, and ground batch glass as well as (2) bulk glasses; gel glass and batch glass were analyzed. These materials were heated for various time periods at 720°C. The crystallization products were identified via x-ray diffraction measurements. The various crystalline species which were found were SiO_2 (α-cristobalite), α-$Na_2O{\cdot}2\ SiO_2$ (NS2), $3Na_2O{\cdot}8\ SiO_2$ (N3S8), $Na_2O{\cdot}3\ SiO_2$ (NS3) and $Na_2O{\cdot}4\ SiO_2$ (NS4).

The results of the study in which the gel glass and ordinary glass crystallization are compared, are given in Table 6. Along the top of the table the heat treatment times are indicated as well as the glass type (M for gel and B for batch or ordinary). On the left the various crystalline phases which were observed are indicated. The designations (S), (G), and (P) indicate a surface slice of bulk glass, ground glass, and gel precursor (powdered), respectively. The numbers 0-5 provide a semi-quantitative scale which indicates the relative amount of crystalline species formed. A "0" signifies the absence of a particular species and a "5" corresponds to a very large amount.

Table 6: Crystal Species Formed After Heating Sodium Silicate Gel and Glasses

Heating Time		5 Hrs		7 Hrs		9 Hrs		16 Hrs		66 Hrs	
Glass	Type	M	B	M	B	M	B	M	B	M	B
α-C											
	S	2	2	-	3	3	3	4	3	3	2
	G	3	3	-	-	-	-	5	4	5	4
	P	5		-		-		5		5	
NS2											
	S	0	0	0	0	0	0	0	0	0	0
	G	0	0	-	-	-	-	0	0	0	0
	P	4		-		-		4		4	
N3S8											
	S	2	0	1	2	1	1	3	2	2	2
	G	1	0	-	-	-	-	3	3	3	3
	P	0		-		-		1		2	
NS3											
	S	1		-	3	3	3	4	3	4	4
	G	4	4	-	-	-	-	0	5	0	4
	P	0	0	-		-		0		0	
NS4											
	S	1	0	-	3	4	3	5	5	4	2
	G	0	0	-	-	-	-	0	0	0	0
	P	0		-		-		0		0	

The overall findings may be briefly summarized as follows: The ordinary and gel glass were found to show qualitatively similar crystallization behavior. Namely, the crystalline phases produced in both glasses for a given heat treatment time were identical. However, the volume fraction of crystallites found in the gel glass was generally larger than that in the batch glass. Such differences, though, were relatively minor. The gel, on the other hand, displayed a unique crystallization behavior, with the appearance of different phases in significantly different amounts. This singular behavior probably reflects the high hydroxyl content as well as altered structural features of the gel.

SUMMARY AND CONCLUSIONS

During the past several years there has appeared mounting evidence[1,13,16,34-36] that gel glasses processed at high temperatures could be different from ordinary glasses. The most dramatic example of this disparity was the seemingly large elevation in the immiscibility temperature of an 18.5 mol percent soda glass in the Na_2O-SiO_2 compositional system.[13] This result as well as others[16,22,31] pertaining to the phase transformation anomalies of gel glasses has been attributed to the enhanced hydroxyl content of gel glasses.

However, our recent findings as well as those of Scherer et al[37] seem to confirm that gel glasses produced by high temperature processing of gels behave similarly to ordinary glasses. In both the structural relaxation studies reported in Reference 37 and our most current studies of liquid immiscibility it was noted that very small changes in glass composition dramatically alter the experimental results. Hence, it appears likely that all previously noted anomalies in the behavior and properties of gel glasses, *which were prepared by high temperature processing,* are attributable to small shifts in composition. This is a satisfying explanation since, as discussed previously, it is difficult to accept that a glass melt prepared from a gel precursor will not equilibrate in a short while when held at a temperature well in excess of the liquidus.

On the other hand, there exists significant differences in structure and crystallization behavior between gels and their corresponding glasses. As gels are heated to increasingly higher temperatures, their structure (and properties) begin to approach those of the glass, if crystallization does not intervene. However, it is still unclear whether or not monolithic glasses produced by low temperature processing of gels are equivalent in all respects to the corresponding glasses produced by melting.

REFERENCES

1. Mukherjee, S.P., Zarzycki, J. and Traverse, J.P., *J. Mater. Sci.* 11, 341-355 (1976).
2. Neilson, G.F., *Phys. Chem. Glasses* 13, 70-76 (1972).
3. Neilson, G.F., *Discuss. Faraday Soc.* 50, 145-154 (1970).
4. Goganov, D.A. and Porai-Koshits, E.A., *Soviet Physics Doklady* 11, 335-337 (1966).

5. James, P.F., *J. Mater. Sci.* 10, 1802-1825 (1975).
6. Uhlmann, D.R. and Kolbeck, A.G., *Phys. Chem. Glasses* 17, 146-158 (1976).
7. Lupis, C.H.P., *Chemical Thermodynamics of Materials,* New York, Elsevier Science Publishing Co., Inc. (1983).
8. Haller, W., Blackburn, D.H. and Simmons, J.H., *J. Amer. Ceram. Soc.* 57, 120-126 (1974).
9. Macedo, P.B. and Simmons, J.H., *J. Res. Nat. Bur. Stand.,* Sec. A, 78, 53-59 (1974).
10. Simmons, J.H., *J. Amer. Ceram. Soc.* 56, 284-285 (1973).
11. Hammel, J.J., *Proceedings of the VII International Congress on Glass,* Vol. 1, Institut National du Verre, Charleroi, Belgium (1966).
12. Andreev, N.S., Goganov, D.A., Porai-Koshits, E.A. and Sokolov, Y.C., in *Structure of Glass* (E.A. Porai-Koshits, ed.) Vol. 3, pp 47-52 (1964).
13. Weinberg, M.C. and Neilson, G.F., *J. Mater. Sci.* 13, 1206-1216 (1978).
14. Levene, L. and Thomas, I., U.S. Patent 3,640,093; February 8, 1972; assigned to Owens-Illinois, Inc.
15. Scholze, H., Gases and Water in Glass, Part Two, *Glass Ind.* 47, 622-628 (1966).
16. Neilson, G.F. and Weinberg, M.C., Anomalous Phase Separation Behavior of Gel-Derived Soda Silica Glasses, *Materials Processing in the Reduced Gravity Environment of Space* (G.E. Rindone, ed.) pp 333-342, Elsevier Science Publishing Co. Inc., New York (1982).
17. Topping, J.A. and Murthy, M.K., *J. Amer. Ceram. Soc.* 56, 270-275 (1973).
18. Tomozawa, M., *Advances in Nucleation and Crystallization in Glasses* (I.L. Hench and S.W. Freiman, eds.) pp 41-50, American Ceramic Society, Inc., Columbus, Ohio (1971).
19. Tomozawa, M. and Obara, R.A., *J. Amer. Ceram. Soc.* 56, 378-381 (1973).
20. Simmons, J.H., Napolitano, A. and Macedo, P.B., *J. Chem. Phys.* 53, 1165-1170 (1970).
21. Kreidl, N.J. and Maklad, M.S., *J. Amer. Ceram. Soc.* 52, 508-509 (1969).
22. Faber, K.T. and Rindone, G.E., *Phys. Chem. Glasses* 21, 171-177 (2980).
23. Adams, R.V., *Phys. Chem. Glasses* 2, 39-49 (1961).
24. Yasui, I., Husegawa, H. and Imaoka, M., Parts 1 and 2; *Phys. Chem. Glasses* 24, 65-78 (1983).
25. Mackenzie, J.D., *Modern Aspects of the Vitreous State* (J.D. Mackenzie, ed.) pp 1-37, Butterworth & Co., Washington (1960).
26. Sakka, S., *Treatise on Materials Science and Technology,* Vol. 22, (M. Tomozawa and R.H. Doromus, eds.) pp 129-167, Academic Press, New York (1982).
27. Scherer, G., *Relaxation in Glass and Composites,* John Wiley & Sons (in press).
28. Rekhson, S., *Sov. J. Phys. Chem. Glass* 1, 417-421 (1975).
29. Mazurin, O.V., *J. Non. Cryst. Solids* 25, 130-169 (1977).
30. Neilson, G.F., Weinberg, M.C. and Smith, G.L., Effect of OH Content on Phase Separation Behavior of Soda-Silica Glasses, presented at, *Third International Workshop "Glasses and Glass-Ceramics from Gels,"* University of Montpellier, France (September 12-14, 1985).
31. Mukherjee, S.P., Zarzycki, J., Badie, J.M. and Traverse, J.P., *J. Non. Cryst. Solids* 20, 455-458 (1976).

32. Phalippou, J., Prassas, M. and Zarzycki, J., *J. Non. Cryst. Solids* 48, 17-30 (1982).
33. Neilson, G.F. and Weinberg, M.C., *J. Non. Cryst. Solids* 63, 365-374 (1984).
34. Yoldas, B.E., *J. Non. Cryst. Solids* 51, 105-121 (1982).
35. Brinker, C.J., Roth, E.P. and Scherer, G.W., *Bull. Amer. Ceram. Soc.* 64, 476 (1985).
36. Yoldas, B.E., *J. Non. Cryst. Solids* 63, 145-154 (1984).
37. Scherer, G.W., Brinker, C.J. and Roth, P.E., *J. Non. Cryst. Solids* (in press).

Part II

Coatings, Thin Films and Surface Treatment

4

Thin Films from the Sol-Gel Process

Helmut Dislich

Schott Glaswerke

Mainz, Federal Republic of Germany

(Translated By: Brigitte Schuegraf)

INTRODUCTION AND HIGHLIGHTS OF THE SOL-GEL PROCESS

The first, although incidental, observation of the sol-gel process dates back to 1846.[1] It covered the hydrolysis and polycondensation of silicic acid under humidity, which progressed to the point of a silicate glass formation.

The second publication in 1939[2] aimed specifically towards SiO_2 layers and, after the Second World War, resulted in the development of rear view mirrors for automobiles,[3] which have been in production since 1953. Anti-reflective coatings[4] followed, which have been in production since 1964, and solar reflecting coatings,[5] deposited on flat-glass, in production since 1969. Several other sol-gel layer products followed.[6,7] These layers basically consist of SiO_2 and TiO_2, as well as mixtures thereof and a variety of other single oxides.

An extension of the chemical principles involved was shown in 1969-1971,[8,9,10] pointing out the importance of reactions of several metal alkoxides in solution under the formation of metal(I)–oxygen–metal(II) bonds. This made possible the production of defined multi-component oxide glasses, glass-ceramics and crystalline layers. This research effort into the involved basic chemistry followed the path used in the study of reactions of metal-organic compounds. It was supplemented with a continuous industrial development for the formation of thin layers,[3-6] which also resulted in a process for the fabrication of transparent granulates, which can be formed in molding presses.[10]

The importance of simple solid-state reactions, e.g. at lower temperatures, was already recognized in 1948 for the formation of co-precipitated hydroxides and oxalates, sometimes combined as powder of multi-component oxides[11] and led to a focused research effort for reactions in solution and those taking place at the solution-substrate interface. This was an absolutely necessary prerequisite.

After an incubation period of several years, the sol-gel process gained worldwide interest. Fibers and unsupported films were developed.[12,13] Other products such as preforms for fiber-optics and other monolithic structures were addressed, sometimes using a super-critical drying step.[15,16] Hollow glass microspheres found applications as containers for nuclear fuels, and deuterium and tritium were encapsulated for laser fusion experiments.[17] Synthetic minerals, which do not exist in nature, were fabricated,[18] the interest in aerogels was re-activated and many other uses were found for this technology.[19]

The knowledge in this field and the importance of the chemistry involved is best documented in a publication "Better Ceramics Through Chemistry."[20]

This short introduction is neither complete nor does it attempt to judge the value of this technology. It is intended to show the widely different directions and applications of the "sol-gel-families," as shown in the various chapters of this book. It should be recognized that several chapters of this book address the subject of thin layer formation. This is not just incidental. In the opinion of the author it is based on the following reasons:

- Thin films, deposited on substrates, are to this date the only sol-gel products produced in large quantities.
- Thin films existed already before the general interest in the sol-gel technology developed.
- With thin films, the advantages of the sol-gel process are relatively easy to attain and its disadvantages can be minimized.
- The latitude in chemistry for the sol-gel process offers the best application basis.

For quite trivial reasons, the surface of solid bodies represents a simple target for the application of economic coatings. This is much easier to accomplish than to design new bulk materials for specific applications.

This chapter attempts to analyze the basis for the above statements. Emphasis is given to review existing products in order to illustrate the state-of-the-art of the sol-gel technology. Secondly, future opportunities for sol-gel products will be discussed. Tolerance from the reader is requested for speculative assumptions, mixed with personal opinions of the author. This chapter can not give a complete subject coverage, but rather attempts to show expected trends. For a comparison of the different thin film coating technologies, the reader is referred to a publication by the author entitled, *Glass: Science and Technology,* edited by D.R. Uhlmann and N.J. Kreidl, Vol. 2, Chapter 8, Academic Press, Inc., 1984. Some duplication in this review of the sol-gel process can however not be avoided.

PRINCIPLES OF THE SOL-GEL DIP PROCESS

The sol-gel dip process is almost exclusively applied for the fabrication of transparent layers, primarily for the deposition of oxide films on float glass as a transparent substrate with a high degree of planarity and surface quality. Other substrates are possible, provided they can withstand the required curing temperature of about 500°C. Film thicknesses up to 1 μm can be deposited, preferred are well defined thicknesses within the wavelength range of visible light. Several additive layers can be superimposed.

The chemical reaction involved is based on metal compounds in alcoholic solutions which can be readily hydrolyzed. The oxides are formed through polycondensation at about 500°C.

Process Technology

Figure 1 shows as example the schematic representation of the IROX-process as used by the Schott-Glaswerk.[5,21] The IROX-process is a single layer coating consisting of TiO_2. Large float glass plates, up to sizes of 3 x 4 meters are carefully cleaned with rotating brushes and transferred in a continuous process through several baths. This is very important since the solution must wet the glass surface. Glass surfaces have a "life" of their own and can be quite different in their behavior. The cleaning step establishes a defined starting condition. After a drying step, the glass plates are dipped into the solution and uniformly withdrawn into a water vapor containing environment. The liquid film runs off the glass plate, adheres to the glass surface and solidifies rapidly through the evaporation of the solvent. At this stage, the chemical formation of the coating layer starts which will be discussed in more detail (Evaporation of solvent, hydrolysis, and condensation). The glass plate enters a furnace zone with increasing and decreasing temperatures of up to 400°-500°C. The chemical formation of the oxide layer continues during this step. The limitation of the process speed is given by the maximum cooling rate of the 4-12 mm thick glass plate.

The thickness of the coating layer is determined by the speed of withdrawal from the coating solution. The coating thickness increases with the withdrawal speed approximately at $dv^{2/3}$. The thickness also depends on the concentration of the solution, its viscosity and on the angle of withdrawal, which is normally 90°.[6] For other angles, the coating thickness is different on both sides of the plate. Other parameters affecting the coating thickness are the surface tension of the solution, vapor pressure and relative humidity above the coating bath. Precise control of the air velocity is required.[7] The clamps holding the plates during the dipping and transport should not enter the solution. After the withdrawal a ridge forms at the lower edge of the plate. This is discarded after the plates are cut to the desired size.

Although the process and the chemical reaction initially appear simple, the dependency of the various process parameters is complex. A careful study of all of those influences is required and guarantees reproducible coating thicknesses and layer quality, without the corrective measures normally required for vacuum coating processes.

Process Advantages

Major advantages of this process are the high degree of uniformity obtained and simple thickness control. Since most applications fall into a thickness range comparable to the wavelength of visible light, even small thickness variations can be easily detected visually. The excellent coating uniformity can be readily observed on many highrise office buildings with solid reflective coated windows.[5]

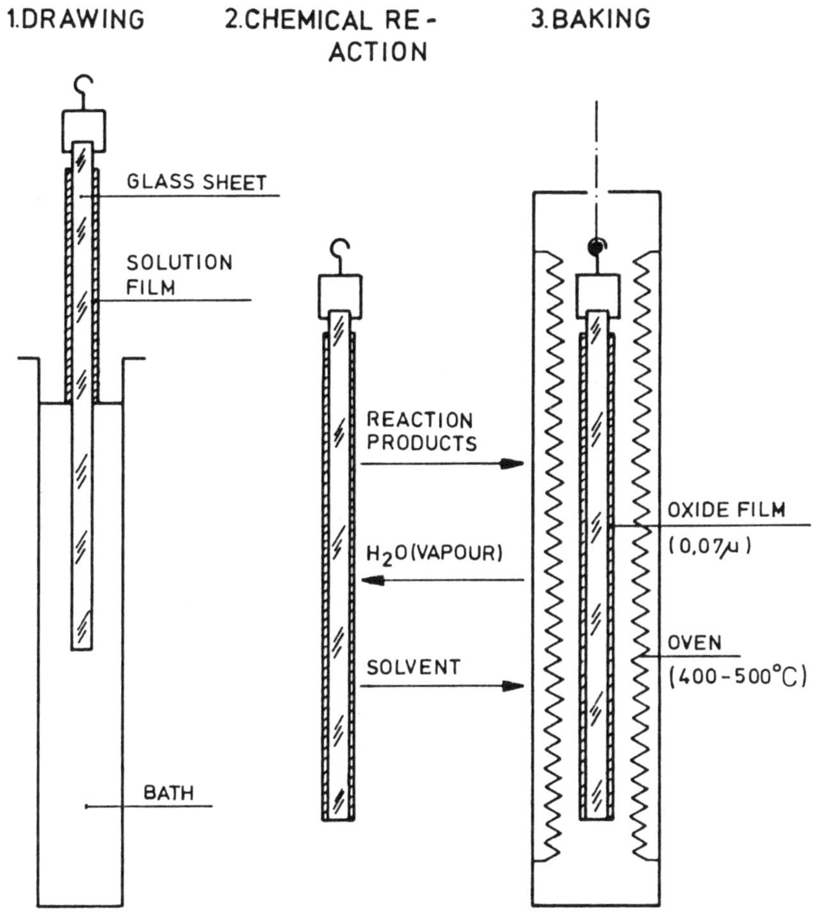

Figure 1: Production in dipping process.

Another advantage is the size of the plates or substrates that can be coated, since in many instances, the economy of the process increases with substrate size, combined with computer-optimized plate cutting to minimize cutting losses. However, the IROX-process, as used at the Schott Glaswerke, is fully automated and requires a large capital investment.

An additional advantage is the ability to produce multi-layer coatings, permitting the fabrication of layers with widely varying optical characteristics, such as transmission and reflection. Production equipment for the deposition of up to 15 layers has been installed. On the other hand, vacuum coating processes, unlike dip processes, permit continuous adjustment of the film parameters during deposition.

For the same optical filter characteristics the simultaneous coating of both plate surfaces requires fewer layers than for a single-sided coating process.[22]

The dip process can also be used to coat tubes, pipes and rods and is superior to the vacuum coating process in this application. Fibers can be coated as well, using a continuous transport of the fibers through the coating bath and curing furnace.

The dip process is limited by runoff uniformity of the excess solution after withdrawal from the coating bath. The previously mentioned disturbances in the runoff make the coatings of cubic or prism-shaped parts impractical.

Other advantages, sometimes significant, are based on the chemistry involved and will be discussed later. For the dip process technology, the chemistry and the actual process form an intricate, interrelated combination. The dip process is the most favored sol-gel process.

Other Coating Techniques

Besides the dip coating process, other techniques have been developed. Instead of withdrawing the parts from the solution, the level of the tank solution can be lowered and the substrates to be coated remain stationary. In a second variation, the coating solution is dripped onto a spinning substrate and spreads evenly. The spinning process is most suitable for the coating of small disks or lenses, but is not very economical.

It is also possible to spray on the coating solution. This does not result in the uniform layer thicknesses required for antireflective optical coatings. It has, however, been successfully applied for the coating of thick layers.[23]

CHEMISTRY AND PHYSICAL PRINCIPLES OF THE SOL-GEL DIP PROCESS

General Comments

The basis for the coating process is the hydrolysis of metal compounds in alcoholic solutions. They are converted into oxides by hydrolysis and polycondensation during and after contact of the substrate with the coating solution. In principle, any solution of a suitable compound which can be hydrolyzed, could be used. From a practical view, the reaction rate for the hydrolysis and condensation for the gel formation should be higher than the crystallization rate from the solution, in order to obtain uniform, transparent coatings.[7] For this reason alkoxides are preferred.

Good wetting of the substrate is important and makes ethanol the preferred solvent. Other alcohols and solvents can be used as well. The radical R in the alkoxide can be, but does not have to be, the same as in the alcohol used. Eventual and partial esterification reactions only occasionally play a detailed role, because the reactions can be controlled in such a way that all radicals R are split off after the formation of the final oxide layer in the heat treatment. The reaction should take place by hydrolysis and not by pyrolysis.

The water content for the hydrolytic reaction can be added to the coating solution, depending on the hydrolytic reaction rate for the oxide formation. This leads to prehydrolyzed and precondensed solutions. Such solutions have a limited useful life, because the reaction proceeds continuously. The coating solutions can be kept active for several months.

During the layer formation, i.e. during the evaporation of the solvent, water vapor diffuses from the tightly controlled atmosphere into the deposited layer and keeps the hydrolysis going. At the same time the polycondensation process continues until the pure oxide layer has been formed at higher temperatures.

The described diffusion-controlled sol-gel process favors the formation of thin layers rather than the formation of massive structures. Minor problems have been observed for layer thickness as low as 0.5 μm. This is one reason why products coated with thin layers have become commercially available whereas solid structures formed by the sol-gel process do not yet exist.

Single Oxides

The formation of single oxide layers is used as an example for the description of the chemical reactions involved. The most frequently used description oxide formation sequence is shown in Figure 2.

$$Si(OCH_3)_4 + 4H_2O \longrightarrow Si(OH)_4 + 4CH_3OH\uparrow$$

$$Si(OH)_4 \longrightarrow SiO_2 + 2H_2O\uparrow$$

Figure 2: Synthesis of SiO_2 (schematic).

It is very schematic and does not describe the actual steps involved. A more accurate process sequence is shown in Figure 3, describing the process steps for the formation of Al_2O_3.[21]

Figure 3 illustrates the possibilities for lattice structure and density modification through variation of the process parameters, resulting in different refractive indices. Film-forming oxides are primarily Group 3 through Group 8 elements of the periodic table, such as Al, In, Si, Zr, Ti, Sn, Pb, Ta, Cr, Fe, Ni and several rare earth elements. They can exist as amorphous or (mostly) crystalline layers. For example, SiO_2 layers are glass-like amorphous. TiO_2 layers exist as anatase and brookite, and in special cases as rutile. This can be influenced by the heat-curing temperature[24] or by the nature of the substrate, especially the surface-content of alkali ions.[25]

Commercially available coatings are often multi-phase structures. Initially the layers are porous and become denser at higher temperatures.

Good film adhesion on carefully cleaned glass substrates has been observed and can be explained by the reaction of the Si–OH groups on the glass surface with the metal–OH or –OR groups of the coating precursors, as shown in Figure 4. Analog considerations can be applied for the adhesion on oxidized metal surfaces, containing OH–groups.

The most commonly used layers are SiO_2 coatings with a low refractive index of $n = 1.45$, and TiO_2 with a high $n = 2.2$. Many filter characteristics, in reflection and transmission mode can be obtained through their combinations. This will be discussed later. Besides the refractive index, the absorption coefficient is an important criterion for the selection of the oxide. For example, ZrO_2 is used if lower absorption in the UV region is desirable. ThO_2 would be even more suitable, but has been discarded because of its radioactive nature.

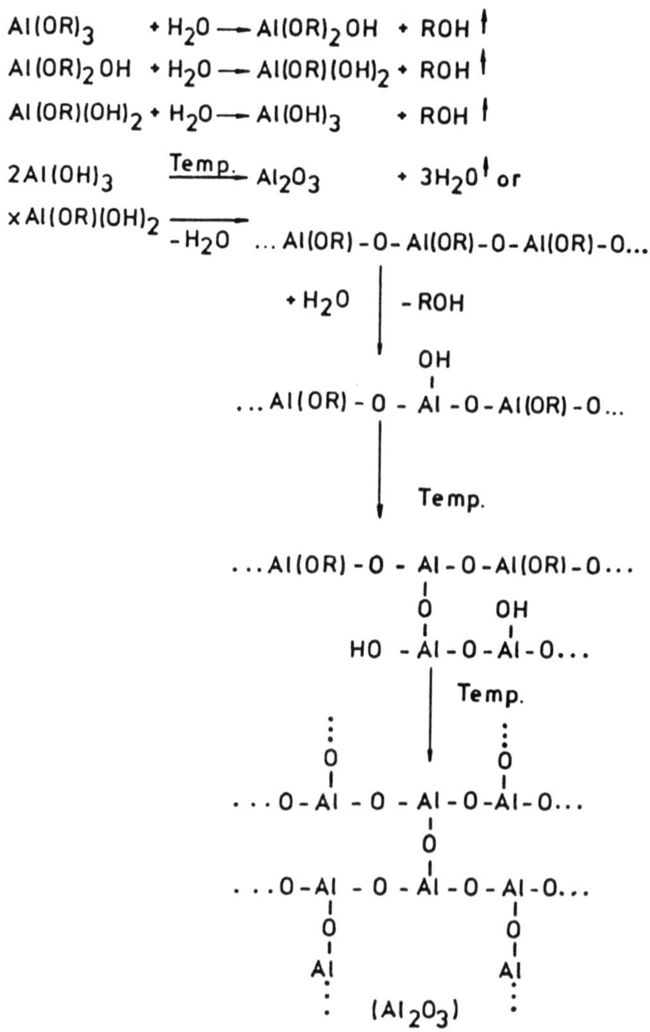

Figure 3: Hydrolysis and polycondensation of $Al(OR)_3$ (schematic).

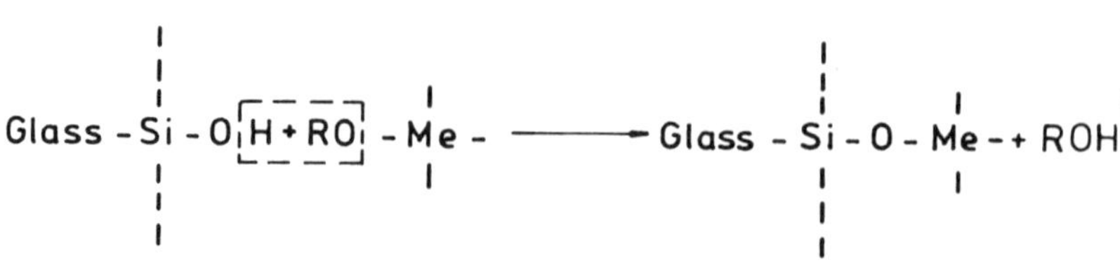

Me = metal

Figure 4: Adhesion of a metal oxide film on glass.

Mixed Oxides

A continuous selection of any index of refraction of n_D = 1.45 to 2.2 can be made from mixtures of Ti and Si alkoxide solutions, as shown in Figure 5.[21]

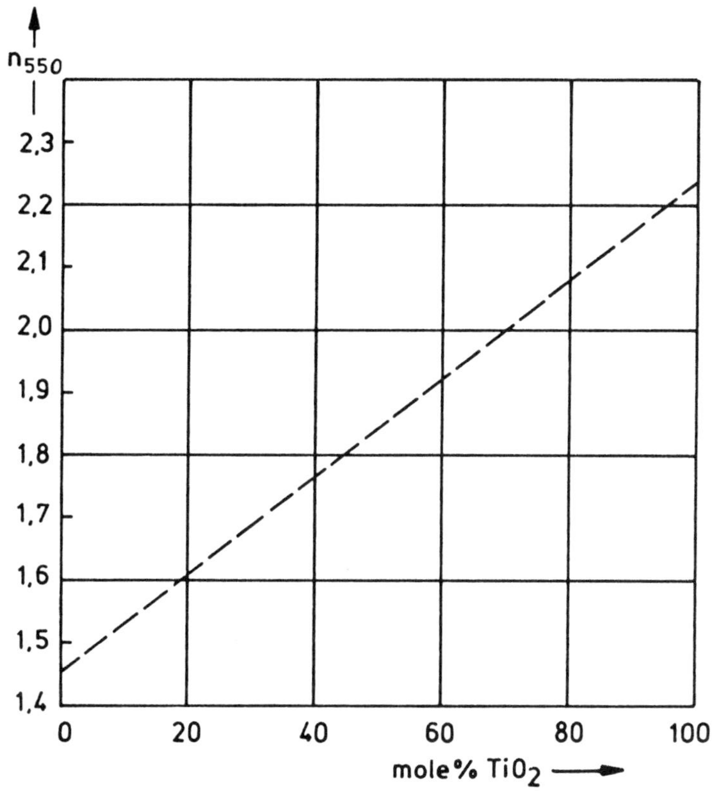

Figure 5: Refractive index of SiO_2–TiO_2 films as a function of the molar proportion of TiO_2 (λ = 550 nm).

This feature is a major advantage of the sol-gel dip technology and is used for the fabrication of many commercial products.[4] By adding absorbing oxides, a wide range of specific optical absorption characteristics can be obtained.[7,21]

In summary, one has the very attractive possibility to design interference filter characteristics, even extending into the UV range.[21]

Cermets

The formation of cermets (2 phases of ceramic and metal) as transparent layers has been accomplished with precious metals some time ago for the TiO_2 (Pd) and TiO_2 (Au) system. According to published patent disclosures,[26] the following process sequence as shown in Figure 6, can be defined:

$$Ti(OC_3H_7)_4 + PdCl_2 + HAuCl_4 \xrightarrow[+H_2O]{in\ ROH} \xrightarrow{400-500°C} TiO_2(Pd,Au)$$

Figure 6: Synthesis of TiO_2 (Pd, Au)-cermet.

The reaction takes place over several unknown intermediate steps and has been significantly modified for major industrial applications.[27]

By increasing the content of $HAuCl_4$, one reaches a very sharp transition point at which the initially optically transparent layer becomes optically dense, gold-colored and electrically conducting.

Non-Oxide Layers

By substituting ammonia, hydrogen sulfide or thiourea for water in the sol-gel process, the chemistry is changed from hydrolysis to the analog ammonia- or sulfide-reaction (ammonolysis or sulfidolysis). This leads to the formation of nitride[28] – and sulfide–layers,[29] sometimes to oxinitrides or oxisulfides.

Multi-Component Oxide Layers

Figure 3 showed the formation of an Al_2O_3 layer from $Al(OR)_3$. By adding of magnesium-alkoxide to the aluminum-alkoxide solution, a reaction product, as shown in the first line of Figure 7 is obtained.

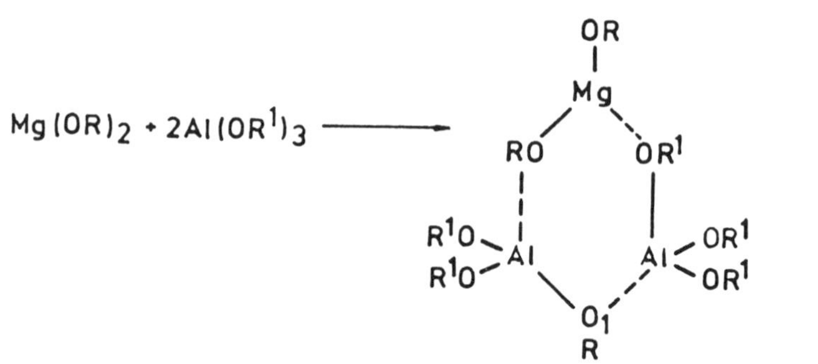

$R = CH_3 \qquad R^1 = CH(CH_3)_2$

$$Mg[Al(OR)(OR^1)_3]_2 + 8H_2O \longrightarrow Mg[Al(OH)_4]_2 + 2ROH + 6R^1OH$$

$$Mg[Al(OH)_4]_2 \longrightarrow Mg[AlO_2]_2 + 4H_2O$$

Figure 7: Spinel-synthesis.

By dip-coating in this solution a *polycrystalline* spinel film $MgO + Al_2O_3$[10] is directly obtained, not by a solid state reaction.

From a solution with a glass composition according to Figure 8,

$$m\,Si(OMe)_4 + n\,Al(O\text{-}sec.Bu)_3 + o\,P_2O_5 + p\,Mg(OMe)_2 \xrightarrow[\text{Temp.}]{H_2O} Si_m Al_n P_o Mg\, O_{\frac{4m+3n+5o+2p}{2}}$$

Figure 8: Synthesis of a silicate-phosphate glass film.

a phosphorus-silicate *glass*-layer is formed, which can protect hydrolytic sensitive glasses from water vapor attack at a thickness of only 0.2 μm, as shown in Figure 9.

Figure 9: Protection of an optical glass against climatic attack by coating with phosphate silicate glass. Top: uncoated (corrosion); bottom: (coated).

A glass crystallizing partially at temperatures between 620°C and 830°C, quite similar to the *glass-ceramic* material Zerodur, can be obtained from an eight component system as shown in Figure 10.

$$m\,Si(OMe)_4 + n\,Al(O\,sec.Bu)_3 + o\,P_2O_5 + p\,LiOEt +$$
$$q\,Mg(OMe)_2 + r\,NaOMe + s\,Ti(OBu)_4 + t\,Zr(OPr)_4$$
$$\xrightarrow{\text{complexation}}$$
$$[(Si_m Al_n P_o Li_p Mg_q Na_r Ti_s Zr_t)(OR)_{4m+3n+p+2q+r+4s+4t}(OH)_{3o}]$$
$$\xrightarrow{\text{hydrolysis}}$$
$$[(Si_m Al_n P_o Li_p Mg_q Na_r Ti_s Zr_t)(OH)_{4m+3n+p+2q+r+4s+4t+3o}]$$
$$\xrightarrow{\text{condensation}}$$
$$[Si_m Al_n P_o Li_p Mg_q Na_r Ti_s Zr_t\, O_{4m+3n+3o+p+2q+r+4s+4t}]$$

Figure 10: Synthesis of a glass ceramic.

The previous three examples show for the coating of glass substrates the principles of very simple multi-component sol-gel processes and the associated wide variety in chemical reactions.

The sol-gel process can be broken down into three basic steps:

1. The combined reaction of several metal- and non-metal alkoxides resulting in a chemical bond of the components. The residual –OR radicals facilitate their solubility.
2. The hydrolysis, which results in a partial transition of the residual –OR and –OH groups, and permits the condensation into two and three dimensional polymer networks, or in the gel formation.
3. The polycondensation, which removes the no longer required –OR and –OH residues and the transformation into oxides.

The technology has the potential for a wide range of applications, such as the controlled addition of impurities, the synthesis of analog structures, such as cadmium-stannate Cd_2SnO_4[31] and indium-tin-oxide[32] for electrically conductive layers. Barium-titanate layers[61,67] and colored SiO_2-layers, with oxides of Cr, Mn, Fe, Co, Ni and Cu[61] are other examples. A survey is given in References 30 and 62. Future trends for such layers will be discussed later.

Organic-Inorganic Layers

A maximum network formation is achieved for the sol-gel synthesis from alkoxides, as shown in Figure 11.

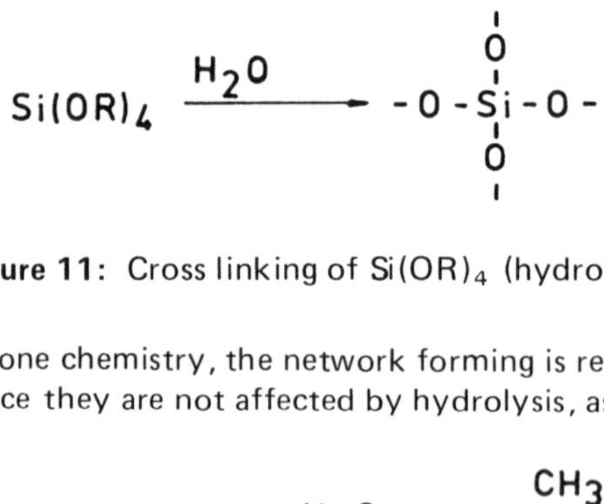

Figure 11: Cross linking of $Si(OR)_4$ (hydrolysis).

For the classic silicone chemistry, the network forming is reduced by the number of Si–C bonds, since they are not affected by hydrolysis, as shown in Figure 12.

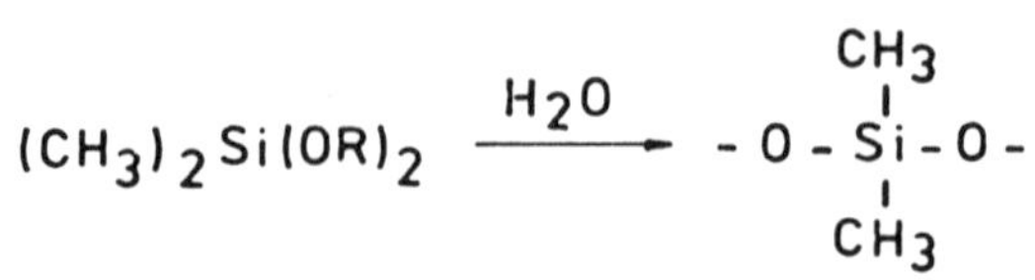

Figure 12: Reaction of $(CH_3)_2Si(OR)_2$ (hydrolysis).

The influence of the methyl groups can be compared to the effect of network-modifiers in glasses. They basically reduce the network density.

The hydrolysis of a thin film of monomethyltrialkoxisilane, formed by dip-processing on silica-glass rods or fibers, leads to glassy films after a heat treatment to 135°C. These films, with a refractive index of n_D = 1.418 and high transmission in the UV region can be considered a monomethylated silica-glass as shown in Figure 13.

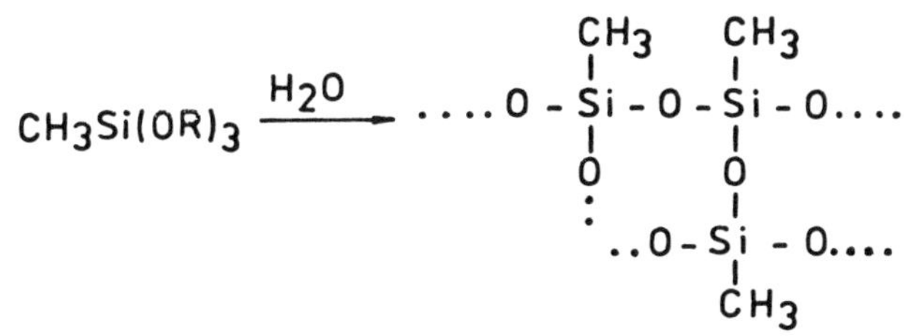

Figure 13: Synthesis of methylpolysiloxane.

With a refractive index lower than the substrate, the coating acts like an optical cladding. This application for UV fiber-optics was developed in the mid 1960's,[33,34] (IR 100 Award 1972). A completely new field for UV optics opened, eliminating the need for bulky lenses and prisms.

Oxide formation based on the principles of alkoxide sol-gel chemistry (Figure 11), can be combined with the basic chemistry of methyl(phenyl)-alkoxides for the formation of silicones (Figure 12), and leads to organic-modified silicates as shown in Figure 14.[35]

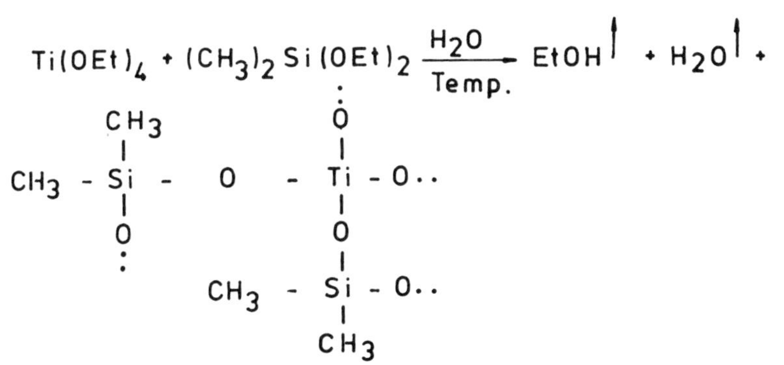

Figure 14: Example of a synthesis of organically modified silicates.

Since essentially all inorganic sol-gel processes can be combined with the processes for the formation of silicones, a wide range of possible thin film structures has become available. Several examples will be given below.

The organic substituted silicates correspond to the known alkyl(aryl)-hetero-poly-siloxanes. However, in the opinion of the author, only the full inclusion of the sol-gel principles allows their synthesis into tailor-made structures. Besides the poly-condensation reaction, the polymerization and poly-addition reactions should be included. Examples are given in Figure 15. The residual radical "R" remains always with the silicon atom, because the Si–C bond is hydrolytically and thermally much more stable.

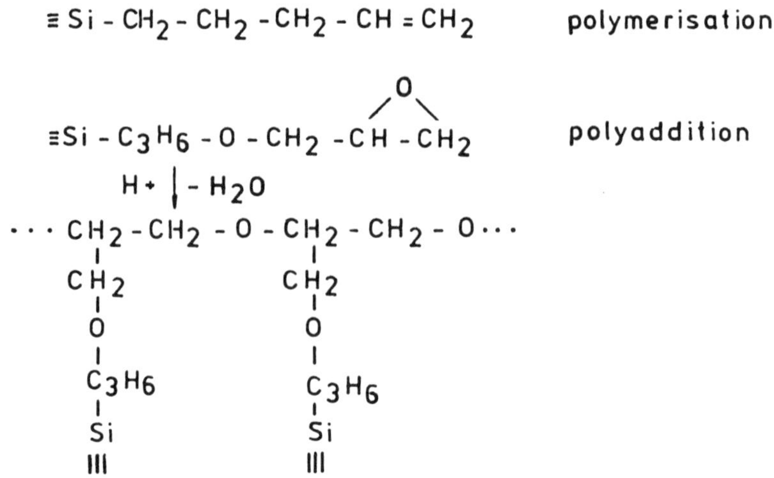

Figure 15: Precursors for polyreactions in organically modified silicates.

Finally, free and pure organic monomers can be added and incorporated into the films, allowing a continuous transition into purely organic polymers. Figure 16 shows an arbitrary example of the possibilities.

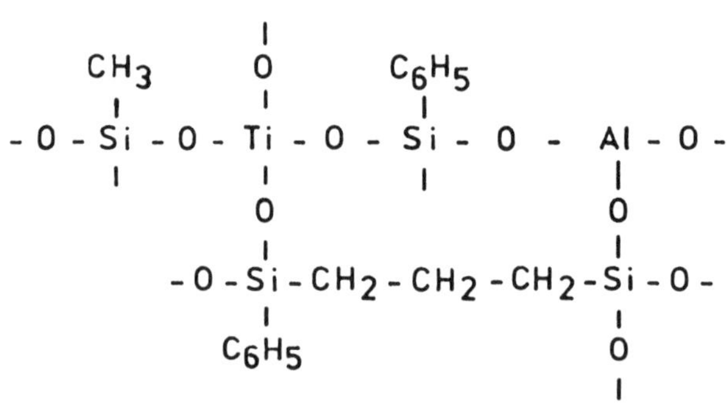

Figure 16: Illustrative example of an organically modified silicate.

Besides the glassy oxide networks, one finds here the network-forming through $-CH_2-CH_2-CH_2-$chains (which could also be a polymer chain). The network modifiers such as alkalis, are replaced by methyl-R and phenyl-R radicals. The "R" radical could also be active in the sol-gel process and participate in additional reactions. The research into the actual structure and nature of those compounds will be, in the opinion of the author, more difficult and time consuming than the practical realization of new synthetic structures.

The extremely large variety of possible combinations in sol-gel chemistry opens the variety of physical properties for these structures, such as optical transmission, refractive index, dispersion, as well as electric, magnetic and mechanical characteristics. Only first approximations can be established.

COATED PRODUCTS BASED ON SOL-GEL TECHNOLOGY

As the heading indicates, only products manufactured in reasonable quantities will be described. This reduces the number of products to coatings for optical filters. The association of the author with the Schott Glaswerke and their connection with the Deutsche Spezialglas AG. explains the selection of the products. Antireflective coatings under a Schott license are also produced by Denton Vacuum Co. Solar-reflective TiO_2 coatings are produced by the Asahi Glass Co., Ltd. For the Asahi coatings the Pd (as used by Schott) is replaced with Au in the TiO_2 film.[26] The chemistry used for these products was described earlier and will not be repeated here.

Rear View Mirrors for Automobiles

The design objective for this product was to increase the contrast and reduce the glare which is especially useful for quartz-halogen headlights.[3] A multilayer coating consisting of TiO_2–SiO_2–TiO_2 has been applied. The reflectivity of this coating is shown in Figure 17.

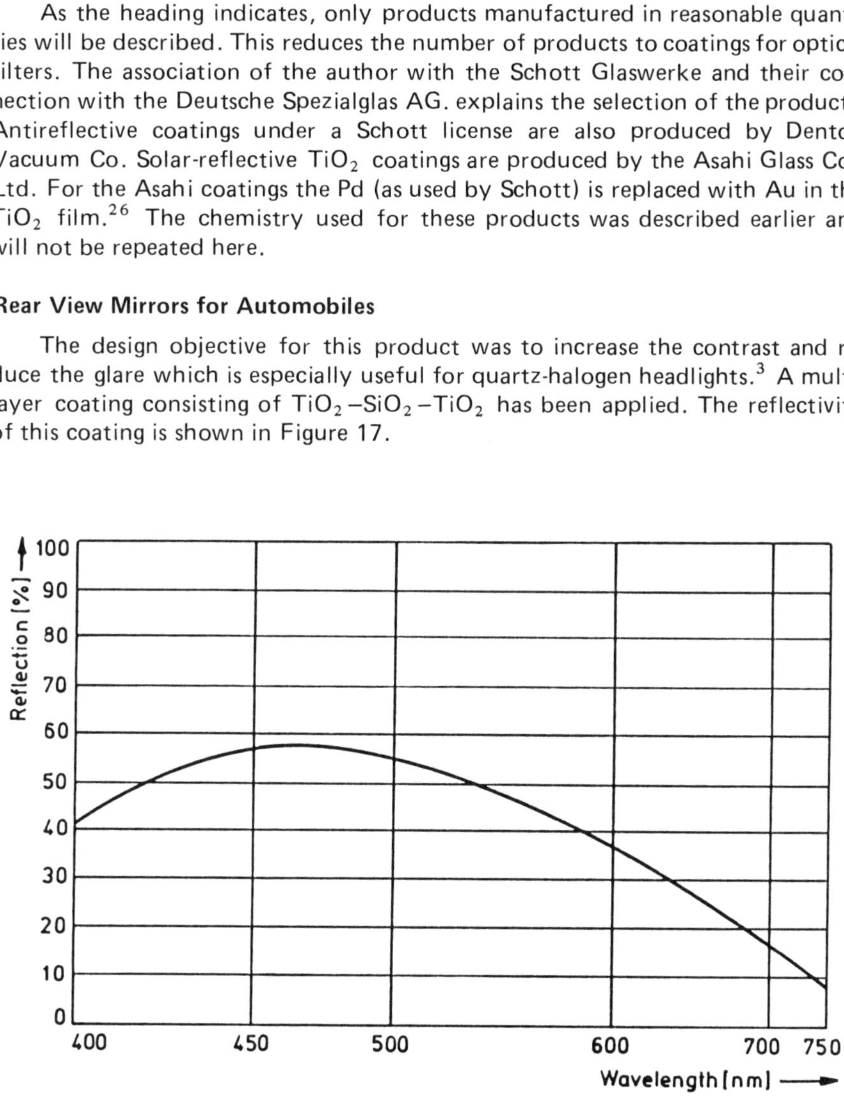

Figure 17: Reflection of a vehicle rear view mirror in the visible region.

The reflection characteristics of these mirrors in the visible spectral range have been selected to enhance the reflection of the blue, shorter wavelengths of the daylight. The quartz lamp spectral distribution is shifted towards the longer, red wavelengths. The coating combines the reduction in glare with a spectral conversion towards higher color temperatures. This results in a neutral reflection of the headlights.

The coatings are stable at temperatures of up to the softening point of the glass. Therefore, coated plane glass plates can be formed into convex shapes at higher temperatures, to increase the field of view.

Automobiles of leading manufacturers in Europe and Japan are incorporating these mirrors.

Solar Reflecting Glass (IROX)

Double sided, single layer coatings of TiO_2 (Pd) are used as solar reflecting coatings. The Pd content determines the desired absorption. Cooling costs for buildings in summer, frequently higher than the heating costs in winter, are considerably reduced. The coatings are thermally stable and permit a high temperature hardening and bending of the glass. A neutral color and the uniformity in coating thickness are the main advantages of the process.

Figures 18-23 demonstrate the characteristics of three different types of IROX coatings.[5]

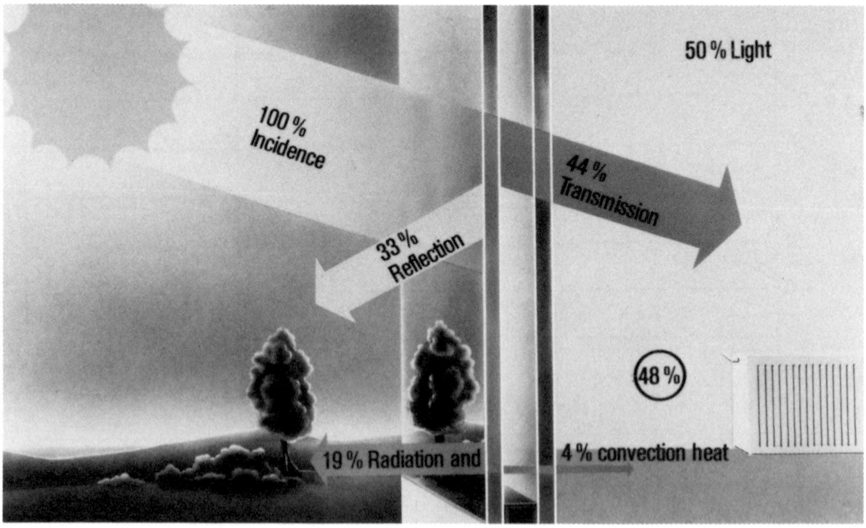

Figure 18: IROX A0, bright. Specially good light transmission. To be used preferably for small windows and single rooms.

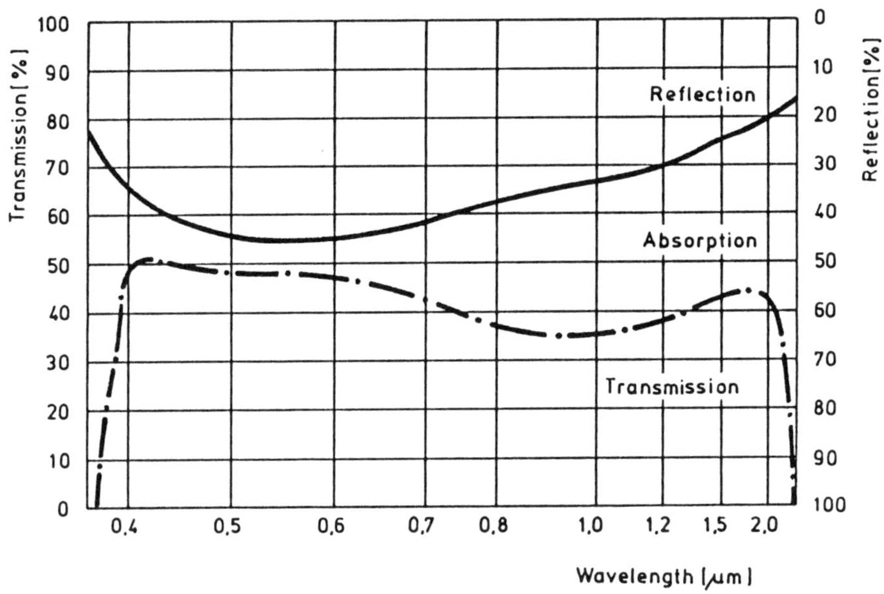

Figure 19: IROX A0. Values for reflection (R), transmission (T) and absorption (A) with IROX insulating glass 2 x 6 mm and the angle of incidence, $\alpha = 0°$.

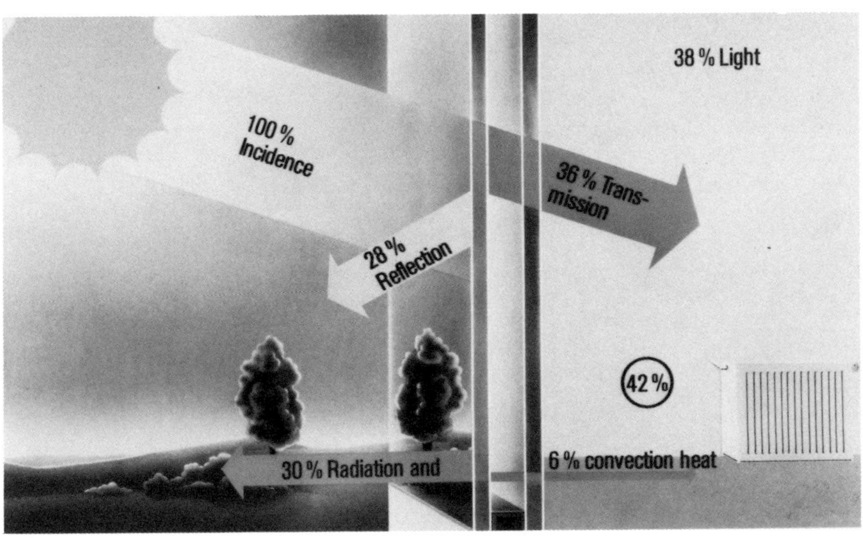

Figure 20: IROX A1, medium. Balanced properties of reflection and transmission. IROX facade panels particularly suited for all-glass facades.

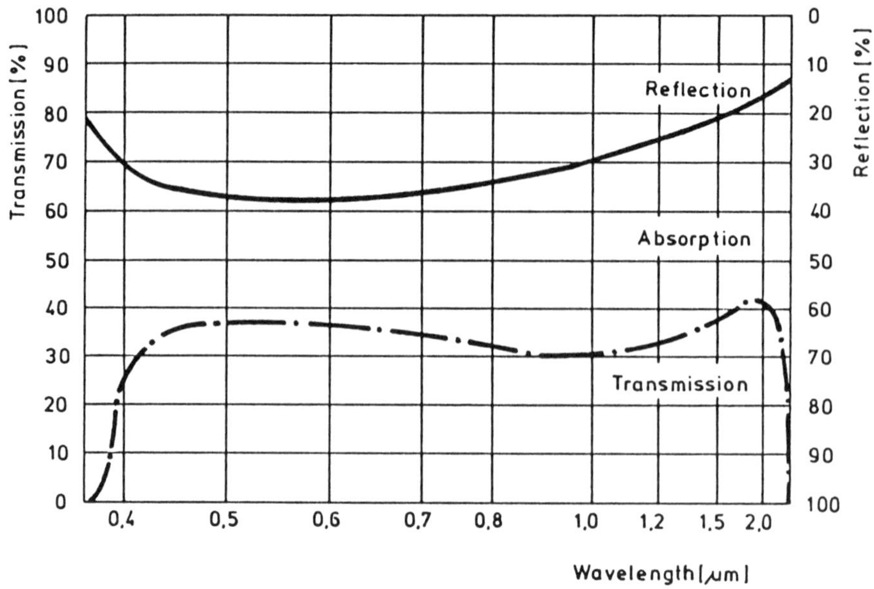

Figure 21: IROX A1. Values for reflection (R), transmission (T) and absorption (A) with IROX insulating glass 2 x 6 mm and the angle of incidence, $\alpha = 0°$.

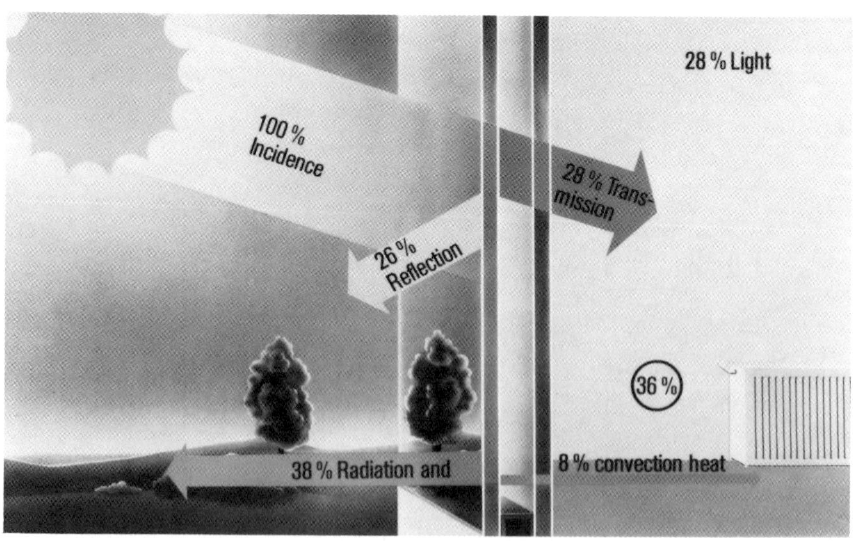

Figure 22: IROX A2, dark. Distinct attenuation of light transmission. Recommended for large-area glazing, open-plan offices and sun-rich countries.

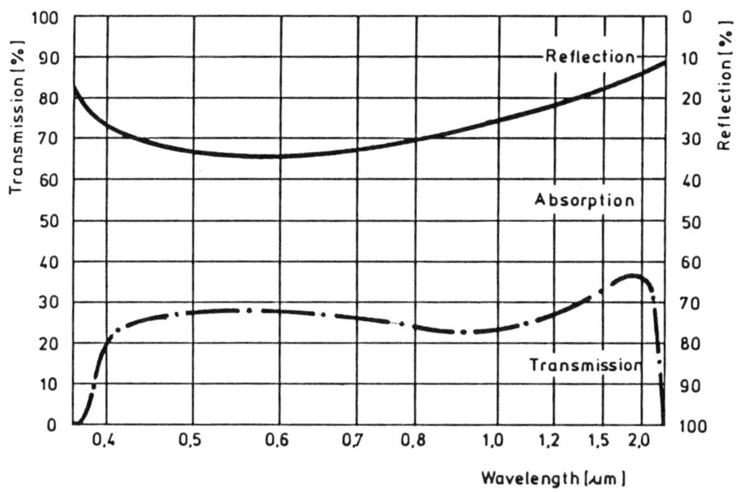

Figure 23: IROX A2. Values for reflection (R), transmission (T) and absorption (A) with IROX insulating glass 2 x 6 mm and the angle of incidence, $\alpha = 0°$.

Anti-Reflective Coatings

Regular flat glass panels for the protection of valuable paintings and for display cases reflect about 4-5% of the incident light, undesirable for glass covers. Figure 24 shows the reduction in reflection for glass plates coated with 2 or 4 antireflective layers.[4]

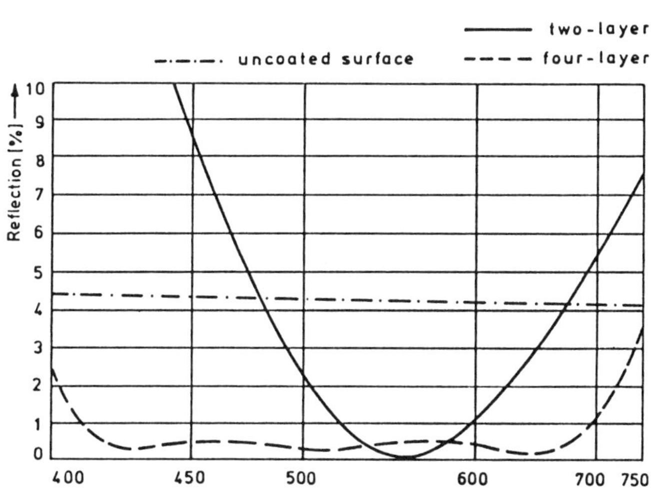

Figure 24: The reflection of a two-layer (–) and a four-layer (- - -) antireflection coating compared with that of an uncoated (- · -) surface.

Other Surface Coated Glasses

The Schott Glaswerke produces and sells about another fifty different optical filter products, produced with the sol-gel dip technology.[6,21] This represents only a fraction of the technically feasible products and shows the multitude of capabilities for this technology. A description of the filter applications would by far exceed the volume of this chapter. Therefore, only the main areas of application will be mentioned here.

- Filters with high light transmission for projection and illumination lamps.
- Cold mirrors with high reflection for visible light and low reflection for infrared radiation.
- Selective reflecting mirrors (edgefilters), such as color separation filters required by the printing trade for the preparation of colorprints, light-splitters for video cameras and lighting equipment in photography, film and TV studios.
- UV mirrors for the separation of the UV radiation in gas discharge lamps and reflectors for illumination testing equipment.
- Achromatic, intensity sensitive light splitters for the separation and combination of light beams in optical systems, semi-transparent mirrors for observation chambers.
- Conversion filters for changing the color-temperature of light sources and recording equipment in TV, film and photography studios.
- Absorption filters such as UV sharp edge filters for the separation of specific spectral ranges in the UV region, such as UV lamps, flashlamps and filters for radiation equipment.
- Antireflection coated glasses with high transmission, such as highly refractive, radiation protection filters, reduced transmission mirrors.
- Glasses with electrically conductive layers to eliminate electrostatic discharge.
- Spectral selective filters, such as dichroic filters with low absorption in the passband range, for light sources in TV, photography and film studios as well as for recording equipment.

SOL-GEL LAYERS UNDER DEVELOPMENT

This section describes layers which are still under development or in pilot production. Some of these developments are close to maturity and ready for introduction into industrial production. Others might fall by the wayside due to other, more economical production technologies. However, they served a purpose by stimulating developments into new directions. This is closely related to the wide range of potential applications for the sol-gel processes.

The final section intends to describe one or another sol-gel layer product with a high level of future importance. The successful product will be determined by the extent of its product or process related advantages.

Antireflective Coatings

The development is driven by application specific requirements. The dominating considerations are the reduction in reflectivity and transmission losses of about 8-10% for standard glass plates. Layers with high, medium and low refractive indices are required. In this respect, the sol-gel process offers decisive advantages. Another important advantage is the high degree in thickness uniformity over large areas because any deviation in uniformity is clearly visible.

Contrast Enhancing Filters for Data Display Screens

Dipped sol-gel layers of TiO_2, SiO_2 and mixed layers of TiO_2/SiO_2 are fabricated as multilayers on absorbing float glass substrates to achieve the desired spectral reflection and transmission as shown in Figure 25. The development is at the pilot plant stage.[36] The absorption can be adjusted to the desired requirements by selecting the proper glass and thickness and the composition of the absorbing coating,[5] altogether a definite advantage of the sol-gel process. A preliminary acceptance test for this product is quite promising.

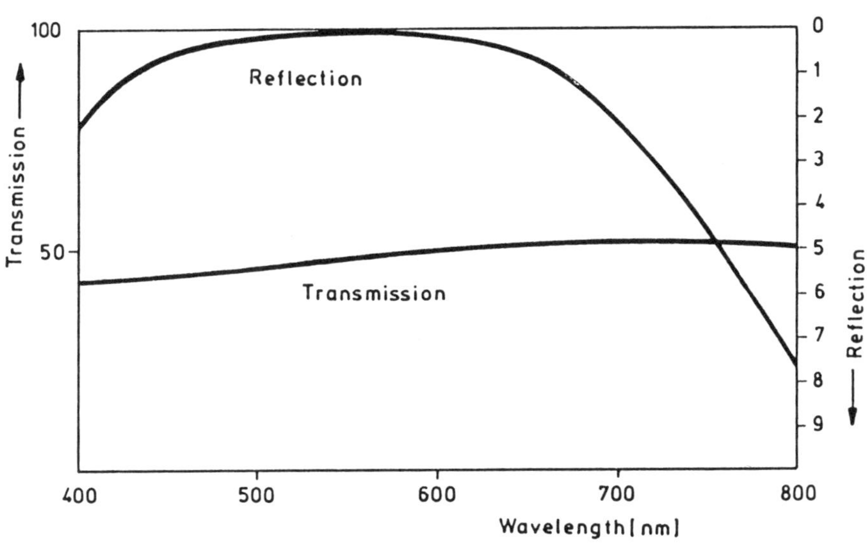

Figure 25: Reflection and transmission curves of an antireflected contrast enhancement filter, coated on both sides.

Porous Antireflective Coatings in the UV-Range

Antireflective coatings in the UV-range can be realized with highly refractive, dense layers of HfO_2, La_2O_3, ZrO_2, ThO_2 and a low refractive layer of SiO_2,[37] with a refractive index $n_D = 1.45$.

Lower refractive indices of $n_D \leqslant 1.45$ can be accomplished through porous layers. A porous, single SiO_2 layer can satisfy the antireflective requirements. The necessary pore size is considerably smaller than the wavelength of the light. These layers are transparent and have a reduced refractive index, lower than the refractive index of dense inorganic materials such as MgF_2 ($n_D = 1.38$).

After exposure to the atmosphere, water absorption in the pores can cause problems.[38,39] An improvement is the deposition of films with a gradient in porosity, equal to the gradient in refractive index, with a refractive index range from the substrate material to the ambient air. The layer is deposited from a $Si(OR)_4$ solution with the addition of a defined amount of water. This film remains porous when it is not heated to temperatures of 400° to 500°C. The pores are subsequently enlarged by etching to the desired refractive index gradient.[40] Such silica glass lenses with broadband antireflective coatings for the visible and UV range are used for the beam concentration in laser fusion experiments. These layers are hardened against laser radiation and protected from the atmospheric water vapor attack and can find many other applications as well.

Antireflective Coating of Silicon Solar Cells

Identical procedures and processes are applied for the coating of glass plates with dense layers. A defined refractive index, e.g. of $n = 2.0$ can be achieved by mixing two solutions, as described in Reference 4 and in use for 40 years. A continuous coating line for silicon substrates appears to be feasible.[41,46]

Leaching of Multicomponent Oxide Layers

This process is based on the phase separation after high temperature treatment of multicomponent oxide glasses, such as borosilicate glass and others. A highly soluble phase is leached out in a suitable etch solution and a skeleton like, porous SiO_2 layer remains. This SiO_2 layer shows a gradient in its refractive index which can be modified to some degree. This coating and leaching technique can be used for the formation of antireflective coatings on a variety of substrates.[42] Certain glass compositions of the system SiO_2, Na_2O, B_2O_3 and Al_2O_3 led to the development of radiation resistant antireflective coatings for high power laser optics, used in laser fusion experiments.[43] The resistance to 1.06 μm, 1 nsec laser pulses was four times higher than for antireflective coatings consisting of vacuum deposited SiO_2-TiO_2 multilayers.[44] It should be mentioned that SiO_2-TiO_2 sol-gel layers, deposited under clean conditions, showed much higher resistance to laser radiation than similar, vacuum deposited layers.[45]

Spray-Coated Diffusor Layers[23]

Besides the previously mentioned antireflective filters for data display screens,[34] cost effective techniques for contrast enhancement and glare reduction of cathode ray tubes (CRT) are needed. These layers should be applied directly to the CRT screen. Expensive lapping and polishing renders a "silk-mat" effect. Etching with hydrofluoric acid causes considerable environmental problems. Dip coatings[58] are not possible due to the shape of the tube. A spray coating[23] of a mixture of Aerosil, SiO_2, and a silicate ester, $Si(OR)_4$, is applied to the CRT front surface and fired at 400°C. A rough layer of Aerosil particles embedded in a glassy SiO_2 film results, acting as diffusor, similar to etched surfaces as shown in Figure 26.

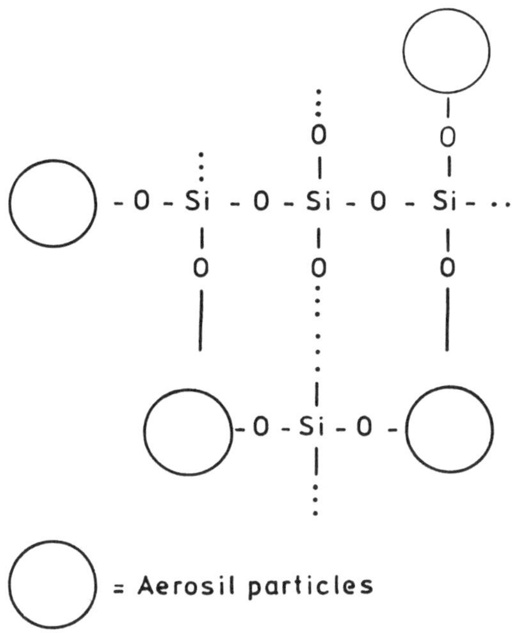

Figure 26: SiO_2-particles in SiO_2 (schematic).

These cost-effective coatings are also suitable for slip casting and can be applied to polymethylmetacrylate to obtain a silky-mat surface, Figure 27.

Figure 27: Surface of a light-scattering coating.

The embedding of Si_3N_4 particles in transparent SiO_2 layers can be achieved in a similar sol-gel spin-on process.[59]

Transparent, Electric Conducting, IR-Reflecting Layers

The sol-gel dip process lends itself to the fabrication of electrically conducting oxide layers used for displays, windows, antistatic coatings and solar cells. A combination of characteristics for those coatings is required, such as good electrical conductivity, high infrared reflection, high transmission of solar radiation in the visible range, ease of etching in acids, high environmental stability, high scratch resistance and good adhesion on glass. In general, indium-tin-oxide layers are preferred, but other coatings have been used for special applications.

Indium-Tin-Oxide Layers

This subject has been covered by the author and his co-workers in recent publications[32,47,48] and a brief coverage should suffice here.

After special cleaning of float-glass plates, an alkali-diffusion barrier layer is applied in a dip process. The plates are subsequently immersed in a tin and indium compound solution. After withdrawal, the coating is heat-treated at 400° to 500°C in a reducing atmosphere and cooled to 200°C. Some typical characteristics of these coatings, variable to a large degree, are shown in Table 1. The transmission and reflection characteristics are shown in Figure 28.

Table 1: Properties of ITO-Films Optimized for Application in Displays (a, c) and for Window-System (b)

	(a)	(c)	(b)
Film thickness [nm]	20-30	∿ 80-100	270
Charge carrier density [cm^{-3}]	4×10^{20}	$5\text{-}6 \times 10^{20}$	$5\text{-}6 \times 10^{20}$
Mobility [$cm^2\ V^{-1}\ s^{-1}$]	15-20	60-70	35-40
Conductivity [$\Omega^{-1}\ cm^{-1}$]	800 - 1 200	5 000-6 000	3 000-3 500
Sheet resistance [Ω/□]	500	∿ 25	10

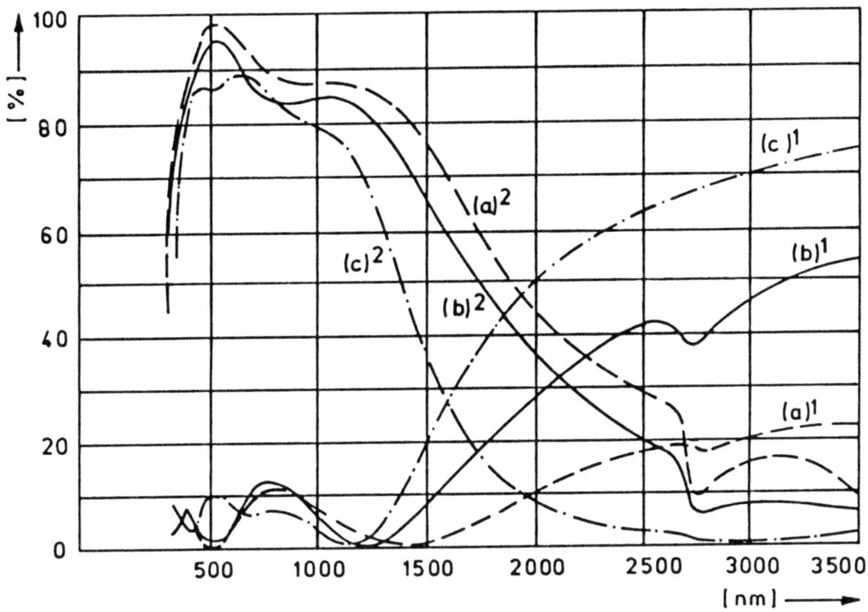

Figure 28: Transmission and reflection curves of ITO-films ( )^1 = reflection; ( )^2 = transmission.

The stability of those layers has been verified in three years of environmental testing. Doped indium oxide layers combine many advantages of the sol-gel technology. By modifying the process chemistry, film characteristics can be tailored to meet specific requirements. For example, by changing the impurity type and concentration, the electrical conductivity can be changed over several orders of magnitude. A high solubility in hydrochloric acid is desirable for the pattern etching required for displays. By incorporating TiO_2 as a dopant, highly acid resistant layers can be obtained. Interesting applications as electrodes in corrosive electrolytes[49] can be expected. The usage of indium, which is relatively expensive, is quite efficient in this process.

Despite the good prospects for the application of ITO (Indium-Tin-Oxide) coatings, other processes, like the SnO_2 on-line coating of float glass or the Ag sputter-coating (between dielectric layers) might be more economical in the mass production of certain products (window panes). The high degree of flexibility of the sol-gel dip coating process for the fabrication of ITO layers offers a real chance.

Cadmium Stannate

The synthesis of cadmium-stannate[31] is characterized by the rigid stoichiometric reaction typical for the sol-gel process. In the process as shown in Figure 29 SnO_2, CdO or $CdSnO_3$ are not formed.

$$2Cd(OAc)_2 + Sn(n\text{-}OBu)_4 \xrightarrow[\text{Temp.}]{H_2O} Cd_2SnO_4 + 4AcOH + 4n\text{-}BuOH$$

Figure 29: Synthesis of cadmium stannate.

Opto-Electronic Films

Layers of PLTZ (Pb, La, Ti, Zr) oxides have been produced successfully with the sol-gel process. This development is still in progress.[50]

Magnetic Films

Iron-oxide layers on borosilicate glass substrates have been deposited and subsequently annealed at 500° to 800°C in a hydrogen atmosphere. ESR (Electron Spin Resonance) and FMR (Field Magnetic Resonance) signals were analyzed and the surface characterized by SEM (Scanning Electron Microscope).[51]

Barrier Films

Barrier layers can serve different applications. The out diffusion of alkali ions from glass substrates into the coating layer can be harmful and should be prevented. A protective barrier layer may be required as protection from atmospheric corrosion. This application covers a fairly wide field.

SiO_2 based layers have been used successfully as barriers, even at elevated temperatures, to retard the out diffusion of alkali ions. A SiO_2 layer under a conductive coating prevents the alkali ions from "poisoning" the conductive film,[32,54] SiO_2–GeO_2 layers, matched to the thermal expansion of the substrate, can protect non-oxide based ceramics against oxidation at elevated temperatures.[52]

Unstable glasses, as used in a laser protective sandwich filter can be attacked in humid environments. They can be successfully protected with a 0.2 μm thick, multicomponent phosphorus-silicate layer.[30]

Pure aluminum phosphate layers can be produced with the sol-gel process and should be resistant against fluorine and fluorides,[53] but have not yet found any industrial applications.

Sol-gel layers made from $SiO_2/B_2O_3/Al_2O_3/Na_2O/BaO$ improve the temperature stability of black chromium solar selective films at 300°C in air by a factor of two.[56]

Regular glass fibers for communication applications embedded in cement are not stable under alkali attack. ZrO_2–SiO_2 glass fibers, drawn from sol-gel deposits, are stable against alkali attack but too expensive for fiber communication applications.[57] Low-cost E-glass fibers (alkali-free Ca–Al–B–silicate glass) coated with ZrO_2–SiO_2 films showed an improvement in stability in initial experiments.[58]

The number of applications is expected to grow in the future.

Sulfide Films

One of the few published papers on this subject describes the formation of photosensitive, oxygen-free Sb_2S_3 layers[29] from $SbCl_3$ and thiourea. A survey is given in "Chemical Solution Deposition of Inorganic Films"[55] which covers not only sulfides. A reference to the sol-gel process is given.

Glassy Thick Films

As mentioned before, sol-gel layers can be produced up to about 1 μm thickness.[59] For the fabrication of thicker layers, up to about 100 μm on glass, ceramic or metal, a different process was developed.[60] A low melting glass powder was dispersed in a metal alkoxide solution. The alkoxide serves as the binder between the substrate surface and the glass powder. A thick, homogeneous film is formed[61] in a subsequent heat treatment exceeding the softening point of the glass powder.

Unsupported Glass Films

Unsupported glass films in the thickness range of 1 to 100 μm can be fabricated with the sol-gel process. Films of SiO_2, SiO_2-TiO_2 and SiO_2-ZrO_2 have been reported.[13,62,63] The most promising technique among several possibilities appears to be the drawing of the film from a solution with suitable viscosity through a slotted nozzle at 800°C to 1200°C.

New Oxide-Based Gel Films

Gel layers not subjected to a high temperature treatment contain oxides, hydroxides, oxyhydroxides and hydrated oxides as recently reported.[64]

Semiconducting V_2O_5 layers are expected to give photographic films antistatic protection,[65,66] and reduce moisture sensitivity when compared to polymer films. V_2O_5 gel layers were reported to show switching and memory effects. WO_3 gel layers exhibit electrochromic characteristics.

Organic Modified Silicate Films

Organic, modified silicates undergo, at least partially, the same hardening mechanism as oxides through polycondensation, and should have the potential for coatings in a partially condensed and soluble state. The reaction will be completed by subjecting the layer to a final heat treatment. This process should enable the design of coatings with specific characteristics.

Scuff Resistant Layers

A coating solution of $Ti(OR)_4$, $Si(OR)_4$, and $(RO)_3Si(CH_2)_3OCH_2CH_2CH_2O$ was prepared and applied as coating for transparent organic polymers and cured at 120°C. The scuff resistance of the polymers was significantly improved by these coatings, highly desirable for many transparent organic polymers, such as polymethylmethacrylate.[35]

Solid Phase System for Radio Immuno Assay

For radio immuno assay analysis a covalent bond of an antibody in a hydrophilic environment was desired. By incorporation of amino-, anilino- and aldehyd-groups,[35] the covalent bond was achieved as well as the hydrophilic characteristics by the cocondensation of alcohol residues (Si–C bonds). In addition, it was desirable to have good adhesion on glass surfaces (inside of small glass bottles). This example shows how different requirements can be successfully combined in the sol-gel technology.

Protective Coatings

Glass-ceramic coverplates for cooking ranges can be attacked in rare instances by hot molten sugar. This attack takes place in a very narrow temperature range; however, it results in a very tenacious coating. The protection can be accomplished by a coating with a special alkyl-heteropolysiloxane, a classic name for an organic modified silicate.[67]

CONCLUSIONS

This chapter has attempted to review trends in the area of thin films from the sol-gel process in a brief and quantitative manner. Some new reports, often with comprehensive original citations have been covered. In addition, the author has tried to show the abundance of variations offered by this technology. Among the few basic principles of the sol-gel chemistry, this is probably the most important characteristic.

Only fundamental principles could be covered here, sufficient for the development of new products, but insufficient in detail for gaining full insight into the intricate reaction mechanisms. This could become a limit for sol-gel technology and causes some discomfort. From discussions and applicable publications, to mention only R.G. Dosch,[68] J.D. Mackenzie[69] and D.R. Uhlmann,[70] it is apparent that this deficiency has been recognized and remedial action has been initiated. This is not a short term undertaking, considering the degree of complexity and variability. Nevertheless, a concerted research effort into sol-gel chemistry will gain new insights into the basic principles, provide us with innovative ideas for its capabilities and will lead us to new and useful applications in the future. The synergy of research and application technology, long proven in many fields, promises a fruitful period for years to come.

REFERENCES

1. Ebelmen, *Annales de Chimie et de Physique,* Ser. 3, Bd 57: 319-55 (1846).
2. Geffcken, W. and Berger, E., Deutsches Reichspatent 736 411; May 6, 1939; assigned to Jenaer Glaswerk Schott & Gen., Jena.
3. Prospect Rückspiegel auf interferenzoptischer Basis. Deutsche Uhrglasfabrik GmbH, Grunenplan, FRG.
4. Prospect Glas ohne Reflexe. Deutsche Spezialglas AG, Grünenplan, FRG.
5. Prospect Calorex®. Das absolut farbneutrale Sonnenreflexionsglas von Schott Glaswerke, Mainz, FRG (1983).
6. Prospect Gläser mit Oberflächenschichten. Schott Glaswerke, Mainz, FRG (1981).
7. Schroeder, H., in: *Physics of Thin Films* (G. Hass, ed.), Vol. 5, pp 87-141, Academic Press, New York (1969).

8. Dislich, H., Hinz, P. and Kaufmann, R.; FRG-Patent 19 41 191; Aug. 13, 1969; assigned to Jenaer Glaswerk Schott & Gen., Mainz, FRG.
9. Levene, L. and Thomas, I.M.; FRG-Patent 20 08 653; Feb. 2, 1970; assigned to Owens-Illinois, Inc., USA.
10. Dislich, H., *Angewandte Chemie International Edition* 10, No. 6: 363-70 (1971).
11. Roy, R., *Journal of the American Ceramic Society* - Discussions and Notes 52, No. 6: 344 (1969).
12. Kamiya, K. and Sakka, S., *Res. Rep. Fac. Eng. Mie Univ.* 2: 87-104 (1977).
13. Mishima, M., Yamamoto, Y., Sakka, S., Kamiya, K. and Makita, K.; FRG-Patent 31 23 205; June 11, 1980; assigned to Central Glass Co., Ltd., Ube, Yamaguchi, Japan.
14. Matsuyama, I., Sagamihara, K., Susa, K., Satoh, S., Iruma, S., Suganuma, T. and Tokorozawa, S.; FRG-OS 31 16 883; April 4, 1981; assigned to Hitachi, Ltd., Tokyo, Japan.
15. Prassas, M., Phalippou, J. and Zarzycki, J., *Journal of Materials Science* 19: 1656-65 (1984).
16. Hench, L.L., Role of chemical additives in controlling sol-gel processes. 2nd International Conference on Ultrastructure Processing of Ceramics, Glasses and Composites. Palm Coast, Florida, Feb. 25-March 1 (1985) to be published by John Wiley & Sons.
17. Mogami, M., Morita, Y., Hayakawa, J. and Komiyama, T., *Bull Gort. Ind. Res. Inst. Osaka* 32(2): 111 (1981).
18. Pentinghaus, H., *Journal of Non-Crystalline Solids* 63: 193-99 (1984).
19. Klein, L.C., *Glass Industry* 63, No. 5, 27-9 (1982).
20. Brinker, C.J., Clark, D.E. and Ulrich, D.R. (eds) *Materials Research Society Symposia Proceedings,* Vol. 32, New York: Elsevier Science Publishing Company, Inc. (1984).
21. Dislich, H. and Hussmann, E., *Thin Solid Films* 77: 129-39 (1981).
22. Hussmann, E., *Chemie für Labor and Betrieb* 12: 463-66 (1975).
23. Hinz, P. and Dislich, H., Anti-reflecting light-scattering via the sol-gel procedure. 3rd International Workshop "Glasses and Glass-Cermics from Gels"Montpellier, France, Sept. 12-14 (1985) to be published by *Journal of Non-Crystalline Solids.*
24. Haisma, J., Heller, P., Pasmans, J.M.M. and Biermann, U.K.P.; European Patent Application 0 136 751; Sept. 2, 1983; assigned to N.V. Philips' Gloeilampenfabriken, Eindhoven, The Netherlands.
25. Bach, H. and Schroeder, H., *Thin Solid Films* 1: 255-76 (1967/68).
26. Furnuchi, S. and Aikawa, K.; FRG-AS 19 31 936; June 24, 1969; assigned to Asahi Glass Co., Ltd., Tokyo, Japan.
27. Dislich, H., Hinz, P., Klein, W. and Baucke, F.G.K., not published, Schott Glaswerke, Mainz, FRG.
28. Brinker, C.J., Formation of oxynitride glasses by ammonolysis of gels. *Communication of the American Society:* C-4-C-5 ((1982).
29. Nayak, B.B., Acharya, H.N., Chandhuri, T.K. and Mitra, G.B., *Thin Solid Films* 9: 309-14 (1982).
30. Dislich, H., *Journal of Non-Crystalline Solids* 57: 371-88 (1983).
31. Dislich, H., Hinz, P. and Wolf, G., U.S. Patent 4,229,491; Dec. 13, 1977; assigned to Jenaer Glaswerk Schott & Gen., Mainz, FRG.

32. Arfsten, N.J., Kaufmann, R. and Dislich, H., European Patent Application 0 114 282; Dec. 8, 1983; assigned to Carl-Zeiss-Stiftung trading as Schott Glaswerke, Mainz, FRG.
33. Dislich, H. and Jacobsen, A., FRG-Patent 14 94 872; July 7, 1965; assigned to Jenaer Glaswerk Schott & Gen., Mainz, FRG.
34. Dislich, H. and Jacobsen, A., *Angewandte Chemie International Edition* 12, No. 6: 439-44 (1973).
35. Schmidt, H., in: *Materials Research Society Symposia Proceedings* (C.J. Brinker, D.E. Clark, D.R. Ulrich, eds), Vol. 32, pp 327-35, New York: Elsevier Science Publishing Company, Inc. (1984).
36. Prospect Kontrastanhebende Vorsatzscheiben für Datensichtgeräte. Schott Glaswerke, Mainz, FRG (1985).
37. Hussmann, E.K. and Schnabel, R., in: *SPIE The International Society for Optical Engineering* (S. Musikant, J. Dupuy, eds), Vol. 400, pp 107-14, SPIE The International Society for Optical Engineering, Bellingham, WC (1983).
38. Yoldas, B.E., *Applied Optics* 19, No. 9: 1425-29 (1980).
39. Yoldas, B.E., *Applied Optics* 23, No. 9: 1418-24 (1984).
40. Yoldas, B.E. and Partlow, D.P.; European Patent Application 0 130 801; June 29, 1983; assigned to Westinghouse Electric Corporation, USA.
41. Yoldas, B.E. and O'Keeffe, T.W., *Applied Optics* 18, No. 18: 3133-38 (1979).
42. Boling, N.L. and McCollister, H.L.; FRG-OS 30 45 635; Dec. 3, 1979; assigned to Owens-Illinois, Inc., USA.
43. Lowdermilk, W.H. and Mukherjee, S.P., *Lawrence Livermore Lab. Report DE 82-006778* (1981).
44. Mukherjee, S.P. and Lowdermilk, W.H., *Applied Optics* 21: 293-96 (1982).
45. Hussmann, E., not published, Schott Glaswerke, Mainz, FRG.
46. Brinker, C.J. and Harrington, M.S., *Solar Energy Materials* 5: 159-72 (1981).
47. Arfsten, N.J., Kaufmann, R. and Dislich, H., in: *Ultrastructure Processing of Ceramics, Glasses, and Composites* (L.L. Hench and D.R. Ulrich, eds), pp 189-96, John Wiley & Sons, New York (1984).
48. Arfsten, N.J., *Journal of Non-Crystalline Solids* 63: 243-49 (1984).
49. Arfsten, N., Kaufmann, R. and Schubert, D.; European Patent Application 0 131 827; July 8, 1983; assigned to Carl-Zeiss-Stiftung trading as Schott Glaswerke, Mainz, FRG.
50. Novoselova, A.V., Turova, N. Ya., Turevskaya, E.P., Yanovskaya, M.I., Kozlova, N.I., *Izvestiya Akademii Nauk SSSR, Neorganicheskie Materialy* 15, No. 6: 1055-67 (1979).
51. Kordas, G. and Arfsten, N., Magnetic thin films produced by sol-gel processes. 2nd International Conference on Ultrastructure Processing of Ceramics, Glasses and Composites. Palm Coast, Florida, Feb. 25-March 1 (1985), to be published by John Wiley & Sons.
52. Schlichting, J. and Neumann, S., *Journal of Non-Crystalline Solids* 48: 185-94 (1982).
53. Rothon, R.N., *Thin Solid Films* 77: 149-53 (1981).
54. Mizuhashi, M. and Gotoh, Y., *Reports Res. Lab. Asahi Glass Co., Ltd.* 32, [2] : 79-86 (1982).

55. Chopra, K.L., Kainthla, R.C., Pandya, D.K. and Thakoor, A.P., in: *Physics of Thin Films* (G. Hass, ed.) Vol 12, pp 167-235, Academic Press, New York (1982).
56. Pettit, R.B. and Brinker, C.J., in: *SPIE Optical Coatings for Energy Efficiency and Solar Application* Vol. 324, pp 176-83, SPIE The International Society for Optical Engineering, Bellingham, WC (1982).
57. Kamiya, K., Sakka, S. and Tatemichi, Y., *Journal of Materials Science* 15: 1765-71 (1980).
58. Guglielmi, M. and Maddalena, A., *Journal of Materials Science Letters* 4: 123-4 (1985).
59. Schroeder, H. and Kaufman, R., FRG-Patent 11 83 413; Dec. 23, 1960; assigned to Jenaer Glaswerk Schott & Gen., Mainz, FRG.
60. Martinsen, J., Figat, R.A. and Shafer, M.W., in: *Materials Research Society Symposia Proceedings* (C.J. Brinker, D.E. Clark, D.R. Ulrich, eds), Vol. 32, pp 145-51, New York: Elsevier Science Publishing Company Inc. (1984).
61. Brinker, C.J. and Scott, T.R.; U.S. Patent 4,476,156; Mar. 10, 1983; assigned to the United States Department of Energy, Washington, D.C.
62. Sakka, S., Kamiya, K., Makita, K. and Yammamoto, Y., *Journal of Non-Crystalline Solids* 63: 223-35 (1984).
63. Sakka, S. and Kamiya, K., *Journal of Non-Crystalline Solids* 42: 403-22 (1980).
64. Livage, J., in: *Solid States Chemistry, Proceedings of the Second European Conference* (R. Metselaar, Heijligers, J.M. and Schoonman, J., eds) pp 17-32, The Netherlands: Elsevier Scientific Publishing Company (1983).
65. French Patent Application 2 318 422; Feb. 11, 1977; assigned to Kodak-Pathe, France.
66. French Patent Application 2 429 252; Jan. 18, 1980; assigned to Kodak-Pathe, France.
67. Steinbach, H.H., Schnurrbusch, K. and Rieder, M.; FRG-OS 29 52 756; Dec. 29, 1979; assigned to Bayer AG, Leverkusen, FRG.
68. Dosch, R.G., in: *Materials Research Society Symposia Proceedings* (C.J. Brinker, D.E. Clark, D.R. Ulrich, eds), Vol. 32, pp 157-61, New York: Elsevier Science Publishing Company, Inc. (1984).
69. Mackenzie, J.D., *Journal of Non-Crystalline Solids* 73: 613-24 (1985).
70. Uhlmann, D.R., Zelinski, B.J.J. and Wnek, G.E., in: *Materials Research Society Symposia Proceedings* (C.J. Brinker, D.E. Clark, D.R. Ulrich, eds), Vol. 32, pp 59-70, New York: Elsevier Science Publishing Company, Inc. (1984).

5

Antireflective Films from the Sol-Gel Process

Richard B. Pettit, Carol S. Ashley, Scott T. Reed and C. Jeffrey Brinker

Sandia National Laboratories
Albuquerque, New Mexico

INTRODUCTION

Antireflection (AR) coatings are being utilized in many applications at Sandia National Laboratories. Several applications support the various solar energy related programs currently carried out by the Department of Energy, including solar thermal and photovoltaic systems. In solar thermal systems, antireflection coatings have been used in parabolic trough collectors to increase the transmittance of protective glass envelopes that are positioned around the receiver tube. In parabolic dish collectors, several receiver designs utilize a transparent window over the receiver aperture to isolate gases, fluids and materials inside the receiver from the outside atmosphere. Antireflecting the glass surface in these applications, using sol-gel derived films, can increase the solar transmittance from initial values of 0.91 to values above 0.96, an improvement of ~5%.

In photovoltaic systems, AR coatings are required on all solar cells in order to reduce significant reflectance losses. For silicon cells, solar reflectance losses of 36% for uncoated material have been reduced to ~12% using single layer TiO_2 sol-gel films. By applying double layer sol-gel coatings of SiO_2 and TiO_2 to silicon, the reflectance losses can be reduced to about 3%. In concentrator systems, acrylic Fresnel lenses are used to focus sunlight on small area solar cells. Porous, AR coatings, which can be applied to the polymeric lens surfaces, allow the lens transmittance to be increased by ~5%. Finally, AR coatings are needed to reduce reflectance losses associated with a thick glass lens element that is positioned next to a solar cell in a concentrating collector. The lens provides no direct concentration, but merely redirects radiation onto the cell when the concentrator experiences some tracking errors or concentrating lens misalignment. An AR coating limits the outer surface reflectance loss to less than 1%.

OPTICAL PROPERTIES OF THIN FILMS

When radiation is incident on the interface between two materials (which can include air or vacuum) with different refractive indices, a certain fraction of the radiation is reflected, while the remainder is transmitted (absorbed). Depending on the application, one may wish to maximize or minimize the amount of reflected radiation. For two materials with complex indices of refraction equal to $\tilde{n}_1$ and $\tilde{n}_2$, the fraction of radiation reflected from their interface at normal incidence is given by (1)

$$R_{1,2} = \frac{[\tilde{n}_1 - \tilde{n}_2]^2}{[\tilde{n}_1 + \tilde{n}_2]^2} . \quad (1)$$

The subscript on R indicates that the radiation passes from material 1 to material 2. The complex index of refraction can be written as $\tilde{n}_1 = n_1 - ik_1$ and $\tilde{n}_2 = n_2 - ik_2$ (where $i = \sqrt{-1}$). The real part of the index represents the ratio of the velocity of radiation in vacuum to that in the material, while the imaginary part of the index is related to energy absorption in the material. The absorption coefficient, α, is related to the imaginary index k through the equation,

$$\alpha = \frac{4 \pi k}{\lambda} , \quad (2)$$

where λ is the wavelength of the radiation. Substituting for $\tilde{n}_1$ and $\tilde{n}_2$ in Equation 1, the reflectance can be written as

$$R_{1,2} = \frac{(n_1 - n_2)^2 + (k_1 - k_2)^2}{(n_1 + n_2)^2 + (k_1 + k_2)^2} . \quad (3)$$

From this equation, it is evident that the reflectance is symmetrical, since switching material 1 and 2 gives the same result, and thus independent of the incident beam direction.

In order to reduce the reflectance at an interface, both n_1 must be approximately equal to n_2 and k_1 must be approximately equal to k_2. On the other hand, high reflectance values result when either n_1 or n_2 is large or k_1 or k_2 is large. For almost all non-metals (this includes insulators, dielectrics, and transparent materials) the effect of k in this equation can be ignored since k is small ($<10^{-2}$). Values for k approaching 0.1 and larger occur primarily for metals. Thus, in the case of non-metals, low reflectance values occur when $n_1 \approx n_2$, independent of the k_1 and k_2 values, while for metals, high reflectance results from the intrinsically large k values for these materials.

If one of the two materials at the interface is air ($n_1 = 0.1$; $k_1 = 0$), Equation 3 reduces to

$$R_{1,2} = \frac{(n_2 - 1)^2 + k_2^2}{(n_2 + 1)^2 + k_2^2} . \tag{4}$$

For a typical transparent dielectric material, like glass, $n_2 \approx 1.5$ and $k_2 \approx 0$. Therefore, the reflectance at an air/glass interface is 0.040 reflectance units or 4.0% (1.00 reflectance units equals 100% reflectance). In a lens or window application, the total reflectance (ρ) for the two surfaces (neglecting any bulk absorption) is equal to

$$\rho = \frac{2 \cdot R}{1 + R} = 0.077 = 7.7\ \%. \tag{5}$$

This reflectance can be almost completely eliminated by modifying the glass surface so that the index of refraction varies smoothly from the value of the substrate (n_2) to the value for air (1.0). Because fully dense (nonporous) materials with refractive indices below about 1.3 do not exist, graded indices are generally achieved by incorporating porosity either in the substrate surface or in a film applied to the surface. Minot[2] produced graded index surfaces on Pyrex by a phase separation/etching process. The Pyrex was heated near its softening point causing the glass to separate into interconnected silica-rich and alkali-borate-rich phases of small characteristic dimensions (~10 nm) much less than the wavelength of visible light. Due to the large difference in solubilities of the two compositional phases, etching in dilute acids causes preferential removal of the alkali-borate-rich phase. Under optimum conditions the resulting porosity is graded normal to the surface, increasing in volume as the outer surface is approached. Thus the index of refraction varies continuously from that of the bulk glass to a value near 1.0 at the outer surface. The graded layer reduces the reflectance over a broad wavelength range. For example, experiments with Pyrex have shown that the reflectance losses for two surfaces can be reduced to values below 0.007 from 300 nm to 2,000 nm.[3] An additional advantage offered by graded index films is that their optical properties are quite insensitive to film thickness as long as the optical thickness (defined as the product of the index and physical thickness) exceeds one-half the wavelength of interest.

Major disadvantages of the phase separation approach include: (1) few substrates exhibit phase separation of the required dimensional scale and connectivity, (2) phase separation often requires long heat treatments at temperatures near or above T_g which may cause deformation of the substrate, and (3) the phase separation process offers limited control over the final porosity which may result in scattering of the transmitted beam.

In order to avoid some of the problems itemized above, McCollister[4] applied phase separable films to glass substrates by a sol-gel process. Heating to 750°C followed by selective etching resulted in an AR surface which exhibited low reflectance over a broad wavelength region. More recently, Yoldas and Partlow prepared graded index surfaces by etching a porous silica film deposited by a sol-gel process.[5] Graded index surfaces may also be formed by neutral solution etching;[6] however, this process is also largely substrate specific.

An alternative and more common technique for antireflecting a surface involves applying thin layers that have index of refraction values that vary in discrete steps between the index of the substrate and air. Using a single coating of the correct thickness and index of refraction, the reflectance at a single wavelength can be reduced to a value near zero. This condition is met when the film has an index of refraction, n_f, given by

$$n_f = \sqrt{n_2 n_1} \tag{6}$$

while the film thickness, d, is adjusted to give a minimum reflectance at a wavelength λ_o defined through the equation

$$\lambda_o = 4 n_f d . \tag{7}$$

For example, to antireflect an air/glass interface at a wavelength of 500 nm requires a film with an index equal to $\sqrt{1.5 \times 1.0} = 1.22$ and a thickness d = 500/(1.22 x 4) = 103 nm.

Better results can be obtained using two thin layers instead of one. In this case there are four independent parameters (n_{f1}, d_1, n_{f2}, d_2) that can be varied. Typically, the parameters are chosen to yield a reflectance value near zero at two separate wavelengths, so that the reflectance is minimized over a broad wavelength range between these wavelengths which, for example, can extend over most of the visible spectrum. An example calculation is shown in Figure 1 for a double layer film on both the upper and lower surface of a thick glass plate with minimum reflectance values at 450 nm and 660 nm. Note that the reflectance is ≤0.02 from 430 nm to 770 nm as compared to an uncoated reflectance of 0.105. As we discuss later, sol-gel processing can be used to obtain both single- and multilayer interference films.

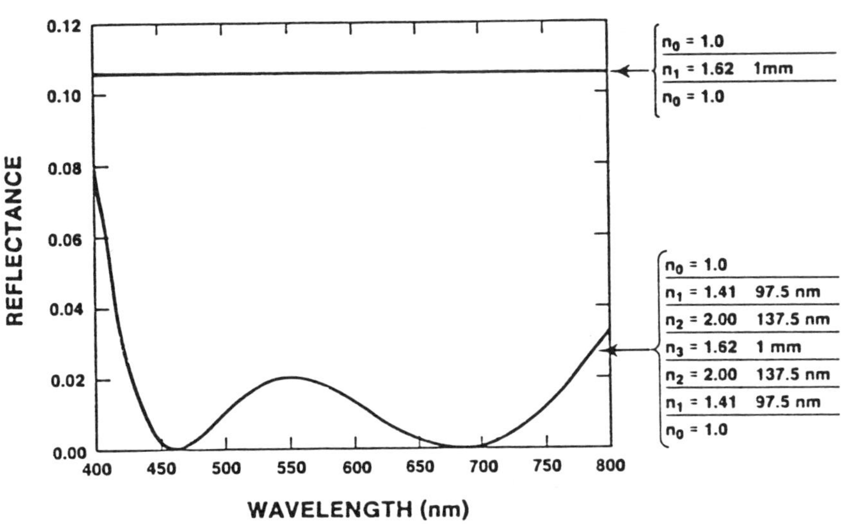

Figure 1: Calculated reflectance versus wavelength for a double layer film on a glass surface. The thickness of the top and bottom layer are 97.5 and 137.5 nm. The refractive indices are 1.41, 2.00, and 1.62 for the top layer, bottom layer, and glass substrate, respectively.

SOL-GEL PROCESSING

Many of the optical requirements of AR coatings can be met by sol-gel processing. The most common sol-gel process of glass preparation uses monomeric compounds [normally alkoxides $M(OR)_n$] of network forming elements (e.g., M = Si, B, Ti, Al) as glass precursors where R is an alkyl group. In alcohol/water solutions the alkoxide groups are removed step-wise by acid or base catalyzed hydrolysis reactions, and are replaced by hydroxyl groups (Equation 8). Subsequent condensation reactions involving the hydroxyl groups yield networks composed of inorganic oxide (M–O–M) linkages (Equations 9 and 10).

$$2\ Si(OR)_4 + 2\ H_2O \rightarrow 2\ (RO)_3Si\text{-}OH + 2\ ROH \quad (8)$$

$$2\ (RO)_3Si\text{-}OH \rightarrow (RO)_3Si\text{-}O\text{-}Si(OR)_3 + H_2O \quad (9)$$

$$(RO)_3Si\text{-}OH + Si(OR)_4 \rightarrow (RO)_3Si\text{-}O\text{-}Si(RO)_3 + ROH \quad (10)$$

Depending on solution conditions, e.g., pH, water concentration, temperature, solvent, etc., continued condensation reactions result in polymers exhibiting a complete spectrum of structures ranging from primarily linear, entangled chains to discrete clusters to fully condensed colloidal particles.[7]

At any point in the polymerization process, the solution can be deposited on a substrate by conventional methods such as dipping, spinning, or spraying. Solution requirements for achieving optical films by dipping have been extensively reviewed by Schroeder.[8] Briefly summarized, it is necessary for the solution to wet the substrate. The solution should be relatively stable with respect to aging and should show a minor tendency towards crystallization. In order to fabricate multilayer coatings, the previously applied layer must be insoluble in the solution and be wettable by subsequent solutions. To achieve this result it may be necessary to heat treat the deposited film prior to subsequent dipping operations.

During the deposition process, the polymers are concentrated and sheared as the solvent flows and evaporates on the substrate surface. Concurrently, the polymers continue to condense as unreacted terminal groups come in contact. Condensation reactions take place both within and between polymeric clusters as well as between the polymers and the substrate. Depending on the relative rates of shear, evaporation and condensation, as well as the surface tension of the solvent, gelation may occur at an early or late stage of the drying process. Once gelled, further evaporation of solvent from the film causes shrinkage and, finally, the formation of porosity. As discussed later, tailored porosity is an important advantage of sol-gel derived AR coatings.

Heat treatment of the porous film causes desorption of physisorbed water and alcohol, further condensation of the skeleton, and at higher temperatures, complete consolidation by viscous sintering. Since the film is constrained in the plane of the substrate, all of the shrinkage occurs in the direction normal to the substrate surface. This imposes a certain amount of tensile stress in the plane of the film which retards consolidation.[9] Since the sintering rate is proportional to the surface energy divided by the product of viscosity and pore size, the shrinkage rate is enhanced by high surface area and small pore size.[10] The viscosity decreases exponentially with increasing temperature and increases with the degree of crosslinking of the skeleton. Thus, the sintering rate is further enhanced by heat treatments which raise the temperature without increasing the degree of crosslinking. This is practically achieved by rapid heating.[10]

AR Coatings

Of interest to the problem of forming AR surfaces is the immense control over refractive index and thickness inherent in the sol-gel processing. The effective refractive index, n, of a mixture of two materials (a and b) in a porous structure is related to the pore volume and composition according to the Lorentz-Lorenz relationship:[11]

$$\frac{n^2 - 1}{n^2 + 2} = V_a \frac{n_a^2 - 1}{n_a^2 + 2} + V_b \frac{n_b^2 - 1}{n_b^2 + 2} \; ; \quad V_a + V_b + V_p = 1.0 \qquad (11)$$

where V_a, V_b, V_p and n_a, n_b, n_p (n_p = 1.0) represent the volume fraction and refractive index of phase a, phase b and the porosity, respectively. Thus, the refractive index of the composite coating (phase a, phase b and the porosity) may be controlled either by appropriate mixtures of two end member phases or by the inclusion of porosity. As discussed by Schroeder,[8] the thickness is controlled by several intrinsic solution parameters including the viscosity, surface tension or vapor pressure. In dipping, thickness is further controlled by speed of withdrawal and angle of inclination of the surface to the liquid interface. The thickness of spun films depends on the spin speed.[12]

An example of refractive index tailoring resulting primarily from composition control is shown in Figure 2 where n is plotted as a function of mol fraction TiO_2, x, in the $xTiO_2-(1\text{-}x)SiO_2$ system.[12] Both Schroeder's and our data show an approximately linear dependence of n on x. This demonstrates that the refractive index in a largely amorphous binary system roughly obeys the Lorentz-Lorenz rule of mixtures (Equation 11). The constant difference between our results and those of Schroeder may reflect a larger volume fraction of porosity in our films or, alternatively, a larger volume fraction crystallinity in Schroeder's films (thus, introducing a fourth phase V_c, rutile or anatase, with a corresponding higher refractive index n_c = 2.6). Both of these hypotheses are consistent with the fact that Schroeder's films were heated to 500°C compared to 450°C in our case.

Figure 2 shows that, for relatively dense films, the refractive index can be continuously varied between about 1.41 and 2.24. In Figure 3 the refractive index and thickness are plotted for a 90% TiO_2·10% SiO_2 composition as a function of temperature after 15 min heat treatments.[12] This figure demonstrates that, in addition to composition, the refractive index also depends on the extent of consolidation of the film. Thus, by independently controlling the composition and extent of consolidation of the film, a wide range of refractive indices is accessible. The refractive index may be further modified by subsequent crystallization (increased n) or etching (reduced n).

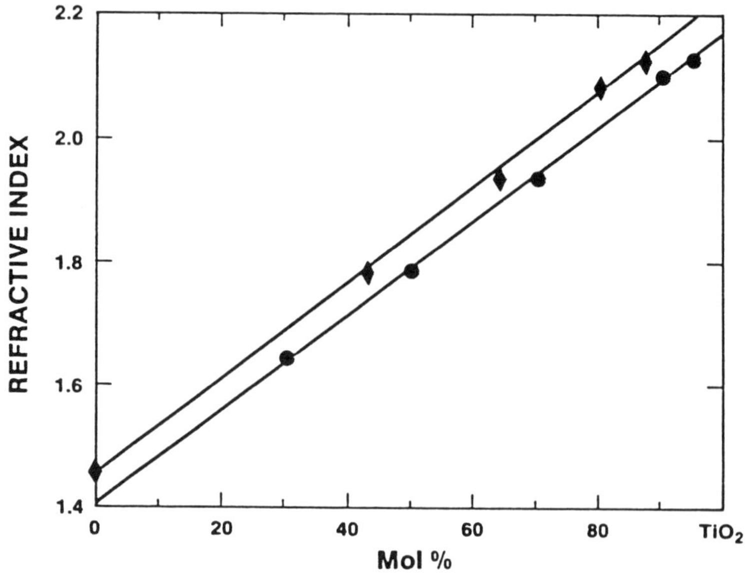

Figure 2: Refractive index as a function of x(mol %) for gel-derived films in the binary system $xTiO_2 \cdot (100 - x)SiO_2$. ♦ are data of Schroeder[8] for films densified at 500°C. ● are data of Brinker and Harrington[12] for films densified at 450°C.

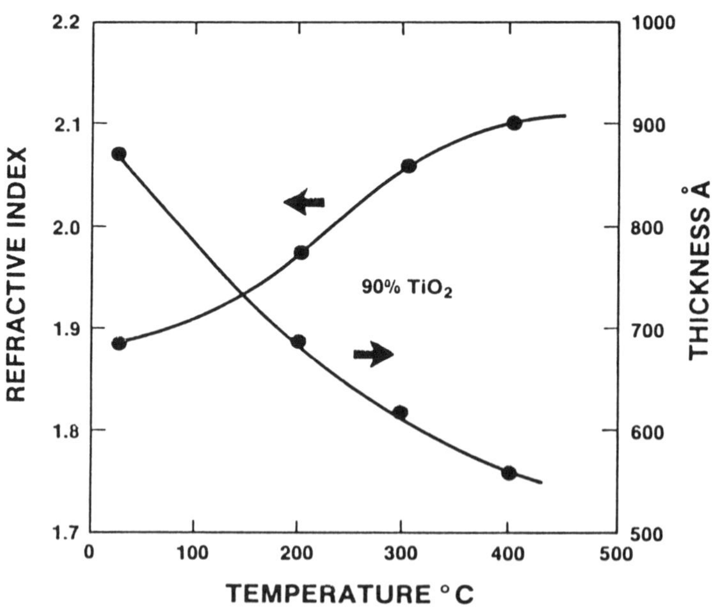

Figure 3: Refractive index and thickness for $90TiO_2 \cdot 10SiO_2$ films after 15 min. heat treatments at the indicated temperatures.[12]

Microstructure Tailoring

In practice, refractive indices in the range 1.4 to 2.3 are generally achieved using dense films in silicate binaries such as $SiO_2 \cdot TiO_2$, $SiO_2 \cdot Al_2O_3$ or $SiO_2 \cdot ZrO_2$. In order to obtain refractive indices below 1.4, it is necessary to introduce porosity. According to Equation 11, the addition of 50 volume % porosity reduces n to 1.25 for SiO_2 and 1.50 for TiO_2. Because refractive indices above 1.4 are easily obtained with fully densified films in binary and multicomponent silicate systems, it is generally of interest to add porosity only to the lowest refractive index end member, e.g., SiO_2 or $SiO_2 \cdot B_2O_3$.

A distinct advantage of the sol-gel processing over conventional film forming processes, e.g., sputtering and chemical vapor deposition, is the ability to tailor the pore volume (and thus, refractive index), pore size and surface area of the deposited film.[13] Because the pore structure reflects to some extent the size and topology of the solution grown polymers (to be discussed below), microstructure tailoring is achieved by controlled polymer growth in solution prior to film deposition. Further refinements are made by controlling the environment of the deposition process or by etching the fired film.

Depending on solution conditions, e.g., pH, water concentration and solvent composition, the polymer growth may result in structures ranging from linear or randomly branched chains to discrete porous clusters to fully dense colloidal particles.[14] Solution conditions also affect the relative rates of condensation and evaporation during deposition. Both polymer topology and the relative rates of condensation and evaporation influence the evolution of structure during film formation.

In silicate solutions prepared at low pH (<2.5) and with understoichiometric additions of water (H_2O/SiO_2 mol ratio <2 in Equation 8), polymer growth is biased towards linear or randomly branched chains.[7] In addition, under these conditions the rate of condensation (Equations 9 and 10) is very low.[7] There are several important consequences of these phenomena: (1) Due to the low condensation rate, highly concentrated solutions can be prepared which are stable with respect to gelation. These concentrated solutions can be used to obtain rather thick films (500-1,000 nm). (2) Linear polymers will tend to align, due to the shear stress imposed during dipping or spinning, which may result in anisotropic optical properties. (3) Because the condensation rate is low, evaporation of solvent combined with shear allows the polymers to become highly compacted and aligned without initially causing gelation. At the latter stages of evaporation, the film may be sufficiently crosslinked to form a gel. However, because the silicate network is still weakly crosslinked at the gel point, further removal of solvent continues to compact the film. When the polymeric skeleton comprising the gel is sufficiently rigid to withstand the compressive stress exerted by liquid menisci on the exterior surface, any additional solvent removal results in the formation of a continuous porous phase which invades the original, solvent-filled regions. Because porosity (if any) is created at the final stages of the drying process, the pore volume is extremely low[14] causing the refractive index to approach that of conventional vitreous SiO_2 (see Figure 4). These films are useful for fabricating multilayer interference coatings and single-layer AR coatings on high index substrates, e.g., silicon.

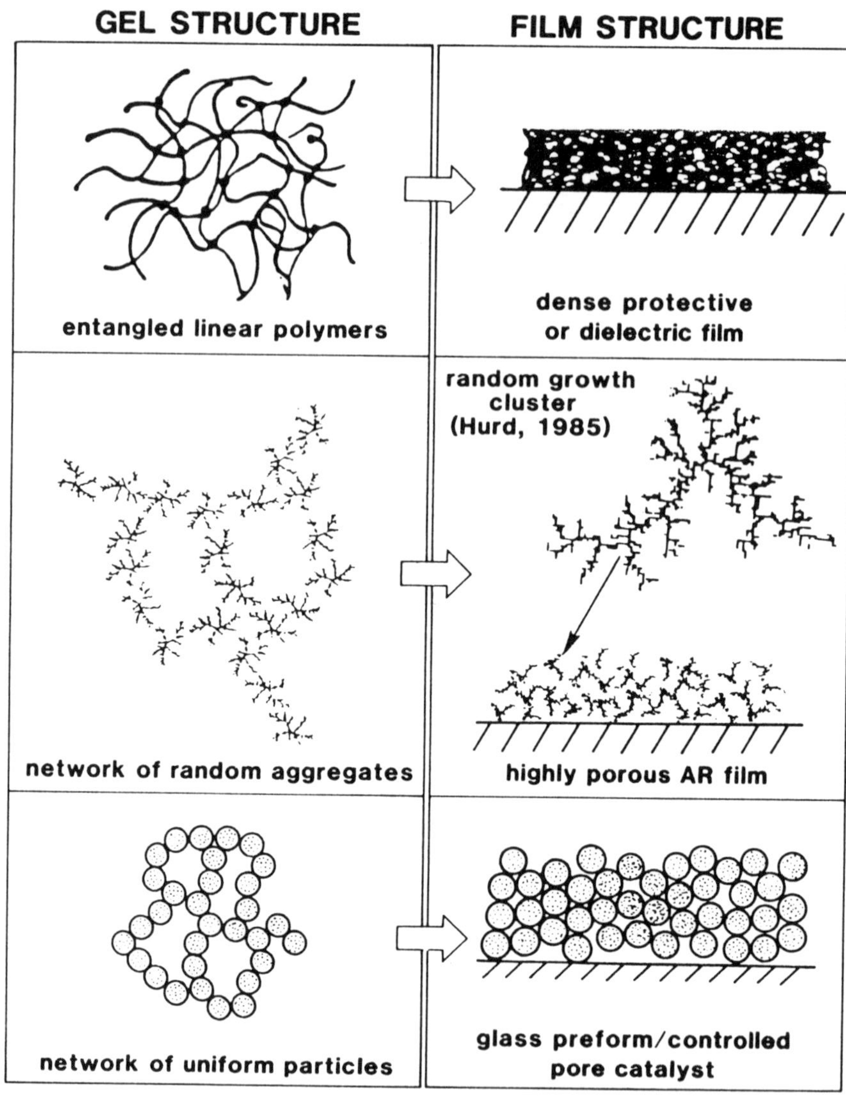

Figure 4: Schematic representation of gel structure and film structure for solution-derived polymers synthesized under different conditions.[16]

If the pH is increased slightly above 2.5 (the isoelectric point of silica), the condensation rate is greatly increased. The consequences of increased condensation rates are as follows: (1) Dilute solutions are required to avoid premature gelation, and thus film thicknesses are limited to ≅200 nm from a single application (2) Growth in dilute neutral solutions may occur by a process analogous to random particle cluster aggregation.[15] In this process, low molecular weight species randomly diffuse and condense on peripheral sites of a growing embryo[16] (see

Figure 5). This results in the formation of weakly condensed, spherically-expanding clusters whose density decreases radially from the center of mass according to: $m \propto r^d$, where m is the cluster mass, r, the radius, and d, the fractal dimension, is less than 3.[15] (3) During film deposition, gelation occurs at an early stage of evaporation, and subsequently, during continued solvent removal, the modulus of the gelled film increases rapidly. Thus, porosity is created at an early stage of the deposition process, and, as shown schematically in Figure 4, the resulting pore volume is quite large (refractive index is low).[14]

Figure 5: Simulation of growth cluster formed by random particle-cluster aggregation under neutral conditions, courtesy of Alan Hurd.[16]

An intriguing aspect of film formation from dilute, neutral systems is that the refractive index is not only low but completely controllable. This is demonstrated in Figure 6 which plots refractive index (measured by ellipsometry) and hydrodynamic radius (measured by light scattering) for borosilicate films deposited from solutions aged at 50°C for up to 30 days. The refractive index and hydrodynamic radius are inversely related and each varies as lagging time. This behavior is consistent with the polymer growth model described above. As the cluster becomes larger it incorporates a larger volume fraction of solvent. If the solvent space is converted to porosity during film formation, the refractive index of the film will decrease in a continuous manner with increasing polymer radius. Because modulus of a porous growth cluster is less than that of a fully condensed particle, the cluster should be compacted somewhat during the deposition process. This may limit the extent of porosity achievable by this method. Despite this possible limitation, a wide range of indices is directly attainable which permits the formation of quarterwave AR coatings on a number of substrates, e.g., glass, acrylic and polycarbonate.

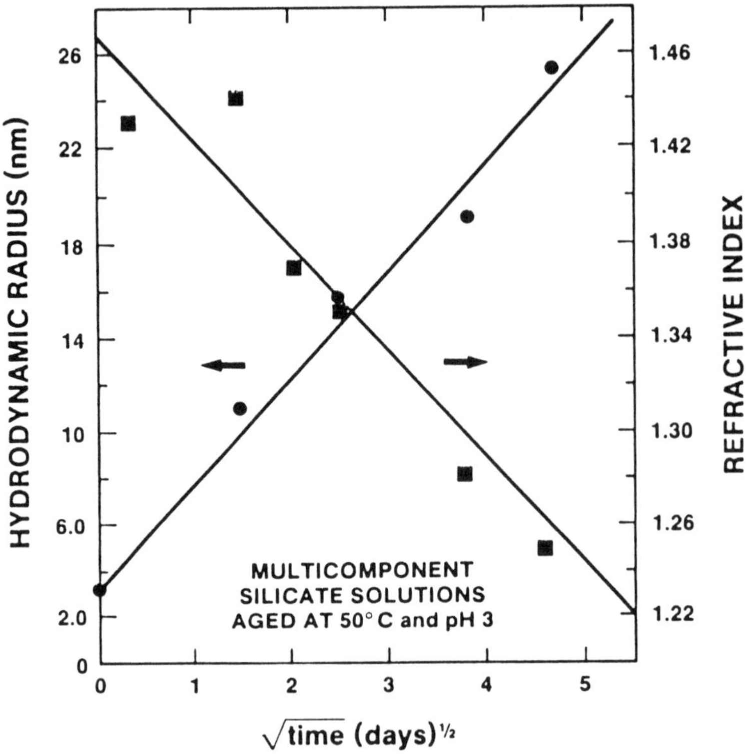

Figure 6: Hydrodynamic radius (●) measured in situ by dynamic light scattering (courtesy of Dale Schaefer) and refractive index (■) of films deposited from aged solutions (measured by ellipsometry) as a function of (solution aging time)$^{1/2}$.

Further increases in pH and/or over stoichiometric additions of water increase the rate of depolymerization in silicate systems (reverse of Equations 9 and 10). This leads to the formation of highly condensed species through bond breakage and reformation (processes which favor formation of the lowest energy configurations, viz. colloidal particles of essentially anhydrous SiO_2).[17] Cathro, et al.[18] prepared AR coatings from commercially available colloidal silicate solutions. Films deposited from solutions stabilized above pH 8 or below pH 2 were porous (n = 1.37). The pore volume was significantly increased (thereby decreasing n to 1.25) by aging the solution at pH 7 prior to film deposition. We expect that at this pH, SiO_2 colloids aggregate to form porous structures similar in form (but larger in scale) to the clusters depicted in Figure 5. However, Cathro, et al. do not present experimental details which would allow us to confirm this hypothesis.

Two further comments can be made concerning the potential for microstructural tailoring in AR coating applications. First, the deposition method (e.g., spinning versus dipping) influences the film structure. If identical solutions are deposited by dipping and spinning, the dipped film will often be denser (higher n) than the corresponding spun film. This is probably a consequence of the increased rate of evaporation that occurs during spinning. Structural arrangements, which allow the polymeric network to rearrange in response to the applied compressive stress, occur slowly with respect to the rate of evaporation, and the network is "frozen" in a more porous state. This effect may be compensated for (or overwhelmed by) the effect of shear on the polymers themselves. Highly ramified structures (such as those depicted in Figure 5) are fragile and may break apart under the high shear rates imposed by spinning, causing n to increase.

Secondly, the microstructure can be further modified by heating and/or etching. Heating initially densifies the skeleton and ultimately causes viscous sintering and complete densification.[10] In the simplest geometrical models, the result of heating is a uniform reduction in pore size (and volume) causing an increase in n. Thus, heating can be used to tailor n as described previously. (Figure 3). Heating may also cause devitrification of the film.

Etching is used to reduce n. Depending on the temperature, dilution and chemistry of the etchant as well as the pore size, composition and thickness of the film, etching porous gel-derived films may result in graded or uniform index films. For example, Yoldas employed a combination of heating and etching to prepare a graded index SiO_2 film used to coat laser optics.[5] In these experiments, heat treatments were used to reduce the pore size prior to etching in order to eliminate scattering at shorter wavelengths.

APPLICATIONS

It is apparent from the above discussion that sol-gel films are highly versatile and eminently suited for the preparation of AR coatings. By proper control of composition and microstructure, refractive indices between about 1.25 and 2.30 are easily obtained. Heating and/or etching provides a further level of refractive index tailoring. In the following sections, we report on several applications of sol-gel processed AR coatings.

AR Coatings on Silicon Solar Cells

Silicon solar cells have a high index of refraction which leads to a solar averaged reflectance value equal to about 36%. This large loss can be significantly reduced by coating the silicon with a thin, transparent dielectric layer with an appropriate dielectric constant and thickness. Single-layer AR coatings have included SiO, SiO_2, TiO_2, Si_3N_4, Ta_2O_5 and Al_2O_3, which leads to average reflectance values of from 10–12%. Further reduction in the reflectance can be accomplished with double-layer AR coatings. Therefore, double layers of SiO_2 and TiO_2, produced by sol-gel processing, were used to antireflect silicon solar cells.[19]

The sol-gel SiO_2 and TiO_2 solutions were applied to 5 cm diameter, polished, n-type, phosphorus doped silicon wafers using a photoresist spinner. The thickness of the fired coatings could be adjusted by varying the spin speed. The thickness of TiO_2 required for optimum optical properties could be applied through a single coating operation, whereas two coating operations were required for the SiO_2 film. Multiple coats were applied with only heat lamp drying between coatings. Samples were then fired at 450°C for 5 minutes to densify both layers. Samples of the individual coatings were also prepared for measurement of the index of refraction and thickness using ellipsometry at 632.8 nm. These values were then used in an optical modeling code (see Appendix) to calculate the spectral reflectance properties.

Calculated and measured reflectance results for single coatings of SiO_2 and TiO_2, and a double-layer coating of the same films are shown in Figure 7. Note that the measured index for the SiO_2 coatings of 1.414 is slightly lower than the published value for fused silica of 1.457 (both at 632.8 nm). This indicates that the film is probably slightly porous. The agreement between the measured and calculated curves is excellent in all cases. For the double-layer coating, the solar averaged reflectance is only 0.049. Because of the excellent agreement between the measured and calculated results, the optical modeling code was used to determine the optimum thickness of each coating required to minimize the solar reflectance. Calculations were performed using the measured index values and the resulting solar averaged properties were fit with a contour plot as shown in Figure 8. Note that the minimum reflectance value of 0.038 is about 0.08 lower than a single-layer AR coating. This optimum sol-gel double-layer coating has a SiO_2 thickness of 100 nm and a TiO_2 thickness of 62 nm. It can also be seen from the figure that the reflectance is not very sensitive to thickness variations in the two films. Thus an increase in reflectance of only 0.01 above the optimum value (e.g., from 0.038 to 0.048) results from SiO_2 thickness variations of ± 25 nm and from TiO_2 thickness variations of ± 10 nm.

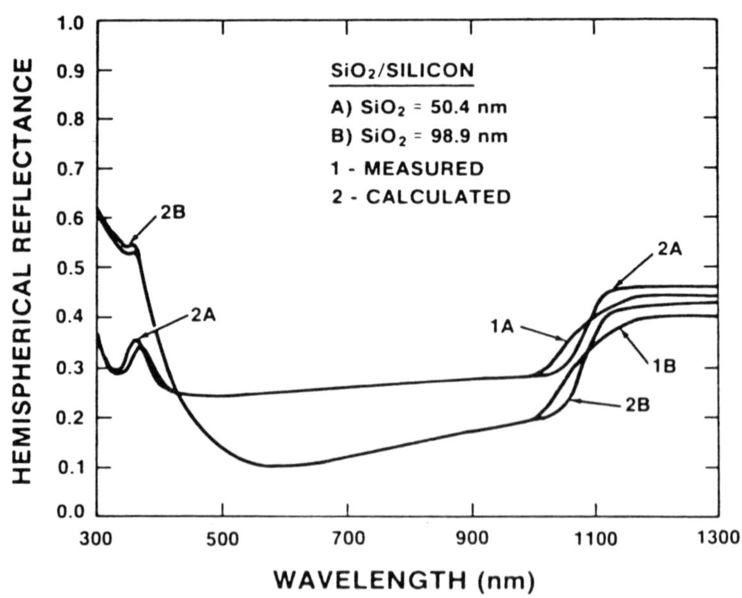

Figure 7A: Spectral hemispherical reflectance properties of single (50.4 nm) and double (98.9 nm) coatings of SiO_2 on silicon: 1A, measured, single; 1B, calculated, single; 2A, measured, double; 2B, calculated, double. Calculated curves use the listed thickness values determined from ellipsometric measurements.

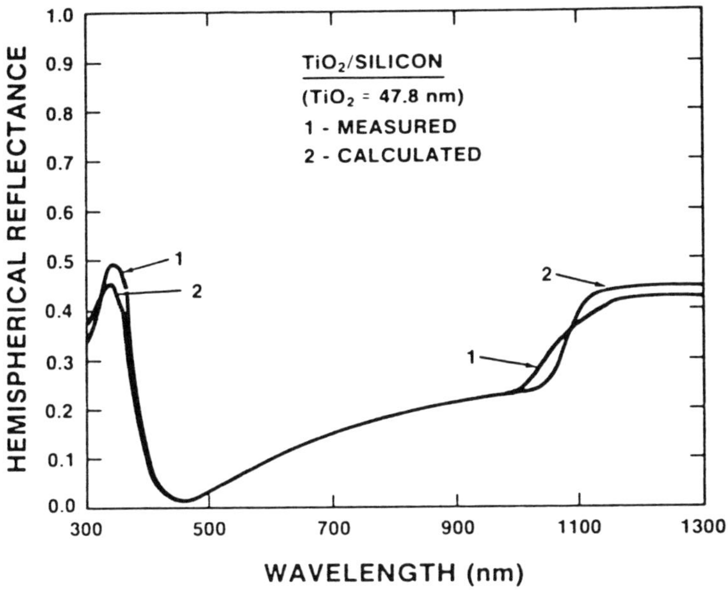

Figure 7B: Spectral hemispherical reflectance properties of a single 47.8 nm coating of TiO_2 on silicon: 1, measured; 2, calculated. The calculated curve uses the listed thickness value determined from ellipsometric measurements.

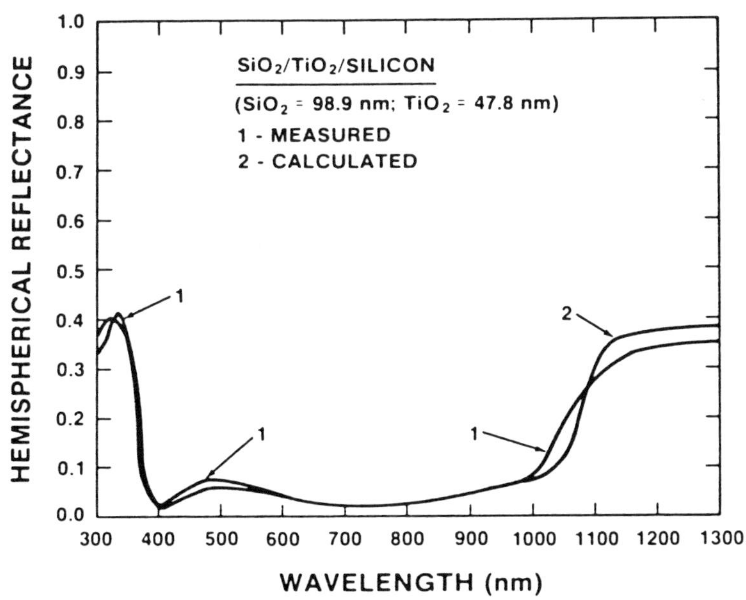

Figure 7C: Spectral hemispherical reflectance properties of a double-layer coating comprising 98.9 nm SiO_2 on 47.8 nm TiO_2 on silicon: 1, measured; 2, calculated. The calculated curve uses the listed thickness values for each film determined from ellipsometric measurements.

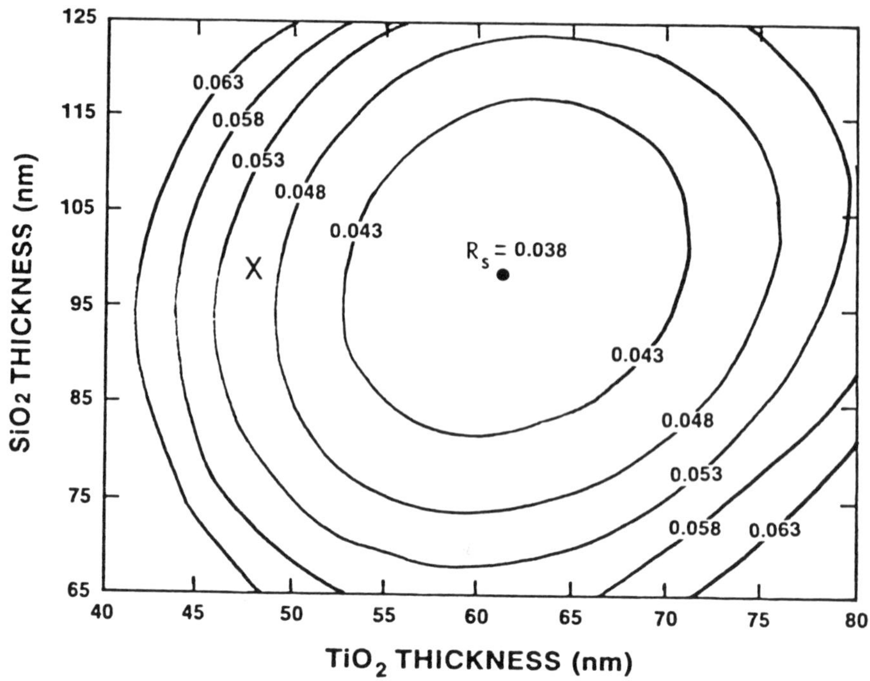

Figure 8: Contours of constant solar averaged reflectance for double-layer coatings on silicon as a function of the thicknesses of the SiO_2 and TiO_2 layers. The point of minimum reflectance • corresponds to 100 nm of SiO_2 and 62 nm of TiO_2. X locates the double-layer coating discussed in the text.

In order to determine if the decrease in reflectance corresponds to an equal increase in the cell efficiency, the performance of a Sandia designed solar cell was determined both before and after coating with the sol-gel double-layer coating. The reflectance properties of an uncoated and coated cell are shown in Figure 9. For the coated cell, the solar reflectance is 0.079, whereas the uncoated cell has a solar reflectance of 0.374. Thus the coating increased the cell solar absorptance from 0.626 to 0.921, or an increase of 47%. The measured cell efficiency increased from 12.1% to 17.4% for an increase of 44%. Because of the excellent agreement between the increase in absorptance and cell efficiency values, it appears that the 450°C firing temperature had no detrimental effect on the cell performance. Based on the measured reflectance of uncoated silicon of 0.36 and the minimum reflectance value possible of 0.038, it should be possible to increase the efficiency of a solar cell by slightly over 50% by using a sol-gel double-layer coating.

Antireflection Coatings on Glass

Most high temperature solar collectors utilize a glass envelope over the receiver surface to reduce convection and conduction losses. Reflection at the

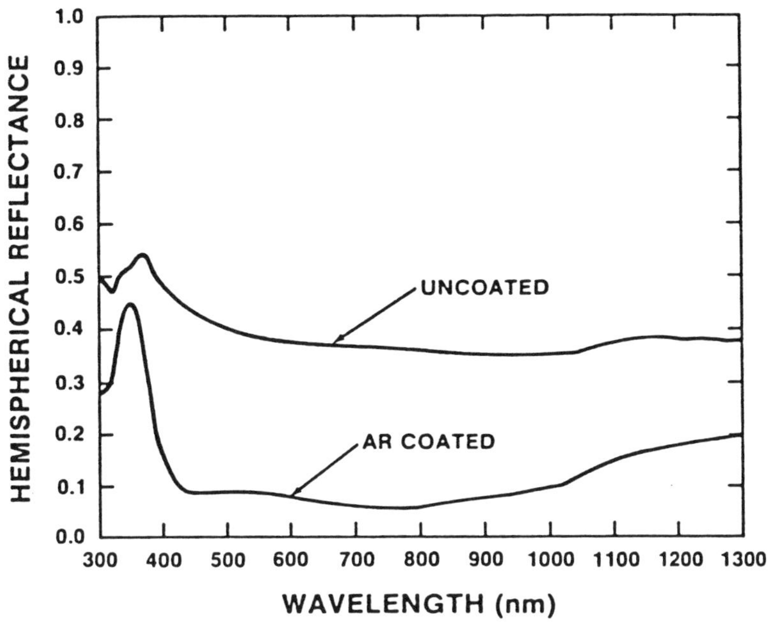

Figure 9: Measured spectral hemispherical reflectance properties of uncoated (as received) and AR coated solar cells.

two glass/air interfaces results in a reduction in the solar transmittance by approximately 0.07 transmittance units. By antireflecting the glass, the solar transmittance can be increased from 0.91-0.92 to 0.97-0.98. As discussed previously, one very effective way this can be accomplished with certain glass compositions is by the formation of a graded refractive index surface. For Pyrex glass, it is necessary to heat treat the glass at 575°C for 24 hours in order to cause the proper phase separation.[3] Because this temperature is close to the softening point of the glass, deformation of the Pyrex envelope is a problem. Therefore, we investigated the use of porous sol-gel coatings to antireflect the Pyrex surface. In the initial laboratory study it was of interest to answer two questions: (1) Do porous or etched gel-derived films provide the required low refractive index ($n_f \cong 1.22$)? (2) Do uniform or graded refractive indices result? To address these issues we formulated solutions in the multicomponent system: $71SiO_2-18B_2O_3-7Al_2O_3-4BaO$ (wt %).[20] Films were deposited on Pyrex microscope slides by a dipping process (3 coats) after various periods of solution aging. The deposited films were then heated for 1000 minutes at 500°C, further solidifying the films in order to increase their chemical and environmental durability.

We initially discovered that the refractive index and etching behavior of the as-deposited (and heated) films depended to a large degree on the solution aging conditions. Figure 10 shows the refractive index, n, as a function of etching time for films deposited from an unaged solution or a solution aged at room temperature for 60 days prior to deposition. Note the differences in refractive index of the heated but unetched films and their completely different etching behavior.

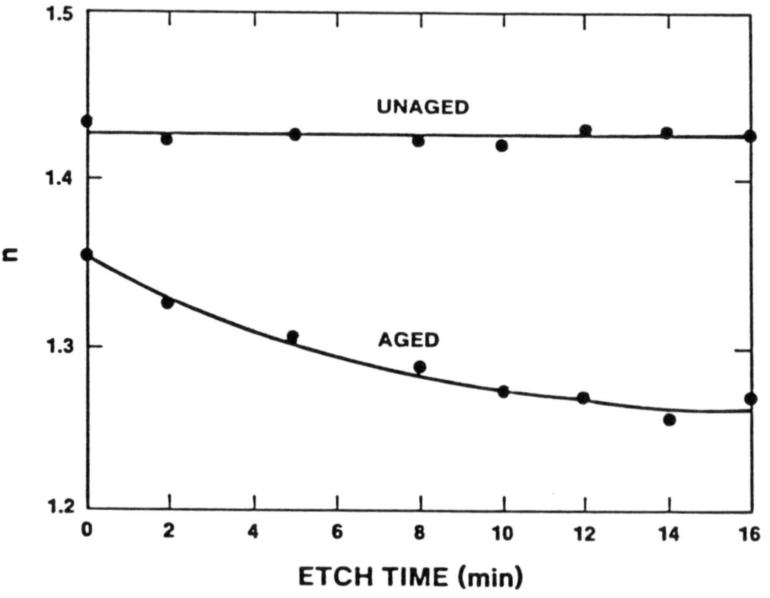

Figure 10: Refractive index of multicomponent silicate films as a function of etching time in NH_4HF_2, $H_2Si_2F_6$. Unaged refers to films deposited from freshly prepared solutions. Aged films were deposited from the identical solution after aging for seven days at 50°C. Both aged and unaged films were heated for 15 min. at 500°C prior to etching.

After heating, the film deposited from the unaged solution has a refractive index which is close to that expected for a melted borosilicate glass and which is unaffected by etching. This suggests that for the film deposited from the unaged solution the 500°C heat treatment resulted in a completely densified film. The film deposited from the aged solution remains porous after heating ($n < 1.47$) and it is possible to reduce the refractive index further by etching.

In Figure 11 the spectral reflectance as a function of etching time is plotted for films deposited from a solution aged at 50°C (multiple coats). Note that the as-deposited film has a minimum reflectance of ≈0.02 at a wavelength of 900 nm, but as the etching progresses, the minimum reflectance decreases to values below 0.005 and shifts to shorter wavelengths. After an etching time of 12 minutes, the solar averaged reflectance (R_s) reaches a minimum value of 0.022 when the reflectance minimum is centered near 600 nm. Optical modeling indicates that the behavior of all curves is adequately described by a film of uniform refractive index and thickness. In Figure 12 are shown the index and thickness as a function of etching time. The as-deposited film has an index of 1.35 and a thickness of 175 nm. As etching begins, the index decreases rapidly and then remains relatively constant at a value near 1.27, while the thickness initially remains constant and then begins to decrease. Therefore it appears that for these films the pores initially enlarge reducing the effective index, and then the thickness decreases shifting the reflectance minimum to lower wavelengths. Fourier transform infrared spectroscopy was used to monitor

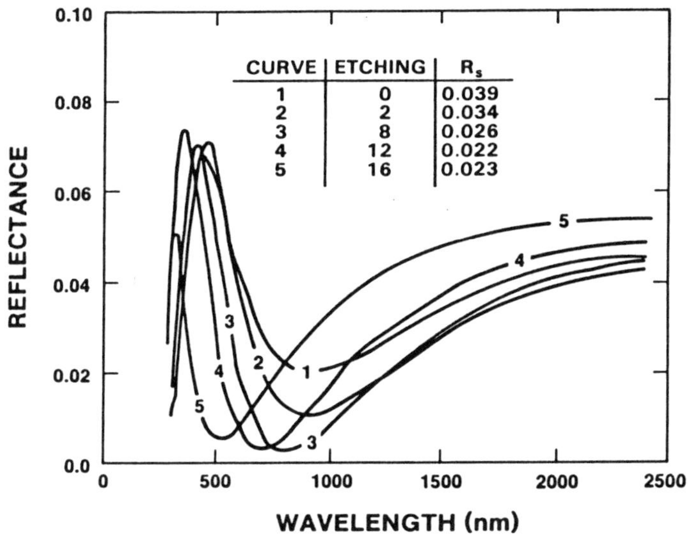

Figure 11: The spectral reflectance properties of a sol-gel coated Pyrex slide both as-coated and after etching for times up to 16 minutes.

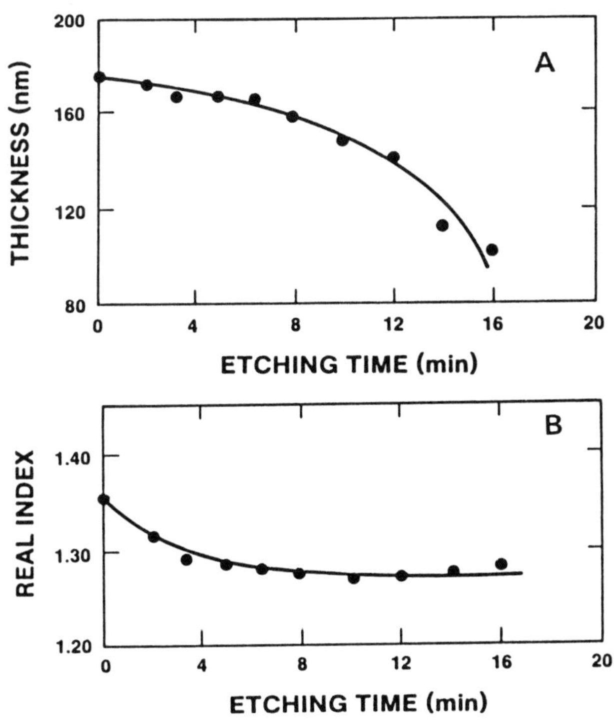

Figure 12: The thickness (A) and index of refraction (B) of a sol-gel film on a Pyrex slide as a function of etching time as determined from spectral reflectance measurements.

compositional changes accompanying etching.[21] The results clearly indicate that selective etching of the borate composition occurs. Apparently, selective etching in this case does not result in a graded refractive index within the film.

Full Scale Process Development. AR films on glass are necessary for a variety of solar applications including Pyrex envelopes used in parabolic trough collectors, low iron glass for flat plate collector glazing and fused silica for high temperature receiver apertures. To determine the applicability of porous sol-gel derived films for use in the solar industry, a full-scale process was developed by Ashley and Reed[22] to antireflect 3.0 m long by 6.0 cm diameter Pyrex tubes. These tubes are used in modular, line-focus parabolic trough collectors to protect the black chrome-coated receiver tube and to substantially reduce convective thermal losses. Antireflecting the inner and outer surfaces of the envelope can substantially increase the solar averaged transmittance of the glass, thus increasing collector efficiency.

In order to qualify the laboratory process for industrial use, first the aging and dipping procedures were substantially simplified and then appropriate modifications were employed to adapt the process to a tubular configuration.

Aging and Etching Conditions. It was established in the laboratory scale process that ~2 months of room temperature aging are required to produce films of the appropriate porosity. The "window" for room temperature aging exceeds 6 months. To reduce processing time, aging was accelerated by increasing the solution temperature to 50°C. Our objective was to find a minimum aging time which gave acceptable AR properties and to establish a "window" for 50°C aging. AR properties improved significantly after a few days of aging at 50°C and an optimum AR effect was achieved after two weeks. The window for 50°C aging extends to ~3 weeks when gelation of the solution occurs. To minimize continued polymer growth once the AR window was reached, the solutions were stored at 4°C when not in use. The solution was warmed to ~23°C in a water bath prior to coating. Films applied from cold, more viscous, solutions were nonuniform and too thick for optimum AR film formation, as has been noted by others.[5] Solution viscosity increased with increasing solution age and decreased with increasing temperature. By proper control of age, temperature, and draining speed, sufficiently thick (~135 nm) coatings were applied with a one-step process. During the course of the full-scale coating operation (~3 mo) negligible increases in viscosity occurred.

After application and heating, sol-gel coatings require acid etching to reduce the thickness and refractive index (increase porosity) for optimum AR film formation. A final coating, 123 nm thick with a refractive index of 1.22, must be obtained in order to produce a (¼)λ interface film with minimum reflectance at 600 nm (Equations 6 and 7). An etchant concentration of 0.015% NH_4HF_2/0.26 N H_2SiF_6 resulted in good AR film formation within 3-5 minutes at room temperature.

Adaptation to Tubular Geometries. Pyrex envelopes, 3.0 m long x 6.0 cm diameter, were drain coated using 20 liters of borosilicate composition, aged at 50°C for 6 days. Prior to coating, the tubes were cleaned as described previously.[22] Particulates on the inner and outer surfaces were removed mechanically as the final step in the cleaning procedure. The tubes were lowered into a 3.3 m x 7.6 cm polyvinyl chloride (PVC) coating tank, thus requiring ~7 meters of vertical clearance. The solution was pumped out of the coating tank at a constant rate and into a glass reservoir using an air-drive, positive displacement pump. A study of experimental variables indicated a correlation between seasonal variations in ambient relative humidity and coating quality. Coatings produced at less than 20% relative humidity were satisfactory, while high relative humidity conditions (~60%) resulted in opaque films with poor AR properties. Therefore in order to standardize drying conditions, the tubes were coated and dried in flowing dry nitrogen. The diameter of the coating tank was selected to minimize the volume of solution required for coating. However, the resulting small clearance between the outer tube surface and the inner wall of the tank caused additional drying problems, i.e., the flow of nitrogen was considerably restricted through this space resulting in poor drying of the outer tube surface. To promote uniform drying on both tube surfaces, a flow restrictor was inserted into the top of the tube prior to coating.[22]

After drying, the coated tubes were heated at 500°C in a specially modified conveyor belt furnace with a total travel time of about 2 hours per tube. A multiple bay PVC tank, equipped with heaters and circulators, was used for etching and rinsing. Initial processing included a hot water rinse after etching to promote rapid drying of the tubes, however tubes rinsed in this manner were consistently over-etched. The elevated temperature of the rinse may have caused accelerated, localized etching within the pores prior to etchant dilution. Therefore tubes were etched for 3-5 minutes in 0.015% NH_4HF_2/0.26N H_2SiF_6, rinsed at room temperature and force dried, in order to obtain the desired AR properties. Spectral reflectance properties of flat, lab-scale samples were determined using a Beckman 5270 spectrophotometer equipped with an integrating sphere accessory. Solar averaged transmittance values were calculated by averaging over the solar spectral distribution of Thekaekara for an air mass of 1.5.[23] Measurements on coated tubes were made using a portable solar reflectometer (Model SSR, Devices & Services Co.) and correlated with spectrophotometer data.[22]

Twelve tubes were processed for installation in the Modular Industrial Solar Retrofit (MISR) test facility at Sandia National Laboratories, Alburquerque, NM. Solar averaged transmittance values ranged from 0.956 to 0.968, compared with 0.915 for uncoated Pyrex, with maximum end-to-end variations of ~0.006. After processing, a slight decrease in transmittance was noted with time, which was attributed to adsorption of water in the porous surface. However, as has been noted for similar processes,[5] brief exposure to low temperatures, e,g, 150°C, returned the transmittance to original values. Reflectance measurements after 4 months of outdoor operation in the MISR facility indicated no decrease in transmittance within the reproducibility (~0.002) of the measurements. (See Figure 13.) Apparently, the operating temperature of the collectors was sufficiently high to minimize AR loss due to adsorption of contaminants (water, etc.) on to the pores. The success of sol-gel processing in coating large-scale substrates, which exhibit no measurable loss in transmittance after extended outdoor exposures, demonstrates the suitability of this method for commercial solar thermal applications.

Sol-Gel AR Films on Plastics

It was desirable to modify the process developed for glass surfaces for use on thermoplastics of interest to the solar industry: acrylic and polycarbonate.

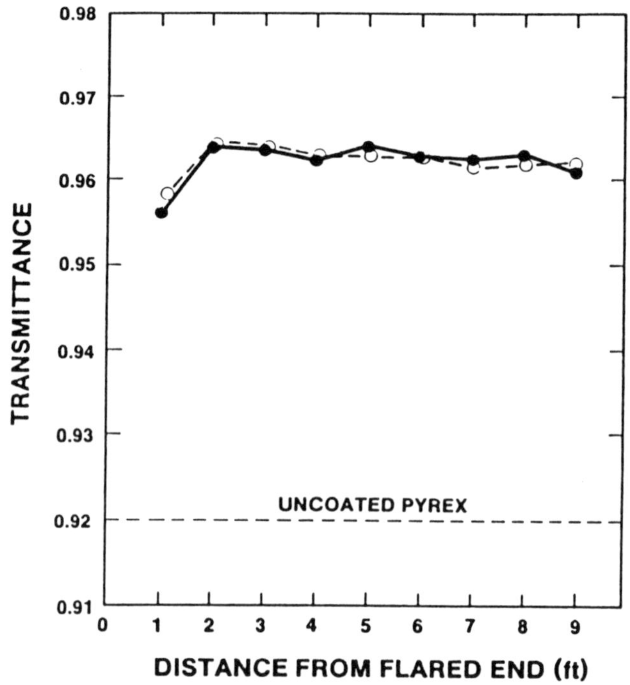

Figure 13: Solar averaged transmittance (averaged values for nine tubes) measured along the length of a 10 ft Pyrex tube coated on both sides with a porous borosilicate film. Open circles as coated, closed circles after 16 weeks of outdoor exposure.

The utility of these materials for solar applications is a result of their low cost, light weight (densities ~40% those of glass), outdoor stability and design versatility due to a variety of manufacturing techniques, i.e., injection molding, compression molding, casting and machining. Acrylic can be precision molded and is the primary candidate for use as a Fresnel lens material in solar photovoltaic applications using concentrators. Polycarbonate, although more expensive and less easily molded than acrylic, has greater dimensional stability and has been used as a glazing for large, flat-plate collectors.

The most serious restriction to the development of a process for AR film formation on plastics is the upper temperature limit of the substrates. Maximum continuous service temperatures of acrylic and polycarbonate are ~80°C and ~115°C, respectively. A second limitation is that unheated sol-gel films do not appear to etch in a manner similar to films heated to 500°C;[24] thus it is difficult to optimize the thickness and refractive index of the film by etching. However, the low index of refraction could be obtained through careful tailoring of the solution, while the proper thickness was obtained through precise control of the deposition process. Thus, the required AR films were deposited on polymers without the need for high temperature curing or acid etching.

As shown in Figure 6, the refractive index of sol-gel films deposited from aged solutions decreases with increasing solution age. However, in the refractive index range of interest for AR film formation on these plastics (<1.25), the aged solution is approaching gelation and thus has a limited shelf-life. It was found that if an aged solution that had gelled was ultrasonically agitated for a short time, it reliquefied and could again be used to produce films with index of refraction values equal to or lower than those obtained before the solution gelled. To further reduce the solution viscosity and increase its stability, the solution was diluted to an equivalent oxide concentration of ~1.5 wt %. With precise control of coating conditions, a film of the correct thickness and refractive index for obtaining a reflectance minimum at 600 nm could be applied to acrylic and polycarbonate substrates. Minimum reflectance values of ~0.005 were obtained for these films. This suggests that ultrasonic treatment of the gel primarily affects the polymer linkages responsible for gelation, while maintaining the requisite microstructure for low index film formation. These ultrasonically treated and diluted solutions are stable for a period of several months before regelation occurs. Additionally, gels that were permitted to "ripen" for an extended period of time after the initial gelation could not be reliquefied by ultrasonic treatment.

Figures 14 and 15 show reflectance spectra for sol-gel AR films applied to acrylic and polycarbonate and heated to ~90°C under a heat lamp. These spectra, which are nearly identical to heated (500°C) and etched sol-gel AR films on Pyrex, exhibit the characteristics of single-layer interference films. Note that the reflectance decreases beyond 1000 nm due to absorption in the polymeric substrate (thickness ≈1.5 mm). A slight decrease in transmittance (0.901 to 0.884) was observed after 3 months outdoor exposure; however, a greater decrease in transmittance (0.854 to 0.833) was observed for uncoated plastics. Transmittance values of both coated and uncoated samples could be significantly increased by rinsing, suggesting that one mechanism of transmittance loss is contamination of the surface. In concurrence with findings of other workers,[5] it appears that although these films may not be extremely abrasion resistant, their weatherability is acceptable for many applications.

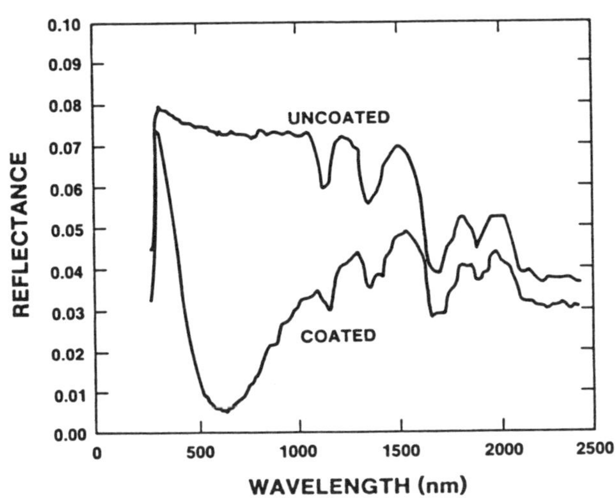

Figure 14: The spectral reflectance of an uncoated and AR coated acrylic substrate.

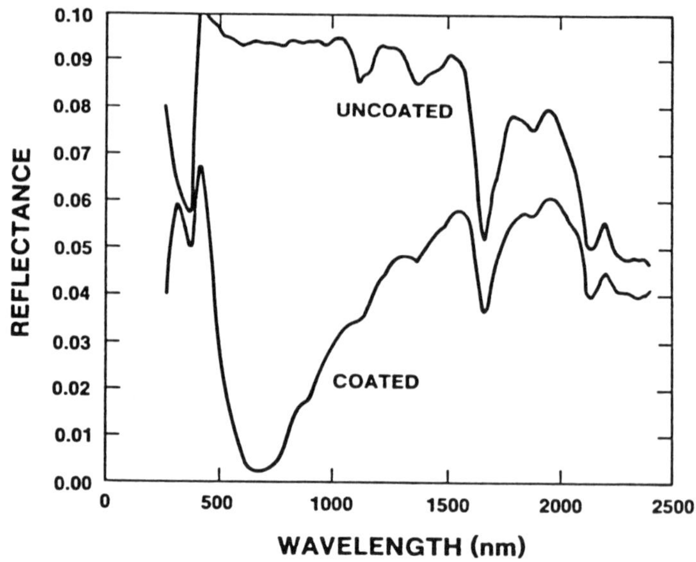

Figure 15: The spectral reflectance of an uncoated and AR coated polycarbonate substrate.

SUMMARY

In summary, we have demonstrated that sol-gel thin film processing permits independent tailoring of both chemical composition and microstructure of the deposited film. Implications of this technology for AR coating applications are: (1) The refractive index can be varied continuously between about 2.30 (relatively dense, partially crystallized TiO_2) and about 1.22 (porous, amorphous SiO_2). Further modifications are made possible by etching (reduced n), consolidation and/or crystallization (increased n). (2) Optical quality films with controlled thicknesses may be deposited on a wide variety of substrates used in solar thermal and photovoltaic technologies including glasses, ceramics, semiconductors, metals and plastics. (3) Multiple layers with differing compositions and/or microstructures may be deposited sequentially with little interdiffusion permitting easy optical simulation and design. (4) It has been demonstrated in several cases that etching phase separated and/or porous sol-gel derived films can result in graded refractive indices.

Compared to more traditional coating methods, e.g., CVD, sputtering and evaporation, the sol-gel approach is certainly more flexible with respect to compositional limitations and substrate size/shape constraints. Perhaps more importantly, it permits microstructural as well as compositional tailoring which is not possible by other coating methods. The possibility for new materials, perhaps with anisotropic optical properties, also exists. Thus the potential for sol-gel derived AR coatings appears bright especially for coating large surfaces with complex shapes or topographies.

APPENDIX: OPTICAL MODELING

In this section, the procedures used to calculate the reflectance and transmittance properties of a generalized stack made up of both thick and thin films are presented. The procedure follows the thin film treatment as described by O.S. Heavens in Reference 1A, but also includes an extension that incorporates thick films as well. A thin film is defined as having a thickness that is comparable to the wavelength of radiation of interest and therefore phase relationships between reflected and transmitted beams at the interfaces are important. This phase information leads to interference effects in the optical properties. On the other hand, a thick film is defined as having a thickness much greater than the wavelength and thus phase information is lost. Thus, when measuring the optical properties of thick films, the finite bandwidth of the measuring instrument, the finite collimation or variation in incident angles, and minor variations in the film thickness all result in averaging of the reflectance and transmittance that eliminates interference (phase) effects. Mathematically, the difference between a thick and thin film will be clearly defined later in this section.

The starting point for the calculation of the optical properties of a stack of thick and thin films involves a derivation of a general formula for the effective reflectance and transmittance properties of a single film as shown in Figure 1A. As shown in this figure, the film has an index of refraction $\tilde{n}_2$ and thickness d_2 and lies between two semi-infinite materials with indices of refraction of $\tilde{n}_1$ above and $\tilde{n}_3$ below. Consider a beam of radiation in material 1 that is incident on the film at an angle θ_1 and is reflected and transmitted into the thin film. The transmitted beam, traveling at an angle θ_2, is modified as it passes through the film and is then incident on the lower interface. At this point, part is reflected and part transmitted. The reflected part is modified as it again passes through the film before it is incident on the top interface for a second time. While this process repeats indefinitely, the individual reflected and transmitted beams can be summed in order to determine an overall transmitted and overall reflected beam.

For this summation, it is important to keep track of the direction of the beam at each interface. Thus a subscript indicates the material identification (1, 2, or 3) with the first number indicating the incident medium and the second number indicating the transmitting medium. In addition, when traversing the film in either direction, a fractional transmittance, T_a, must be included; the value used for this factor depends upon the film index of refraction and thickness and will be defined later. With this nomenclature, the terms in the overall reflected beam, R, are given by the sum (see Figure 1A).

$$R = R_{1,2} + T_{1,2}\, T_a^2\, R_{2,3}\, T_{2,1} + T_{1,2}\, T_a^4\, R_{2,3}^2\, R_{2,1}\, T_{2,1} + \ldots \quad (1A)$$

Rearranging we have

$$R = R_{1,2} + T_{1,2}\, T_{2,1}\, T_a^2\, R_{2,3} \left[1 + T_a^2\, R_{2,3}\, R_{2,1} + T_a^4\, R_{2,3}^2\, R_{2,1}^2 + \ldots\right] \quad (2A)$$

The series inside the brackets can be summed so that

$$R = R_{1,2} + \frac{T_{1,2}\, T_{2,1}\, T_a^2\, R_{2,3}}{1 - T_a^2\, R_{2,3}\, R_{2,1}} \quad . \tag{3A}$$

Similarly, the terms in the overall transmitted beam, T, are given by the sum (see Figure 1A).

$$T = T_{1,2}\, T_{2,3}\, T_a + T_{1,2}\, T_a^3\, R_{2,3}\, R_{2,1}\, T_{2,3} + T_{1,2}\, T_a^5\, R_{2,3}^2\, R_{2,1}^2\, T_{2,3} + \ldots \tag{4A}$$

Rearranging, we have

$$T = T_{1,2}\, T_{2,3}\, T_a \left[1 + T_a^2\, R_{2,3}\, R_{2,1} + T_a^4\, R_{2,3}^2\, R_{2,1}^2 + \ldots\right] . \tag{5A}$$

Summing the series gives

$$T = \frac{T_{1,2}\, T_{2,3}\, T_a}{1 - T_a^2\, R_{2,3}\, R_{2,1}} \quad . \tag{6A}$$

The two Equations 3A and 6A can now be applied separately to the situation of a thin film and to a thick film.

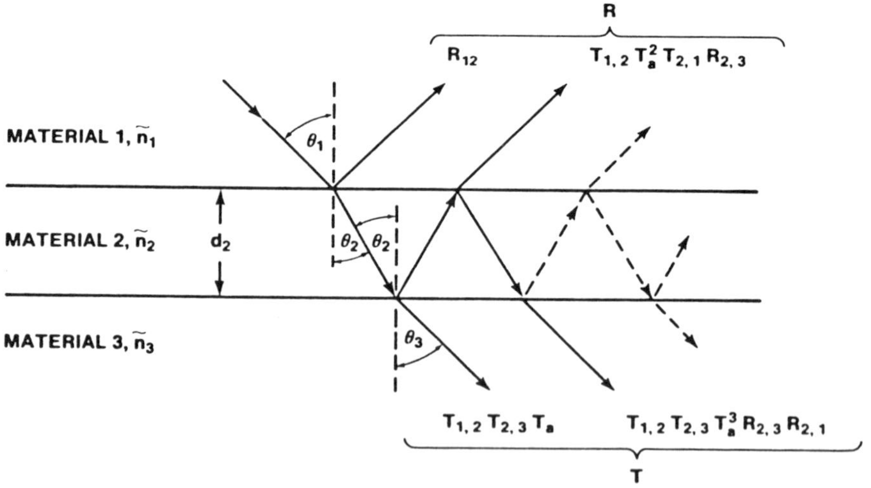

Figure 1A: Schematic showing the multiple reflected and transmitted beams that make up the overall reflected beam (R) and transmitted beam (T).

For a thin film, the Fresnel coefficients determine the reflection and transmission amplitudes that are used at the interfaces. For an incident beam traveling from material 1 to material 2, the Fresnel coefficients for s- and p-polarization are given by (7A)

$$r^s_{1,2} = \frac{\tilde{n}_1 \cos\theta_1 - \tilde{n}_2 \cos\theta_2}{\tilde{n}_1 \cos\theta_1 + \tilde{n}_2 \cos\theta_2} = - r^s_{2,1}$$

$$r^p_{1,2} = \frac{\tilde{n}_1 \cos\theta_2 - \tilde{n}_2 \cos\theta_1}{\tilde{n}_1 \cos\theta_2 + \tilde{n}_2 \cos\theta_1} = - r^p_{2,1} \qquad (7A)$$

and (8A)

$$t^s_{1,2} = \frac{2\,\tilde{n}_1 \cos\theta_1}{\tilde{n}_1 \cos\theta_1 + \tilde{n}_2 \cos\theta_2} = \frac{\tilde{n}_1 \cos\theta_1}{\tilde{n}_2 \cos\theta_2}\, t^s_{2,1}$$

$$t^p_{1,2} = \frac{2\,\tilde{n}_2 \cos\theta_1}{\tilde{n}_1 \cos\theta_2 + \tilde{n}_2 \cos\theta_1} = \frac{\tilde{n}_2 \cos\theta_1}{\tilde{n}_1 \cos\theta_2}\, t^p_{2,1} \qquad (8A)$$

where the complex angles θ_1 and θ_2 are determined from Snell's law of refraction:

$$\frac{\sin\theta_1}{\sin\theta_2} = \frac{\tilde{n}_2}{\tilde{n}_1} . \qquad (9A)$$

From Equations 7A and 8A, it can be shown that

$$1 - \left[r^s_{1,2}\right]^2 = t^s_{1,2} \cdot t^s_{2,1}$$

$$1 - \left[r^p_{1,2}\right]^2 = t^p_{1,2} \cdot t^p_{2,1} \qquad (10A)$$

Since identical relationships between the Fresnel coefficients hold for s- and p-polarization, the s or p superscript will be omitted in the following equations. Substituting the relationships between the r and t coefficients derived in Equations 7A and 9A into Equation 3A for the reflectance amplitude, we have

$$r = r_{1,2} + \frac{(1 - r^2_{1,2})\, r_{2,3}\, T^2_a}{1 + T^2_a\, r_{1,2}\, r_{2,3}} \qquad (11A)$$

or

$$r = \frac{r_{1,2} + r_{2,3}\, T^2_a}{1 + r_{1,2}\, r_{2,3}\, T^2_a} . \qquad (12A)$$

Similarly, the transmittance amplitude from Equation 6A is determined to be

$$t = \frac{t_{1,2}\, t_{2,3}\, T_a}{1 + r_{1,2}\, r_{2,3}\, T_a^2} \quad . \tag{13A}$$

It is important to note that, in deriving Equations 12A and 13A, only relationships between $r_{1,2}$, $r_{2,1}$, $t_{1,2}$ and $t_{2,1}$, which all involve the (1,2) boundary, were required to simplify the general equations. No assumptions were made about relationships between the reflection and transmission coefficients at the (2,3) boundary. This fact will become important later. In addition, it should be emphasized that the resulting r and t coefficients are complex and therefore have both an amplitude and phase factor. The quantity T_a for a thin film represents the change in amplitude and phase associated with traversing the film and is given by 14A

$$T_a = \exp(-2\,\iota\,\delta_1) \tag{14A}$$

where

$$\delta_1 = \frac{2\,\pi}{\lambda}\,\tilde{n}_1\, d_1 \cos\theta_1 \quad . \tag{15A}$$

Equations 12A and 13A represent recursion relations that allow a thin film with its two interfaces to be replaced with a single interface and effective Fresnel coefficients.

Application of the above relations to determine the overall reflection and transmission values of a stack of mixed thin and thick films is straightforward. Consider the generalized stack shown in Figure 2A. First the complex Fresnel coefficients (r_i and t_i) are calculated at all the interfaces in the multifilm stack using Equations 2A, 8A, and 9A. Next each group of thin films that is positioned between thick films (three such thin film layers are shown in Figure 3A) is collapsed into a single interface with "effective" reflection and transmission Fresnel coefficients. This is accomplished by starting with the bottom two interfaces in each thin film group and, using the recursion relations of Equations 12A and 13A, replacing these interfaces with a single interface that has associated with it the calculated effective Fresnel coefficients. Next, this interface and the one above are collapsed into a single interface with new effective Fresnel coefficients. This process is repeated until the complete group of thin films is collapsed to a single interface with effective Fresnel coefficients. This procedure results in the correct reflectance and transmittance amplitude coefficients since the recursion relation relies only upon relationships between the Fresnel coefficients at the upper interface and this interface has not been previously collapsed. Thus this procedure works only if one starts with the bottom most thin film in a group. Starting at the top of a thin film group and working downward is not possible using the recursion relations derived here, but may be possible using different recursion relations. The alternative recursion relations are more complicated and thus are not as useful as those derived here.

In order to collapse the remaining thick films, the Fresnel amplitude coefficients are converted to intensity coefficients. The reflection and transmission

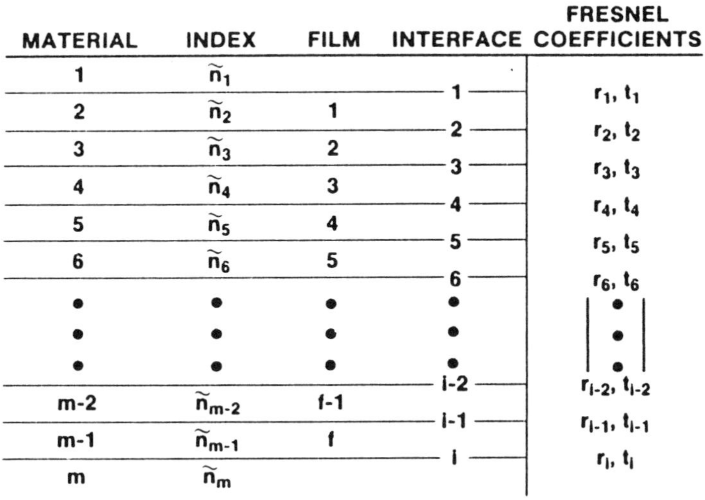

Figure 2A: A generalized thin/thick film stack showing the labeling of the materials (m), films (f), and interfaces (i).

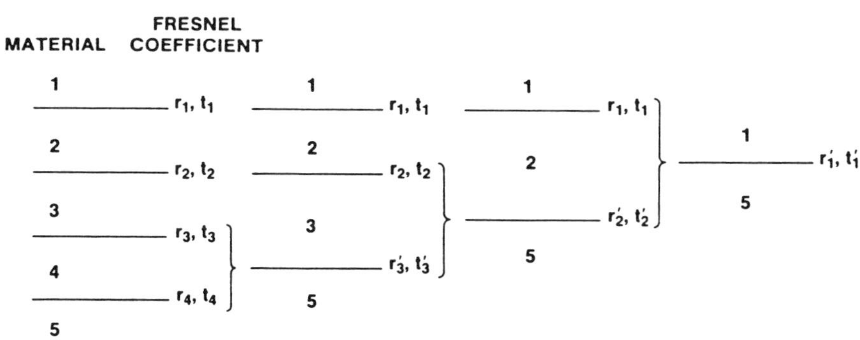

Figure 3A: Schematic showing the procedure followed to collapse interface pairs into a single, effective interface. When completed, the Fresnel coefficients (r_1', t_1') represent the complete 3 thin film stack.

intensity coefficients are proportional to the square of the corresponding Fresnel coefficients and will be represented by R and T, respectively. The reflection intensity for both s- and p-polarization at a boundary is given by

$$R_{1,2} = r_{1,2} \cdot r^*_{1,2} \qquad (16A)$$

where r^* is the complex conjugate of r. For the transmission intensity coefficient, T, a factor must be included which takes into account propagation in a medium of refractive index $\tilde{n}$ (see Reference 4A). Thus the transmitted beam intensity for s- and p-polarizations is given by

$$T^s_{1,2} = \frac{\mathrm{Re}\,[\tilde{n}_2 \cos\theta_2]}{\mathrm{Re}\,[\tilde{n}_1 \cos\theta_1]}\, t^s_{1,2}\, [t^s_{1,2}]^*$$

(17A)

$$T^p_{1,2} = \frac{\mathrm{Re}\,[\cos\theta_2/\tilde{n}_2]}{\mathrm{Re}\,[\cos\theta_1/\tilde{n}_1]}\, t^p_{1,2}\, [t^p_{1,2}]^*$$

After calculating the intensity coefficients, multiple reflections between the remaining interfaces must be taken into account. This is accomplished using the generalized recursion relation of Equations 3A and 6A. Again, the collapsing process is initiated with the bottom pair of remaining interfaces. From the Equations for R and T, it can be seen that the reflection and transmission coefficients of the upper interface of the pair of interfaces are required for radiation passing from below the interface to above the interface ($R_{2,1}$ and $T_{2,1}$ in Equations 3A and 6A). If the upper interface represents a group of collapsed thin films, then the reflection and transmission coefficients are different for radiation passing from top to bottom through the interface as compared to radiation passing in the reverse direction. Therefore, effective Fresnel coefficients need to be calculated for each thin film group for radiation passing from the bottom to the top of the interface. The same iterative procedure described above and shown in Figure 3A is used. With the necessary forward and reverse intensity coefficients, successive interfaces are collapsed and effective intensity coefficients are calculated using Equations 3A and 6A each time. Again it is important that this calculation starts at the bottom of the remaining interfaces and proceeds upward through the stack. The value used for T_a in the case of a thick film is given by

$$T_a = \exp\left[-\frac{4\pi\, kd \cos\theta}{\lambda}\right] = \exp(-\alpha d \cos\theta) \tag{18A}$$

where α is the absorption coefficient. When the last interfaces are collapsed, the resulting reflection and transmission coefficients represent the overall reflection and transmission intensities for the original multilayer thick and thin film stack.

In situations where the stack involves only thick films, simplified recursion relations for R and T can be derived similar to the relations determined for the Fresnel coefficients, r and t. Thus from Equations 16A and 17A, it can be shown that

$$R_{1,2} = R_{2,1} \quad \text{and} \quad T_{1,2} = T_{2,1}\,. \tag{19A}$$

Using these equalities, Equations 3A and 6A can be simplified for the case of thick films to

$$R = R_{1,2} + \frac{T_a^2\, T_{1,2}^2\, R_{2,3}}{1 - T_A^2\, R_{1,2}\, R_{2,3}}$$

(20A)

$$T = \frac{T_{1,2}\, T_{2,3}\, T_a}{1 - T_a^2\, R_{1,2}\, R_{2,3}}\,.$$

REFERENCES

1. Heavens, O.S., *Optical Properties of Solid Thin Films,* (Academic Press, NY) 1955, Chapter 7.
2. Minot, M.J., *J. Opt. Soc. Am.* 66 (1976) 515.
3. McCollister, H.L. and Pettit, R.B., *Solar Energy Eng.* 105 (1983) 425.
4. McCollister, H.M. and Boling, N.L., U.S. Patent 4,273,826, June 1981.
5. Yoldas, B.E. and Partlow, D.P., *Applied Optics* 23 (1984) 1418.
6. Cook, L.M., Lowdermilk, W.H., Milam, D. and Swain, J.E., Lawrence Livermore Nat. Labs Report, UCRL-86909 (1981).
7. Brinker, C.J., et al, *J. Non-Cryst. Solids* 63 (1984) 45.
8. Schroeder, H., in *Physics of Thin Films: Advances in R&D,* George Hass and Rudolf E. Thun, eds. 5 (1969).
9. Scherer, G.W., unpublished results.
10. Brinker, C.J., Drotning, W.D. and Scherer, G.W., in *Better Ceramics Through Chemistry,* C.J. Brinker, D.E. Clark and D.R. Ulrich eds. (Elsevier, NY) 1984 p. 25.
11. Born, M. and Wolf, E., *Principles of Optics,* (Pergamon, NY) 1975, p. 87.
12. Brinker, C.J. and Harrington, M.S., *Solar Energy Mat.* 5 (1981) 159.
13. Brinker, C.J., Lenahan, R.J. and Keefer, K.D., U.S. Patent Appl. #
14. Brinker, C.J. and Scherer, G.W., *J. Non-Cryst. Solids* 70 (1985) 301.
15. Witten, T. and Sander, L.M., *Phys. Rev. Lett.* 47 (1981) 1400.
16. Hurd, A.J., unpublished results.
17. Iler, R.J., *The Chemistry of Silica,* (Wiley, NY) 1979.
18. Cathro, K., Constable, D. and Solaga, T., *Solar Energy* 32 (1984) 573.
19. Pettit, R.B., Brinker, C.J. and Ashley, C.S., *Solar Cells* 15 (1985) 267.
20. Brinker, C.J. and Pettit, R.B., Sandia Nat. Labs Report, SAND 83-0137 (1983) 68.
21. Brinker, C.J., Haaland, D.M. and Pettit, R.B., Annual Meeting of the American Ceramic Society, Chicago, IL (April 1983).
22. Ashley, C.S. and Reed, S.T., Sandia Nat. Labs Report, SAND 84-0662, (1984).
23. Lind, M.A., Pettit, R.B. and Masterson, K.D., *J. of Solar Energy Eng.* 102 (1980) 34.
24. Reed, S.T., unpublished results.

APPENDIX REFERENCES

1A. Heavens, O.S., *Optical Properties of Solid Thin Films,* (Academic Press, NY) 1955, Chapter 4.

2A. Heavens, O.S., *Optical Properties of Solid Thin Films,* (Academic Press, NY) 1966, p. 51.

3A. Heavens, O.S., *Optical Properties of Solid Thin Films,* (Academic Press, NY) 1955, p. 56.

4A. Born, M. and Wolf, E., *Principles of Optics,* (Pergamon, NY) 1975, p. 631.

6

Oxynitride Thin Films from the Sol-Gel Process

Carlo G. Pantano, Richard K. Brow and Lee A. Carman

Department of Materials Science and Engineering
The Pennsylvania State University
University Park, Pennsylvania

INTRODUCTION

There has been a considerable amount of interest in the development of the sol/gel process for the deposition of thin films.[1,2] Although the optical properties required for anti-reflection or high reflection coatings have received the most attention, one also recognizes the possible use of the sol/gel process for the preparation of scratch resistant coatings, diffusion and oxidation barriers, and dielectric films. In fact, it is often suggested that this method may be especially applicable for the low-temperature processing and fabrication of microelectronic devices; e.g., to deposit field and gate oxides, passivation coatings, or interlayer dielectrics.

Silicon oxynitride films (SiO_xN_y), in particular, have received much attention because of their unique dielectric, chemical and thermal properties. In contrast to silicon dioxide films (SiO_2), they exhibit higher dielectric constants and dielectric breakdown strengths, they are excellent impurity diffusion masks, and when applied to semiconductor or metal substrates they can provide high-temperature oxidation resistance. Altogether, these characteristics render silicon oxynitride films important materials for microelectronic processing and device applications. Moreover, the optical properties—especially the refractive index—can be tailored through control of the O-to-N composition ratio; thus, these films also have potential applications in optoelectronic devices and packages.

The two methods most often used to fabricate silicon oxynitride films are (1) chemical vapor deposition,[3-5] and (2) direct thermal nitridation of silicon surfaces or silicon-dioxide thin films.[6–10] Since the primary applications of these materials are in the realm of silicon-device technology, the direct nitridation of silicon and silicon-dioxide thin films has been studied most extensively. This approach, if successfully developed, would eliminate the extra, often complicated processing steps associated with CVD, and could be incorporated directly into the processing schemes already utilized in silicon technology.

A number of studies have already shown that the creation of oxynitride films is more easily controlled by thermal nitridation of silicon-dioxide thin films than by direct nitridation of silicon.[8–10] Thus, most of the effort in this area has focused upon the nitridation, in ammonia, of thermally-grown silicon dioxide films on silicon substrates. But it is quite clear that this process is limited by the transport of reactants (NH_x) and reaction products (H_2O) through the dense oxide/oxynitride film. Long reaction times (of the order of hours) and high temperatures (in the range of 1100° to 1200°C) are required to carry out the process, and in virtually all cases, a graded nitrogen content is observed through the film.

In the work reviewed here, silica sol/gel films were deposited onto silicon wafers, and then, were converted to dense oxynitride films by thermal treatment in ammonia. The microporosity and reactivity of the deposited gel film facilitates the incorporation of higher nitrogen contents at lower temperatures, and more importantly, the nitrogen concentration is uniform through the film. In general, it is possible to produce films with controlled compositions in the range SiO_2 to Si_2N_2O. This chapter first discusses the thermochemistry associated with the Si–O–N–H system and then goes on to describe the chemical characteristics of the nitrided sol/gel films. Finally, the optical properties, dielectric properties and oxidation resistance of the materials are reviewed.

THERMOCHEMISTRY IN THE Si–O–N–H SYSTEM

It has long been known that ammonia can react with silicon dioxide (and silicate glasses) at high temperature to form oxynitride materials. For example, several percent of tricoordinated nitrogen was incorporated in porous glasses exposed to ammonia at temperatures in excess of 700°C,[11] and in silicate glass melts bubbled with ammonia around 1400°C.[12,13] More recently, it has been shown that oxynitride films, containing up to 40 to 50 mol % nitrogen, can be produced by reacting thin (<100 nm) thermally-grown silicon dioxide films in ammonia at high temperature (>800°C) ammonia. For example, Ito, et al[6] reacted a 50 nm SiO_2 film in ammonia for one hour at 1200°C to produce a material that contained approximately 40 mol % nitrogen. Similarly, Hayafuji and Kajiwara[8] reacted 10 nm SiO_2 films in various pressures of ammonia at 1100°C and reported that they could incorporate up to 50 mol % nitrogen at the surface of the film.

In order to understand the reactions that produce such silicon oxynitride materials, and to predict how changes in the reaction conditions affect their development, it is useful to examine the equilibrium thermodynamics. Due to the reactivity of silica sol/gel thin films, it is quite likely that equilibrium may, in fact, be achieved during the high temperature nitridation. Thus, the high temperature thermodynamic equilibria in the Si–O–N–H system were evaluated using a version of the SOLGASMIX computer program, based on the work of Eriksson.[14] The program calculated equilibrium compositions in systems containing a gaseous phase, condensed mixtures and condensed phases of either invariant or variable stoichiometry. The equilibrium calculations are based upon a minimization of the summation of the free energies of all possible chemical species in the system of interest. The system is defined by the initial input conditions which include temperature, total pressure, the elemental mol fractions of reactants, and the Gibbs free energies of formation and entropies of each possible chemical species. The concentration of these chemical species, at equilibrium, are obtained after a series of iterative calculations.

For the present study, 38 gaseous species and four condensed phases (Si, SiO_2, Si_2N_2O and Si_3N_4) were considered in the calculations. The enthalpy and entropy of formation of each of the 42 species were used as part of the input data. Most of this thermodynamic data was obtained from standard sources. The gaseous species are, of course, assumed to mix ideally. In the case of the solid phases SiO_2, Si_2N_2O and Si_3N_4, the SOLGASMIX code allows one to treat these as solid solutions. The initial concentration ratios of the elements in the system are also required as part of the input data. A range of NH_3:SiO_2 ratios was used in the initial calculations to determine what input parameters produced calculated nitrogen concentrations that best matched the experimental results (see below). Once the optimum input parameters were established, further calculations were performed to determine how variation in the other reaction parameters affected the extent of nitridation.

Figure 1 shows the effect of temperature on the relative abundance of silicon-containing phases at equilibrium as a result of the reaction between one atmosphere of ammonia and silicon dioxide. Below about 400°C, only silicon dioxide is calculated to be thermodynamically stable. As the temperature is increased, the formation of silicon oxynitride is favored over that of silicon dioxide, until at around 1000°C, the solid phase is predicted to consist solely of Si_2N_2O. As the temperature is further increased, the formation of silicon monoxide gas becomes significant. In fact, at temperatures in excess of 1200°C, no solid phase remains stable as all of the silicon originally present as silicon dioxide vaporized to various gaseous species (but, primarily silicon monoxide). Clearly, the production of Si_3N_4 is not possible under these conditions.

Figure 1 indicates that the optimum temperature range for the incorporation of significant amounts of nitrogen into silicon dioxide would be between 900° and 1100°C; the lower temperature limit is set by the thermodynamic stability of silicon dioxide and the upper limit is set by the evolution of silicon monoxide. This is the same temperature range generally used for the thermal nitridation of silicon and silicon dioxide during semiconductor device processing,[10] and as shown later, this temperature range leads to optimum nitridation of the silica sol/gel films. The calculated composition-temperature relationship in Figure 1 also agrees quite well with other reported results on the interactions between ammonia and silicon dioxide. Marchand and Lang[15] reported that small amounts of nitrogen were incorporated into silicon dioxide powders heated in ammonia at temperatures as low as 550°C, and that the powder could be completely converted to Si_2N_2O at temperatures approaching 1000°C. Similarly, the evolution of SiO(g) from silicon dioxide heated at >1200°C in reducing atmospheres is well known.[16] The results of the thermodynamic calculations thus provide a realistic description of the products of the reaction between ammonia and silicon dioxide.

It should be noted that although the SOLGASMIX results shown in Figure 1 predict the simultaneous formation of SiO_2 and Si_2N_2O below 1000°C, this does not imply that oxynitride materials made by reacting ammonia with silicon dioxide are necessarily two-phase. The experimental data (presented later) shows that the oxynitride films made by reacting ammonia with the sol/gel derived silicon dioxide are actually single-phase amorphous materials similar in structure to CVD-silicon oxynitrides. These latter materials have been described as solid solutions of silicon dioxide and silicon nitride,[17] and therefore, the separate oxide and oxynitride phases shown in Figure 1 can also be treated as the end-components of a single phase solid solution.

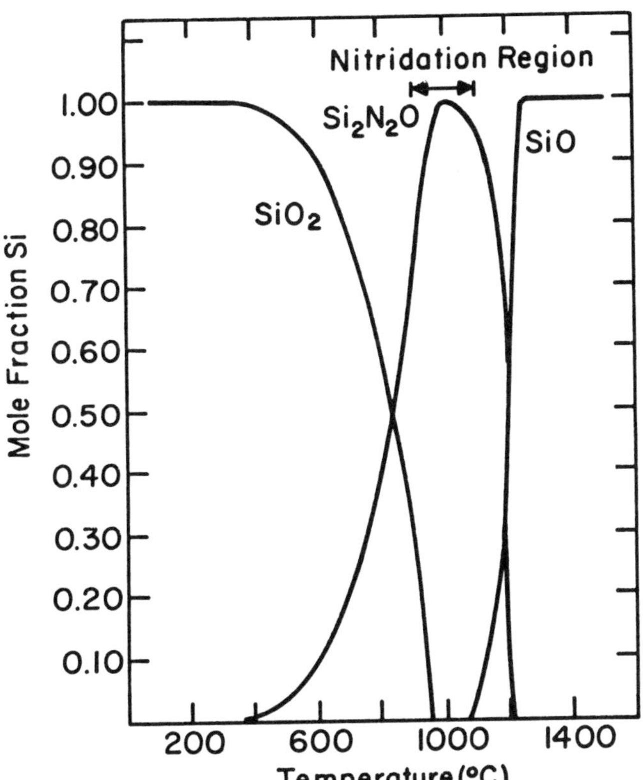

Figure 1: The effect of temperature upon the relative abundance of silicon-containing solid phases at equilibrium in the system NH_3–SiO_2; calculated using SOLGASMIX.

The effects of changing the total oxygen content in the nitridation atmosphere can also be evaluated using the SOLGASMIX program. The calculated oxygen partial pressure in the pure ammonia:silicon dioxide system was on the order of 2×10^{-24} atmospheres at 1000°C. Adding oxygen to the initial input concentrations increases the calculated equilibrium oxygen partial pressure, and reduces the nitrogen content of the stable solid phase as shown in Figure 2. [The nitrogen contents in Figure 2 were calculated assuming a solid solution (SiO_xN_y) of SiO_2 and Si_2N_2O.] The reduced nitrogen contents in the solid phase are due to the more favorable formation of SiO_2 relative to Si_2N_2O at the higher P_{O_2}. These latter calculations have been confirmed experimentally. For example, Ito et al[6] have shown that the direct nitridation efficiency of silicon is reduced when even trace levels of oxygen are present in the ammonia stream. This is the reason why nitridation of silicon to produce Si_3N_4 is so difficult.

Another practical consequence of the result shown in Figure 2 is that the final nitrogen content of an oxynitride material can be controlled by intentionally doping the ammonia atmosphere with oxygen. For example, if a film with a specific refractive index is required for an optical application the desired film composition, and thus the refractive index, will be fixed at equilibrium by the oxygen activity. The refractive indices of CVD-silicon oxynitride films have been controlled in a similar manner.[18,19]

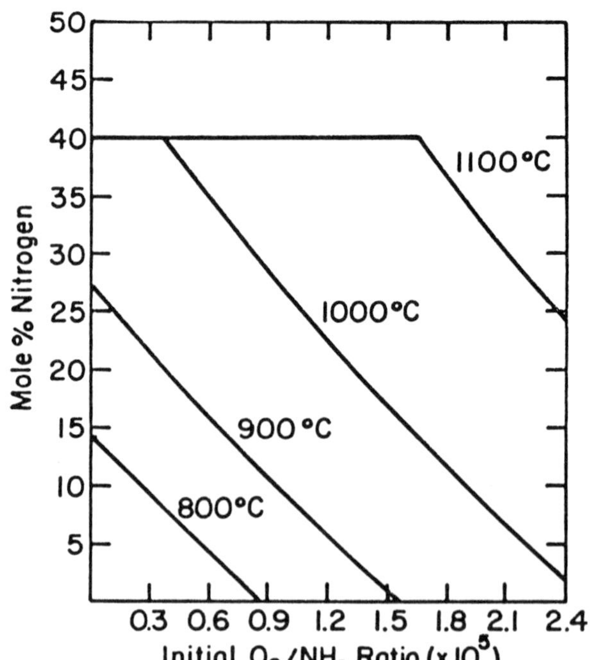

Figure 2: The effect of oxygen in the ammonia atmosphere upon the equilibrium nitrogen content in the SiO_xN_y solid phase during the SiO_2–NH_3 reaction; calculated using SOLGASMIX.

The effect of hydrogen in the nitridation atmosphere can also be illustrated by the thermochemical calculations. The hydrogen content is fixed when ammonia is used for nitridation due to the thermal decomposition of NH_3, but in these calculations, the nitridation in various N_2/H_2 mixtures was examined. Figure 3 shows the results of increasing the hydrogen content in the initial input reactants on the calculated temperature dependency of the nitrogen contents of the stable solid phase. Less than 0.1 mol % nitrogen is calculated to be in the stable solid product of a reaction between 100% N_2 and silicon dioxide. An increase in the hydrogen content of the ambient increases the calculated nitrogen content of the solid phase. These results were qualitatively confirmed by experiment. Sol/gel derived silicon dioxide thin films heated in excess of 1000°C in N_2 contained no detectable nitrogen, while those heated under the same conditions in ammonia were found to contain up to 40 mol % nitrogen. Interestingly, though, there is always more nitrogen found in ammonia treated films (experimentally) than would be predicted by the corresponding mixture of hydrogen and nitrogen (i.e., 25% N_2-75% H_2). This suggests that the decomposition products of NH_3, presumably NH_x radicals, are involved in the reaction; these cannot be included in the calculations due to lack of thermodynamic data.

In summary, the SOLGASMIX computer program can be used to predict the effects of treatment conditions on the development of oxynitride materials by the reaction between the nitridation atmosphere and silicon dioxide. The most efficient nitridation atmosphere is seen to be ammonia at 900° to 1100°C. Oxygen and water must be eliminated from the system in order to maximize the amount of nitrogen in the solid phase. These general processing conditions were used to successfully form oxynitride films by reacting ammonia with sol/gel derived silica films.

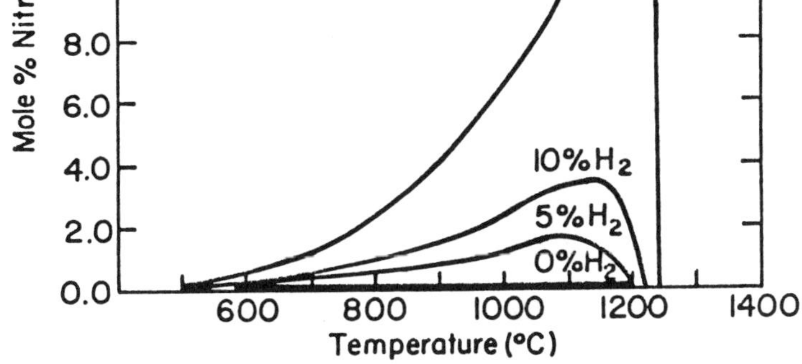

Figure 3: The equilibrium nitrogen concentration in the SiO_xN_y solid phase during the reaction between SiO_2, N_2, and H_2; calculated using SOLGASMIX.

FILM FORMATION

The techniques for the deposition of the silica sol/gel thin films were developed by Glaser.[20,21] Solutions with varying amounts of tetraethoxysilane (TEOS), ethanol and water were prepared with a small amount of 1 M HCl added to promote hydrolysis. The solutions were mixed for 30 minutes at 60°C in a reaction vessel equipped with a reflux condensor. The mixtures were then cooled to 40°C and diluted with ethanol (generally four parts ethanol to one part solution).

The substrates for the sol/gel thin films were polished <100> single crystal silicon wafers. The wafers were cleaned using electronic grade reagents by the RCA method. Approximately 250 μl of the diluted solution was uniformly distributed over the surface of the wafers using a photoresist spinner.

Figure 4 shows how a number of the processing parameters influence the thickness of the films.[21] In all cases, the film thickness was measured ellipsometrically at 632.8 nm. The open circles in the upper righthand corner of the figure represent films formed from solutions with a constant H_2O/TEOS = 4. The volume percent TEOS was varied by the addition of ethanol to the initial solution. Because these curves extrapolate through zero, it is apparent that for a constant H_2O/TEOS ratio, the film thickness is dependent only upon the oxide content of the solution. An increase in the speed of the photoresist spinner during film deposition is seen to decrease the final film thickness. It may be especially noteworthy, though, that the spinning rate also influences the refractive index of the films. Figure 5 presents this dependence where an increase in the deposition rate is shown to decrease the refractive index. This is probably due to the decreased time available for condensation/polymerization at the higher deposition rate. Thus, the film gels at a lower density. Assuming that equilibrium is, in fact, achieved during the nitridation in ammonia, though, the extent of nitridation will not be influenced by the initial density of the silica film.

The crosses (X) in Figure 4 represent films prepared from solutions with a constant ethanol concentration but with increasing H_2O/TEOS ratios. These films were all deposited at 3,000 rpm. The dotted line extrapolation in Figure 4 would represent the situation where the additional water simply 'dilutes' the oxides present in the solution. The fact that the experimental points deviate in a systematic way indicates non-ideal behavior wherein the water influences the condensation of the oxide film. Again this effect may be accounted for by changes in the density and/or micromorphology of the films which, in this case, have been hydrolyzed under different conditions.[21] It is apparent though, that these unique features of the gel films (due to differences in the solution hydrolysis) are 'annealed-out' after the films have been densified (see the filled points in Figure 4).

The majority of the oxynitride films discussed in this chapter were prepared from solutions with an H_2O/TEOS ratio of four and were deposited at 2,000 rpm for 15 seconds. Under these conditions, the dried-gel thin films are of the order 100 nm in thickness. After deposition, the thin films were thermally treated in a resistance-heated horizontal silica tube furnace. A pretreatment for one hour at 500°C in flowing O_2 was employed to remove residual organic species from the films. After purging the furnace with dry nitro-

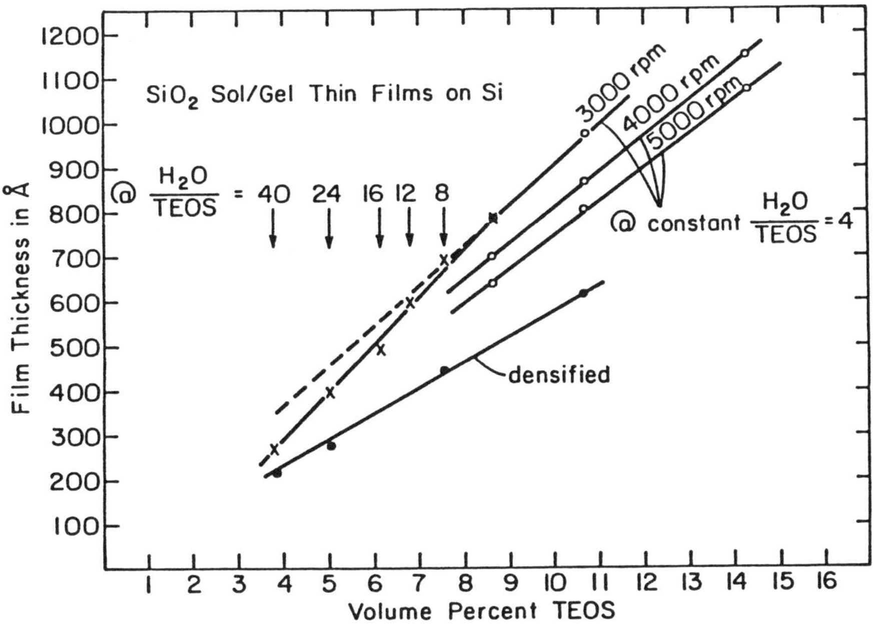

Figure 4: The dependence of film thickness upon the sol(ution) composition, the spinning-rate during deposition, and the densification in nitrogen.

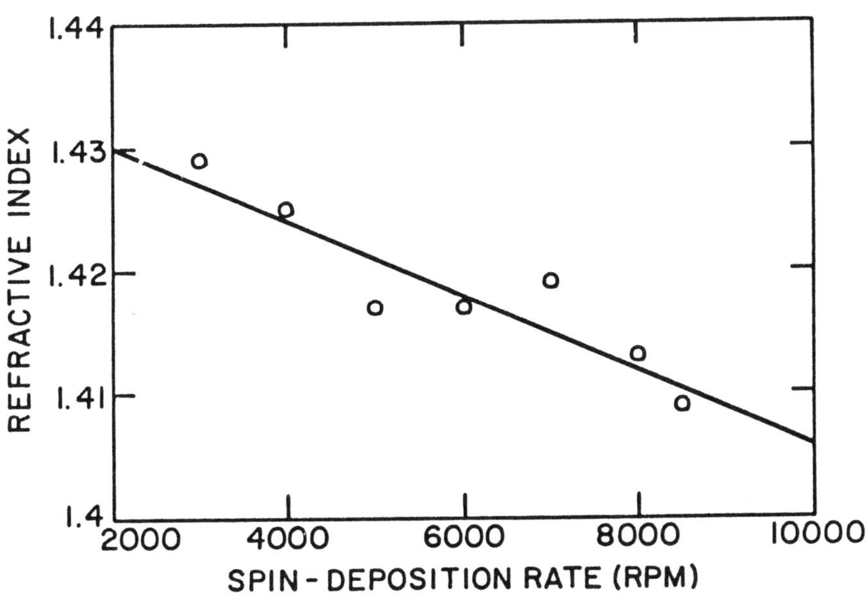

Figure 5: The influence of spin-deposition rate, i.e., gelling time, upon the refractive index of silica films (H_2O/TEOS = 4).

gen, the samples were heated at 10°C/min under flowing (0.5 ℓ/min) anhydrous ammonia to the treatment temperature. They were held for a specified time at the treatment temperature and were then cooled to room temperature. These ammonia treatments were carried out at between 500° and 1100°C. The final thicknesses of the initially ~100.0 nm thick, porous-gel films are presented in Figure 6; clearly, most of the film shrinkage occurs in the range of >750°C. Since nitridation of the films becomes appreciable at T ⩾750°C (see below), the shrinkage observed at T ⩾750°C is believed to be the result of both nitridation (due to changes in composition and structure) as well as, the elimination of microporosity (due to viscous sintering). It was verified through optical measurements (see below) and transmission electron microscopy that the films are fully dense at T ⩾900°C. Although the oxynitride films described here are all between 50.0 to 60.0 nm, one should note that by controlling the initial thickness of the deposited films and/or varying the film preparation parameters as shown in Figure 4, the final thickness of the dense oxynitride films can also be controlled. In general, oxynitride films ranging in thickness from 10.0 nm to 60.0 nm can be prepared in this manner.

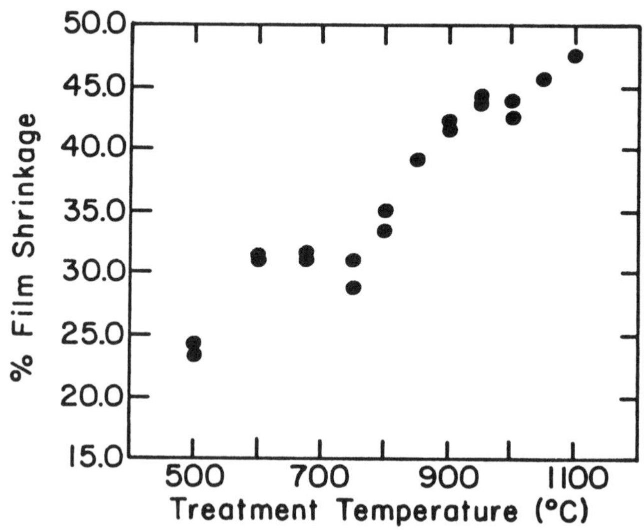

Figure 6: The temperature dependence of film thickness after a two-hour nitridation in ammonia; the initial thickness of the silica sol/gel film was ~100 nm.

FILM COMPOSITION AND STRUCTURE

The effects of the ammonia treatments upon the film compositions were monitored by a number of surface sensitive techniques. X-ray photoelectron spectroscopy (XPS) has been used to quantitatively determine the composition and chemical structure of these oxynitride films.[22] The analyses of the compositions were based on the relative areas of the Si 2p, N ls and O ls high resolution photoelectron spectra. A set of bulk oxynitride glasses, Si_2N_2O powder, and CVD-Si_3N_4 films, whose nitrogen and oxygen contents were independently verified, were used to calibrate the photoelectron intensity ratios. Figure 7 shows the measured nitrogen concentration as a function of the treatment temperature. The nitrogen content increases significantly as a result of ammonia treatments over 750°C. The nitrogen concentration approachs 40 mol % (the nitrogen content of Si_2N_2O) for treatments at 1000°C (the error bars shown represent the uncertainty of quantitative XPS). It should be noted that because of the surface sensitivity of XPS, the compositions shown in Figure 7 are representative of only the outer ten volume percent of the films. However, it is shown below that these films exhibit relatively uniform film composition depth profiles, and thus, the XPS data should provide good quantitative analyses of the overall film compositions.

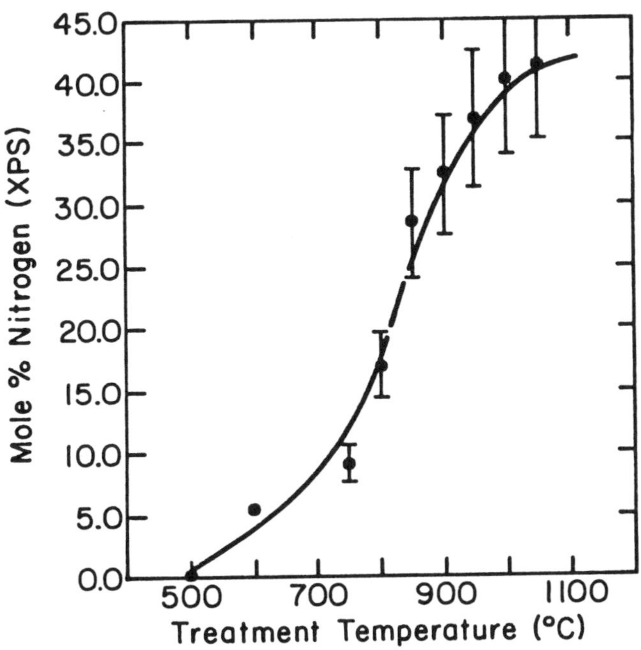

Figure 7: The nitrogen content of SiO_xN_y films prepared by the thermochemical nitridation of sol/gel silica films in ammonia; measured using x-ray photoelectron spectroscopy (XPS).

An analysis of the high resolution binding energies from the ammonia treated sol/gel films has been used to describe the chemical structure of the films, and in a sense, to verify that the silica gel has been 'nitrified.' This analysis has been described rather extensively in other publications.[23] Some of those results are summarized in Table 1 where the Si 2p, N 1s and O 1s binding energies of crystalline Si_2N_2O powder are compared with those for a sol/gel film treated in ammonia at 1000°C which contains approximately 40 mol % nitrogen. The Si 2p, N 1s and O 1s binding energies in both materials are seen to be equivalent within the resolution of the experiment (± 0.2 eV); thus one may conclude that these materials have comparable chemical structures. Figure 8 shows that the N 1s binding energy decreases from around 398.5 eV in samples treated below 800°C to about 397.5 eV in samples treated at higher temperatures. A comparable set of data shows a systematic decrease in the Si 2p binding energy with increasing overall nitrogen contents. This binding energy shift has been explained using a simple Pauling charge model that shows that the average silicon bond becomes more covalent with increasing nitrogen content and thereby leads to the decrease in the average Si 2p binding energy.[23] Altogether, it seems clear that as a result of the ammonia treatments, nitrogen replaces oxygen on the silica tetrahedra in the sol/gel film network and the nitrogen is coordinated by three silicon atoms; i.e., an oxynitride structure is formed.

Table 1: High Resolution Binding Energies (eV)

Sample	Si 2p	N 1s	O 1s
Si_2N_2O Powder	101.7	397.3	532.0
Si_2N_2O Sol/Gel Film	101.9	397.5	532.1

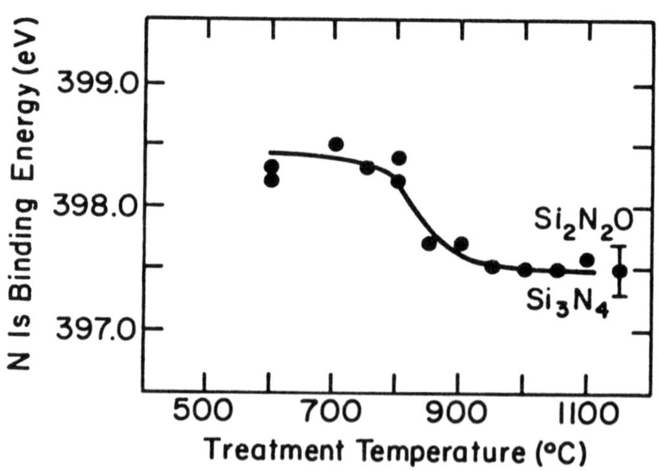

Figure 8: The N 1s binding energy, measured with x-ray photoelectron spectroscopy, in SiO_xN_y films prepared by the nitridation of silica; note that the N 1s state in the nitrided film is indistinguishable from that in stoichiometric Si_3N_4 and Si_2N_2O.

Structural information about the films can also be obtained from infrared absorption measurements. In the present study, the Fourier transform infrared specular reflectance spectra shown in Figure 9 were obtained from a series of films treated for two hours in ammonia at increasing temperatures. The primary reflectance consists of a single peak that decreases in frequency and increases in width as the nitrogen content of the films increases. For example, the reflectance maximum for a film treated at 600°C, with a refractive index of 1.472, was at 1,060 cm^{-1}, while that for a film treated at 1050°C, with a refractive index of 1.712, was at 880 cm^{-1}. In the case of SiO_2 films, the reflectance maximum has been reported to occur at 1,060 to 1,086 cm^{-1}, whereas in CVD-Si_3N_4 films, the maximum is found at 830 to 870 cm^{-1}.[3,24-26] Thus, it seems clear that the observed shift in position of the maxima in Figure 9 is due to the incorporation of nitrogen in silica.

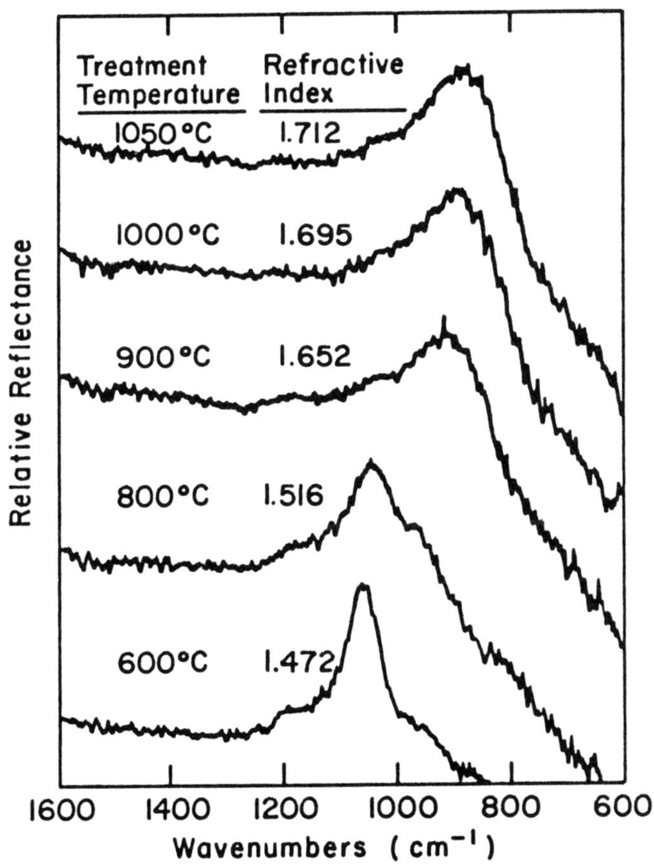

Figure 9: The infrared reflectance spectra for silica sol/gel films nitrided at various temperatures in ammonia; the corresponding refractive indices are also presented.

The infrared reflectance peak maxima for the nitrided sol/gel films (in Figure 9) are plotted as a function of their respective refractive indices in Figure 10. This plot also includes data reported in the literature for silicon oxynitride films prepared by CVD.[3,27] It will be shown in a later section of this chapter that the refractive index is a measure of the nitrogen content in the films. Thus, the correlation presented in Figure 10 verifies that the location of the IR absorption maxima is related to perturbations of the Si–O lattice vibrations due to the network substitution of nitrogen. The continuous shift in the maxima (towards the position for Si_3N_4) indicates that the incorporation of nitrogen is random and independent of the total nitrogen content (at least in the range between SiO_2 and Si_2N_2O). The fact that both the CVD and sol/gel films exhibit this behavior is further verification that nitridation of silica-gel in ammonia leads to an amorphous, chemically-homogeneous silicon oxynitride material.

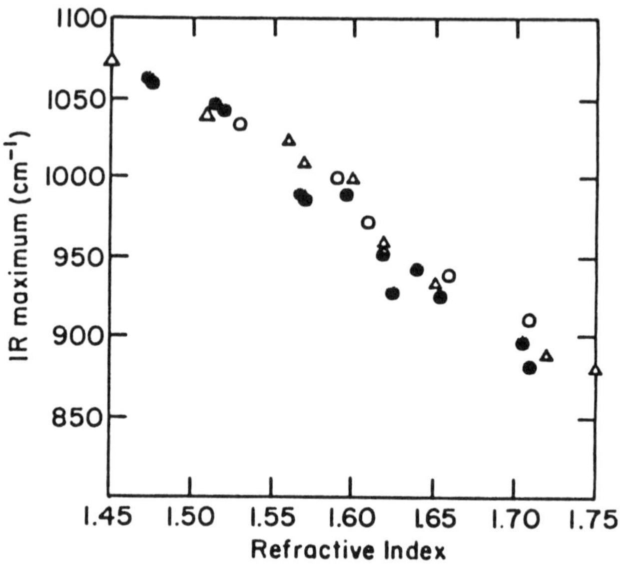

Figure 10: The relationship between the IR reflectance maxima (see Figure 9) and the refractive index of sol/gel-derived SiO_xN_y films (●); the data for CVD-- SiO_xN_y films is also shown (○ from Reference 3, △ from Reference 27).

Finally, profiling with Auger electron spectroscopy (AES) and secondary-ion mass spectroscopy (SIMS) was used to examine the in-depth compositional homogeneity of the films. These data, too, have been presented elsewhere.[28] In general, they show uniform distributions of oxygen and nitrogen throughout the bulk of films. This is in contrast to the profiles observed after ammonia treatment of thermally-grown oxide on silicon (i.e., dense silica films). Figure 11 shows the N/Si Auger signal ratios measured versus depth in two nitrided silica films, one was obtained by nitridation of a dense thermal oxide and the other a deposited microporous sol/gel film. In both cases, the films were about 50.0 nm after being treated in ammonia at 1000°C for four hours. It is quite clear that the nitridation is limited in the case of the thermal-oxide, and this leads to a significant nitrogen gradient. The profile is considerably more uniform, in general, in the case of the nitrided gel film.

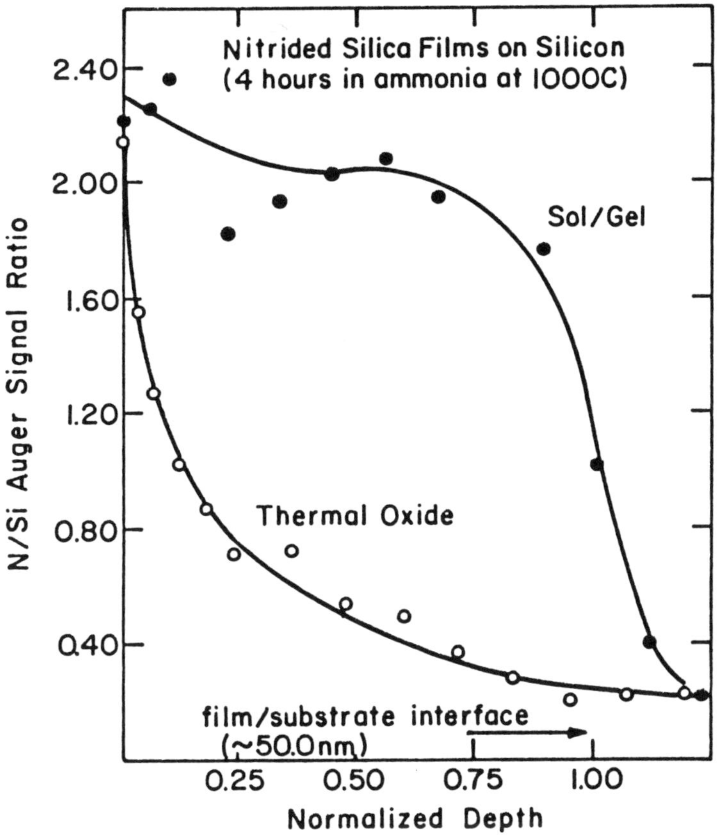

Figure 11: Auger sputter depth-profiles for nitrided silica films using sol/gel vs thermal-oxide precursors.

OPTICAL PROPERTIES

Figure 12 shows how the temperature of a two-hour ammonia treatment affects the measured refractive index of the films; the indices were ellipsometrically determined at 632.8 nm. The indices of the films treated below 600°C were generally between 1.40 and 1.42. This indicates that a substantial degree of microporosity is present in these films. The treatments carried out at 600° to 750°C produced films with refractive indices which were not very different from that of silicon dioxide (~1.46). For treatments in excess of 750°C, there is a significant increase in the refractive indices of the films. The index approached 1.70 for treatments at 1000°C but did not change appreciably for treatments over 1000°C. The correlation between film thickness (Figure 6) and refractive index (Figure 12) shows quite clearly that the composition and structure changes in the film are responsible for the enhanced film shrinkage at $T \geqslant 750°C$.

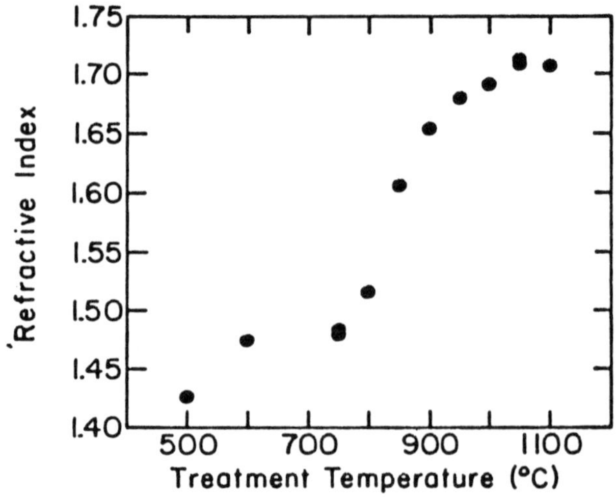

Figure 12: The refractive index of silica sol/gel films after a two-hour nitridation in ammonia; measured ellipsometrically at 632.8 nm.

Figure 13 shows the effect of treatment time on the refractive indices of films reacted at 800°, 900° and 1000°C. Most of the increase in the film refractive index occurs during the first two hours of the ammonia treatment. The high refractive indices obtained after 'zero' treatment times indicates that the reaction between the film and ammonia occurred quite rapidly; i.e., while the films were being heated to the treatment temperature.

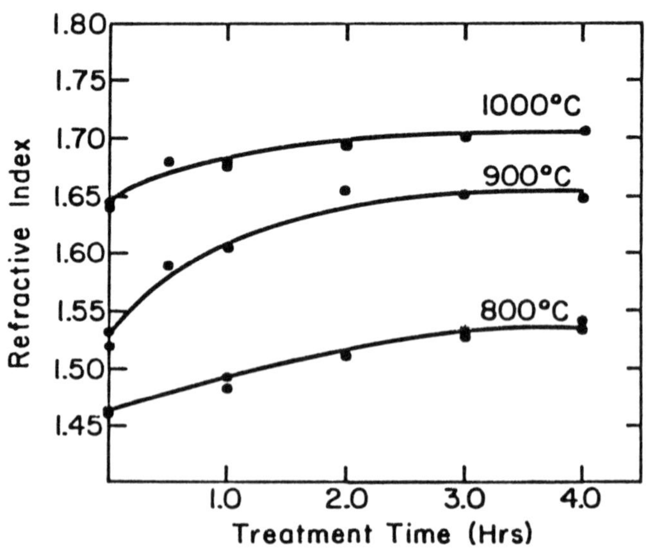

Figure 13: The time-dependence of refractive index, i.e., nitridation, for ~100 nm silica sol/gel films in ammonia.

It is quite evident that a relationship exists between the refractive indices and the compositions of the oxynitride films. Indeed, a number of theoretical models and effective medium approximations exist that do relate the optical properties of a multicomponent material to its composition. Here, it is shown that a linear approximation based upon the well-known Lorentz-Lorenz relationship[29] provides a reasonable description of the refractive index-composition relationship (the relationship based upon an effective medium approximation is reported in Reference 22). If one assumes that the optical properties of the oxynitride films can be represented by the optical properties of a composite of silicon dioxide and silicon nitride, the weight fraction of nitrogen in the film (W_N) can be written in terms of its refractive index (n_f) using:

$$W_N = \frac{\dfrac{n_f^2 - n_o^2}{\rho_o}}{\dfrac{n_f^2 - n_o^2}{\rho_o} + \dfrac{a_N}{a_o}\,\dfrac{n_N^2 - n_f^2}{\rho_N}} \tag{1}$$

where n_o and n_N are the refractive indices of silicon dioxide and silicon nitride, respectively, ρ_o and ρ_N are their respective densities, $a_o = (n_o^2)^{-1}$ and $a_n = (n_N^2)^{-1}$.

The ellipsometrically measured refractive indices (in Figure 12) and the XPS composition data (in Figure 7) for the ammonia treated sol/gel derived thin films are plotted against one another in Figure 14. A comparable set of data was obtained for oxynitride films prepared by CVD and these, too, are presented in Figure 14. The solid line in Figure 14 is the calculated relationship based upon Equation (1). A number of things should be noted. First, the data for the nitrided sol/gel films and the CVD-oxynitride films are quite similar, again implying a similarity in the chemical structures of the different CVD and sol/gel films. Second, since the ellipsometrically determined refractive indices reflect a 'bulk' composition for the film, the sol/gel films exhibit an in-depth homogeneity which is comparable to the CVD materials. Thirdly, the approximation based upon the Lorentz-Lorenz equation accurately predicts the compositional dependency of the refractive indices. And finally, Figure 9 indicates that the refractive indices of silicon oxynitride films may be used for a quick and convenient estimate of their overall nitrogen content.

Figure 15 shows the dispersion characteristics of the films over the wavelength range 400 to 650 nm. These, too, were ellipsometrically determined. They show that the dispersion in these materials increases with the extent of nitridation. These data were also used to verify the ellipsometric measurement of film thickness (already presented in Figure 6); i.e., the film thickness calculated using the ellipsometer readings at 400 nm, 550 nm, and 628 nm were all identical (not shown). This verifies that these films satisfy the single-layer ellipsometric model used for the refractive index measurements reported in this work.

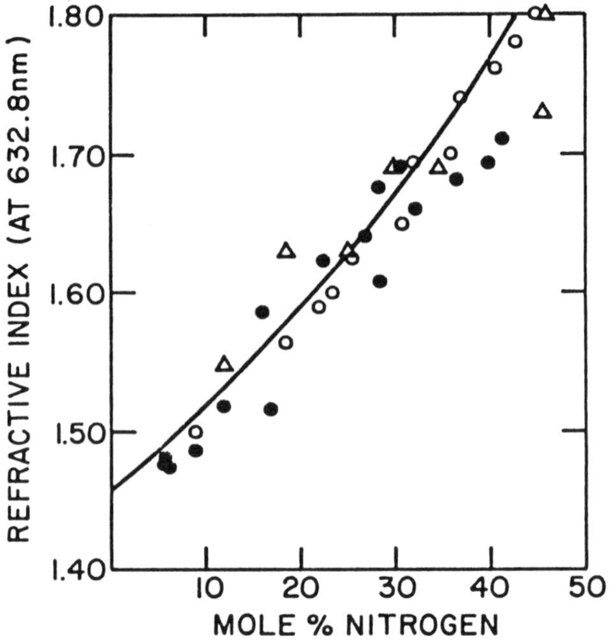

Figure 14: The refractive index of SiO_xN_y films as a function of the nitrogen concentration (y); the closed points are data for nitrided silica sol/gel films, the open points are data for CVD oxynitride films,[3,27] and the solid line is the Lorentz-Lorenz approximation.

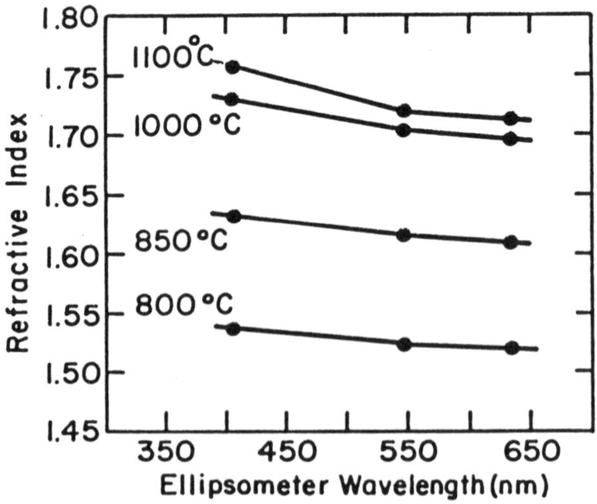

Figure 15: The refractive index of SiO_xN_y films—prepared by nitridation of silica in ammonia—as a function of wavelength; measured using ellipsometry.

ELECTRICAL PROPERTIES

The electrical properties of silica and oxynitride sol/gel films have been described in detail elsewhere.[30] Here, only the current-voltage (IV) or current field (IE) characteristics, dielectric constants and dc-dielectric breakdown strengths of the nitrided silica sol/gel films, which were deposited on silicon wafers, are described. In these studies, an array of Al electrodes was evaporated onto the densified films through a metal mask; the electrodes were approximately 200 nm thick and ~7 x 10^{-3} cm^2. This essentially produces an array of metal-insulator-semiconductor (MIS) diode capacitors whose electrical properties can be obtained very conveniently. Usually, about 30 of the individual diodes on each wafer were tested. The IV curves were obtained by incrementally increasing the voltage and then recording the steady-state current after a few minutes stabilization. The time increment required to attain a steady-state current was determined by first monitoring the current-time (IT) characteristic of each diode. The breakdown field was measured by applying a ramp voltage (~0.5 V/s), and then recording the maximum voltage drop across the diode. The dielectric constants were determined using the maximum capacitance obtained in the accumulation regime of the high frequency CV curve.

Figure 16 presents a comparison of IE curves for sol/gel oxide and oxynitride films, sputtered Si_3N_4, pyrolytic Si_3N_4 and oxynitride, and thermal SiO_2 films. The sol/gel films were prepared with a H_2O/TEOS ratio of 10 and treated to 1000°C in vacuum and 1000°C in ammonia, respectively. The vacuum treatment has resulted in lower leakage currents than the ammonia treatments, and this is consistent with the IE behavior of sputtered and pyrolytic oxide and oxynitride films;[24,31,32] i.e., nitrides and oxynitrides usually exhibit higher conductances than pure oxides. Nevertheless, the leakage currents through the sol/gel films are higher, in general, than the vacuum-deposited oxides and oxynitrides. [The sol/gel oxynitride and pyrolytic oxynitride films shown here have nearly identical refractive indices (~1.7), and therefore, would be expected to contain comparable concentrations of nitrogen.) The higher conductance through the sol/gel films is probably related to the higher ionic impurity levels associated with the spin-on film deposition versus vacuum film deposition, as well as, to the differences in the ultrastructure of the films.

The sol/gel films exhibit ohmic behavior at very low fields (not shown in Figure 16); this is presumably an ionic conductivity due to the presence of alkali and other mobile impurities. Perhaps of more significance, though, is the observed linear relationship between log I and $E^{1/2}$ (Figure 16). This dependence is also observed for vacuum deposited thermal SiO_xN_y and SiO_2 films. It indicates that the electrons which give rise to these leakage currents are injected into the conduction band of the oxide or oxynitride from the silicon substrate. Considering the rather low applied fields at which these currents are observed, it implies that there are rather high local electric fields at the silicon/sol-gel oxide/oxynitride interface. This situation may arise due to the presence of heterogeneities and protuberances at the interface which focus the field lines applied across the film, or due to a lowering of the interface barrier by the interfacial adsorption or immobilization of positively charged ionic impurities (see below).

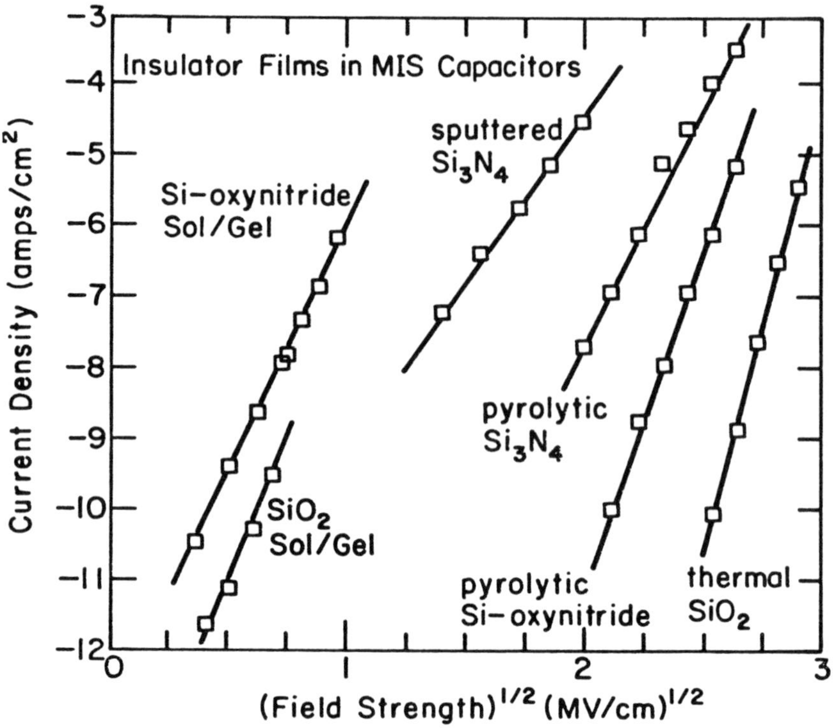

Figure 16: The current-field½ relationship for silicon oxide, oxynitride and nitride thin films prepared by a variety of techniques.

The influence of film thickness upon the IV characteristics of sol/gel films treated to 1000°C in ammonia is represented in Figure 17; all of these films were prepared from solutions with a H_2O/TEOS ratio of 10. Not surprisingly, the thicker oxynitride films exhibit lower leakage currents and can support much higher voltages than the thinner films. Thus, one can prepare more reliable dielectric insulators with the sol/gel method by depositing thicker layers. In fact, the films with thicknesses greater than 100 nm exhibit low leakage currents and acceptable dielectric strengths even after lower temperature treatments; consequently, they may be reasonable dielectric insulators even without complete densification at high temperature. Unfortunately, there is a limit to the thickness that can be deposited with the sol/gel method, and where thickness uniformity is critical, the maximum film thickness is further reduced by the need to use dilute solutions. Hence, silica sol/gel films, whether fired or unfired, will be useful insulators only where leakage currents of the order nanoamps to picoamps can be tolerated.

Of course, the most challenging applications for these materials, especially in microelectronics, are within the realm of very thin films, and thus, it is necessary to understand and control the properties of films $<$100 nm in thickness. Here, the dielectric breakdown strength is, perhaps, the most important

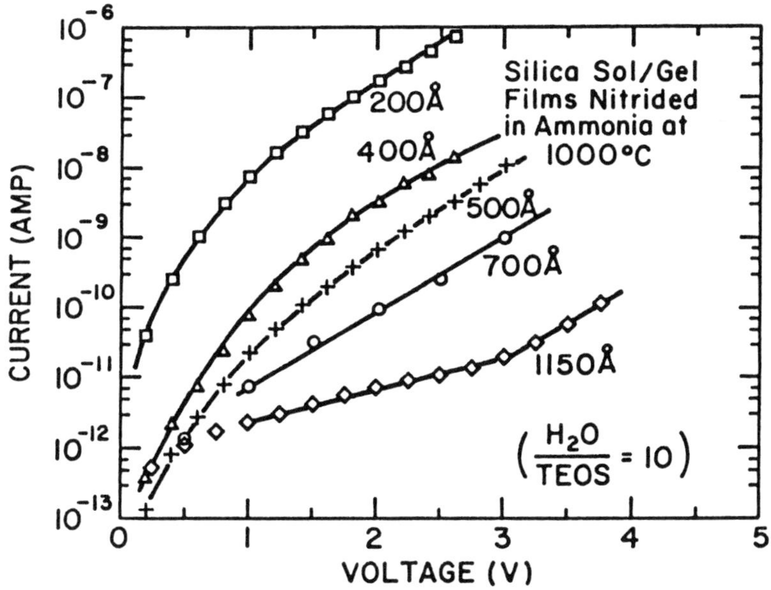

Figure 17: The current-voltage relationship for nitrided silica sol/gel films of various thicknesses.

characteristic. Thus, Figure 18 shows the dependence of the breakdown field upon film thickness and heat-treatment temperature in ammonia and in vacuum. One notes that the breakdown field increases with decreasing film thickness. This behavior is consistent with breakdown behavior for thermal silica films, where the breakdown field for intrinsic breakdown increases with decreasing oxide thickness.[33] On the other hand, defect density is reported to increase with decreasing thickness,[34] thus giving rise to higher probabilities of low field breakdown in very thin insulator films. This trend and interpretation is consistent with the behavior of most systems in the thickness range less than 100 nm.[35] The higher dielectric strengths in the films heat-treated to higher temperatures correlate directly with the IE behavior (not presented here); i.e., a reduction in the conductance of the film occurs with increasing treatment temperature and this leads to an increase in the breakdown strength.

It is quite surprising, though, that the dielectric strengths of the films treated at 1000°C in ammonia are indistinguishable from those of the vacuum treated films. The oxynitride is expected to exhibit better dielectric strength than the pure oxide, and these films (treated at 1000°C in ammonia) should contain of the order 30 to 40 atomic percent nitrogen. One interpretation to their comparable dielectric strengths is that the breakdown of these MIS devices is not intrinsic to the dense oxide or oxynitride material itself, but rather, is limited by microstructure in the film or defects in the device. This interpretation is also consistent with the low values of dielectric strength for any sol/gel deposited film (E_{max} = 0.2 – 1.0 MV/cm) at least when compared to the thermal silicon oxide (E_{max} = 10 MV/cm).

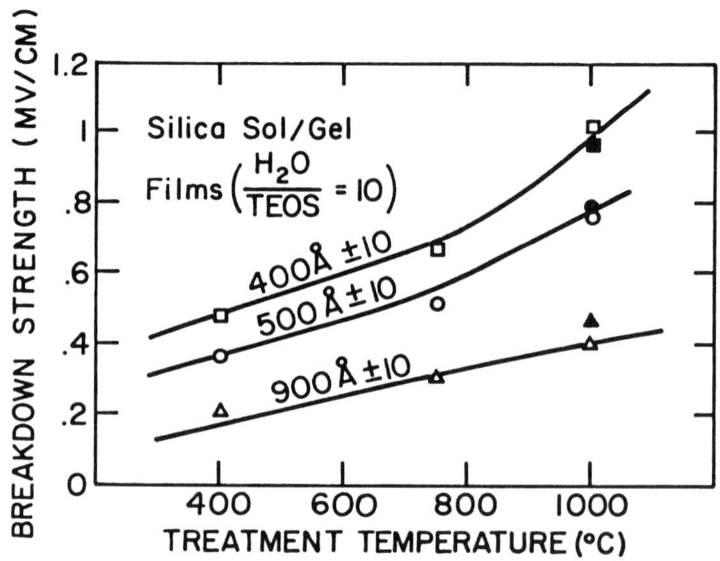

Figure 18: The dielectric strength of silica sol/gel films of various thicknesses after heat-treatment in vacuum (open points) and in ammonia (closed points).

Although the nitridation treatment in ammonia did not measurably enhance the dielectric strength, the dielectric constants of the sol/gel films did increase with the extent of nitridation in ammonia. Figure 19 shows the dielectric constant versus the measured nitrogen content. All of these films were approximately 50 nm in thickness after densification. The dielectric constant increases linearly with nitrogen concentration. This is due to the more atomically dense network in the oxynitrides and the additional polymerization modes provided by the 3-coordinated nitride species. Clearly, the ability to incorporate up to 40% nitrogen in the silica sol/gel films provides a wide degree of latitude in the control of dielectric constant; that is, films with dielectric constants between 3.8 and 5.5 can be produced depending upon the treatment temperature. It is noteworthy that the nitridation of dense thermal SiO_2 films at 1000° to 1200°C yields oxynitrides with dielectric constants of no more than 4.5.[6]

Although the dielectric constant can be enhanced and controlled to a great extent through nitridation of sol/gel films, it is apparent that the sol/gel films exhibit high leakage currents and low dielectric strengths, in general, when compared with thermal SiO_2 and vapor deposited oxides or oxynitrides. The high leakage currents and low dielectric strengths in the sol/gel films are probably enhanced by the presence of impurities and defects. Spectrochemical analyses of the sol/gel solutions and SIMS depth profiles of the deposited films, have verified the presence of sodium, potassium and boron impurities. The commercial, reagent grade TEOS was also subjected to spectrochemical analysis with results indicating that the TEOS itself may be acting as a source of impurities in the films (particularly the boron). The presence of these impurities in the

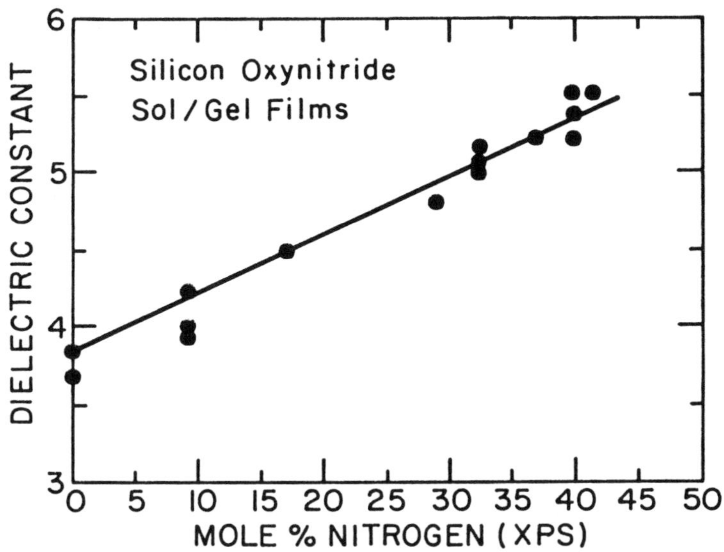

Figure 19: The dependence of the dielectric constant upon nitrogen concentration in SiO_xN_y films prepared by nitridation of silica sol/gel films in ammonia.

oxide film can enhance the injection of charge carriers into the oxide film due to a lowering of the interface barrier. Using a bias-temperature stress test, the CV characteristics of these films (not reported here) verify the presence of mobile positively charged impurity species. The injection of electrons into the conduction band of the oxide would be further enhanced due to the high local fields at or near the semiconductor/oxide interface. These high local fields could be associated with microstructural features intrinsic to the film/ substrate interface (e.g., micropores, impurity crystallites, pyrolysis products, stress lines, etc.), as well as with any extrinsic defects (e.g., pinholes, dust, etc.). Thus, the development of higher purity alkoxides, sols and process environments will be required before the intrinsic electrical properties of sol/gel films of any kind are realized.

OXIDATION RESISTANCE

Silicon oxynitride films, deposited or formed on silicon substrates, are known to limit substrate oxidation much more effectively than comparable silicon dioxide films.[8,36] The oxidation resistance of the ammonia treated sol/ gel derived silica films was studied by using an ellipsometer (at 632.8 nm) to measure the change in the apparent film thickness as a function of oxidation time and temperature.[37] Several films with different initial nitrogen contents were examined, along with a thermally grown oxide film used as a reference. The oxidation treatments were performed in a closed fused silica tube furnace, between 800° and 1000°C, in both wet and dry oxygen flowing at 0.5 ℓ/min.

The samples were periodically removed from the oxidation furnace to be analyzed on the ellipsometer and then returned to the furnace for further oxidation.

Figure 20 shows how the ellipsometrically determined thicknesses of several nitrided thin films, along with a thermally grown silicon dioxide film, change with time when oxidized at 900° and 1000°C. The thickness of the thermal oxide increased steadily with oxidation time due to the reaction between the oxidant, which diffuses through the oxide film, and the underlying silicon substrate. In contrast, the ammonia treated films exhibit little increase in thickness during the initial oxidation period, but abruptly begin to grow at a rate similar to the thermal oxide after longer oxidation times. Similar curves were obtained at 800° and 950°C. The time (t_{ox}) before this rapid change in the oxynitride film thickness is summarized in Table 2. The film with an initial refractive index of 1.68 (and an initial thickness of about 50 nm) prevented significant substrate oxidation for nearly 100 hours at 800°C; at 1000°C, this film resisted substrate oxidation for about 8 hours. Since the films with the higher initial refractive index contain more nitrogen, the data in Table 2 shows that increasing the initial nitrogen content of the oxynitride films increases their oxidation resistance.

Several nitrided sol/gel films were also oxidized with wet oxygen (bubbled through H_2O at 95°C). The oxynitride films oxidized much more rapidly in wet oxygen than in dry oxygen; this is comparable to the behavior of oxide and oxynitride films prepared by other methods. For example, a film with an initial refractive index of 1.72 had a t_{ox} of approximately 100 min when oxidized in the wet ambient at 800°C, compared to a t_{ox} of approximately 6,000 min when oxidized in the dry ambient.

The rapid increase in film thickness after t_{ox} is assumed to be due to oxidation of the underlying silicon substrate. Thus, the nitrided films were analyzed by secondary ion mass spectroscopy (SIMS) at various stages of the oxidation process to determine how the oxidation process affected the elemental compositional profiles through the films. The primary ion beam used in the SIMS analyses consisted of mass filtered $^{40}Ar^+$ accelerated to 7 keV with a current of 0.5 μA. The depth profiles were created by plotting the $^{42}SiN^+$ and $^{44}SiO^+$ secondary ion intensities, referenced to the $^{30}Si^+$ signal intensity, versus the normalized film depth. The normalized film depth was taken to be equal to the absolute sputtering time divided by the total time required to sputter through the film to the silicon substrate.

The film with an initial refractive index of 1.58 was oxidized for various times at 1000°C; the corresponding nitrogen depth profiles are shown in Figure 21. The unoxidized film had a relatively uniform nitrogen distribution. After exposure to the high-temperature oxygen, however, nitrogen was removed from the film surface. The thickness of this nitrogen depleted surface layer is seen to increase with increasing oxidation time, while the nitrogen signal intensity at the film-substrate interface remains relatively constant. The fact that the nitrogen was not uniformly removed from the oxidized film, but instead formed diffusion-limited profiles, indicates that the original oxynitride film was dense and impeded the migration of the oxidant to the silicon substrate.

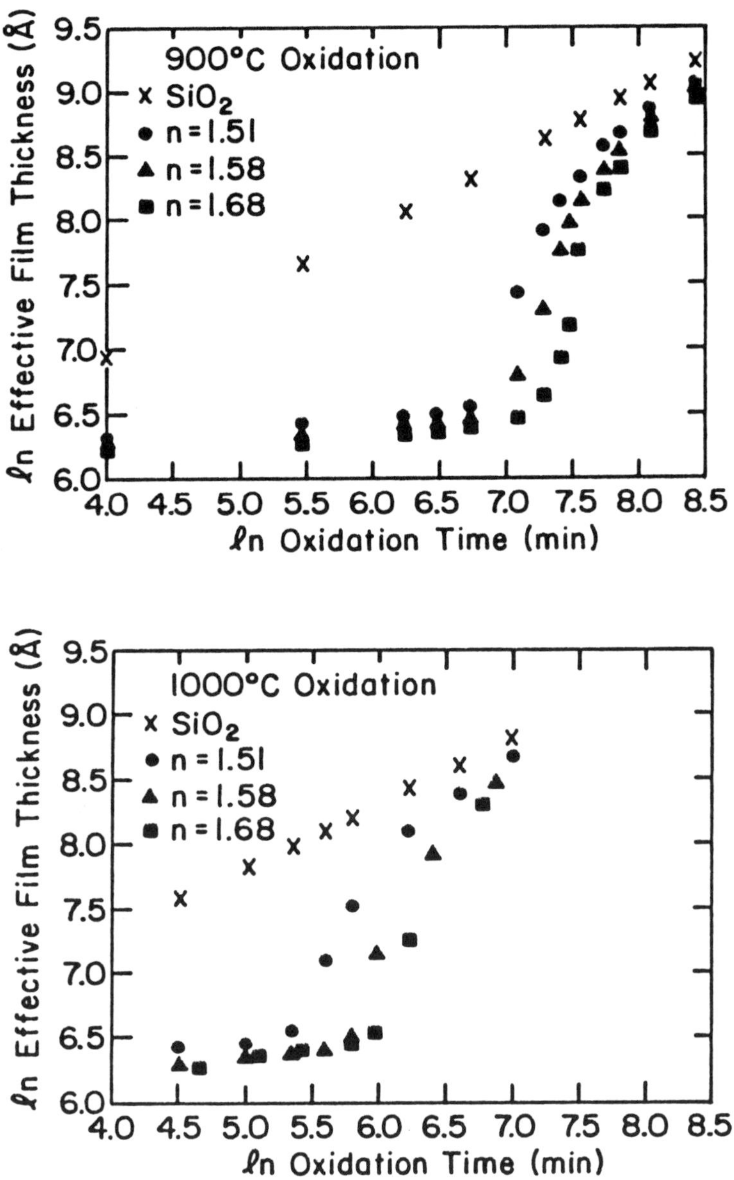

Figure 20: The change in thickness of SiO_2(X) and various SiO_xN_y (closed points) thin films during exposure to dry oxygen at 900°C (upper) and 1000°C (lower); the SiO_xN_y materials were prepared by nitridation of silica sol/gel films in ammonia while the SiO_2 was obtained by oxidation of silicon; the film thickness was measured with ellipsometry.

Table 2: Oxidation Time, t_{ox} (min), Prior to Rapid Film Growth

	Initial film refractive index		
Temperature	1.51	1.58	1.68
800°C	4500	5400	5750
900°C	1085	1410	1445
950°C	450	790	1125
1000°C	205	320	465

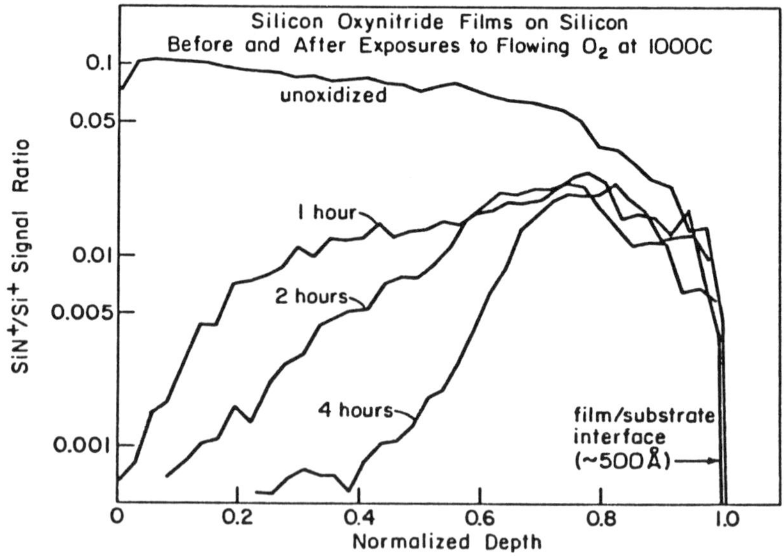

Figure 21: The time-dependent depth distribution of nitrogen—determined by secondary ion mass spectroscopy (SIMS)—after oxidation of sol/gel derived SiO_xN_y films.

Figure 22 shows the SiO^+/Si^+ and SiN^+/Si^+ depth profiles of a film, with an initial refractive index of 1.70, that had been oxidized for 8 hours at 1000°C. The oxygen signal is high and the nitrogen signal is very low at the film surface. This is taken as evidence for the conversion of the surface of the film to silicon dioxide as a result of the oxidation treatment. Near the film-substrate interface, the nitrogen signal is increased and the oxygen signal is reduced,

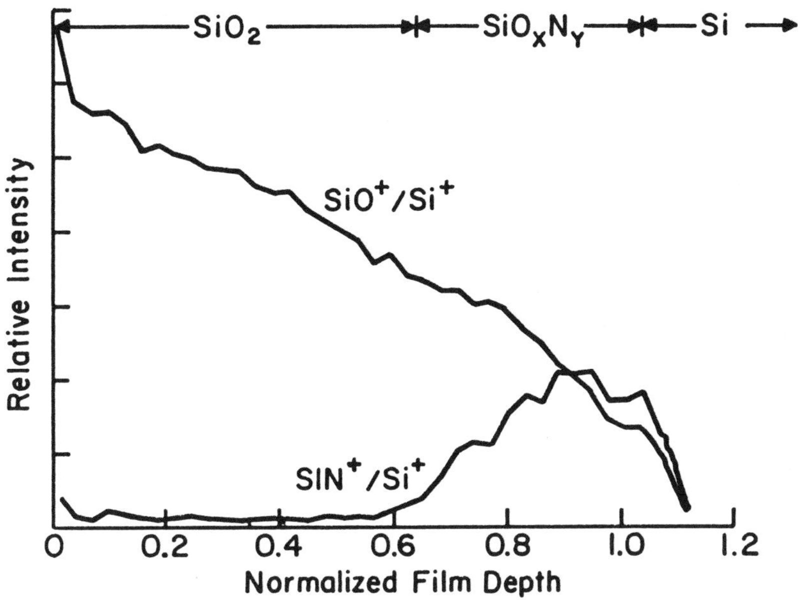

Figure 22: The relative depth distributions of oxygen and nitrogen in a sol/gel derived SiO_xN_y film after 8 hours of oxidation.

indicating that some of the original oxynitride material remains. There is no evidence for a buildup of oxygen in this interfacial region, so it is unlikely that the substrate begins to oxidize before the film.

The depth profiles shown in Figures 21 and 22 illustrate the mechanism through which the sol/gel derived oxynitride films resist substrate oxidation. The oxynitride films are denser than silicon dioxide, and so they restrict oxidant diffusion to the substrate. In order for the oxidant to reach the substrate, it must first react with the oxynitride film to convert it to an oxide. This reaction is then given by:

$$SiO_xN_y(s) + \frac{(2-x)}{3} O_2(g) \rightarrow SiO_2(s) + \frac{y}{2} N_2(g). \tag{2}$$

Reaction (2) occurs during the first stage of oxidation; i.e., for times shorter than t_{ox}. The conversion to a less dense silicon dioxide then accounts for the slow increase in film thickness for $t < t_{ox}$, as seen in Figure 20. Once the oxynitride has completely converted to silicon dioxide, the barrier is eliminated and the film growth rate rapidly increases as the substrate reacts with the diffusing oxidant ($t > t_{ox}$).

Figure 20 and Table 2 indicate that increasing the nitrogen content of the initial film (designated by the higher initial refractive index) results in an improvement in the oxidation resistance of the film. There are several possible explanations for this. For example, the oxidation reaction rate constant for reaction (2) may decrease with increasing nitrogen content. Also, it is known that the evolution of nitrogen during the oxidation of silicon nitride inhibits the transport of oxygen through the oxide surface layer.[38] Therefore, the increased amount of molecular nitrogen liberated in the oxidation of the more extensively nitrided films can reduce the permeation of the oxidant to the oxide-oxynitride interface. In either case, one can expect a more oxidation resistant SiO_xN_y film with increasing y/x.

In summary, silicon oxynitride films produced by a reaction between ammonia and sol/gel derived silicon dioxide are shown to resist the oxidation of silicon substrates. Increasing the amount of nitrogen initially incorporated in the films increases their oxidation resistance. This oxidation resistance stems from the ability of the film to prevent the diffusion of oxygen to the underlying film substrate. Instead, oxygen reacts with the oxynitride to convert it to silicon dioxide. Once this conversion is complete, the film no longer acts as an oxidation barrier. These studies also verify the density (or absence of continuous microporosity) of the oxynitride thin films prepared by the nitridation of silica gels.

SUMMARY

Sol/gel derived silicon dioxide films appear to be an effective precursor material for the development of silicon oxynitride thin films by a high temperature reaction with ammonia. Because of their initial microporosity, these films are rapidly incorporated with high concentrations of nitrogen (up to 40 mol %) to produce a compositionally homogeneous oxynitride material. The chemical structure of the films was thoroughly investigated and it was verified that the substitution of 3-coordinated nitrogen species occurs during this nitridation treatment. The reaction mechanism responsible for this nitridation are reported elsewhere,[28] but essentially, they involve the chemisorption of amine species within the microporosity of the gel at low temperatures and their subsequent condensation to nitrides at temperatures in excess of 800°C. Although further development of the processing will be required to produce films with acceptable electrical transport properties, it was shown that the refractive index, dielectric constant and oxidation resistance could be controlled through the ammonia treatment temperature. These properties were related directly to the corresponding concentration of nitrogen in the resulting SiO_xN_y films.

REFERENCES

1. See Chapters 13, 14 and 15 in *Ultrastructure Processing of Ceramics, Glasses and Composites,* Larry L. Hench and Donald R. Ulrich, Eds. (Wiley & Sons, New York, 1984) for reviews and complete lists of references.
2. See other chapters in this volume.
3. Rand, M.J., and Roberts, J.F., *J. Electrochem. Soc.,* 120(3), 446 (1973).
4. Nguyen, V.S., Burton, S., and Pan, P., *J. Electrochem. Soc.,* 131, 2348 (1984).

5. Lucovsky, G., Richard, P.D., Tsu, D.V., Lin, S.Y., and Markunas, R.J., *J. Vac. Sci. Techn.,* 4(3), 681 (1986).
6. Ito, T., Nozaki, T., and Ishikawa, H., *J. Electrochem Soc.,* 127(9), 1053 (1980).
7. Habraken, F.H.P.M., Kuiper, A.E.T., Tamminga, Y. and Theeten, J.B., *J. Appl. Phys.,* 53(10), 6996 (1982).
8. Hayafuji, Y. and Kajiwara, K., *J. Electrochem. Soc.,* 129(9), 2102 (1982).
9. Nemetz, J.A., and Tressler, R.E., *Solid State Techn.,* February, 79 (1983).
10. Moslehi, M.M. and Saraswat, K.C., *IEEE J. Solid-State Cir.,* SC-20(1), 26 (1985).
11. Elmer, T.H., and Nordberg, M.E., Nitrided Glasses, Paper No. 30, *Proc. VII International Congress on Glass,* Brussels, Belgium, 1965; 11 pp. Institut National de Verre, Charleroi, Belgium (1965).
12. Mulfinger, H.O., and Franz, H., *Glastechn.,* Ber., 38(6), 236 (1965).
13. Mulfinger, H.O., *J. Amer. Ceram. Soc.,* 49(9), 462 (1966).
14. Eriksson, G., *Chemica Scripta,* 8, 100 (1975).
15. Marchand, R. and Lang, J., *C.R. Acxad. Sc. Paris,* Serie C, 264, 969 (1967).
16. Wagner, C., *J. Appl. Phys.,* 29(9), 1295 (1958).
17. Tombs, N.C., Sewell, F.A., Comer, J.J., *J. Electrochem. Soc.,* 116(6), 862 (1969).
18. Chu, T.L., Szedon, J.R., and Lee, C.H., *J. Electrochem. Soc.,* 115(3), 318 (1968).
19. Raider, S.I., Flitsch, R., Aboaf, J.A., and Pliskin, W.A., *J. Electrochem. Soc.,* 123(4), 560 (1976).
20. Pantano, C.G., Glaser, P.M., and Armbrust, D.J., Sol/Gel Thin Films, in *Ultrastructure Processing of Ceramics, Glasses and Composites,* Larry L. Hench and Donald R. Ulrich, Eds., (John Wiley and Sons, New York, 1984) pp. 161-177.
21. Glaser, P.M. and Pantano, C.G., *J. Non-Cryst. Solids,* 63, 209 (1984).
22. Brow, R.K., and Pantano, C.G., in *Better Ceramics Through Chemistry,* Mat. Sci. Res. Soc. Symp. Proc., Vol. 32, Eds. C.J. Brinker, D.E. Clarkland, D.R. Ulrich, North-Holland, New York (1984) 361-367.
23. Brow, R.K. and Pantano, C.G., *J. Amer. Ceram. Soc.,* 169, 314 (1986).
24. Brown, D.M., Gray, P.V., Heumann, F.K., Phillip, H.R. and Taft, E.A., *J. Electrochem. Soc.,* 115(3), 311 (1968).
25. Phillip, H.R., *J. Phys. Chem. Sol.,* 32, 1935 (1971).
26. Taft, E.A., *J. Electrochem. Soc.* 118(8), 1341 (1971).
27. Gaind, A.K., Ackermann, G.K., Lucarini, V.J., and Bratter, R.L., *J. Electrochem. Soc.,* 124(4), 599 (1977).
28. Brow, R.K. and Pantano, C.G., Thermochemical Nitridation of Microporous Silica Films in Ammonia, *Journal of the American Ceramic Society* (1986).
29. Born, M., and Wolf, E., *Principle of Optics,* 3rd ed., Pergamon Press, Oxford (1965) 84-90.
30. Carman, L.A., and Pantano, C.G., Dielectric Properties of Silicon Dioxide and Silicon-Oxynitride Sol/Gel Thin Films, in *Science of Ceramic Chemical Processing,* Larry L. Hench and Donald R. Ulrich, Eds., (John Wiley & Sons, New York, 1984).

31. Deal, B.E., Fleming, P.J., and Castro, P.L., *J. Electrochem. Soc.,* 115, 300 (1968).
32. Gritsenko, V.A., Dikavskaja, N.D., and Mogilnikov, K.P., *Thin Solid Films,* 51, 353 (1978).
33. Lai, S.K., *Silicon Processing,* ASTM STP 804, D.C. Gupta, Ed., American Society for Testing and Materials (1983) 260-272.
34. Adams, A.C., Smith, T.E., and Chang, C.C., *J. Electrochem. Soc.,* 127, 1787 (1980).
35. Solomon, P., *J. Vac. Sci. Technol.,* 14, 1122 (1977).
36. Murarka, S.P., Chang, C.C., and Adams, A.C., *J. Electrochem. Soc.,* 126, 996 (1979).
37. Brow, R.K., and Pantano, C.G., *Appl. Phys. Lett.,* 48, 27 (1986).
38. Goursat, P., Lortholary, P., Tetard, D., and Billy, M., in *Proceedings of the 7th International Symposium on the Reactions of Solids,* eds. J.S. Anderson, M.W. Roberts and F.S. Stone (halsted, New York, 1972), 315-326.

Part III

Continuous, Discontinuous and Woven Fibers

7

Fibers from the Sol-Gel Process

Sumio Sakka
Institute for Chemical Research
Kyoto University
Uji, Kyoto-Fu, Japan

INTRODUCTION

Conventionally, glass and ceramic fibers have been prepared by drawing or blowing the high temperature melts through the orifice. This melting method requires conversion of raw materials to a homogeneous, high temperature melt, making it difficult to prepare fibers of compositions which are high melting or immiscible in the liquid state. In the case of glassy and amorphous fibers uncontrollable crystallization may inhibit fiber drawing from the melt. The sol-gel tech technique[11] applied to fiber preparation may be free from these difficulties. In the sol-gel technique, fibrous gels are drawn from the sols around room temperature, and then are converted to glass or ceramic fibers by heating at several hundred to one thousand degrees centigrade.[2] Recently, various kinds of glassy fibers, such as silica, alumina-silica and zirconia-silica fibers, and ceramic fibers, such as SiC, zirconia, alumina and titania fibers, have been prepared by this type of sol-gel technique, that is, the method based on the conversions sol → gel fiber → glass or ceramic fiber.

In this chapter, glass fiber formation from metal alkoxide will be described, in which gel fibers, drawn from the sol near room temperature, will be converted to glass fibers by heating. The underlying science in this process is described using an example of SiO_2 glass. Properties and applications are also given. Besides this type of fiber formation, formation of zirconia, titania and alumina fibers based on unidirectional freezing of gel, a modification of sol-gel technique, will be described.

VARIATIONS OF SOL-GEL FIBER PREPARATION

There are several variations or modifications in the sol-gel fiber preparation, as shown in Table 1. The classification shown in Table 1 is rather arbitrary. Here, loose definitions of sol, gel and sol-gel process are used.

Table 1: Modifications of Fiber Preparation Through Sol-Gel Process

Method	Characteristics	Starting materials	Fiber-forming temperature	Product: Composition	Product: State
1	Low temperature drawing. (Hydrolysis-polycondensation of alkoxides)	Metal alkoxides.	Room temp.	SiO_2, SiO_2-ZrO_2, SiO_2-TiO_2.	Glass
2	Low temperature drawing. (High concentration sol, Addition of drawing reagent)	Inorganic salts, Metal alkoxides.	Room temp.	Al_2O_3, Al_2O_3-SiO_2, TiO_2, ZrO_2.	Amorphous, Crystals.
3	Low temperature drawing.	Carboxysilanes.	Room temp.	SiC.	Amorphous, Crystals.
4	High temperature drawing.	Metal alkoxides, Fumed silica	2200°C(SiO_2)	SiO_2.	Glass
5	Unidirectional freezing of gel	Inorganic compounds (Chlorides, Oxychlorides)	0°C	SiO_2, TiO_2, ZrO_2, Al_2O_3.	Amorphous, Crystals.

In this paper methods 1 and 5 are described in some detail. Method 1, especially, is stressed, since this method represents the essence of fiber formation through the sol-gel process.

In method 1, metal alkoxides in solution are hydrolyzed and polymerized, so that the solution might become viscous and drawable without any help or addition. The composition of the starting solution must be chosen so that it may produce the drawable sol. The gel fibers become glass on heating at high temperatures.

Method 2 is similar to method 1. In method 2, drawing is accomplished by high concentration of fiber materials, use of a complex reaction with a large concentration of aliphatic acids, or addition of a suitable drawing reagent such as polyvinyl alcohol to the solution for drawing after the hydrolysis and polymerization reaction.[3-5] Inorganic metal salts, such as metal chlorides, metal nitrates and metal sulfates, as well as metal organic materials, such as acetates and metal alkoxides, are used as starting materials. When gel fibers are heated, volatiles, such as water, chlorine, sulfur dioxide, nitrogen oxide and carbon dioxide are lost, and fibers become compact. Usually, amorphous fibers are obtained at low temperatures. They become microcrystalline upon heating at higher temperatures. Thus, alumina, titania and zirconia fibers are produced.

Horikiri et al.[5,6] prepared heat resistant Al_2O_3 and Al_2O_3–SiO_2 amorphous and crystalline fibers. Aluminum alkoxide was changed to aluminoxane by partial hydrolysis. Polyvinyl alcohol or some other organic polymer solution was added to the aluminoxane for the Al_2O_3 composition and to the mixture of aluminoxane and organosiloxane compound for the $Al_2O_3-SiO_2$ compositions, in order to adjust the viscosity of the solutions for fiber drawing. Fibers were drawn and heated to the pertinent temperatures for the final products. Addition of SiO_2 to the Al_2O_3 composition makes stronger aluminous fibers.

In method 4 for obtaining SiC fibers,[7,8] dodecamethyl cyclohexasilane was converted to polycarbosilane polymer by treatment in an autoclave at 400°C for 48 hours. The fraction with molecular weight of about 1,500 was dissolved in benzene, to make a viscous solution. Fibrous gels drawn from this solution were converted to black SiC fibers with metallic luster by heating in vacuum at 1000°C. Heating these fibers at 1200°C to 1500°C resulted in mechanically strong, oxidation-resistant fibers consisting of β-SiC grains of 33-69 Å diameter.

Method 4 presents a new method of fabricating silica or silica-based optical fibers. Large rods of transparent silica glasses called preform are first prepared through the sol-gel technique and fibers are drawn at temperatures of 2200°C to 2300°C. Accordingly, the essential process in this method is fabrication of large rods of silica glass. Susa[9] made silica glass rods as large as 3 cm in diameter and 10 cm in length by heating silica gel rods at 1100°C to 1200°C. Tetramethoxysilane $Si(OCH_3)_4$ was hydrolyzed and polycondensed in a solution with or without a small concentration of ammonia as catalyst to make a gel cylinder. Rabinovich et al.[10] and Scherer and Luong[11] used colloidal silica sols prepared from fine, fumed silica particles. They could make silica glass cylinders of larger size than Susa's, indicating that bulk glass formation is easier in this so called colloid method. However the temperature employed for preparing silica cylinders (>1350°C) was higher than in the fabrication from metal alkoxides.

Method 5 is a modification of the sol-gel method. In this method, a hydrogel cylinder is unidirectionally frozen by slowly immersing in a cold medium from the bottom toward the top. Gel fibers several tens of microns in diameter are separated as a bundle upon thawing the frozen cylinder at room temperature, when conditions were satisfactory for sol and gel formation and freezing as described in section 4.

FIBERS THROUGH LOW TEMPERATURE DRAWING FROM METAL ALKOXIDE SOLS

Significance of Fiber Drawing at Low Temperature

Low temperature direct preparation of fibers based on the sol-gel technique is described. The fiber formation in this method (method 1 in Table 1) is illustrated in Figure 1, for the formation of SiO_2 glass from $Si(OC_2H_5)_4$ as an example.

A homogeneous $Si(OC_2H_5)_4$ solution containing water, alcohol and hydrochloric acid is prepared at room temperature. When this solution is kept standing under reflux at temperatures from room temperature to 80°C, the hydrolysis and polycondensation reactions proceed, leading to an increase in viscosity of the alkoxide solution. When the composition of the starting $Si(OC_2H_5)_4$-H_2O-C_2H_5OH--HCl solution is appropriate, fibers can be drawn from the viscous solution with viscosities higher than about 10 poises. Fibers, drawn near room temperature, are gel fibers, which can be converted to silica glass fibers by heating at 800°C, for example.

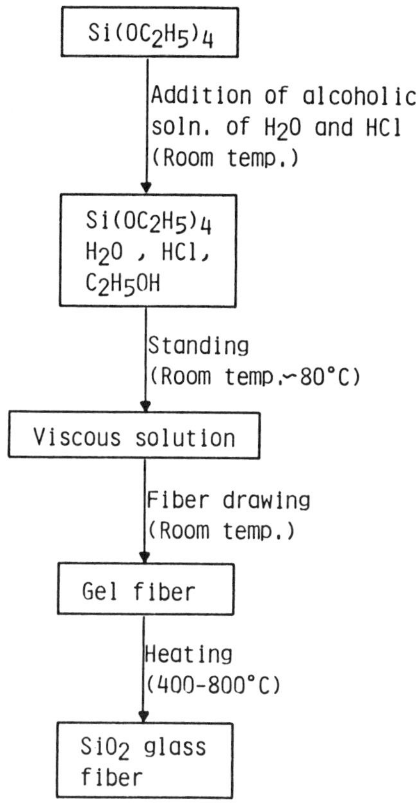

Figure 1: Low temperature glass fiber formation through the sol-gel technique. Formation of SiO_2 fiber from $Si(OC_2H_5)_4$ is given as an example.

One will notice that in this method, the maximum heating temperature throughout the processes is much lower than that required for SiO_2 fibers drawn from alkoxy-derived or melt-derived bulk glass rod. Thus, this technique is really the low temperature process for fiber formation. Another advantage can be seen in that continuous refractory fibers can be obtained by the present method. For instance, 3M Company[12] manufactured silica-alumina fibers by continuously drawing from metal alkoxide solution. Thus, for the first time, refractory silica-alumina fibers have been made continuously to be woven into textiles.

It should be noted that, in order to prepare fibers, the metal alkoxide solution has to exhibit spinnability in the course of hydrolysis-polycondensation and the drawn gel fibers have to be quickly solidified and dried on fiber drawing. The spinnability may appear when some kind of viscosity-increasing reagent is added to the solution. If such reagent is used, however, drawn fibers have to be immediately fixed by some special means, for example, by passing the fibers through the setting solution. Otherwise, a fiber once formed may shrink in the direction perpendicular to the fiber axis until it is separated into upper and lower parts and changes to a round drop of solution because of the surface tension. Therefore, the condition has to be found in which the solution becomes spinnable as a result of the progress of hydrolysis-polycondensation, that is, the condition for the occurrence of spinnability without addition of any viscosity-increasing reagent. In this case, the fibers drawn from the viscous, spinnable solutions are quickly solidified during drawing due to the progress of hydrolysis and polycondensation caused by the moisture in the air, because the reaction has already progressed before drawing to a sufficiently high degree and only a small degree of extra reaction is required for fixing the fibers.

Conditions for Gel-Fiber Drawing

Possibility of Fiber Drawing. The conditions in which silica fibers can be drawn from a tetraethoxysilane solution without addition of any foreign viscosity-increasing reagent will be discussed. It should be noted that all starting solutions do not necessarily become drawable. The composition of the solution must be appropriate, in order for fiber drawing to be possible. Generally, it has been shown[13] that the alkoxide solution becomes spinnable when the content of water used for hydrolysis is low and the catalyst is acid. When the water content is high or an alkali-like ammonia is chosen as the catalyst, the solution does not exhibit spinnability even when the viscosity reaches more than 10 poises.

Figure 2 is a diagram showing the fiber drawing behavior versus composition relationship for the formation of SiO_2 glass fibers from the $Si(OC_2H_5)_4$–H_2O–C_2H_5OH solution.[14] The ratio $HCl/Si(OC_2H_5)_4$ is kept at 0.01. The hydrolysis reaction was carried out at 80°C. The triangular diagram can be divided into four areas. The components of the solution are not miscible with each other in area I,

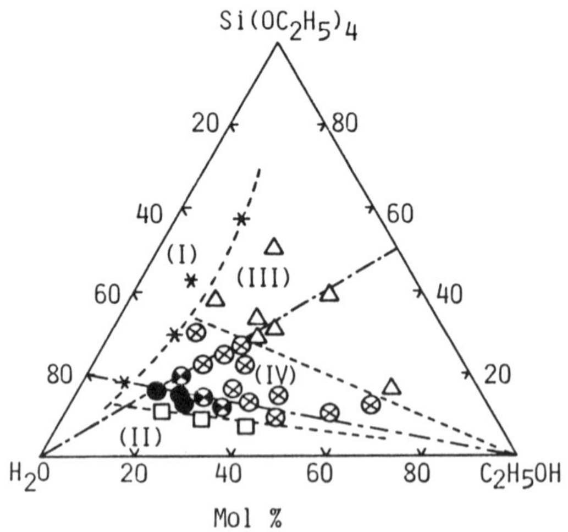

Figure 2: Relationship between fiber drawing behavior and composition of $Si(OC_2H_5)_4$–H_2O–C_2H_5OH solution with $HCl/Si(OC_2H_5)_4$ = 0.01 hydrolyzed at 80°C; (*) immiscible (area I), (□) not spinnable (area II), (△) no gel formation (area III), (●) circular cross section (area IV), (⊗) noncircular cross section (area IV), (⊖) circular and noncircular cross section (area IV).

where the C_2H_5OH content is low. In area II, where the $H_2O/Si(OC_2H_5)_4$ ratio r is over 5, the solutions become elastic in nature just before solidification to a gel, exhibiting no spinnability. In area III, in which r is less than 1.5, too much time is required for the solution to become viscous and solidify. In area IV, where r ranges from 1.5 to 4, the solutions exhibit spinnability and, therefore, fiber drawing is possible. On heating to 500°C to 900°C the drawn fibers become SiO_2 glass fibers.

Besides the possibility of fiber drawing, there is a problem that the cross-section of the fibers thus prepared are not necessarily circular, unlike ordinary glass fibers drawn from a high temperature melt. A circular cross-section can be obtained in some limited compositions of area IV. This problem will be discussed later.

Time Period Required for the Reaction Leading to Occurrence of Spinnability. It has been shown[13] that fibers can be drawn from the solutions of pertinent compositions when their viscosity reaches about 10 poises as a result of progress of the hydrolysis-polycondensation reaction. The time required for a solution to reach the drawable state is a function of ambient temperature—more than several days for room temperature, but two or three hours for 80°C. Two or three hours may be reasonable for the practical application of the procedure.

Another requirement for practical application is that the length of time available for fiber drawing should be long. This possibility has been found[15] by examining the time dependence of viscosity for the alkoxide solutions of different water contents shown in Figure 3. The reaction temperature was fixed at 80°C, while the measurements of viscosity were made at 25°C. It is seen in Figure 3 that in the solution with the $H_2O/Si(OC_2H_5)_4$ = 1.7, the increasing rate of viscosity reaches about 10 poises. This is thought to be caused by the exhaustion of the water added to the solution for the hydrolysis-polycondensation reaction. Accordingly, it is possible in this case to continue drawing for a prolonged time. Limiting the water added to the original solution, controlling the relative humidity of the surrounding atmosphere,[16] lowering the solution temperature during fiber drawing and covering the solution tightly may prolong the available time period for fiber drawing so that this problem has been solved.

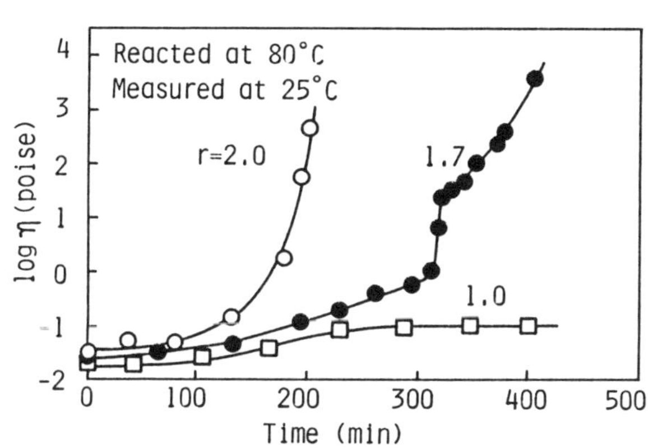

Figure 3: Change in viscosity of $Si(OC_2H_5)_4$–H_2O–C_2H_5OH solution at 80°C; $r = H_2O/Si(OC_2H_5)_4$.

Shape of Fiber Cross-Section. Figure 4 shows that some fibers have a circular cross-section and some have a noncircular cross-section depending on the composition of the starting tetraethoxysilane solution.[14] This never happens in fiber drawing from a glass melt, which contains no volatile constituents. It has been shown that the compositions which are characterized by a smaller shrinkage on sol-to-gel conversion at fiber drawing give a circular cross-section. For instance, a composition which shows 40% shrinkage gives fibers of noncircular cross-section, while a composition which shows 20% shrinkage gives fibers of circular cross-section.

It should be stressed here that the composition of the starting solution appears to be the only important factor that affects the shape of the cross-section. The same shape is obtained with different drawing methods, that is, by pulling with a rod and by extruding through an orifice. Whether the rod has a pointed or round end and whether the orifice is round or triangular do not affect the shape of the cross-section. The viscosity at drawing does not affect the shape, either. The only effect of the viscosity is that drawing at high viscosities gives thicker fibers and drawing at low viscosities gives thinner fibers.

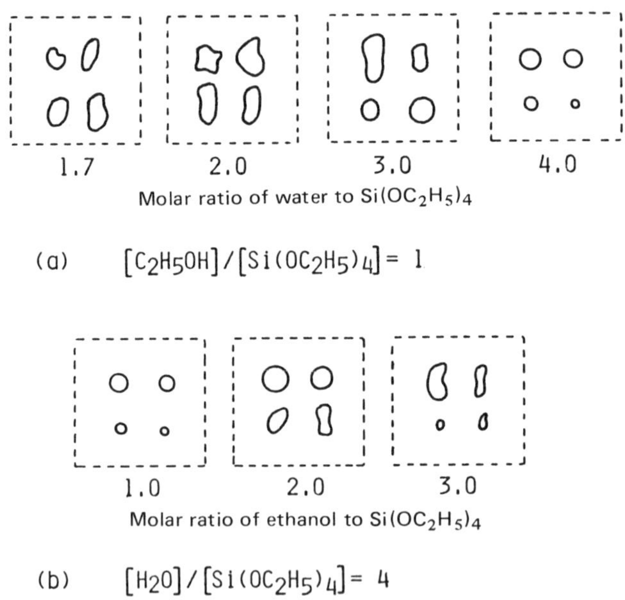

Figure 4: Schematic showing the change in shape of cross-section of SiO_2 fibers with composition of starting $Si(OC_2H_5)_4$ solution. (a) Change with water content at molar ratio $C_2H_5OH/Si(OC_2H_5)_4 = 1$. (b) Change with C_2H_5OH content at molar ratio $H_2O/Si(OC_2H_5)_4 = 4$.

Process of Hydrolysis-Polycondensation

Reduced Viscosity. It has been shown that the lower water content of the alkoxide solution is favorable for fiber drawing, while no spinnable state appears when the water content is high. It may be assumed that linear polymers are formed in the low water content solutions which show spinnability in the course of progress of hydrolysis-polycondensation reaction. It is also assumed that three-dimensional networks or colloidal particles may be formed in the high water content solutions and ammonia-catalyzed solutions which show an elastic nature before gelation and no spinnability.

In order to confirm these, the molecular weights and intrinsic viscosities of the solution taken in the course of gelation of the tetraethoxysilane (TEOS) solutions have been measured and discussed in terms of the shape of the polymers produced in the solutions. For this purpose, Sakka and Kamiya[13] determined the relation between the reduced viscosity and concentration for five different $Si(OC_2H_5)_4$ solutions; two solutions acid-catalyzed with r being unity, one solution with r being 20 and two solutions catalyzed with ammonia. r is the $H_2O/Si(OC_2H_5)_4$ mol ratio. The SiO_2 concentration of the solution was varied by diluting the solution with alcohol.

According to Huggins,[17] for linear polymer solutions, the reduced viscosity η_{sp}/C is expressed as a function of the concentration C by the relation

$$\eta_{sp}/C = (\eta) + k(\eta)^2C \tag{1}$$

where η_{sp}, (η) and k denote the specific viscosity, the intrinsic viscosity and a constant, respectively. This suggests that the plot of η_{sp}/C - C is a straight line with a slope. On the other hand, the solution containing spherical particles follow the Einstein relation

$$\eta_{sp}/C = K/\rho \tag{2}$$

indicating that the plot is parallel to the concentration axis.

The experimental results showed that the two acid-catalyzed solutions with r being 1 exhibit slope, indicating that linear polymers are formed in the solution. For other solutions, no slope was seen for the plots, indicating that the particles in the solution are round, and this is the reason for the absence of spinnability of these solutions.

Intrinsic Viscosity. The shape of particles formed in the solutions can be also examined from the intrinsic viscosities and molecular weights. The compositions of the solutions used in the measurement are shown in Table 2. Solutions 1 and 2 are characterized by lower water content, solution 3 by intermediate water content and solution 4 by higher water content. The solutions have been kept at 30°C for the hydrolysis-polycondensation reaction. A portion of the solution was taken at various times in the course of the reaction. The silicon alkoxide polymers in the solution have been trimethylsilylated[18] for stabilization dissolved in benzene and subjected to the measurement of molecular weight and viscosity.

Table 2: Compositions and Properties of $Si(OC_2H_5)_4$ Solutions

Solution*	$Si(OC_2H_5)_4$ (g)	$[H_2O]/[Si(OC_2H_5)_4]$	C_2H_5OH (ml)	Concentration of SiO_2 (wt %)	Spinnability	Time for gelling at 25°C(h)
1	169.5	1.0	324	33.3	Yes	233
2	178.6	2.0	280	42.3	Yes	240
3	280.0	5.0	79	61.0	No	64
4	169.5	20.0	47	33.5	No	138

* $[HCl]/[Si(OC_2H_5)_4]$ = 0.01 for all glasses

Figure 5 shows the variation of number-averaged molecular weight $\bar{M}_n$ with the reduced reaction time t/t_g, where t_g is the geling time. Figure 6 shows the η_{sp}/C - C plots, where η_{sp} is the reduced viscosity and C is the concentration, for the alkoxide polymers taken from solution 1 at various reaction times. The numbers attached to the lines in the figure denote the measured values of $\bar{M}_n$. It is seen that the slopes of the straight lines and the values of (η) are larger for larger $\bar{M}_n$ values. The occurrence of the slope in the lines indicates that linear polymeric particles are found in the solution.[13] Figure 7 shows the variation of intrinsic viscosity (η) with t/t_g.

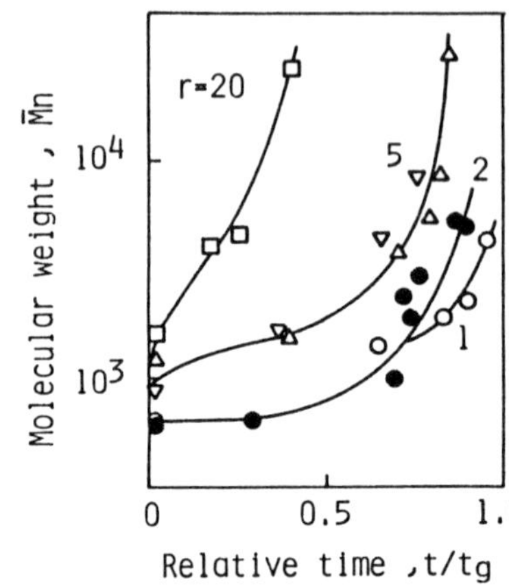

Figure 5: Change of number-average molecular weight $\bar{M}_n$ of the trimethylsilylated siloxane polymer with relative time t/t_g (t_g is the gelling time) for $Si(OC_2H_5)_4$ solutions with different r's. The marks ▽ and △ for the solution with r of 5.0 correspond to the polymers trimethylsilylated with trimethylchlorosilane (TMC) and hexamethyldisiloxane (HMDS), respectively.

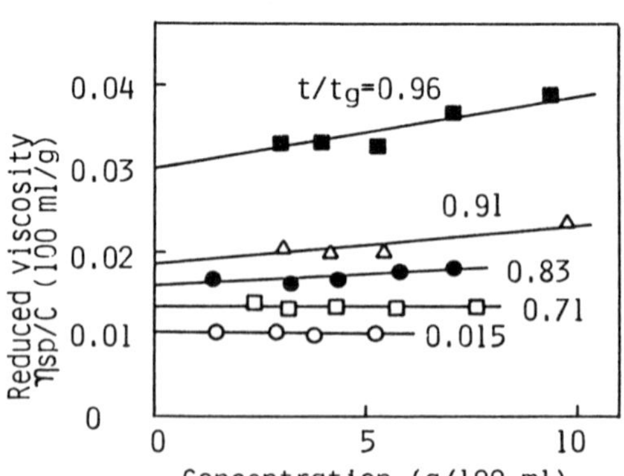

Figure 6: Relationship between reduced viscosity η_{sp}/C and concentration of the trimethylsilylated siloxane polymers for solution 1 with r of 1.0.

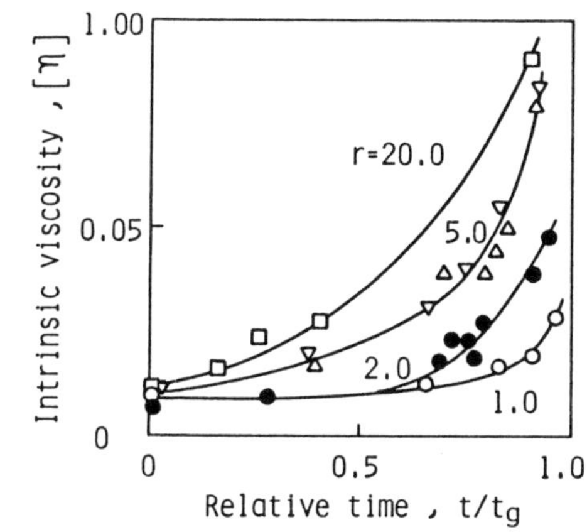

Figure 7: Change of intrinsic viscosity (η) with relative time t/t_g for $Si(OC_2H_5)_4$ solution with different r's.

Figure 8 shows the log $\overline{M}_n$ versus log (η) plots. The slope of the plot is larger than 0.5, that is, 0.75 and 0.64 respectively for solutions 1 and 2 in which the water content is low. It is less than 0.5, that is, 0.34 for solution 4, in which the water content is high. The slope for solution 3 which has an intermediate water content is about 0.5 at the early stage of the hydrolysis-polycondensation reaction and about 0.2 at the later stage.

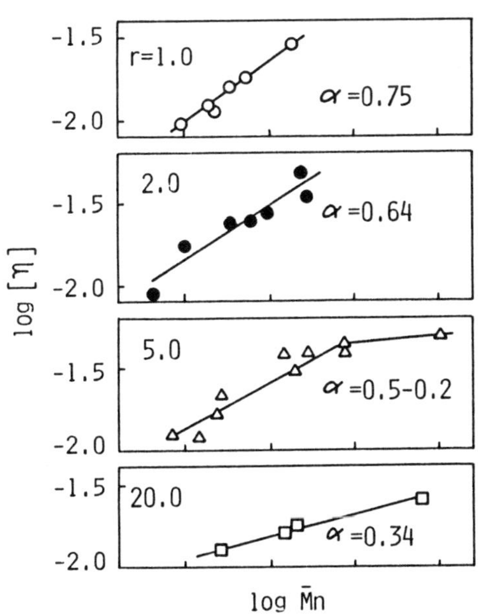

Figure 8: Relative between number-average molecular weight $\overline{M}_n$ and intrinsic viscosity (η) of the trimethylsilylated siloxane polymers for the $Si(OC_2H_5)_4$ solutions with different r's.

It is known that for polymer solutions (η) is related to $\overline{M}_n$ by the expression,[19,20]

$$(\eta) = k\, M_n^{\alpha} \tag{3}$$

where k is a constant depending on the kind of polymer, solvent and temperature. The exponent α, that is, the slope of the log (η) - log $\overline{M}_n$ plot takes a value between 0 to 2.0. The value depends on the shape of polymeric particles: $\alpha = 0$ for rigid spheric particles, $\alpha = 0.5$ to 1.0 for flexible, chain-like or linear polymers and $\alpha = 1.0$ to 2.0 for nonflexible or rigid, rod-like polymers.[20] It is reported that for high polymers containing siloxane bondings, Si–O–Si, $\alpha = 0.5$ for linear polydimethylsiloxane, $\alpha = 0.21$ to 0.28 for branched or cross-linked polymethylsiloxane and $\alpha = 0.3$ for spherical polysilicates.[21]

Referring to these, the present experimental results on the (η) - $\overline{M}_n$ relationships can be interpreted to show the following. In solution 1 with $r = H_2O/Si(OC_2H_5)_4 = 1.0$, the alkoxide polymers are formed in the solution. Linear polymeric particles are also found in solution 2 with particles at the early stage of the reaction but three-dimensional or spherical growth of polymeric particles occur at the later reaction stage close to the onset of gelation. In solution 4 containing a much higher amount of water expressed by r = 20, three-dimensional or spherical growth of alkoxide polymeric particles is predominant.

The above results are summarized in Table 3. Linear polymeric particles are the main reaction products in solutions 1 and 2 which become drawable. Three-dimensional polymeric particles are dominant in solution 4 which does not show spinnability but forms a large bulk gel and glass. Solution 3 is intermediate between the above two cases, resulting in no drawability and no bulk gel formation. This systematic conclusion may prove the author's prediction that fiber drawing is only possible for the alkoxide solutions containing linear polymeric particles.[1,2] With this information, silica fibers can be continuously drawn as shown by Figure 9. This is a picture of continuously drawn silica fibers. A bundle of fibers has been cut from a spinning drum.

Table 3: The Exponent α's for the Alkoxide Polymers and Properties of $Si(OC_2H_5)_4$ Solution

Solution	H_2O(r)	α	Type of polymer	Spinnability
1	1.0	0.75	Linear	Yes
2	2.0	0.64	Linear	Yes
3	5.0	0.5	Branched	No
		0.2	Three-dimensional	
4	20.0	0.34	Three-dimensional, Spherical	No

The Nature of Linear Polymeric Particles. The nature of the polymeric particles is not exactly known yet; whether they are linear polymers with a single chain, those with triple chains, or linear aggregates of fine, round beads is not known.[22] Formation of any of those polymeric particles could take place, since the functionality of the $Si(OC_2H_5)_4$ molecule is four. It is noted that Bechtold

Figure 9: Photograph of continuously drawn silica fibers.

et al.[23] proposed from their chemical reaction studies that the triple-chain models hold.

Fibers of Compositions Other Than Silica

A similar sol-gel method using metal alkoxides has been applied to the preparation of oxide glass fibers other than SiO_2 fibers. Generally, other metal alkoxides are hydrolyzed and polycondensed much more quickly and easily than $Si(OC_2H_5)_4$. Accordingly, preparation of the starting solution requires certain precautions; a water-containing acid such as hydrochloric, sulfuric, nitric or acetic acid cannot be added to the metal alkoxide mixture, because it may cause nonuniform hydrolysis and precipitation of colloids in the solution. The addition of water as an alcoholic solution may often cause nonuniform precipitation.

Then, the air humidity is used to induce hydrolysis and produce a viscous spinnable solution. Since moisture is absorbed through the surface on exposure of the metal alkoxides-alcohol mixed solutions to air, the solution must be vigorously stirred during the reaction to keep it homogeneous. Thus, glass fibers of composition 30% $Al_2O_3 \cdot 70\% SiO_2$ in weight are produced from $Al(O\text{-}i\text{-}C_3H_7)_4$ and $Si(OC_2H_5)_4$.[24,25] Glass fibers of the ZrO_2–SiO_2 and Na_2O–ZrO_2–SiO_2 systems containing 7% to 48% ZrO_2 are produced from $Zr(OC_3H_7)_4$, $Si(OC_2H_5)_4$ and $NaOCH_3$.[26]

It takes a long time for the alkoxide mixture to become spinnable with this method. Forcible introduction of water vapor and CO_2 gas into the solution can improve the process; it remarkably shortens the time required for the solution to become viscous and spinnable for producing TiO_2-SiO_2 glasses.[27] Glass fibers of 10% $TiO_2 \cdot 90\%$ SiO_2 composition can be prepared by adding a mixture of acetic acid and water to the alkoxide mixture,[28] as shown in the previous section.

Properties of Fibers Synthesized by the Sol-Gel Process

Basic Properties. Basic properties, such as density, refractive index, thermal expansion and elastic moduli and their change with heating of fibers drawn from the metal alkoxide solutions may be similar to those of the plates and particles prepared in the same way. Naturally, one must note that fibers drawn near room temperature should be a gel with fine pores, and that the porosity decreases with increasing heating temperature. Such changes have been measured with other forms of gels than gel fibers.[29-31] It is known that gels with 20-60% porosity, in which open and continuous pores of about 100 Å diameter are dispersed, become glasses of compact structure when they are heated to temperatures near or a little higher than the glass transition temperatures of the respective glasses. It is also noted that water, alcohol and other volatile matters in gels are lost by vaporization during heating.

Properties of New Glasses: Alkali-Resistance of Zirconia-Containing Fibers. Glass reinforced concrete is made by incorporating alkali-resistant glass fibers into cement. ZrO_2 is an important component which increases the alkali-resistance of the glass. However, glasses containing more than 20% ZrO_2 are difficult or impossible to be manufactured by the method of melting, because of very high melting temperature and crystallizing tendency during cooling.

Glass fibers of the ZrO_2-SiO_2 and $Na_2O-ZrO_2-SiO_2$ systems containing up to 48% ZrO_2 have been prepared by the sol-gel method from $NaOCH_3$, $Zr(N\text{-}OC_3H_7)_4$ and $Si(OC_2H_5)_4$.[26,31] Figure 10 shows the change of alkali durability as a function of the ZrO_2 content of glass. The alkali durability is expressed by the weight loss of the fiber in 2 N NaOH solution at 96°C. It is seen that the alkali durability increases with increasing ZrO_2 content of glass. It should be noted that this type of study is possible since at present the sol-gel low temperature synthesis can give glasses of very high ZrO_2 concentrations.

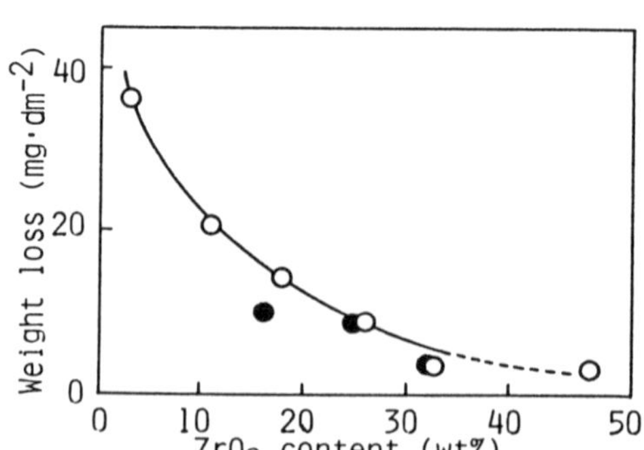

Figure 10: Weight loss in 2 N NaOH solution at 96°C for 4h, plotted against the ZrO_2 content. Weight loss data are calculated based on the decrease in fiber diameter. ○ fibers of the ZrO_2-SiO_2 system, ● fibers of the $Na_2O-ZrO_2-SiO_2$ system.

Appearance of Films. Fibrous gels are easily converted to glass fibers on heating, if no crystallization occurs. However, the glass fibers of SiO_2, SiO_2-TiO_2 and $Na_2O-ZrO_2-SiO_2$ compositions are often yellow, brown and black after being heated to convert them to glass fibers. The coloring or blackening may be attributed to fine carbon particles. It is supposed that fibrous gels become denser as a result of further hydrolysis and polymerization after fiber drawing and that such densification makes the vaporization of organic residues from the fibers difficult. Organic residues would be carbonized at high temperatures, causing blackening of the glass fibers. It is important to note that the diameter of fibers produced by the author and his colleagues is 10-40 μm and that organic substances cannot leave the fibers. On the other hand, if the structure of fibrous gels is loose due to insufficient progress of hydrolysis and polymerization, organic substances would leave more easily. As a matter of fact, fibrous gels heated immediately after fiber drawing produce colorless and transparent glass fibers.[26]

Mechanical Strength of Fibers. It is known that the mechanical strength of glass fibers is generally much higher than that of bulk glasses such as rods and plates, and that the strength is very sensitive to fabrication conditions. Figure 11 shows the tensile strength of SiO_2 glass fibers prepared from $Si(OC_2H_5)_4$ by the sol-gel method.[31] For the abscissa, the cross-sectional area or the calculated diameter is shown to represent the fiber thickness, since the strength measurement was made on fibers with a circular or noncircular cross-section. Anderegg's data of tensile strength of fused silica fibers[39] are also shown for comparison.

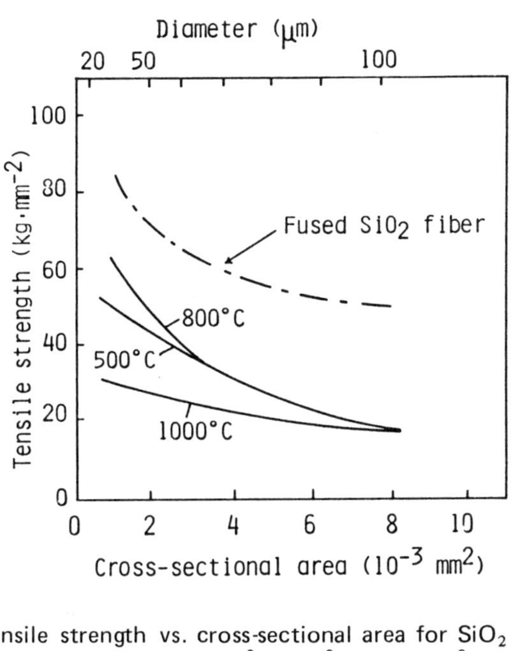

Figure 11: Tensile strength vs. cross-sectional area for SiO_2 fibers prepared by the sol-gel method and heated at 500°C, 800°C and 1000°C, respectively.

It is seen that the strength of gel-derived silica glass fibers is generally lower than that of melt-derived fibers and that the strength does not change with a temperature of heat treatment higher than 500°C. Factors affecting the strength of as-drawn gel-derived SiO_2 fibers and consolidated glass fibers are extensively discussed by LaCourse.[32]

Figure 12 shows the tensile strength of gel-derived zirconia-silica fibers of the composition $20ZrO_2 \cdot 80SiO_2$ in weight percent.[34] It is seen that the strength of this glass is generally higher than that of gel-derived SiO_2 glass and for smaller diameters compares with fused silica fibers.

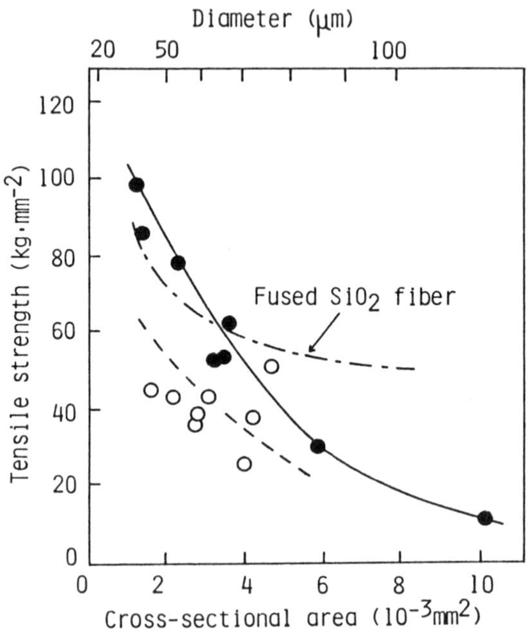

Figure 12: Tensile strength vs. cross-sectional area for fibers prepared by the sol-gel method. ● $20ZrO_2 \cdot 80SiO_2$ (wt %) heated at 800°C, ○ SiO_2 heated at 800°C, — · — · — Fused SiO_2 fiber (Anderegg).[32]

FIBERS FORMED BY THE UNIDIRECTIONAL FREEZING OF GEL

The basic idea of preparing fibers by unidirectional freezing of gel was originated by Mahler and Bechtold.[35] They made silica gel fibers by unidirectionally freezing a silica hydrogel cylinder from bottom to top. In this process, numerous, needle-like ice crystals grow in the cylinder at a temperature slightly lower than 0°C, causing the silica component to be concentrated into spaces between the ice crystals to form silica gel fibers. Thawing of ice crystals at room temperature gives a bundle of silica gel fibers. Thereafter this method was applied to preparing zirconia,[36] titania[37] and alumina[38,39] fibers. Fibers thus prepared have an angular cross-section, since fibers are formed between ice crystals which have hexagonal configuration. Gel fibers as prepared are hydrated, porous and essentially amorphous and, depending on the composition, are converted to crystalline fibers on heating.

Preparation of the Fiber

Table 4 summarizes the conditions of fiber formation, together with measured properties. In order to prepare hard hydrogels, which may successfully produce oxide fibers on freezing and subsequent thawing, electrolytes such as Na^+, K^+ and Cl^+ ions are removed by dialysis. This is carried out by placing the cellulose tube filled with the sol in distilled water for several days. It is assumed that for successful formation of gel fibers, sufficient metal-oxygen-metal bonds have to be formed. This requires a fairly high concentration of oxide components in the sol as shown in Table 4.

For unidirectional freezing, the cellulose tube containing the hydrogel is put into a polyethylene cylinder lined with polyurethane foam, and the cylinder is lowered into a -78°C cold bath of dry ice-ethanol mixture at a rate of 2-9 cm/h. The apparatus is shown in Figure 13. The lowering rate of the gel-containing cylinder markedly affects the formation of fibers. Since it is assumed that the formation of oxide gel fibers in unidirectional freezing may be achieved by the cellular growth of ice crystals in the form of needles, the freezing rate R and the temperature gradient at the freezing front G must have appropriate values.

Tiller et al.[40] express the condition for the plane front growth of the solvent in the unidirectional solidification of the solution as follows.

$$G/R \geqslant \frac{m \cdot C_o (1-k)}{k \cdot D} \tag{4}$$

Here, m is the slope of liquidus line, C_o is the initial concentration of the solution, k is the partition ratio and D is the diffusion coefficient. Similar reasoning of solidification is assumed to hold in the unidirectional freezing of ice, which is considered in this experiment. In freezing of the same gels, the value of the right-hand side of Formula 4 is identical. When G/R is large because of the small lowering rate, the ice crystals grow by the plane front growth mechanism. Then, the oxide component is expelled to the front area of the freezing ice crystals and no oxide fibers may be formed. For higher lowering rates causing small G/R values, ice crystals grow by the cellular growth mechanism as ice fibers. The oxide component is concentrated in the region between ice fibers, forming oxide gel fibers on freezing. The formation of long fibers for larger lowering rates is explained by this reasoning. For lowering rates which are too high and thus cause very low G/R values, however, the occurrence of large constitutional supercooling causes formation of discrete, particulate ice crystals and no long fibers are formed. Thus, fibers are formed when the lowering rates are intermediate.

The optical micrograph of a bundle of fibers, scanning electron micrograph (SEM) of separated fibers and SEM of fiber cross-sections are shown in Figure 14[37] for TiO_2 fibers.

Table 4: Conditions of Fiber Formation in Unidirectional Freezing of Gels and Properties of Fibers

	SiO_2 (34)	ZrO_2 (35)	TiO_2 (36)	Al_2O_3 (37)
Starting chemical	Sodium silicate	$ZrOCl_2 \cdot 8H_2O$	$TiCl_4$	Al metal
Formation of sol	Ion exchange	Hydrolysis	Hydrolysis	Reaction with acid
Oxide content of sol	0.5 - 5 M	1 - 3 M	1 - 2 m	1 - 1.5 M
Shape of particles	-	-	-	Feather-like
Gellation	Addition of NH_4OH	Dialysis	Dialysis	Dialysis
Appearance of gel	Transparent	Transparent	Transparent	Opaque
State of gel	Amorphous	Amorphous	Amorphous	Amorphous
Cylinder for freezing	240mm long, 27mm dia.	150mm long, 12mm dia.	150mm long, 12mm dia.	150mm long, 12mm dia.
Lowering rate on freesing	0.3 - 150cm/h	4 - 6 cm/h	2.2 - 4.5 cm/h	6 - 9 cm/h
Fibers as dried:				
Length	15 cm	10 cm	10 cm	5 cm
Diameter	50 μm	10 - 50 μm	20 - 100 μm	60 - 200 μm
Composition	$SiO_2 \cdot (1/3)H_2O$ (Dried at 150°C)	$ZrO_2 \cdot H_2O$ (Dried at 290°C)	$TiO_2 \cdot 0.3H_2O$ (Dried at 120°C)	$Al_2O_3 \cdot 2.8H_2O$ (Dried at 110°C)
State	Amorpuous	Amorphous	Anatase	Bohemite
Specific surface area	900 m^2/g	–	350 m^2/g	240 m^2/g
Tensile strength	86 ± 35 MPa	19 MPa	–	50 MPa
Fibers after heating				
Tensile strength	510± 216 MPa (Heated at 925°C)	30 MPa (Heated at 290°C)	–	180 MPa (Heated at 1000°C)
State	Glass	Amorphous	–	θ-Alumina

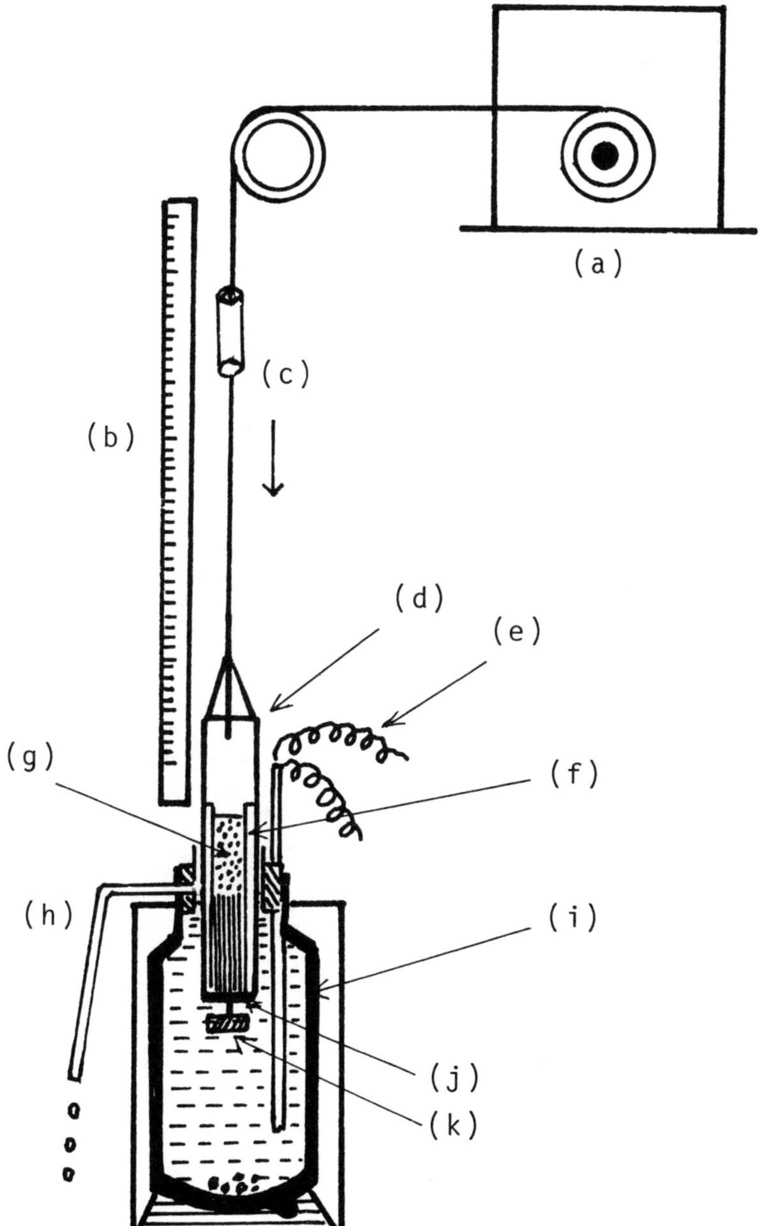

Figure 13: Apparatus used for unidirectional freezing of gel. (a) Motor, (b) Scale, (c) Guide, (d) Polyethylene cylinder, (e) CA thermocouple, (f) Polyurethane foam, (g) Gel, (h) Overflow, (i) Dry ice/Ethanol cold bath, (j) Brass bottom, (k) Weight.

Figure 14: Fibers prepared by unidirectional freezing of TiO_2 gel. The gel is prepared from 2.0 M TiO_2 hydrosol. (a) Photograph of a bundle of fibers, (b) Scanning electron micrographs (SEM) of separated fibers and (c) SEM of fiber cross-section.

Properties of Fibers Made by Unidirectional Freezing

Some of the properties of oxide fibers thus prepared are shown in Table 4. The diameter of the fiber is not constant throughout the fiber axis, because G and R change with the distance from the bottom of the cylinder, when a constant lowering rate is adopted. Figure 15 shows that the fiber diameter is linearly related to (1/GR). Oxide gel fibers have a polygonal cross-section.

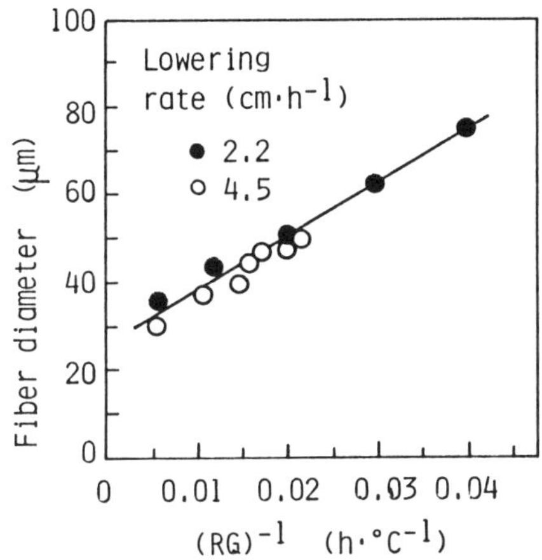

Figure 15: Relation between the fiber diameter and the reciprocal of the product of the freezing rate (R) and the vertical temperature gradient (G) at the freezing front. The fibers were prepared from 1.0 M TiO_2 hydrosol-derived gel.

Fibers are porous, hydrate gels after thawing and drying at low temperatures, as shown in Table 4. Dried gels show very large surface areas–900 m^2/g in silica, 350 m^2/g in titania and 240 m^2/g in alumina. This suggests that fibers can be applied as support for catalysts and enzymes.

Heating of gel fibers give glass or microcrystalline materials. Silica fibers become high strength silica glass fibers.[35] The strength of alumina fibers is the highest at 140 MPa when heated at 100°C. The fibers could be used as reinforcing materials, if higher strength is achieved.

REFERENCES

1. Sakka, S., in: *Treatise on Materials Science and Technology* (M. Tomozawa and R.H. Doremus, ed.), Vol. 22, pp 129-167, Academic Press, New York (1982).
2. Sakka, S., in: *Better Ceramics Through Chemistry.* Mat. Res. Soc. Symp. Proc. Vol 32 (C.J. Brinker, D.E. Clark and D.R. Ulrich, ed.), pp 91-99, North-Holland, New York (1984).
3. Winter, G., Zirngibul, H. and Mansmann, M.; Japan Patent 48-17615; May 30, 1974; assigned to Beyeraktiengesellschaft.
4. Wainer, E.; German Patent 1249832; April 11, 1968; assigned to Horizon Inc.
5. Horikiri, S., Tsuji, K., Abe, Y., Fukui, A. and Ichiki, E.; Japan Patent 49-108325; Oct. 15, 1974; assigned to Sumitomo Chemical Company.
6. Horikiri, S., Tsuji, K., Abe, Y., Fukui, A. and Ichiki, E.; Japan Patent 50-12335; Feb. 7, 1975; assigned to Sumitomo Chemical Company.
7. Yajima, S., Hayashi, J. and Omori, M., *Chem. Lett.* 931-934 (1975).

8. Yajima, S., Okamura, K., Hayashi, J. and Omori, M., *J. Amer. Ceram. Soc.* 59: 324-327 (1976).
9. Susa, K., Matsuyama, I., Satoh, S. and Suganuma, T., *Electronics Lett.* 18: 499-500 (1982).
10. Rabinovich, E.M., MacChesney, J.B., Johnson, D.W., Simpson, J.R., Meagher, B.W., Dimarcello, F.V., Wood, D.L. and Sigety, E.A., *J. Non-Crystal. Solids* 63: 155-161 (1984).
11. Scherer, G.W. and Luong, J.C., *J. Non-Crystal. Solids* 63: 163-172 (1984).
12. 3M Company Catalog, Nextel Ceramic Fiber, and Private Communication.
13. Sakka, S. and Kamiya, K., *J. Non-Crystal. Solids* 48: 31-46 (1982).
14. Sakka, S. and Kamiya, K., in: *Materials Science Research* Vol. 17, Emergent Process Methods for High-Technology Ceramics (T. Davis, H. Palmour and T. Porter, ed.), pp 83-94, North-Holland, New York (1984).
15. Sakka, S., Kamiya, K. and Kato, T., *Yogyo-Kyokai-Shi* 90: 555-556 (1982).
16. LaCourse, W.C., Dahar, S. and Akhtar, M.M., *Commun. Am. Ceram. Soc.* C-200 - C-201 (1984).
17. Huggins, M.L., *J. Am. Chem. Soc.* 64: 2716-2718 (1942).
18. Lents, C.W., *Inorg. Chem.* 3: 574-579 (1964).
19. Budley, W.J. and Mark, H., in: *High Molecular Weight Organic Compounds* (R.E. Burk and O. Grunmitt, ed.), Interscience Publisher, New York, 1949, pp 7-112.
20. Tsuchida, H., *Science of Polymers,* Baihukan Publishing Company, Tokyo, 1975, pp 85-87.
21. Abe, Y. and Misono, T., *J. Polymer Sci. Polymer Chem.* 21: 41-53 (1983).
22. Sakka, S., Tanaka, Y. and Kokubo, T., Hydrolysis and polycondensation of dimethyldiethoxysilane and methyltriethoxysilane as materials for sol-gel process, to be published in *J. Non-Crystal. Solids.*
23. Bechtold, M.E., Vest, R.D. and Plambeck, L., *J. Am. Chem. Soc.* 90:17: 4590-4598 (1968).
24. Kamiya, K., Sakka, S. and Tashiro, N., Preparation of refractory oxide fibers from metal alcoholates–possibility of fiber drawing in the course of hydrolysis of the alcoholates. *Yogyo-Kyokai-Shi* 84: 614-618 (1976).
25. Kamiya, K. and Sakka, S., *J. Mater. Sci.* 15: 2937-2939 (1980).
26. Kamiya, K., Sakka, S. and Tatemichi, Y., *J. Mater. Sci.* 15: 1765-1771 (1980).
27. Sakka, S. and Kamiya, K., in: *Proc. Int. Symp. Factors Densificat. Sinter. Oxide Non-Oxide Ceram.* (S. Somiya and S. Saito, ed.), Tokyo Institute of Technology, Tokyo, 1978, pp 101-109.
28. Kamiya, K., Sakka, S. and Ito, S., *Yogyo-Kyokai-Shi* 85: 599-605 (1977).
29. Nogami, M. and Moriya, Y., *J. Non-Crystal. Solids* 37: 191-201.
30. Yamane, M. and Okano, S., *Yogyo-Kyokai-Shi* 87: 434-439 (1979).
31. Sakka, S., *Hyomen* 19: 430-437 (1981).
32. Anderegg, F.O., *Ind. Eng. Chem.* 31: 290-298 (1939).
33. LaCourse, W.C., *Better Ceramics Through Chemistry.* Mat. Res. Soc. Symp. Proc., Vol. 32 (C.J. Brinker, D.E. Clark and D.R. Ulrich, eds.), pp 53-58, North-Holland, New York (1984).
34. Kamiya, K., Yoko, T., Iwanaka, M., Sakai, A. and Sakka, S., *Zirconia-Ceramics (Japan)*, 5: 39-52 (1984).

35. Mahler, W. and Bechtold, M.F., *Nature* 285: 27-28 (1980).
36. Kokubo, T., Teranishi, Y. and Maki, T., *J. Non-Crystal. Solids* 56: 411-416 (1983).
37. Maki, T., Teranishi, Y., Kokubo, T. and Sakka, S., *Yogyo-Kyokai-Shi* 93: 387-393 (1985).
38. Maki, T. and Sakka, S., Formation of alumina fibers in unidirectional freezing of gel. To be published in *J. Mat. Sci. Lett.*
39. Maki, T. and Sakka, S., *J. Non-Crystal. Solids* 82: 239-245 (1986).
40. Tiller, W.A., Jackson, K.A., Rutter, J.W. and Chalmers, B., *Acta Met.* 1: 428-437 (1953).

8

Alumina-Boria-Silica Ceramic Fibers from the Sol-Gel Process

Harold G. Sowman

3M

St. Paul, Minnesota

INTRODUCTION

Ceramic fibers discussed in this chapter include those comprising compositions from the Al_2O_3–B_2O_3 and Al_2O_3–SiO_2 binary systems as well as fibers which contain all three oxides, Al_2O_3, B_2O_3, and SiO_2. Since it is not within the scope of this chapter to include a dissertation on fibers designated as "alumina" fibers, compositions discussed herein have been limited arbitrarily to those containing at least 10% by weight of B_2O_3 or SiO_2 or of B_2O_3 and SiO_2 combined. The major emphasis is also directed to the preparation and properties of continuous fibers (or filaments). It is recognized, however, that staple fibers and non-woven mats of fibers of this system are useful, especially for high temperature insulation and thermal shielding; this will be expanded upon later in the discussion on applications.

The addition of other oxides to the basic Al_2O_3–B_2O_3–SiO_2 compositions is sometimes beneficial. The purpose of such modification may be to obtain colors for decorative applications, grain growth inhibition, increase in modulus of elasticity, enhancement of chemical resistance to certain agents, or for improvement in mechanical durability of fabric woven from fibers of this system. Alternatively, metal oxides may be added as sources of metals which may be precipitated by heat-treating in reducing atmospheres. The resulting metal-containing fibers, then, will be cermet fibers having a discontinuous metal phase dispersed within a continuous ceramic matrix.

Fibers may also be fired in controlled atmospheres so as to cause cracking of the fugitive organic components and precipitation of carbon.

PROCESSING

The major share of published technical information on ceramic fibers by the sol-gel process is contained in patents. A recent book by Bracke, Shurmans and Verhoest[1] of the European Patent Office, The Hague, Netherlands provides a compilation of recent developments on inorganic fibers including the development of fibers by sol-gel processes.

"Sol-gel" processes are generally considered to include those which utilize solutions, sols, and mixtures of solutions and sols as precursors for ceramic articles. The precursors can be polymerizable materials such as alkoxides. Examples of materials used as sources for Al_2O_3 include aqueous solutions of chlorides, nitrates, formates, and formoacetates as well as aqueous sols of colloidal alpha alumina monohydrate. Boria (B_2O_3) may be derived from boric acid. A basic aluminum acetate having a formula of $Al(OH)_2(OOCCH_3)\cdot \frac{1}{3}H_3BO_3$, available under the trademark "Niaproof" (Niacet Corporation), is a particularly useful material which can serve as a precursor for both Al_2O_3 and B_2O_3 in a molecular ratio of 3:1. An aqueous solution of this precursor can be concentrated by removal of water to a syrupy consistency and formed into fibers, much like the spinning of polymeric or glass fibers. Aluminum formoacetate behaves in a similar fashion. Silica sols are readily available from commercial sources for SiO_2, especially in aqueous systems. Silica sols can be obtained with different colloid sizes, such as, for example, 2 to 20 nm or larger, with SiO_2 solids of 15 to 50% by weight. Well-known trademarks for silica sols include "Ludox" (E.I. du Pont de Nemours & Co., Inc.) and "Nalco" (Nalco Chemical Co.). Although aqueous systems provide for convenience, precursor materials may also be organic-based, requiring non-aqueous solvents. For example, Horikiri, et al[2] has described a process for preparing alumina-silica fibers by spinning a solution of at least one polyaluminoxane and one or more kinds of silicon-containing compounds such as dimethyl polysiloxane in an organic solvent, hydrolyzing, and then calcining to an $Al_2O_3–SiO_2$ fiber.

During spinning, drying, and firing, the control of conditions such as relative humidity, atmosphere, temperature, and temperature/time relationships may require precise control for the preparation of high quality $Al_2O_3–B_2O_3–SiO_2$ fibers. This is also the case for all ceramic fibers prepared via the sol-gel process. It is especially important to avoid ignition during the removal of organics as firing proceeds,[3,4] since over-heating may occur. This can cause uncontrolled release of remaining fugitives and excessive crystal growth, resulting in weak, fragile fibers.

$Al_2O_3–B_2O_3$ FIBERS

The fabrication of continuous $Al_2O_3–B_2O_3$ fibers were described by Sowman.[4] As stated previously, basic aluminum acetate having an equivalent $Al_2O_3:B_2O_3$ ratio of 3:1 was spinnable after concentration by removal of water from an aqueous solution. In a specific example for the preparation of continuous fibers, a 37.5% solution of basic aluminum acetate in water was concentrated in a rotating evacuated flask partially immersed in a 32° to 36°C water bath. After concentration to an equivalent Al_2O_3 content of 28.5%, a viscous liquid with a viscosity of 100,000 to 150,000 cp was obtained. Continuous fibers were formed by pressurizing this solution at 83 to 97 x 10^4 Pa (120 to 140 psi) so as to extrude it through a spinnerette having thirty 100-μm diameter holes. Individual bundles of fibers were fired from room temperature to 600°, 800°, 1000° and 1200°C, respectively, in air. Those fired to temperatures up to 1000°C were shiny, clear, and colorless. Those fired to 1200°C were opaque and very fragile. X-ray diffraction patterns of the fibers fired at 1000° and 1200°C for one-half hour were consistent with that for aluminum borate ($9Al_2O_3 \cdot 2B_2O_3$). Tensile strength of the 1000°C fired fibers was 1.04 GPa (150,000 psi) and modulus of elasticity was 155 GPa (22.5 x 10^6 psi). Fiber diameter was about 12 μm and density was 2.90 g/cm^3. Firing at 1400°C for one-half hour resulted in loss of some of the boria since the x-ray diffraction patterns indicated the major crystalline component to be $9Al_2O_3 \cdot 2B_2O_3$ with a minor amount of alpha (α)-Al_2O_3.

Fibers having the stoichiometric composition of $9Al_2O_3 \cdot 2B_2O_3$ with no excess B_2O_3 were also fabricated from concentrated viscous aqueous solutions of basic aluminum acetate and aluminum formoacetate. It is interesting to note that the compound $2Al_2O_3 \cdot 1B_2O_3$ indicated in published phase diagrams[5] was not found in Al_2O_3–B_2O_3 fibers; only the $9Al_2O_3 \cdot 2B_2O_3$ compound was detected.

Ceramic fibers having Al_2O_3 and B_2O_3 as the only components and having B_2O_3:Al_2O_3 in ratios of 2:9 or greater were found to be inferior, with regard to high temperature applications, to those containing SiO_2. But the role of B_2O_3 when combined with alumina and silica in ceramic fibers is important as noted below.

THE EFFECT OF BORIC OXIDE ADDITIONS

Boria has been reported to be effective in lowering the temperature required for the formation of mullite and for retarding the transformation of alumina into the alpha form.[6,7] The presence of boria in an amount equivalent to or greater than a $9Al_2O_3$:$2B_2O_3$ ratio in Al_2O_3–B_2O_3–SiO_2 compositions was found to prevent the formation of any crystalline alumina,[4] the only crystalline phase during firing, according to x-ray diffraction analyses, being an aluminum borosilicate species, possibly an aluminum borate-mullite solid solution. The formation of such a solid solution is consistent with a published phase equilibrium diagram of the $9Al_2O_3 \cdot 2B_2O_3$–$3Al_2O_3 \cdot 2SiO_2$ system[5] and may explain the lower temperature initiation of the mullite-type structures noted previously. The x-ray diffraction patterns for the mullite ($3Al_2O_3 \cdot 2SiO_2$) and the aluminum borate ($9Al_2O_3 \cdot 2B_2O_3$) compounds are quite similar, notable differences including a 4.35 diffraction line present in the $9Al_2O_3 \cdot 2B_2O_3$ pattern and a reversal of the ratio of intensities of the 5.4 and 3.4 lines. This is shown in Table 1 along with values obtained for Al_2O_3–B_2O_3–SiO_2 fibers.[4,8]

Table 1: Comparison of Relative Intensities of Selected X-Ray Diffraction Line

Diffraction Line	Mullite ($3Al_2O_3 \cdot 2SiO_2$)	Aluminum Borate ($9Al_2O_3 \cdot 2B_2O_3$)	Aluminum Borosilicate ($Al_2O_3-B_2O_3-SiO_2$)
5.4	30-40	100	60-100
4.35	0	60	0-50
3.4	100	40	40-80

The data show that the relative intensities of selected lines from x-ray powder diffraction patterns of the crystalline species of Al_2O_3–B_2O_3–SiO_2 fibers lie between those for mullite and aluminum borate. The ratio of relative intensities of these lines in diffraction patterns of the aluminum borosilicate crystalline material has been found to be somewhat dependent on temperature and time, the ratio changing more toward that of mullite as temperature and time increases, presumably as the development of the solid solution species proceeds.

MICROSTRUCTURE DEVELOPMENT

The correlation of differential thermal analysis (DTA), x-ray diffraction, and electron microscopy of selected Al_2O_3–B_2O_3–SiO_2 fiber compositions provides important information on the process of microstructure development and resultant properties. The results of studies on fibers with Al_2O_3:B_2O_3 molecular ratios of 9:2 and 3:1 follow with the most detailed data being on fibers having a composition of $3Al_2O_3$:$1B_2O_3$:$3SiO_2$. Electron microscopy studies show that the precursor used has a dramatic effect on microstructure, especially prior to the complete reaction of the SiO_2 to form a silicate compound in the system. In the amorphous state, when the SiO_2 is in colloid form, SiO_2 colloids remain as discrete identifiable regions in the microstructure. This is shown in a comparison of transmission electron micrographs (TEM) of microtomed sections shown in Figures 1 and 2.

The $9Al_2O_3$:$2B_2O_3$ fibers were prepared from an aqueous mixture of boric acid-stabilized aluminum acetate and aluminum formoacetate. The $9Al_2O_3$:$2B_2O_3$:$6SiO_2$ fibers were prepared from an aqueous mixture of boric acid-stabilized aluminum acetate, aluminum chloride hexahydrate, and a silica sol.

The $3Al_2O_3$:$1B_2O_3$:$3SiO_2$ unfired fibers were prepared according to the flow sheet of Figure 3. Drying of the fibers was done by drawing through a heated tower. Bundles of the dry fibers were fired in an air-atmosphere furnace from room temperature to 600°C over a period of one hour. Separate portions of these were then refired at 800°, 900°, 1000°, 1200°, and 1400°C, respectively, for an hour.

Transmission electron micrographs of microtomed sections of fibers selected from each fired lot shown in Figures 4 through 9 show a dramatic change in microstructure of the specimens fired to 900°C (Figure 5) as compared to those fired to 800°C (Figure 4).

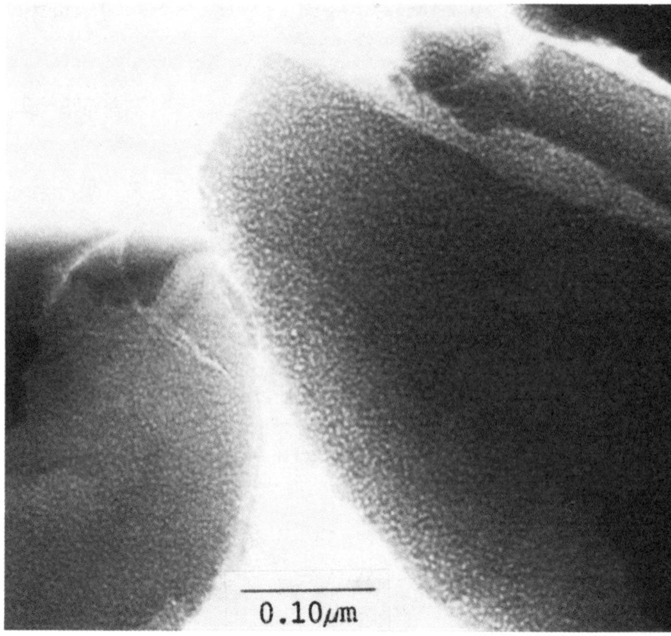

Figure 1: TEM of sections of Al_2O_3:$2B_2O_3$ fibers fired at 800°C.

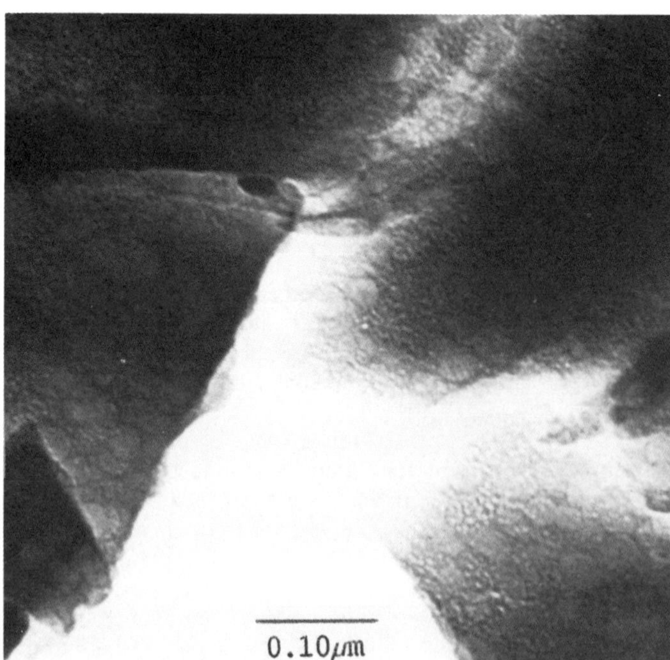

Figure 2: TEM of sections of $9Al_2O_3$:$2B_2O_3$:$6SiO_2$ fibers fired to 800°C.

PREPARE AQUEOUS MIXTURE

Stir Together Basic Aluminum Acetate

(Boric Acid-Stabilized) and Silica Sol

FILTER

CONCENTRATE

Rotating Evacuated Flask

(Viscosity=75000 CPS)

SPIN

Spinnerette

30 - 100 μm Dia. Holes

Pressure

1 - 2.2 MPa

(150 - 180 psi)

DRAW AND COLLECT

Rate

30-100m/min

Figure 3: Preparation of unfired precursor $3Al_2O_3:1B_2O_3:3SiO_2$ fibers.

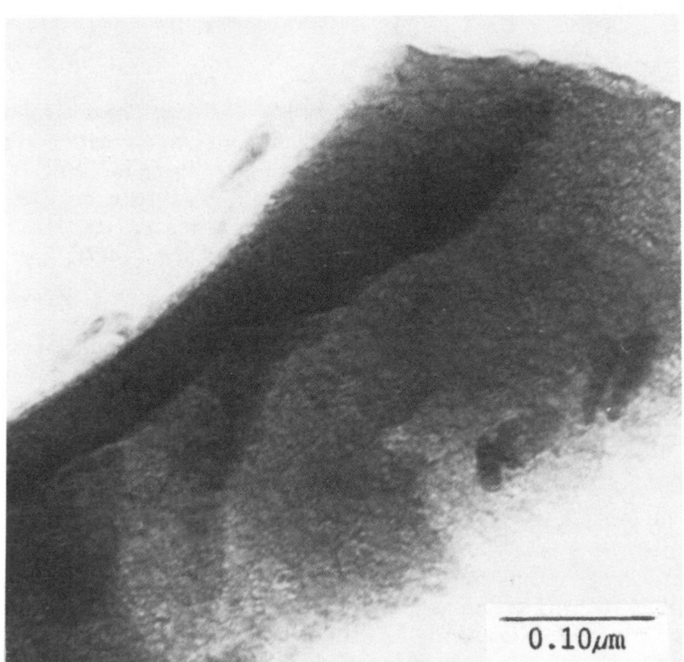

Figure 4: TEM of sections of $3Al_2O_3:1B_2O_3:3SiO_2$ fibers fired to 800°C.

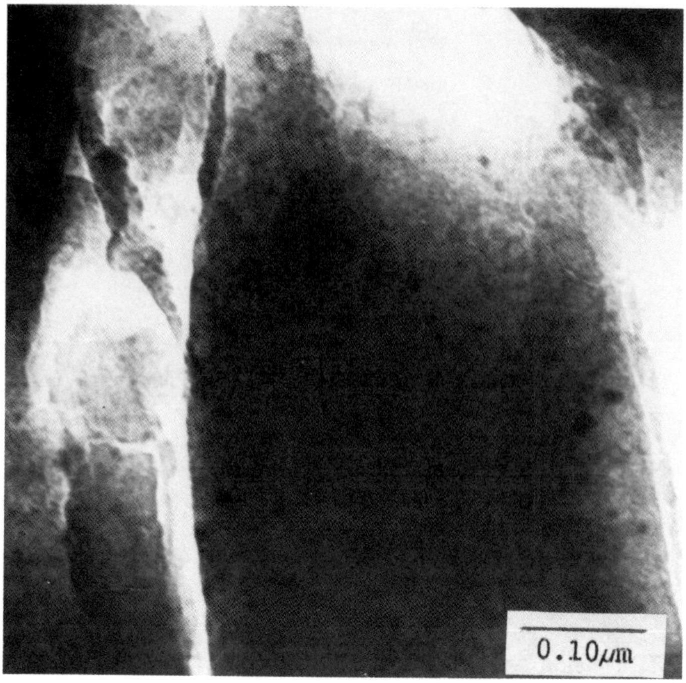

Figure 5: TEM of sections of $3Al_2O_3$:$1B_2O_3$:$3SiO_2$ fibers fired at 900°C for 1 hour.

Tiny crystallites have appeared after the 900°C firing. These are further refined after firing to 1000°C (Figure 6) and significant grain growth and development of well-defined prismatic crystals is shown after firing to 100°C (Figure 7) and 1200°C (Figure 8). In addition to the crystalline component, considerable amorphous material, e.g., B_2O_3 and SiO_2, remains in the $3Al_2O_3$:$1B_2O_3$:$3SiO_2$ fiber. Excessive crystal growth is in evidence after firing to 1400°C as shown in Figure 9.

Indeed, the handleability of fibers of $3Al_2O_3$:$1B_2O_3$:$3SiO_2$ is reduced seriously after exposure to 1400°C for as little time as one hour. At such temperatures B_2O_3 volatilizes rapidly, causing severe damage to the fibers. X-ray data show that at this temperature the loss of B_2O_3 causes a shift in the diffraction pattern toward that of mullite, i.e., the ratio of relative intensities of the 5.4/3.4 diffraction lines becomes 40/100, essentially that of pure mullite. A summary of the x-ray analyses and electron microscopy examinations is provided in Table 2.

Differential thermal analysis (DTA) results on unfired fibers of the $3Al_2O_3$:$1B_2O_3$:$3SiO_2$ compositions are consistent with the sudden change in microstructure shown by x-ray diffraction analyses and electron micrographs of fired fibers. Figure 10 is a tracing of a DTA on fibers heated at a rate of increase of 10°C/min.

The endothermic reactions occurring at the lower temperatures, e.e., below about 300°C, are attributable to the volatilization of water and other fugi-

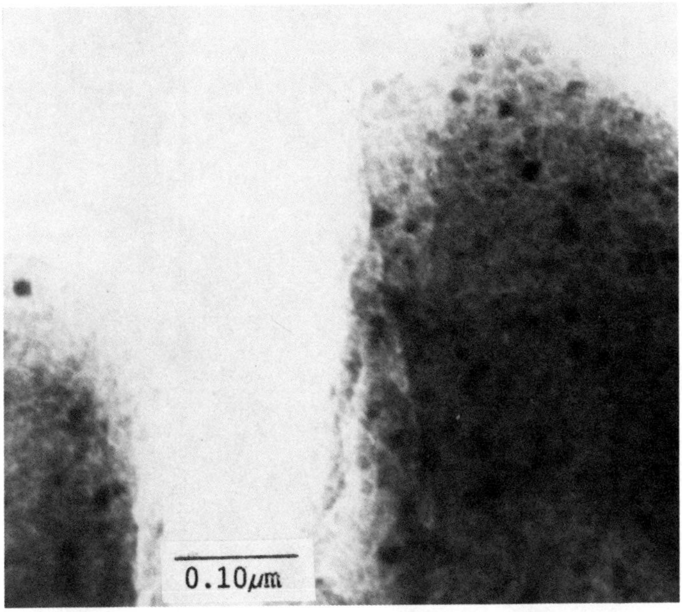

Figure 6: TEM of sections of $3Al_2O_3$:$1B_2O_3$:$3SiO_2$ fibers fired at 1000°C for 1 hour.

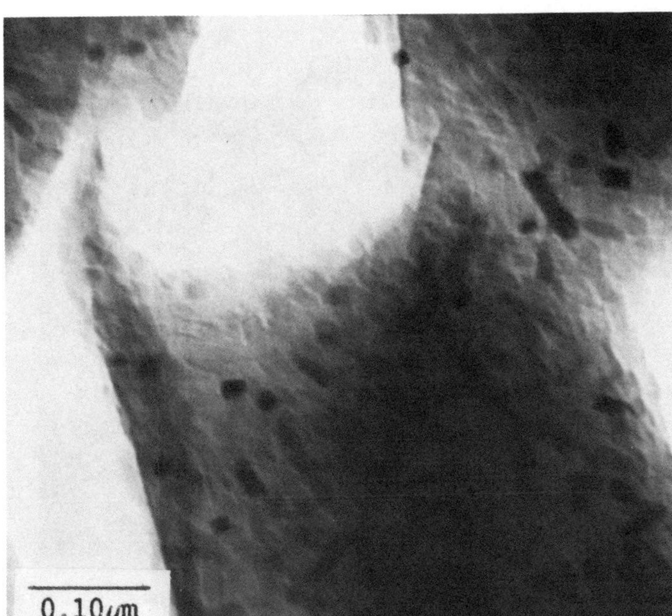

Figure 7: TEM of sections of $3Al_2O_3$:$1B_2O_3$:$3SiO_2$ fibers fired at 1100°C for 1 hour.

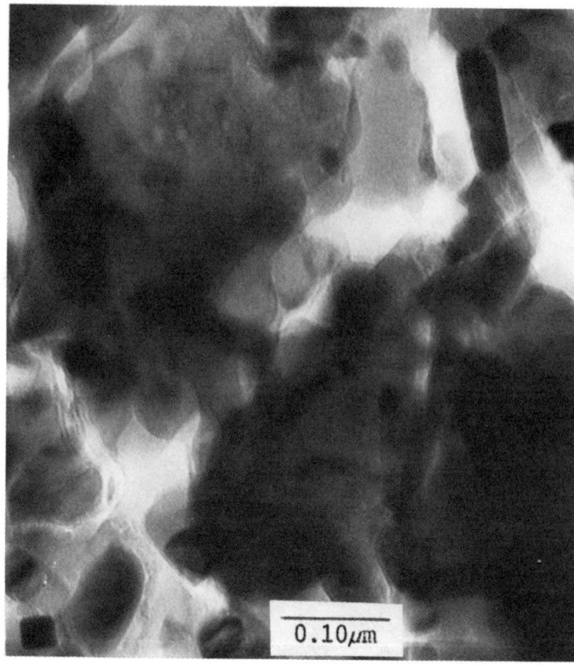

Figure 8: TEM of sections of $3Al_2O_3$:$1B_2O_3$:$3SiO_2$ fibers fired at 1200°C for 1 hour.

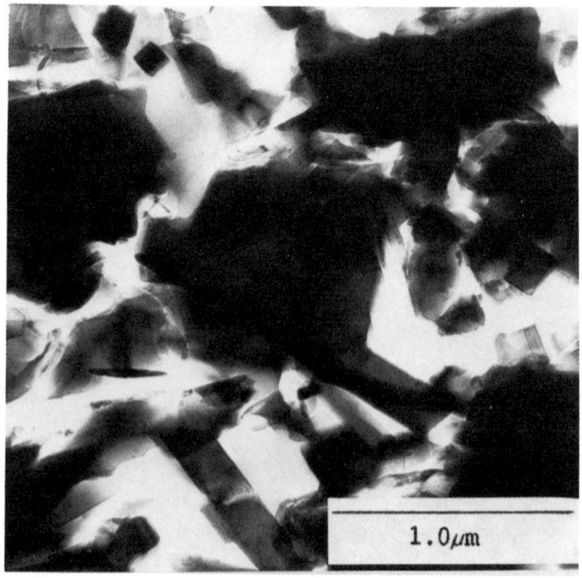

Figure 9: TEM of sections of $3Al_2O_3$:$1B_2O_3$:$3SiO_2$ fibers fired at 1400°C for 1 hour.

Table 2: Microstructure of $3Al_2O_3$:$1B_2O_3$:$3SiO_2$ Fibers

Firing	X-ray Analysis			Electron
	Crystalline	Relative Intensity		
Temp°C	Species	5.4	3.4	Microscopy
600	Amorphous	-	-	SiO_2 Colloids Porous Structure
800	Amorphous	-	-	SiO_2 Colloids Porous Structure
900	Aluminum Borosilicate	100	60	Dense Crystallites 100-180A; Not Porous
1000	"	100	90	Dense, Cube Shaped Crystallites to 270A; Lath-Shaped Crystals up to 150 X 360A
1100	"	100	90	Lath-Like Crystals 90 - 240A X up to 800A Long In Amorphous Matrix
1200	"	100	90	Same as Above with Crystals 330 to 750X up to 2400A long
1400	Mullite	40	100	Lath-Like Crystals up to 1μm in length

tives. The exotherms at temperatures between about 300° and 600°C are associated with the decomposition and oxidation of fugitive components. The sharp exotherm at 885°C is coincident with the formation of the aluminum borosilicate crystalline species of the system; total reaction time for this formation at the 10°C/min heating rate is about one to one and a half minutes. It seems remarkable that the ceramic fibers can survive this without apparent damage during the firing process. Further studies are expected to provide more information on changes which take place during this crystallization process.[9]

A differential thermal analysis of basic aluminum acetate (boric acid stabilized) is shown in Figure 11. A comparison of this graph with that of Figure 10, shows a significant difference in that region of the curve charting the exothermic reactions associated with the decomposition, oxidation, and volatilization of fugitive components. The presence of silica in the Al_2O_3–B_2O_3–SiO_2 precursor fiber apparently has a dramatic effect on the temperature and rate

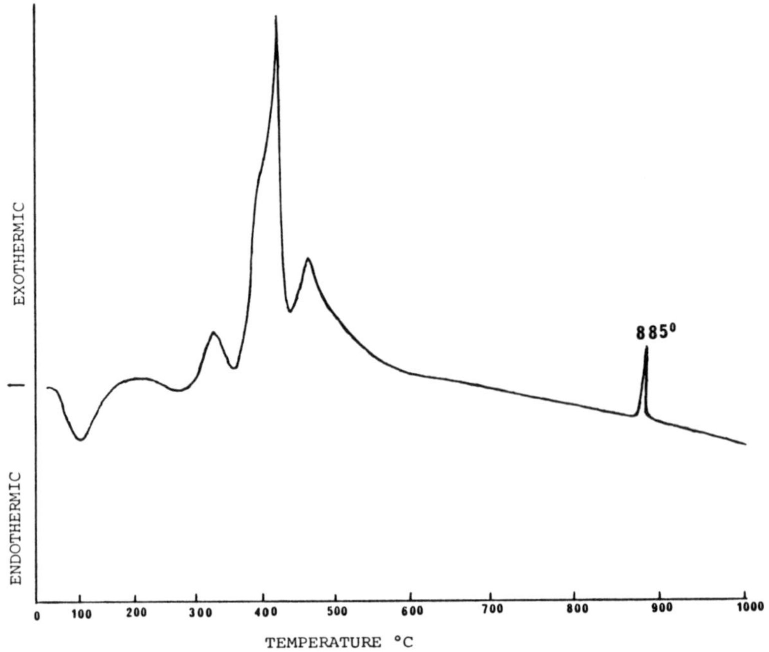

Figure 10: DTA of $3Al_2O_3$:$1B_2O_3$:$3SiO_2$ precursor fibers (rate = 10°C/min).

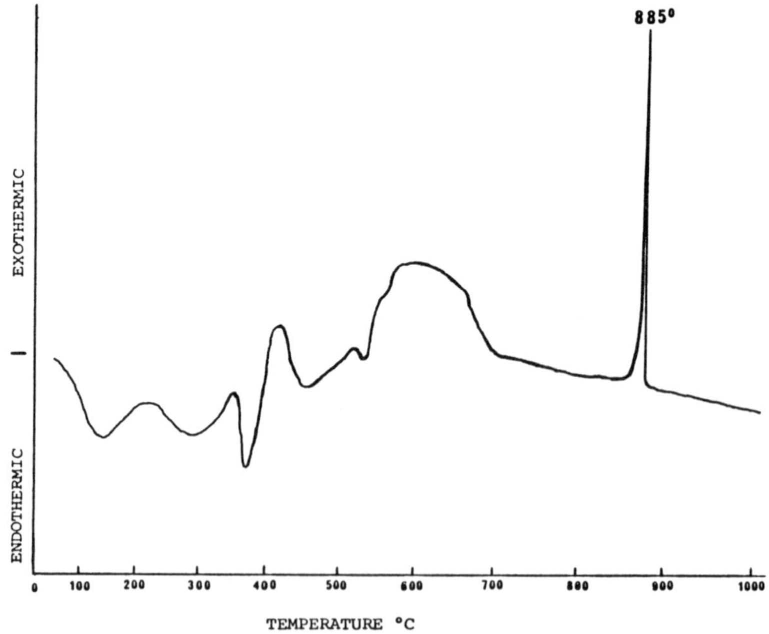

Figure 11: DTA of basic aluminum acetate (boric acid-stabilized; rate = 10°C per min).

of reactions occurring at this stage in the firing process. This is in line with knowledge that fibers of this system containing silica are less difficult to fire satisfactorily than those without.[4] The sharp exotherm at 885°C in Figure 11 is associated with the formation of the $9Al_2O_3 \cdot 2B_2O_3$ compound. The exothermic reaction indicated by the peak at 885°C in Figure 10 for $3Al_2O_3$:$1B_2O_3$:$3SiO_2$ fibers is convincing evidence that the first crystalline species which appears in the Al_2O_3–B_2O_3–SiO_2 fibers may be aluminum borate, followed by a solid solution reaction with SiO_2 to result in the aluminum borosilicate crystalline species of this fiber.

PROPERTIES OF Al_2O_3–B_2O_3–SiO_2 FIBERS

Mechanical and chemical properties of fibers of this system may be affected significantly by composition and by firing to cause crystallization. Amorphous fibers may be fired so as to be essentially nonporous, and such fibers, in general, are as strong as fibers which have been fired to crystallize the aluminum borosilicate species. Modulus of elasticity values are affected by the formation of crystalline materials. For example, the modulus values of fibers having an Al_2O_3:B_2O_3:SiO_2 ratio of 3:1:2 change from about 110 to 124 GPa (16 to 18×10^6 psi) in the dense amorphous form to about 131 to 145 GPa (19 to 21×10^6 psi) after the formation of a crystalline phase caused by firing at about 900°C.

The latitude in using a wide range of compositions for fibers in this system makes it possible to vary properties, e.g., index of refraction and modulus of elasticity. In one study, to determine the effect of SiO_2 content with a fixed Al_2O_3:B_2O_3 ratio of 3:1, it was found that increasing additions of SiO_2 (aquasol), as indicated previously, were coincidental with ease of pyrolysis. Modulus of elasticity values of fired fibers ranged from about 83 GPa (12×10^6 psi) for a composition of $3Al_2O_3$:$1B_2O_3$:$10SiO_2$ to about 155 GPa (22.5×10^6 psi) for $3Al_2O_3$:$1B_2O_3$:$0SiO_2$. Compositions between these SiO_2 limits having the $3Al_2O_3$:$1B_2O_3$ ratio had modulus of elasticity values according to an essentially linear relationship based on the SiO_2 content.

Chemical properties have been found to be changed significantly after crystallization. This is especially important in view of potential applications in high temperature filtration which may require resistance to corrosive gases such as sulfuric acid. For greatest resistance to corrosive environments, fibers of woven or nonwoven textiles of Al_2O_3–B_2O_3–SiO_2 compositions should be heated to obtain maximum combination of the components into the mullite, aluminum borate, or aluminum borosilicate crystalline materials.

Al_2O_3–SiO_2 FIBERS

The fabrication of continuous and noncontinuous fibers of Al_2O_3–SiO_2 with these oxides as the sole components or which contain only small amounts of additives is described in several U.S. patents and publications.[2,10,14] In most of the described processes, aqueous solutions, sols, or mixtures of these are utilized. As mentioned earlier, nonaqueous systems which are polymerizable[2] can also be used as precursors. Information available indicates that the polymerization and pyrolysis steps may require lengthy times and considerable control procedures. Generally, fibers of Al_2O_3–SiO_2 compositions are fired so as to remove carbon, resulting in oxide fibers essentially free of carbon. However, Mansmann, et al[12] prepared fibers containing carbon intentionally, e.g., 0.1 to 4%. Their fibers were prepared by dry spinning a solution of an aluminum carboxylate, hydrolyzed silicic acid ester or organoalkoxy-silane, and polyethylene oxide. Firing was done in air up to 1100°C. However, for conversion to α-Al_2O_3/mullite or γ(gamma)-Al_2O_3/mullite, temperatures greater than 1100°C were required. At these higher temperatures, an inert atmosphere was found to be necessary so as to prevent removal of the carbon which they believed to be necessary for retention of good mechanical properties. Consequently, the higher temperature firing was accomplished in a nitrogen atmosphere.

The fabrication of continuous fibers of Al_2O_3–SiO_2 precursors by dry spinning of a mixture of polyaluminoxane and silicon-containing compounds is outlined by Horikiri.[2] A solution of the mixture is extruded through a spinnerette and wound on a bobbin at a speed of 20 to 400 m/min. The resultant fibers are hydrolyzed with humid air and then calcined.

Continuous strands of Al_2O_3–SiO_2 fibers comprising up to 400 continuous fibers were spun by Karst and Sowman[10] according to a process described by Borer and Krogseng.[3] Briefly, continuous strands of green fibers were made by extruding fiberizable, viscous concentrates of mixtures of sols and solutions through a spinnerette and, at the same time, drawing the extruded fibers continuously, by means of drawing rolls, through a drying tower followed by bringing them together as continuous strands. The resultant strands of green fibers were then accumulated in a relaxed, loose, unrestrained series of loops on a substrate, such as a moving belt, which passed through a furnace, continuously. The continuous ceramic strands were then pulled through a second furnace, such as a tube furnace, for densification and straightening.

During the firing of the Al_2O_3–SiO_2 fibers, one first obtains amorphous fibers after removal of the fugitive components. If carbon is removed and if no colorant oxides or pigments are present, the oxide fibers are preferably transparent, this transparency being generally coincident with the best mechanical properties. As firing proceeds, η(eta)- or γ-Al_2O_3 is formed between 900° and 1000°C. The preferred fiber of Horikiri, et al[2] is that in which the Al_2O_3 is in the γ form. Further heat treatment, e.g., 1100° to 1250°C, causes a reaction between the Al_2O_3 and SiO_2 to form mullite. In compositions containing more than 35% SiO_2, cristobalite may form; with less than about 25% SiO_2, α-Al_2O_3 may be present. In general, these species are not desirable since their formation causes physical defects in the resultant fibers. There appears to be general agreement among those who have developed ceramic fibers with regard to this. The continuous fibers of Karst, et al[10] were free of α-Al_2O_3 and crystalline silica, such as cristobalite, i.e., not discernible by x-ray diffraction analysis. With compositions having alumina and silica in ratios of 67 to 77 parts by weight Al_2O_3 to 33 to 23 parts by weight SiO_2, conversion to mullite at temperatures of 1200° to 1400°C resulted in continuous fibers with excellent resistance to fracturing. These results were confirmed in a later U.S. patent,[14] showing that flexing tests on blankets of fibers, formed by a centrifugal spinning device and having alumina to silica weight ratios of 70 to 75 Al_2O_3 to 30 to 25 SiO_2, calcined at 1250°C, then heat treated at 1500°C for 24 hours, had excellent flexibility.

Alumina-silica fibers which are near the mullite composition and have no B_2O_3 or relatively low levels, e.g., 2%, are especially useful for high temperature applications. The low levels of B_2O_3 coincide with lower weight loss at temperatures significantly greater than 1200°C and, consequently, less shrinkage. In addition, modulus of elasticity values may be as high as 242 GPa (35 x 10^6 psi) which coupled with the high temperature properties, makes such fibers candidates for the reinforcement of metal and ceramic matrices. The microstructure of an Al_2O_3–SiO_2 fiber having a nominal composition of 98% ($3Al_2O_3$:$2SiO_2$) and 2% B_2O_3 is shown in Figure 12. This is an electron micrograph of an ion-milled section of a fiber converted to mullite. Mullite crystallites appear to be somewhat blocky rather than needle-like as shown previously for fibers having considerably more B_2O_3 (see Figures 7 and 8 of sections of $3Al_2O_3$:$1B_2O_3$:$3SiO_2$ fibers).

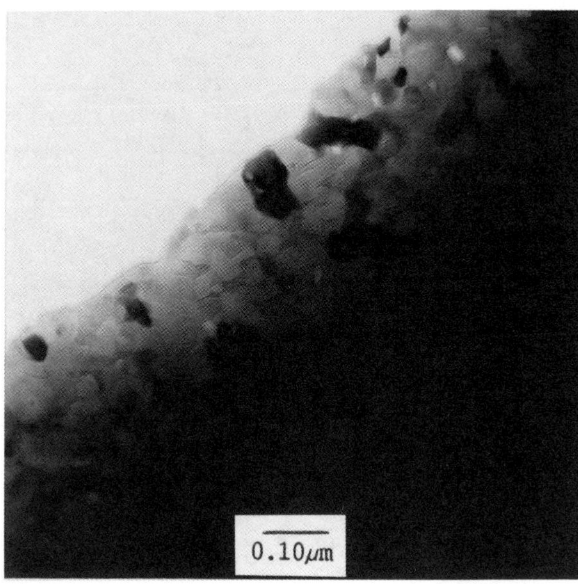

Figure 12: TEM of ion-milled section of mullite fiber (E.A. Richards, 3M).

COMMERCIAL Al_2O_3–B_2O_3–SiO_2 FIBERS

Ceramic fibers of binary and ternary compositions of this system, claimed to be prepared by sol-gel or chemical processes, are available from several companies. Continuous fibers in strand or yarn form are manufactured by Sumitomo Chemical Company, Ltd. and the 3M Company. The fiber product from Sumitomo is generally identified as the Sumitomo Alumina Fiber and is also referred to as a High-Performance Alumina Fiber.[13] Several fiber compositions are available from the 3M Company. These are trademarked as Nextel ceramic fiber and are manufactured as continuous filament strands. Each is available as roving or fabric. They can also be supplied in mat form made up of small diameter, e.g., ~3 μm diameter, intertwined fibers, especially useful for high temperature thermal insulation. Discontinuous alumina-silica fibers in bulk or mat form are also available from Imperial Chemical Industries, Ltd., trademarked as Saffil fibers.[15,16] These are formed by the spinning of solutions containing organic components. The resulting oxide fibers comprise δ (delta)-Al_2O_3 with 5% SiO_2. Property data for continuous filament fibers are shown in Table 3.

Table 3: Properties of Commercially-Available Fibers

PROPERTY	Sumitomo Alumina(13)	Nextel 440 fiber	Nextel 480 fiber	Nextel 312 fiber
Composition				
wt % Al_2O_3	85	70	70	62
" B_2O_3		2	2	14
" SiO_2	15	28	28	24
Predominant Crystalline Species	gamma Al_2O_3	Eta Al_2O_3	Mullite	Aluminum borate-mullite
Density (g/cm^3)	3.2	3.0-3.05	3.1	2.7
Liquidus Temp °C$^{(a)}$		1850$^{(b)}$	1850$^{(b)}$	1800
Diameter(μm)	17	10-11	10-11	11
Tensile Str.				
(10^3 psi)	260	300$^{(c)}$	250-350$^{(c)}$	250-300$^{(c)}$
GPa	1.8	2.1	1.7-2.4	1.7-2.1
Mod. of Elast.				
(10^6 psi)	30.5	27-28	30-35	20-22
GPa	210	190	207-242	138-152
Use Temp $^{(d)}$ (C°)	1250	1425	1425	1200
Manufacturer	Sumitomo Chemical Co. Ltd.	3M	3M	3M

(a) completely liquid

(b) estimated minimum value; see phase diagrams by Aramaki and Roy (17) and Risbud and Pask (18)

(c) 2.5 cm gauge length

(d) short-term use temperature may be higher

APPLICATIONS

The high temperature properties of alumina-boria-silica fibers coupled with continuous filament yarn and the ability to utilize standard textile equipment in processing has considerable influence on applications. Reinforcement of polymers, ceramics, and metals are major areas of interest.

Ceramic Fiber—Metal Composites

The reinforcement of aluminum with the Sumitomo alumina fiber and the development of a process for infiltration of fiber bundles was discussed by Abe, et al.[13] Because of the non-wettability of the fibers by molten aluminum, pressure was found to be necessary to overcome the resistance to infiltration.

Ceramic Fiber—Polymer Composites

The results of an extensive study[19,20] of fibers having a composition of 62% Al_2O_3, 14% B_2O_3 and 24% SiO_2 (Nextel 312 fiber) in polymeric composites as reinforcement has shown that such composites have increased stiffness, improved fatigue life, and dielectric properties as compared to those reinforced with glass fiber. Ceramic fibers having intermediate values of modulus of elasticity, e.g., 138 to 152 GPa (20 to 22 x 10^6 psi), are particularly interesting as hybrids with Kevlar fibers (DuPont) because of the relatively comparable values for stiffness of polymer-fiber composites of each. By combining the ceramic and Kevlar fibers, a hybrid composite results, lighter in weight because of the Kevlar fibers and with greater compressive strength contributed by the ceramic fibers.

Flame Barriers

Special conditions requiring resistance to high temperature have expanded the needs for Al_2O_3–B_2O_3–SiO_2 fibers in polymeric composites. These include reinforcement of resins such as polyimides and epoxies for flame-shielding applications for jet engine compartments. For example, as flame or fire shield components, Nextel fiber-resin composites have been plied over graphite fiber-reinforced composites, the ceramic fiber-containing composites serving as barriers to prevent burning of the combustible graphite fiber-resin structures.

Ceramic-Ceramic Composites

The high temperature resistance of fibers of the Al_2O_3–B_2O_3–SiO_2 system, especially compositions having a high percentage of mullite or alumina, make them candidates for use in ceramic-ceramic composites. Applications may include reinforcement of polycrystalline or amorphous matrices, including glass. They may also be used as fiber components in fibrous composites. An especially important application has been the combination of Al_2O_3–B_2O_3–SiO_2 fibers (Nextel 312 fibers) and high purity SiO_2 fibers (Johns Manville Corp.) which are reacted together to form the fibrous composite tiles used as part of the heat shield for the space shuttles, Discovery and Atlantis. This protective barrier, known as "fibrous refractory composite insulation" (FRCI)[21,22] is used, principally, on the shuttle's underside portion which attains the highest temperatures. This new insulation is stronger and can withstand higher temperatures than the earlier reusable surface insulation (RSI) comprising only silica fibers bonded together. Especially unique to the structure of FRCI is the bonding which develops at the point of contact of the Al_2O_3–B_2O_3–SiO_2 fibers and the silica fibers, apparently caused by reactions between boria and silica. A rigidized, open structure discernible in scanning electron micrographs[23] results. The fiber-to-fiber bonding of Al_2O_3–B_2O_3–SiO_2 and SiO_2 fibers in a fibrous composite is shown in Figure 13.

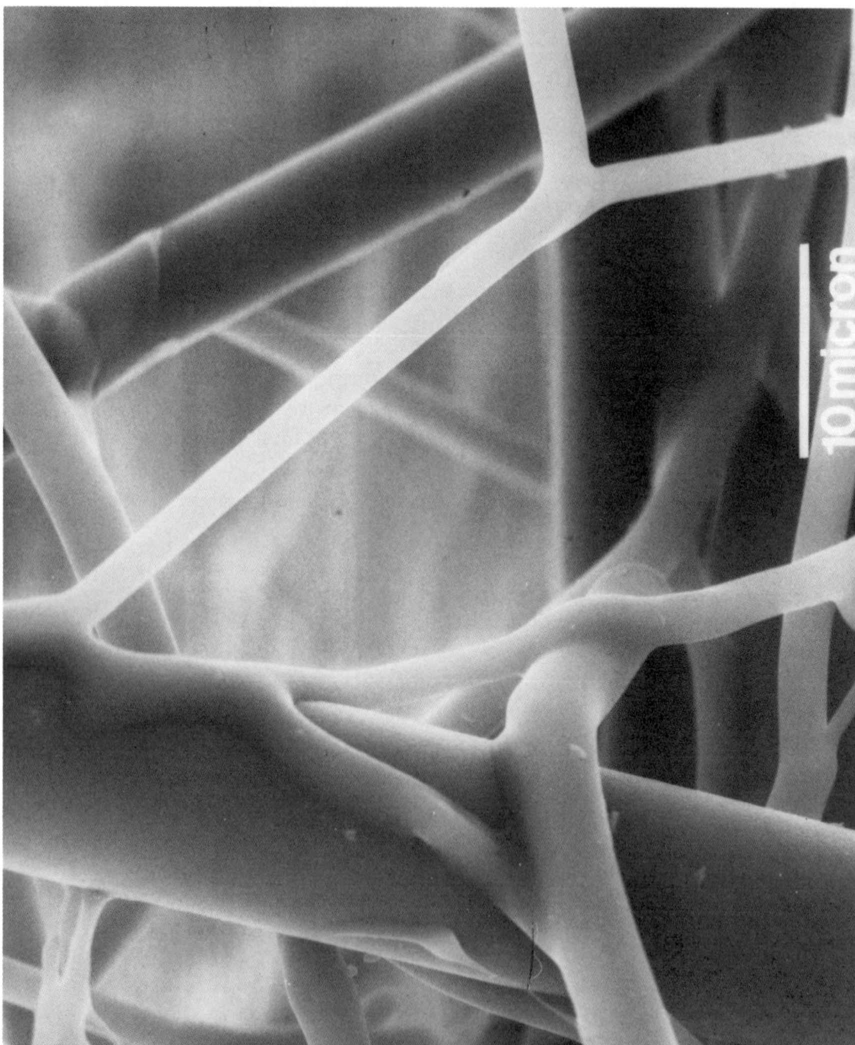

Figure 13: SEM of fibrous composite with 20% $3Al_2O_3{:}1B_2O_3{:}2SiO_2$ fibers and 80% SiO_2 fibers; firing temperature = 1330°C (K.E. Owens, 3M).

High temperature fibers can serve as shaped substrates which also act as reinforcement for chemical vapor deposited materials, e.g., silicon carbide (SiC). Alumina-boria-silica fibrous shapes such as textile tubes of Nextel 312 fibers can be treated so that SiC deposits on the fibers and, with continued reaction, infiltrates and essentially fills the voids so as to result in fiber-reinforced SiC composites. Such composites utilizing Nextel ceramic fibers are known as SICONEX™ Fiber-Reinforced Ceramic. Tubes of this have been tested at temperatures to 1300°C for extended periods of time and are expected to be useful for products such as heat exchangers and radiant gas burner tubes.

High Temperature Fabric

The textile quality of fabric which can be woven with present state-of-the-art Al_2O_3–B_2O_3–SiO_2 fibers made by sol-gel processes make the fabric useful for many applications. For example, fabric of Nextel 312 fibers has been used for furnace belts, flame curtains, seals, and wall insulation. In braided forms, such textile materials protect thermocouples, electrical cables, and fuel lines and serve as high-temperature covering for gaskets used around openings, such as doors, on the space shuttles;[24] as woven fabric it is applied as an envelope-like cover for gap-fillers used between tile on the hot side of the thermal shield.

These new sol-gel fibers may help in the solution to some of our pollution problems and improvement in efficiency of future power plants. High-temperature fabric on oil booms, used to corral oil spills on water, allows the contained pool to be burned, a preferred way for more complete cleanup especially in ice areas where normal collection is impossible. Incinerators, many of which are inoperative because of failure to meet modern standards for effluents, may be able to function with high temperature fabric filters made of Al_2O_3–B_2O_3–SiO_2 fibers.

Testing of filter bags of ceramic fibers has been in process for a number of years. Present filters made of glass fiber are useful to about 290°C. Bags of Nextel 312 fiber[25] have been tested successfully for limited periods of time in pressurized fluidized bed coal combustion systems at temperatures up to 815°C. Should tests show success for sufficient periods of time, such bags may be useful in systems designed to utilize hot gases from the combustion process in the generation of additional power by passing the hot gases directly through gas turbines after filtration to remove particulate materials.

MODIFIED Al_2O_3–B_2O_3–SiO_2 FIBERS

Leached Al_2O_3–B_2O_3–SiO_2 Fibers

Ceramic fibers of Al_2O_3–B_2O_3–SiO_2 comprising an alumina-containing crystalline species and amorphous material can be leached so as to result in a uniform microporous sheath enveloping a substantially nonporous core of the original composition. Specifically, fibers comprising the crystalline aluminum borosilicate species with amorphous boria and silica as well as those comprising crystalline alumina, e.g., η-Al_2O_3, and amorphous silica can be leached in a controlled manner with a hydrofluoric acid leachant. The microporous sheath is very uniform as shown in Figure 14, and may be treated with an infiltrate to further modify the sheathed fiber. The infiltrated fibers may be in fiber, woven, or nonwoven textile forms and heat-treated so as to convert to ceramic or cermet composite fibrous forms. Although numerous applications may be envisioned for these materials, they may be especially useful as supports for catalysts.

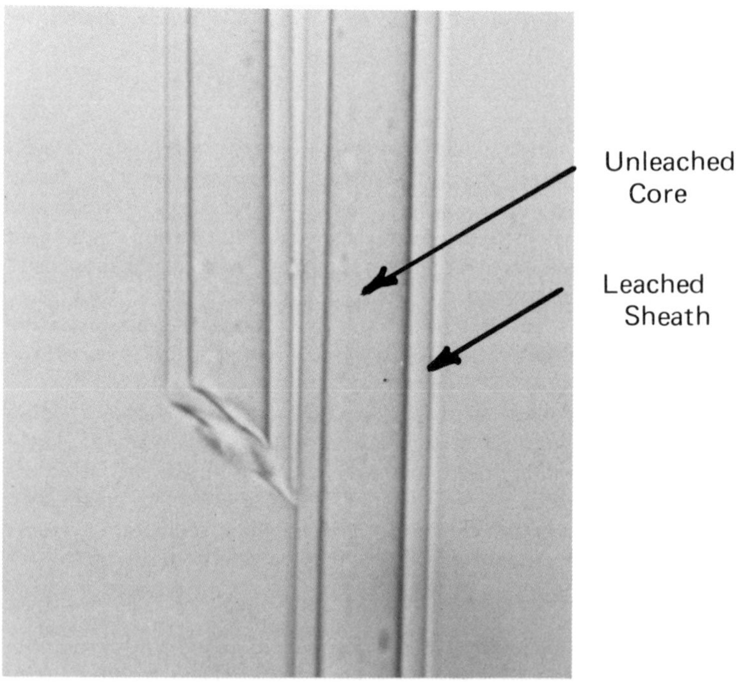

Figure 14: Leached $3Al_2O_3:1B_2O_3:2SiO_2$ fibers; fiber diameters ~11 μm.

Because of the chemical resistance of the aluminum borate, aluminum borosilicate, and aluminum silicate (mullite) crystalline materials of the Al_2O_3–B_2O_3–SiO_2 systems, fibers composed of these species without amorphous B_2O_2 and/or SiO_2 present cannot be leached to form the sheathed fiber described above. Also, fibers which are amorphous are destructively attacked with attempts to leach in hydrofluoric acid.

Cermet Fibers

Fiber compositions of this system may be modified with significant amounts of additive materials, e.g., 25% by weight or more. For example, compositions may contain oxides such as those of copper, zirconium, nickel, chromium, iron, etc. Such additives will affect indices of refraction, color, and physical properties such as modulus of elasticity. Nickel oxide additions will react so as to form a nickel aluminate spinel ($NiAl_2O_4$). Since this reaction in sol-gel systems used for preparation of Al_2O_3–B_2O_3–SiO_2 fibers can take place at temperatures below that required for formation of the aluminum borosilicate crystalline species, the spinel can form at the expense of the silicate and borosilicate materials.

Fibers containing oxide additives which are reducible in atmospheres such as hydrogen or hydrogen-inert gas mixtures can be heat-treated in such atmospheres at high temperatures, e.g., 800° to 900°C, to result in cermet fibers. The metallic components in the cermet fibers are dispersed as discrete metallic globules which, in general, are not in contact with each other, the resulting cermet fiber being electrically nonconductive. In the case of iron or nickel additives the fibers are attracted to a magnet. Depending on the concentration and temperature treatment, the spheroidal or globular metallic particles are distributed in a graded concentration increasing toward the outer surface, and may protrude beyond the ceramic fiber matrix as shown in Figure 15.

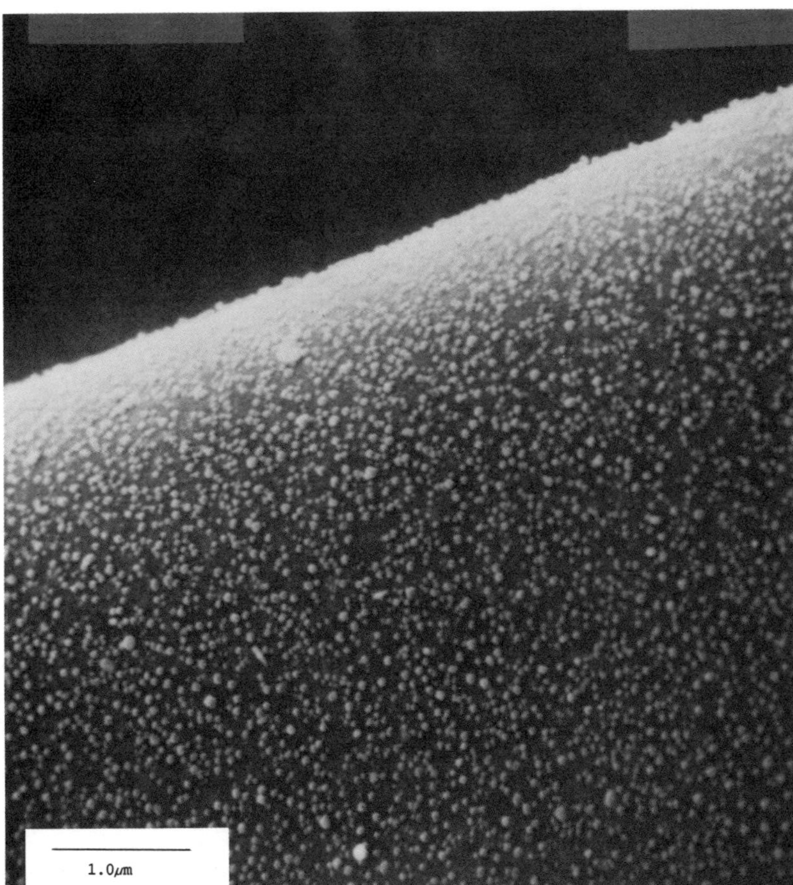

Figure 15: SEM of surface of nickel-Al_2O_3–B_2O_3–SiO_2 cermet fiber.

Such cermet fibers have been found to have useful catalytic properties.[26] These cermet fibers can remain strong after the reduction process and have higher modulus of elasticity values than the initial ceramic fiber, and are useful for reinforcement applications. The reduction process can be done on fibers per se or on woven or nonwoven forms.

SUMMARY

Sol-gel technology makes possible the preparation of ceramic fibers over a broad range of compositions in the Al_2O_3–B_2O_3–SiO_2 system. Such fibers can be modified by heat-treatment to cause crystallization of Al_2O_3, Al_2O_3–B_2O_3, Al_2O_3–B_2O_3–SiO_2, and Al_2O_3–SiO_2 species. Physical, chemical, thermal, and optical properties obtainable make them useful for numerous high temperature and reinforcement applications. Furthermore, compositions may be modified by other additives so as to develop specific desired properties such as modulus of elasticity, color, and index of refraction or to provide compositions which can be converted to cermet fibers comprising metallic particles and a ceramic fiber matrix. Continuous fibers can be of such high quality that yarn prepared with them can be converted into textile-quality fabric on standard equipment. The high temperature properties of such fabric make it useful for many applications, examples of these including the filtration of hot gases from coal-burning power plants, high temperature insulation, thermal shielding, and as substrates or reinforcement components in ceramic-ceramic or metal-ceramic composites.

REFERENCES

1. Bracke, P., Schurmans, H., and Verhoest, J., *Inorganic Fibres & Composite Materials,* Pergamon Press, New York (1984).
2. Horikiri, S., Tsuji, K., Abe, Y., Fukiu, A., Ichiki, E.; U.S. Patent 4,101,615; July 18, 1978; assigned to Sumitomo Chemical Company, Limited.
3. Borer, A., Krogseng, G.P.; U.S. Patent 3,760,049; September 18, 1973; assigned to Minnesota Mining and Manufacturing Company.
4. Sowman, H.G.; U.S. Patent 3,795,524; March 5, 1974; assigned to Minnesota Mining and Manufacturing Company.
5. Kim, K.H., and Hummel, F.A., System Al_2O_3–B_2O_3 (tentative), Figure 308; Gielisse, P.J., and Foster, W.R., System Al_2O_3–B_2O_3–SiO_2, Figure 763; Kim, K.H., and Hummel, F.A., System $9Al_2O_3 \cdot 2B_2O_3$–$3Al_2O_3 \cdot 2SiO_2$ in *Phase Diagrams for Ceramists* (E.M. Levin, C.R. Robbins, and H.F. McMurdie) The American Ceramic Society, Columbus, Ohio (1964).
6. Blaze, J.E.; U.S. Patent 3,503,765; March 31, 1970; assigned to the Babcock & Wilcox Company.
7. Long, W.G., Development of Fine-Diameter Mullite Fiber, NASA CR-134612, August (1974).
8. Thatcher, W.E., 3M Center, 3M Company; Personal Communication.
9. McArdle, J.; Pennsylvania State University; Personal Communication.

10. Karst, K.A., Sowman, H.G.; U.S. Patent 4,047,965; September 13, 1977; assigned to Minnesota Mining and Manufacturing Company.
11. Mansmann, M.; U.S. Patent 3,947,534; March 30, 1976; assigned to Bayer aktiengesellschaft.
12. Mansmann, M., Schmidt, L.; U.S. Patent 3,982,955; September 28, 1976; assigned to Bayer aktiengesellschaft.
13. Abe, Y., Horikiri, S., Fujimura, K., Ichiki, E., High Performance Alumina Fiber and Alumina/Aluminum Composites in: *Progress in Science and Engineering of Composites* (T. Hayashi, K. Kawata and S. Umekawa, ed.) ICCM-IV, Tokyo (1982).
14. Miyahara, K. and Nakayama, N.; U.S. Patent 4,159,205; June 26, 1979; assigned to the Carborundum Company.
15. *SAFFIL FIBRES,* brochure published by ICI, Mond Division, New Ventures Groups, The Heath, Runcorn Cheshire WA74QF.
16. Product Information, Refractories Division, Babcock and Wilcox, January 1, 1976.
17. Aramaki, S., and Roy, R., *Jour. Am. Ceram. Soc.* 45:229 (1962).
18. Risbud, S.H., and Pask, J.A. *Jour. Am. Ceram. Soc.* 61:63-67 (1978).
19. Pipes, R.B., Center for Composite Materials, University of Delaware, Newark DE; Johnson, D.D., and Karst, K., 3M Center, 3M Company, *NEXTEL 312 Ceramic Fiber Polymeric Composites,* a publication of Ceramic Fiber Products, 3M, St. Paul, MN.
20. Pipes, R.B., Johnson, D.D., Karst, K.A., Ceramic Fiber Polymeric Composites in: *Proceedings of the Thirty First Annual Reinforced Plastics Technical Conference,* Washington, DC (1976).
21. Leiser, D.B., and Goldstein, H.E.; U.S. Patent 4,148,962; April 10, 1979; assigned to NASA.
22. Leiser, D.B., Smith, M., and Goldstein, H.E., *Am. Cer. Soc. Bull.* 60 (11) 1201-1204 (1981).
23. Newquist, C.W. Pfister, A.M., Miller, A.D., and Scott, W.D., *Am. Cer. Soc. Bull.* 60 (11) 1205-1209 (1981).
24. Korb, L.J., Morant, C.A., Calland, R.M. and Thatcher, C.S., *Am. Cer. Soc. Bull.* 60 (11) 1188-1193 (1981).
25. White, L.R., O'Brien, D.L., and Schmitt, G.A., Ceramic Fabric for Filtration at High Temperatures, 550° to 1600°F in the *Proceedings of the Third Conference on Fabric Filter Technology for Coal-Fired Power Plants,* (EPRI), Scottsdale, Arizona, November, 1985.
26. Hagen, D.F. Haddad, L.C., and Markevka, J.S., Deuterium Reactions with Microwave Sustained Helium Plasma Detection for Gas Chromatography, presented at the 1986 Winter Conference on Plasma Spectrochemistry, January 2-8, 1986, Kailua-Kona, Hawaii; to be published in *Spectrochimica Acta.*

9

Continuous Filament Fibers by the Sol-Gel Process

William C. LaCourse

Alfred University
Alfred, New York

INTRODUCTION

A major problem associated with sol-gel processing is the difficulty in producing large shapes. Drying stresses associated with capillary forces, and shrinkage stresses during firing cause cracking and fragmentation into small pieces. These effects are minimized in fibers since diffusion distances are reduced and stresses can be relieved by local bending of the fiber. Unfortunately these simplifications are important only after the gel fiber is formed from the sol. Production of high quality fibers from sols remains a difficult task.

In the present discussion of gel fibers we consider only "continuous filament" fibers from alkoxide based sols. Techniques such as directional freezing[1,2] or blown fiber produced from colloidal sols[3] are not considered. The emphasis is on effects of preparation procedures on the sol structure and how it affects fiber formation. Each stage of the overall process will be treated in detail, focusing first on silicate sols, then extending the results to other systems.

SOL STRUCTURE

Sol Requirements for Continuous Filament Formation

In oxide melts fibers are generally drawn in a viscosity range of 10–100 Pa-s. Outside this range fibers tend to "fracture," at the drawing rates employed in industry (20 m/sec). At these high shear rates the structure and rheological behavior of the melts become extremely important.

Sol-derived fibers have not been formed by high shear rate processes. Individual lengths in the range of 1 meter are typically hand drawn, although Sakka[4] has had success in continuous filament formation, collecting the drawn fiber on a spinning drum. Fibers can be drawn in the viscosity range between about 1-1,000 Pa-s.[5,6] During the sol to gel conversion all sols pass through this viscosity range. However, not all can be drawn into fiber. The sol structure must be such that the material can withstand the stresses associated with attenuation. Furthermore, the rate of change of viscosity can be so rapid that fibers can only be drawn for a few minutes. Finally, the shear rate dependence of viscosity will be at least as important in sol processes as it is in oxide glasses.

Initial Sol Structure. To understand these influences it is necessary to briefly review the effect of preparation conditions on sol structure. For simplicity, we will consider first pure silica sols produced from tetraethoxy silane (TEOS).

The structure of silicate sols can vary dramatically, depending on the conditions extant during the hydrolysis and condensation reactions. Sols produced from TEOS, ethyl alcohol and water tend to be composed of individual spherical particles when the reactions are carried out at high pH. Base catalyzed silica sols cannot be drawn into fiber. At high pH (pH = 9-11) the spherical sol particles mutually repel due to a high surface charge from the reaction

$$\text{Si–OH} + \text{OH}^- \rightarrow \text{Si–O}^- + \text{H}_2\text{O} \qquad (1)$$

The sol therefore tends to exist as individual particles. (Figure 1A) Gelation generally occurs only after evaporation of the sol liquid, or growth of the particles to large size, forcing them into close contact. At gelation the viscosity increases rapidly and may reach 100 Pa-s or more.[7] (Figure 2 - Curve B.) Drawing of a continuous filament however, requires interparticle bonding so that the viscous liquid remains intact under the drawing strain. A system of noninteracting or weakly bonded spheres is incapable of withstanding such stresses.

At somewhat lower pH (4-8) particle-particle interactions occur via collision and local condensation reactions which form a neck between particles. (This process has been described in detail by Iler[8] for silicic acid sols and LaCourse et al for alkoxides.[7]) It is therefore possible to create local regions having a three dimensional gel structure of coalesced particles. (Figure 1B) However, these structures are still unable to support the drawing stresses encountered in fiber drawing operations. Evidently, when the sol viscosity reaches a value appropriate for drawing, the microgel regions are only weakly interacting.

It is possible to prepare totally different sol structures when TEOS is reacted at low pH (less than 3.0) and with low water to TEOS ratios. Consider first the hypothetical case in which a sol is reacted with 2 mols of water per TEOS molecule at a pH near the iso-electric point (near pH = 2-2.5 for SiO_2). Here the rate of hydrolysis is much greater than that for condensation since either Si-O^- or $\text{Si–(OH}_2)^+$ sites are required for condensation.[7-9] In such a case the average TEOS molecule will react via

$$\text{Si(OR)}_4 + 2\text{H}_2\text{O} \rightarrow (\text{OH})_2\text{–Si–(OR)}_2 \qquad (2)$$

If one now neglects condensation reactions resulting in edge shared tetra-

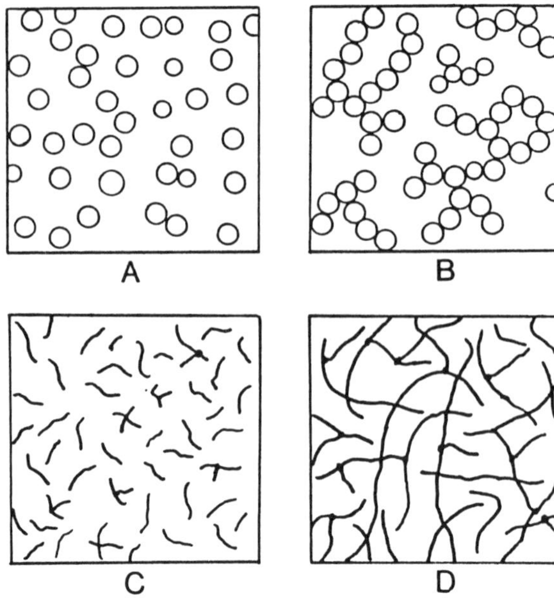

Figure 1: Proposed sol structures. (A) Noninteracting spherical particles at high pH. (B) Particles coalesced into microgel regions: pH 5-8. (C) Early stage of growth in low water content, acid catalyzed sols. (D) Acid catalyzed sol after extensive reaction among sol molecules.

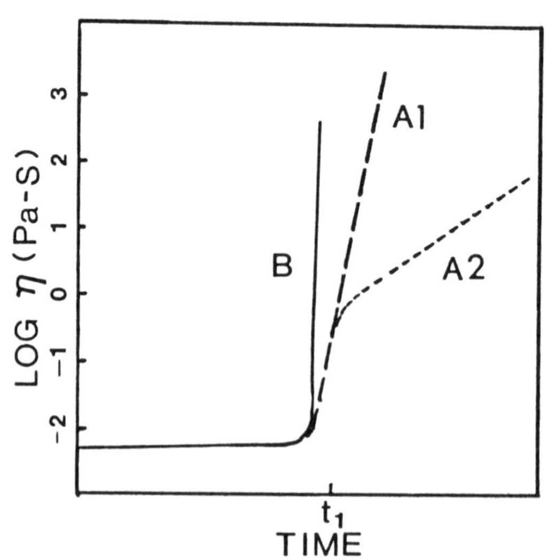

Figure 2: Typical viscosity-time plots for sols. Curve B is for base catalyzed sols composed of spherical particles. Curve A1 is for acid catalyzed, prereacted sols in moist atmosphere. Curve A2 - source of water was removed at time T_1.

hedra (See section on titania sols.) these hydroxylated monomers wll react, via the simplified condensation reaction given below.

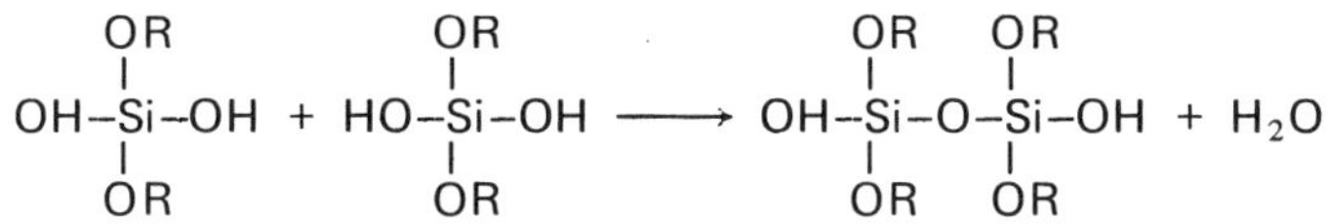

(3)

(As noted above the reaction is actually between ionized sites. For details on the nature of acid catalysis see Reference 9.)

Additional reactions between these dimers and monomers at OH sites eventually leads to linear chains. Eventually, the monomer concentration decreases to zero, and the growth of sol molecules via monomer-chain reactions stops. A structure similar to that shown in Figure 1C will result. It cannot yet be fiberized since aspect ratio of the sol molecules (length/diameter) is small and no extensive network has yet formed.

The nature of the above structure will have a strong influence on gelation since reactions leading to growth of the sol particles must now occur between the preformed chains. Only major additions of water, a strong base, or additional monomer can alter subsequent growth. The prereaction step is therefore crucial to production of sols capable of being fiberized.

This simple analysis requires modification to account for the behavior of "real" sols. First, concurrent hydrolysis and condensation reactions must be allowed. Condensation forms water which may be used for additional hydrolysis of silicate monomers or sites on the growing chains. For a water/TEOS ratio of 2.0 this results in crosslinking of the chain molecules with formation of more complex molecules. Less crosslinking would be present at lower water contents. It is also necessary to account for the tendency of silica sols to form closed rings and spherical particles, rather than crosslinked chains.[8]

With these modifications the additional growth of the sol can be described. As time goes on the average molecular weight of the sol particles increases, by chain-chain reactions, which tend to increase the average length, aspect ratio, and degree of branching of the chains. Eventually large units react and intertwine and the sol viscosity increases. If the concentration of TEOS is high, or if the alcohol is allowed to evaporate, the sol will reach a viscosity sufficient to allow fiber formation. Furthermore, the sol structure will likely be able to withstand at least low drawing stresses due to entanglement of the large, flexible sol molecules illustrated in Figure 1D.

The type of growth illustrated by the above analysis is likely to dominate at pH values near 1-3. Here the condensation rates are slow, due to the absence of ionized hydroxyl groups [$Si{-}O^-$ or $Si{-}(OH_2)^+$]. Furthermore the low charge on sol molecules near the iso-electric point provides only a small driving force for sphere formation. The net result is the formation of many small molecules which tend to grow slowly. They may continue to grow and/or to react to form larger molecules, but formation of large, dense spherical particles is unlikely.

Increasing the pH much above 2 causes two changes. Most importantly the rate of condensation increases due to increased concentration of $Si{-}O^-$ sites. Similar to a nucleation and growth process, fewer, but larger particles tend to form. Repulsion of ionized sites on the same molecule will also increase the tendency for molecules to rearrange into a shape which permits maximum shielding of the charge. This, combined with the increased reaction rates tends to favor spherical particle formation, with the size of the particles increasing with increasing pH.

Similar effects occur if the pH is decreased substantially below 1.0. Positively charged $Si-OH_2$ sites form, leading to an increased rate of condensation and decreased gel times. Importantly however, the rates of reaction in acid sols are orders of magnitude less than in base catalyzed material. Therefore nucleation of small molecules or very small particles still occurs. In silicic acid, $Si(OH)_4$, spherical particles less than 5 nm in diameter form.[8]

Dahar[5,10] was able to draw fiber from sols produced at pH values near 0.0 where the tendency for sphere formation is increased. He suggested that chains, produced from coalescence of extremely small (5 nm) particles rather than polymeric chains, can also lead to sols capable of being drawn into fiber. Chain entanglement can still occur, and due to their small diameter and residual OR groups, substantial chain flexibility would be retained by the coalesced chains.

Summary-Sol Structure. At low water contents and at pH values where reaction rates are slow the sol structure develops in two stages:

1. Small molecules-particles nucleate and grow primarily by reaction with monomers.
2. Reaction between sol molecules occurs creating an extensive network of branched chains.

Processes for Silica Fiber

The analysis given above allows one to design a procedure which will lead reproducibly, to sols capable of being fiberized. One such procedure employed in our laboratories is outlined in Figure 3. A detailed analysis of the function of each step in the process is provided below.

Mixing. The function of the mixing step is simply to combine the chemicals in such a way as to promote homogeneous reaction between the water and TEOS. Since water and TEOS are immiscible, alcohol is generally added as a co-solvent. As pointed out by Yoldas[11] however, alcohol plays a major role in determining the sol structure.

Our general procedure is to prepare two solutions. One contains half of the desired alcohol and all the TEOS, while the other has the remainder of the alcohol and the desired amount of water and HCl. After 15 minutes mixing the two solutions are simply poured together, covered and allowed to react. For the procedure given in Figure 3, an alcohol/TEOS volume ratio of 0.5 or more is suggested. Ratios as low as 0.30 can be employed.[12] However, this leads to an increased tendency for heterogeneous hydrolysis and condensation, and makes the process less reproducible.

Prereaction. Prereaction steps are carried out in a closed container. This is critical for reproducible sols. A major function is to eliminate all low molecular weight silicate polymers so that volatilization losses are minimized during later reactions. Many would refer to this stage of reaction as prehydrolysis. However, as discussed above more than hydrolysis occurs. In fact we now believe that after two hours mixing the basic structure of the sol has been set.

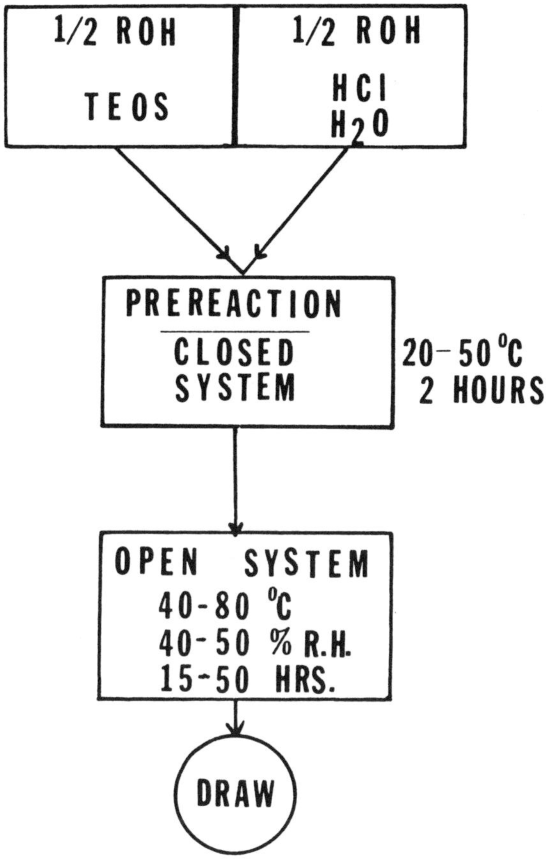

Figure 3: Procedure for producing fiberizable SiO_2 sols.

The particular structure extant at the end of this stage will depend on the concentrations of silicate monomer, water and alcohol, and on temperature. At a given pH, increased water or decreased alcohol content will increase the reaction rates and will lead to fewer, and larger sol species. At very low pH, less than 1.0, we believe it possible for very small, semi-spherical particles to form. The tendency for sphere formation will depend strongly on the pH, as well as the water and alcohol present. At low water contents and reasonable dilution by alcohol it is unlikely that substantial numbers of spherical particles can form due to the lack of highly hydrolyzed monomers. Therefore, even at low pH one expects molecular sol particles to dominate. As the water content increases, more rapid and complete hydrolysis will increase the tendency for spherical particle formation.

Sol Aging. This is generally carried out in an open container, at a temperature somewhat above ambient,[5,6] and often in a controlled humidity environment. Its main function is to allow development of a spacially extended sol network with slowly increasing viscosity. The processing occurring during this step can be summarized as follows:

1. Alcohol evaporates rapidly, decreasing the sol volume to a fairly constant level.
2. The water content of the sol equilibrates with the atmosphere.
3. Continued reaction and condensation occurs, leading to an increased viscosity of the sol. Since monomer concentrations are minimal, molecule-molecule interactions dominate.
4. During later stages of growth the reaction becomes limited by the availability of water, with the most rapid reaction occurring at the sol-air interface.

These processes are complex and interdependent. They are influenced by temperature and relative humidity, as well as surface to volume ratio of the sol container. It has been shown that changes in the above parameters can strongly influence the rate of increase in sol viscosity. However, no obvious differences in drawing characteristics have been noted.

Sakka[6] noted that increased aging temperatures decreased the time required to reach sufficient viscosity for fiberization but noted no change in behavior. LaCourse et al[5] also investigated effects of temperature and atmosphere. It was shown that one can allow rapid viscosity increases until an appropriate value is reached. Then by lowering the temperature, and/or prohibiting additional water from entering the sol, it is possible to slow further increases in viscosity (Figure 2 - Curve A2) to a rate of less than 4% per hour. In this way the sol viscosity remains in an appropriate drawing range for several hundred hours. It was also noted that sols could be frozen for long periods of time, then reheated without loss of drawing ability.[13]

Drawing. Prepared in the above manner, sols having viscosities in the range of 1-100 Pa-s can be drawn into fiber. Even after drawing, however, reactions have not stopped. The first indication of this is the fiber geometry. As noted by Sakka[4] and LaCourse[14] the fiber generally has a noncircular cross-section, with the cross sectional area dependent on the "prereaction" step. Increased water or decreased alcohol contents result in a more circular cross-section. The range of shapes observed and effects of water and alcohol content are shown in Figure 4.

Sakka notes that fibers characterized by less shrinkage during the sol to gel conversion, are more circular. LaCourse noted similar trends in aspect ratios with increasing water content.[14] The same effects are also observed in pure, gel derived titania fibers.[15] It is also noted[16] that the fibers tend to have localized regions of increased diameter, producing a wavy surface as illustrated in Figure 4e.

We believe these geometries to be a natural consequence of the gelation process. During drawing the surface of the fiber gels rapidly to a high volume structure due to evaporation of residual alcohol and to additional reaction with atmospheric moisture. The inside remains liquid however, leading to two post gelation processes:

1. Elongated and oriented molecules relax.
2. Evaporation of internal sol liquid and continued hydrolysis and condensation cause the interior to shrink.

The first process will cause axial contraction of the sol. The interior will be under a tension and the surface will be under a compression. These stresses can be relaxed by localized void formation or by localized increases in the fiber diameter. Evidently the surface has some ability to adjust, and a wavy surface results.

The second process causes additional axial, as well as radial, stresses as the interior attempts to shrink. Radial stresses can be relieved only by formation of a void in the fiber interior or by a change in fiber geometry. If the fiber surface remained fluid a uniform reduction in diameter, with the fiber remaining cylindrical, could occur. However, gelation generally occurs during drawing, and the surface is not able to contract uniformly. It must therefore form a geometry having a greater surface to volume ratio. As the amount of shrinkage increases (decreased water or increased alcohol content), the tendency will therefore be to form an ellipse and, with further internal contraction, a figure 8 shape. Triangular or square shapes are also possible if the initial shrinkage is not symmetric.

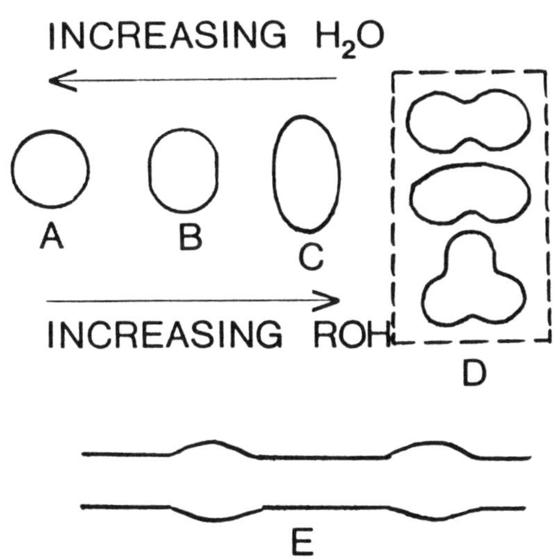

Figure 4: (A-D) Geometry of fiber cross sections, showing general effects of water and alcohol content. (E) Surface geometry.

Silica Fiber Properties

As Drawn Fibers. The strength of as drawn fibers (unconsolidated) has been studied by LaCourse.[14] The initial strength was found to be independent of cross sectional area, but strongly dependent on the initial water content, increasing from an average of 55 MPa for a 1.5 mol ratio of water to TEOS, to 110 MPa for a ratio of 2.2, as shown in Figure 5.

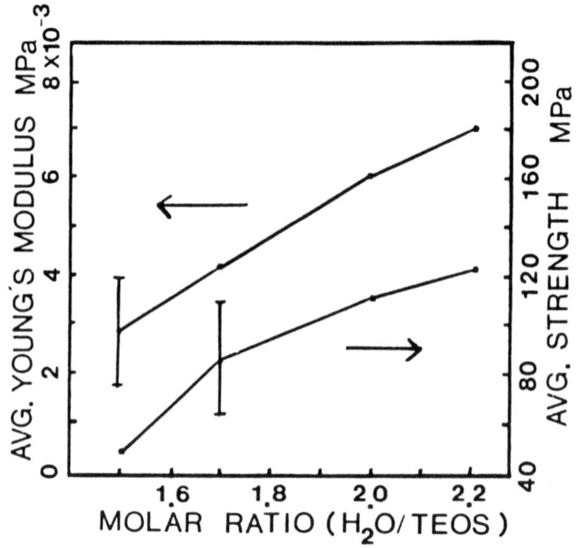

Figure 5: Effect of H_2O/TEOS ratio on strength and modulus of unconsolidated (as drawn) fibers.

The stress-strain behavior of as drawn fibers is also dependent on water content.[17] Those prepared with low water contents have a low Young's modulus (Figure 5) and often exhibit a yield stress. The tendency to exhibit a yield stress decreases with increasing water/TEOS, and those with more than a 2/1 ratio rarely exhibit this effect. Young's Modulus and M.O.R. increase with aging at room temperature, and after long times the strength of large diameter fibers decreases. Maximum average strengths in the range of 150 MPa have been observed for fibers dried for two weeks at room temperature.

Alcohol content and the drawing viscosity also influence strengths of as drawn fibers. Alcohol effects were not statistically significant. At high water contents (2.4 water/TEOS) strengths were higher for fiber drawn from high viscosity sols.[14]

These results are consistent with earlier discussions of sol structure. Higher water contents produce a dense, more crosslinked structure leading to increased strength and stiffness. Viscosity effects are also expected since a higher viscosity corresponds to a more completely reacted sol structure.

Consolidated Fibers. For a given preparation procedure, drying and firing schedules have a major influence on strength. A detailed study of these effects has been carried out by Mohideen.[12,16] Effects of water content and fiber diameter are shown in Figure 6. It was found, in agreement with Sakka,[18] that the highest strengths were obtained with gels prepared with a 1.7 water/TEOS ratio, and fired to 800°C immediately after drying. Mohideen used a heating rate of 300°C/hr. Lower or higher water contents, faster or slower heating rates or increased drying times resulted in lower strengths. Heat treatments above 800°C also produced lower strengths, possibly due to crystallite formation.

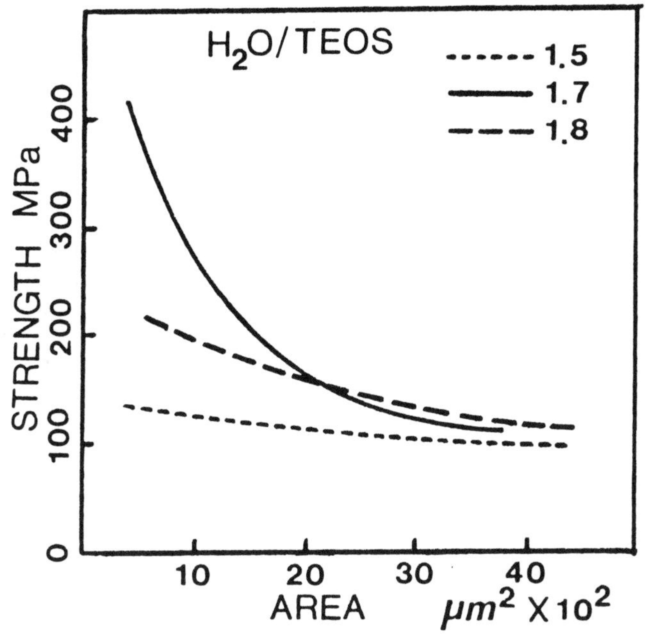

Figure 6: Effect of H_2O/TEOS ratio and cross-sectional area on strength of fibers. All fibers fired at 800°C at a rate of 300°C/hr.

Importantly, if increased aging times were employed it was still possible to obtain high strength fibers. However, a slower heating rate was required. With a 48 hour age at room temperature a 60°C/hr heating rate produced strengths almost equivalent to those obtained for unaged samples heated at 300°C/hr.

The optimum water content is not the same as that which gives the highest "as drawn" strengths. Coupled with the combined effects of drying time and heating rate this suggests that at least two gel parameters control strength.

1. The gel structure must be strong enough to withstand drying and firing stresses without producing major flaws.
2. The pore structure must be open so that water and the byproducts of organic burnout can diffuse out of the system without causing flaw formation.

Sols prepared with high water contents satisfy the first requirement, but their relatively dense structure does not facilitate removal of gasses formed during firing. Samples prepared with 2.4 water/TEOS often had a brown tinge,[14] indicating residual carbon in the structure. Sols prepared at 1.5 or 1.6 water/TEOS presumably do not have a sufficiently crosslinked structure to withstand stresses associated with drying and firing. Furthermore, sols fired after long aging would be expected to have a more dense, brittle structure, and would consequently require a slower heating rate for minimum flaw generation.

TiO_2, ZrO_2, AND BINARY OXIDE FIBERS

There are few literature references to other alkoxide derived fibers. The main problem is that other systems tend to exhibit extremely rapid hydrolysis and condensation reactions leading to precipitation. Since one expects the same general structural requirements for fiber drawing in nonsilicate sols, the increased rates of reaction will clearly make it difficult to grow chain or small particle structures in these systems.

Kamiya et al.[18,19] have formed binary ZrO_2-SiO_2 and ternary $Na_2O-ZrO_2-SiO_2$ fibers using metal alkoxides. Kamiya and coworkers also report formation of 30% Al_2O_3–70% SiO_2,[20] and several TiO_2-SiO_2 fibers with up to 70% TiO_2. The problem of rapid reaction was minimized by first mixing TEOS and the appropriate metal alkoxide with alcohol, in the absence of water. The system was then allowed to hydrolyze slowly in air.

The above process requires long times to reach a drawing viscosity. However, for the TiO_2-SiO_2 system it was found that a mixture of water vapor and CO_2 gas could be bubbled through the sol. This considerably decreased preparation times, but limited the TiO_2 content to 50%.[21]

It is of interest that the resistance of gel derived ZrO_2 containing glasses to dissolution by strong alkaline solutions is equivalent to those prepared by melting.[21] Furthermore, since the durability increases with increasing ZrO_2 content, and since it is possible to prepare gel derived material with greater ZrO_2 contents, gel derived fibers are potentially superior to melted compositions.

Pure ZrO_2 fibers[22] have also been formed from Zr-acetate sols. Viscosity was controlled by evaporation of water then aging at a given water content. The viscosity increased slowly with aging time, and fibers could be blown using a double attenuation process. The sol was first extruded through a hypodermic needle with a subsequent blowing step which further reduced the diameter to approximately 5 μm. Kilinski[23,24] has successfully drawn fiber from similar sols. He also was able to draw fibers from ZrO_2 and ZrO_2-SiO_2 sols prepared by nitric acid catalysis of Zr-alkoxides with low water contents.[24] While the binary silicates drew easily, it was extremely difficult with sols of pure ZrO_2.

Kim and LaCourse[15,25] have recently shown that the problems associated with use of many metal alkoxides can be reduced by choice of raw materials. Fibers of TiO_2 prepared from sols containing titanium-diisopropoxide bisacetylacetonate (TIAA) could be drawn with a wide variety of preparation conditions, and have been fired successfully to above 400°C without crystallization.

The unique properties of these sols are due to the acetyl acetonate groups (AcAc). When water is added to a TIAA-alcohol solution, the isopropoxy groups are rapidly hydrolyzed while the AcAc groups remain. The partially hydrolyzed TIAA (two OH per molecule) is then forced to react to form a linear chain structure consisting of repeating Ti–O–Ti bonds (Figure 7c.) or edge shared $Ti_2O_2(AcAc)_4$ dimers. (Figure 7b.) Both species probably exist in TIAA sols, with the concentration depending on the overall concentration of TIAA and the degree of hydrolysis.

The rate of reaction of both the chain or dimer will be slow. However, the dimer structure will be quite stable since AcAc groups are not easily hydrolyzed, and hydrolysis is required for additional growth. AcAc can be hydrolyzed by addition of either a strong acid, or a strong base. The rate of reaction depends on the pH and water content, and is faster in base than in acid. In both pH ranges molecular growth can be controlled, and structures similar to those found in pure silica sols can be produced.

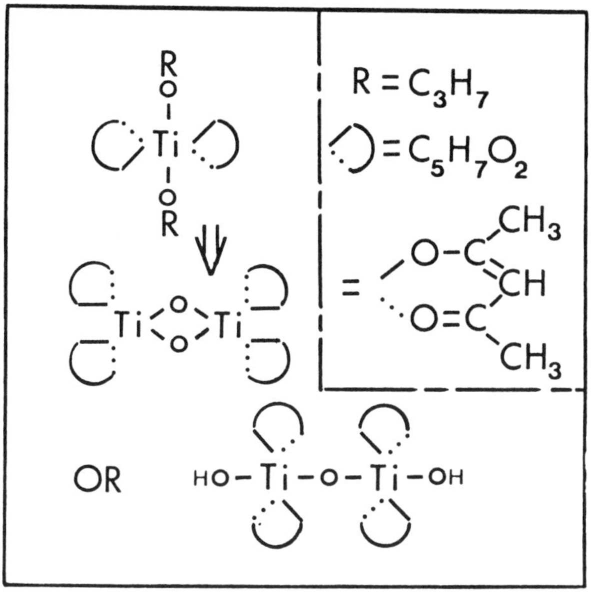

Figure 7: Monomer (top left) of titanium diisopropoxide bisacetyl acetonate. Dimer (middle left) forms by hydrolysis of iPr groups and with condensation between monomers. Chain structures are also possible (bottom).

As in silica sols, the continued growth of molecules with crosslinking causes a slow increase in viscosity. Fibers are most easily drawn under base conditions with water/TIAA ratios in the range of 5. An outline of the procedure is shown in Figure 8.

It is also possible to prepare TiO_2–SiO_2 compositions over the full compositional range. For sols containing less than 50% TiO_2 it is possible to use procedures developed for pure silica, the main difference being that TIAA slows the gelation rate from that of pure TEOS due to its reduced rates of reaction in acid. It is likely that the average molecular weight of the sol molecules is lower in these sols. Slight adjustments in procedure are required at high TiO_2 contents. In particular, a higher water content must be used due to the tendency of TIAA to form extremely small (short) molecules with acid hydrolysis. The structure would be similar to that shown in Figure 1c. This, in turn, leads to less than optimum fiber drawing qualities.

Doyle and LaCourse[26] have shown that zirconium-AcAc raw materials can also be used for production of fibers. Using Zr-dibutyl-$(AcAc)_2$, compositions ranging up to 80 mol % ZrO_2 were successfully fiberized. The preparation procedure is essentially identical to that employed for pure TEOS sols. Control of the process was more difficult with the higher Zr contents and pure ZrO_2 could not be formed using the above procedure.

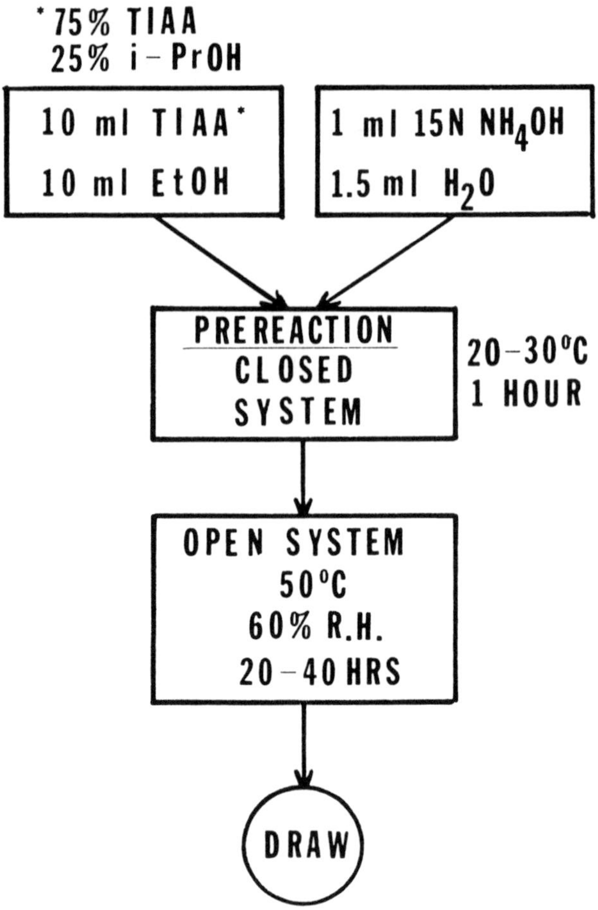

Figure 8: Preparation procedure for pure TiO_2 fibers using TIAA. For TiO_2-SiO_2 fibers follow procedure in Figure 3.

No detailed study of the properties of fibers prepared from AcAc raw materials has been carried out. Kim[25] was successful in firing pure TiO_2 gels to 450°C without crystallization. While the resulting fibers could be handled strengths were not measured. Preliminary results indicate that the binary TiO_2-SiO_2 fibers are similar in strength to gel derived pure SiO_2, but the ZrO_2-SiO_2 fibers are weaker, particularly at higher zirconia contents.

CLOSING COMMENTS

While much has been learned over the past 10 years, suprisingly few studies involving drawn fiber from alkoxide derived sols have been reported. Most of the work has focused on fiber preparation, rather than resulting properties. However, progress should be rapid over the next few years since it is now possible to prepare a number of compositions having potentially useful properties. No doubt additional compositions will be prepared as our understanding of sol chemistry progresses. The use of "modified" alkoxides with slowly hydrolyzing organic groups offers considerable opportunity. A number of materials are available commercially, and many are easily prepared in the laboratory using the pure alkoxide as starting material.

Whether these advances will lead to additional commercial applications remains to be seen. The cost of raw materials is a major drawback. In many cases however, (e.g. TiO_2 and ZrO_2 silicates) the gel route may be the only way to prepare certain compositions. Melt derived fibers cannot be drawn due to high melting temperatures and the tendency to crystallize. Furthermore, the gel processes described above can be carried out on a "batch" or continuous production basis. No major changes, other than batch composition, are required in order to produce SiO_2, ZrO_2, etc. As a result, a single plant might conceivably produce an almost unlimited number of low volume specialty compositions.

In the final analysis, success of gel derived fibers will depend on their properties. If reasonable strengths can be attained in compositions having other useful properties there will likely be at least a limited market. However, even for the simplest systems, characterization studies are barely underway, and virtually no data are available for more complex compositions. Part of the problem has been a necessary focus on producing fiber. Sufficient progress seems to have been made in this area, so that in the next few years an increased emphasis on characterization should be possible. Hopefully these studies will confirm the potential of the sol-gel process for production of unique compositions and useful products.

Acknowledgements

The author would like to thank Dr. Lisa Klein for the opportunity to contribute to this volume. I am indebted to a number of associates, particularly Dr. Sunuk Kim and Dr. Masyood Akhtar, who took part in several long and useful discussions regarding the work. Financial support from the N.Y.S. Science and Technology Foundation and the Kay-Fries Corp. are gratefully acknolwledged.

REFERENCES

1. Mahler, W. and Chowdhry, U., in: *Ultrastructure Processing of Ceramics, Glasses and Composites* (L.L. Hench and D.R. Ulrich, eds.), pp. 207-218, John Wiley & Sons, New York (1984).
2. Sakka, S., in: *Better Ceramics Through Chemistry* (C.J. Brinker, D.E. Clark and D.R. Ulrich, eds.), pp. 91-100, North-Holland, New York (1984).
3. Sweeting, T.B., U.S. Patent 4,277,269, May 22, 1979.
4. Sakka, S., *Am. Ceram. Soc. Bull.* 64: 1463-66 (1985).
5. LaCourse, W.C., Dahar, S. and Akhtar, M., *J. Am. Ceram. Soc.* 67: C200-1 (1984).
6. Sakka, S. and Kamiya, K., *J. Non-Cryst. Sol.* 48: 31-46 (1982).
7. LaCourse, W.C., Akhtar, M., Dahar, S., Sands, R.D. and Steinmetz, J., *J. Can. Ceram. Soc.* 53: 18-23 (1983).

8. Iler, Ralph K., *The Chemistry of Silica,* John Wiley & Sons, New York (1979).
9. Keefer, K.D., in: *Better Ceramics Through Chemistry* (C.J. Brinker, D.E. Clark and D.R. Ulrich, eds.), North-Holland, New York (1984).
10. Dahar, S., M.S. Thesis, N.Y.S. College of Ceramics, Alfred University, May 1983.
11. Yoldas, B.E., in: *Ultrastructure Processing of Ceramics, Glasses and Composites* (L.L. Hench and D.R. Ulrich, eds.), John Wiley & Sons, New York (1984).
12. LaCourse, W.C. and Mohideen, U., in: *Proceedings of the XIV International Congress on Glass,* Vol. II, pp. 63-70, Indian Ceramic Society, Calcutta (1986).
13. LaCourse, W.C., unpublished results.
14. LaCourse, W.C., in: *Better Ceramics Through Chemistry* (C.J. Brinker, D.E. Clark and D.R. Ulrich, eds.), pp. 53-58, North-Holland, New York (1984).
15. Jim, S. and LaCourse, W.C., *Proceedings of the Florida Conference on Sol-Gel Processing,* Feb. 1985 (In Press).
16. Mohideen, U., M.S. Thesis, N.Y.S. College of Ceramics, Alfred University, May 1984.
17. Schlickling, B., B.S. Thesis, N.Y.S. College of Ceramics, Alfred University, May 1985.
18. Kamiya, K. and Sakka, S., *Yogyo-Kyokai-Shi* 85: 308-9 (1977).
19. Kamiya, K., Sakka, S. and Tatemichi, T., *J. Mater. Sci.* 15: 1765-71 (1980).
20. Kamiya, K., Sakka, S. and Tashiro, N., *Yogyo-Kyokai-Shi* 84: 614-18 (1976).
21. Sakka, S., in: *Treatise on Materials Science and Technology* (M. Tomozawa and R.H. Doremus, eds.), Vol. 22, pp. 129-67, Academic Press, New York (1982).
22. Leroy, E., Robin-Brosse, C. and Torre, J.P., in: *Ultrastructure Processing of Ceramics, Glasses and Composites* (L.L. Hench and D.R. Ulrich, eds.), pp. 219-31, John Wiley & Sons, New York (1984).
23. Kilinski, B., M.S. Thesis, N.Y.S. College of Ceramics, Alfred University, August 1986.
24. Kilinski, B. and LaCourse, W.C., *Am. Ceram. Soc. Bull.* (1986).
25. Kim, S., Ph.D. Thesis, N.Y.S. College of Ceramics, Alfred University, August 1986.
26. Doyle, E. and LaCourse, W.C., to be published.

Part IV

Monoliths, Shapes and Preforms

10

Monolith Formation from the Sol-Gel Process

Masayuki Yamane
Department of Inorganic Materials
Tokyo Institute of Technology
Tokyo, Japan

INTRODUCTION

During the past decade the formation of monolithic glass bodies via the sol-gel technique has been studied with great interest because of the many advantages this technique has over conventional glass melting.[1-6]

The sol-gel process begins with the formation of a gel, usually from metal alkoxide starting materials, in a suitable mold at a temperature below 100°C. It is generally agreed that the formation of a gel from metal alkoxides occurs through hydrolysis and polycondensation. The hydrolysis and polycondensation of metal alkoxide dissolved in a proper solvent yields a colloidal suspension of the reaction products called a sol. The sol undergoes a transition to a soft porous mass, a wet gel, which eventually becomes a solid porous aggregate of extremely small particles known as xerogel. If the xerogel contains a sufficient amount of strong metal-oxygen bonds such as Si–O, it can be converted into a monolithic glass consisting of a random three-dimensional network of the bonds by simply sintering or hot-pressing at a temperature well below the liquidus temperature of the system.

Monolithic glasses thus obtained can have compositions which are difficult or even impossible to obtain using melt technology. For example, glasses having an extremely high melting temperature can be made at a substantially reduced temperature. Another example is the possibility of making homogeneous glasses which would ordinarily phase separate during conventional processing. Sol-gel glasses are also free of crucible contamination, which is particularly important for high technology materials.

Monolith formation by the sol-gel process was carried out with the aid of hot-pressing during the early stages of development due to the difficulty of obtaining a monolithic gel.[7-10] Currently, however, a number of glasses can be obtained by directly sintering monolithic gels.[11-33]

GEL PREPARATION

Types of Gels Used for Monolith Formation

In monolith formation by the sol-gel process, the quality of the final glass depends heavily on the quality of the precursor gel. If the precursor has a crack, the final glass will also have a crack. Thus, the preparation of a crack-free monolithic gel is perhaps the most important step in the sol-gel process.

There are two major types of gels used for monolith formation. First is the type obtained from organo-metal compounds mentioned above. It is based on the sol formed by hydrolysis and polycondensation of metal alkoxides such as tetramethoxysilane [$Si(OCH_3)_4$ or TMOS] or tetraethoxysilane [$Si(OC_2H_5)_4$ or TEOS].

The second type of gel is formed from a sol consisting of a colloidal suspension of silica which is obtained by a method other than the direct hydrolysis of organo-metal compounds.[34-41] For example, such sols may consist of dispersed silicic acid or dispersed fumed silica obtained by the flame hydrolysis of $SiCl_4$. The sol-gel transition of the dispersed silicic acid is made by destabilizing the sol with the addition of potassium silicate solution and formamide.[34] The gel formation from fumed silica dispersed in water is carried out by adding HCl or NH_4OH to the sol.[35-39] The sol formed by dispersing fumed silica particles in a non-polar medium is gelled with the aid of gelling agent such as ammonia or amine which deprotonates the silanol groups covering the surface of the particles.[40,41]

Gels of the second type shrink very little during drying, and can be made crack-free in large sizes. Unfortunately, the composition of glasses made with these gels is limited to either pure or very nearly pure silica because of the low reactivity of the dispersed silica particles with other chemicals.[42]

Gels formed from metal alkoxides, on the other hand, are applicable to the formation of both pure silica and multicomponent silicate glasses because of the chemical reactions which occur when the sol is made. Gels for multicomponent glasses are made by hydrolyzing mixed solutions of various alkoxides or by hydrolyzing TMOS or TEOS with an aqueous solution of one or more salts.[24-33] Thus, the alkoxide method has a broad area of application.

The major drawback to the alkoxide method is the problem of fracturing due to large shrinkage during the drying of the gel.[14] This fracturing problem is worse in multicomponent systems than it is in the pure silica system, and thus has been a major factor in inhibiting the development of multicomponent glasses via the alkoxide technique.

Gel Formation from Silicon Alkoxide

Composition of Precursor Solution: Silicon alkoxides are, in general, immiscible with water. The hydrolysis of TMOS or TEOS for the formation of a monolithic gel is, therefore, usually carried out by using methanol or ethanol as a mutual solvent for the two immiscible liquids. Moreover, the amount of water introduced for the hydrolysis of TMOS or TEOS is usually much larger than the theoretical amount[43] shown in the following schematic Equations (1) and (2):

$$nSi(OR)_4 + 4nH_2O = nSi(OH)_4 + 4nROH \tag{1}$$

$$nSi(OH)_4 = nSiO_2 + 2nH_2O \tag{2}$$

where, R is CH_3 or C_2H_5.

In practice, 4 to 20 mols of water containing acid or base catalysts, i.e., HCl, HNO_3, NH_4OH is introduced per mol of TMOS or TEOS.[11-13,18-20,44-48] Table 1 shows some examples of the ratio among TMOS or TEOS, alcohol, and water employed for the monolith formation by various researchers. The introduction of excess water is to assure the complete hydrolysis and the development of crosslinking to form a strong three dimensional network of Si–O–Si bonds throughout the gel.[44-48]

Table 1: Some Examples of the Composition of Precursor Solutions for Monolithic Silica Gels (mol ratio)

TMOS	TEOS	Methanol	Ethanol	H_2O	Catalyst	Ref.
1		4.6		4	——	(11)
	1		3.8*)	16	HNO_3(1N)	(13)
1		1.2-2.4*)		4-4.65*)	NH_4OH(-0.01N)	(16)
	1		4.8	6.5	——	(18)
1		4.5		4	NH_4OH(PH=10.6)	(22)

*) : Estimated from volume.

Alcohol and water are no longer needed after a sol has been formed. A wet monolithic gel formed from an alkoxy-derived sol normally shrinks to some extent while being aged in its mold due to the removal of some of its internal alcohol and water. When the gel is allowed to dry, it shrinks by a rather large amount, and this shrinkage often leads to the build up of a non-uniform stress which fractures the gel.

Stress build up in a drying gel is caused by capillary force, which is proportional to the evaporation rate and inversely proportional to pore size.[14] During the initial stage of drying, these forces are large because the rate of evaporation is large. During the final stages of drying the rate of evaporation is small, but the pore size is considerably reduced and fracture causing capillary force can occur. The primary techniques in eliminating fractures are to reduce the total shrinkage, which leads to a xerogel of low bulk density, and to control the rate of evaporation of liquids trapped in the gel.

The amount of total shrinkage of a gel depends on the chemical and physical conditions during gel formation.[13,48] It depends on the gelling temperature as well as the composition of the precursor sol. Thus, the type of catalyst used and the molar ratio between alkoxide, alcohol, and water are important variables in controlling gel shrinkage. Shrinkage can be reduced by removing alcohol from the sol prior to pouring it into a mold. This is done in practice by using a rotary evaporator or by pulling a vacuum.[13,49,50] Also noteworthy is a method described in a patent of sol formation without the addition of alcohol.[51] In this method, TEOS is vigorously mixed with water and a small amount of a suitable acid catalyst such as HCl, and the result is a clear sol.

Effect of Catalyst: There are many papers on the effects which various catalysts have on the sol-gel transition.[52,60] Acid catalysts such as HCl and HNO_3 are thought to promote hydrolysis through electrophilic reaction.[54-57] The rate of hydrolysis increases monotonically as the pH of the starting mixture decreases from 7. However, the condensation reaction rate has a local minimum at around pH = 2.[57] Thus, sols prepared with an acid catalyst require very long gelling time, and the resulting gel undergoes large shrinkage during drying.[56]

The hydrolysis of TMOS and TEOS using a base catalyst takes place via a nucleophilic reaction.[57] The hydrolysis in this instance takes place slowly and the silicate monomers tend to condense before they are fully hydrolyzed. In general, gels prepared from solutions containing a base catalyst such as ammonia shrink less than those prepared from acid catalyzed solution. The resulting xerogel has a low bulk density.[56,61,62] This is shown in Table 2.

Table 2: Effects of Catalyst on the Gelling Time of TMOS and the Bulk Density of the Resulting Xerogels*[56]

Solution	A	B	C
Catalyst	——	HCl	NH_4OH
pH of Water	7	3	9.5
Gelling Time (h)	16	150	6
Bulk Density (g/cm^3)	1.51	1.56	0.82

*) TMOS : MeOH : H_2O = 1 : 4.6 : 4

It is thought that base catalysts accelerate the coagulation of colloidal particles in a sol by deprotonating the silanol groups covering the surface of the particles.[62] This results in the formation of secondary particles. When the secondary particles come into contact with each other by Brownian motion they tend to stick together, thus creating ternary and quaternary particles. If the formation of the strong Si–O–Si bonds occurs during this coagulation the resulting ternary and quaternary particles will not have spherical symmetry. When these irregular particles coagulate to form a gel, they will tend to do so in a loose rather than a dense packing, and the dried gel will have a fairly low bulk density.

Effect of Temperature: In a sol-to-gel transition carried out in a well sealed container, the following gelling time goes as given by Equation (3);

$$\frac{1}{t_{gel}} = A\exp(-E^* / RT) \tag{3}$$

where, t_{gel} is gelling time, A is the usual Arrhenius constant and E^* is an "apparent" activation energy.[63] In general, higher temperatures cause shorter gelling time and reduced shrinkage during drying.[11] This is shown in Table 3 in terms of the bulk density of the xerogel.

Table 3: Effect of Gelling Temperature on the Bulk Density of Xerogels Prepared by the Hydrolysis of TMOS*[11]

Sample	A	B	C	D
Gelling Temperature (°C)	54	65	68	70
Bulk Density (g/cm^3)	1.46	1.13	1.02	0.98

*) TMOS : MeOH : H_2O = 1 : 4.6 : 4

One of the reasons for these temperature effects is that the dehydration polycondensation reaction between silanols on the surface of the sol particles is accelerated as the temperature is increased. Another reason is the increased rate of particle collision at higher temperatures. Higher particle collision rates imply higher rates of bond formation through dehydration polycondensation. At elevated temperatures the particles are able to bond together to form a fairly rigid structure which hinders the shrinkage of the wet gel to a densely packed dry gel.

Drying: In addition to producing gels with low bulk density, drying of wet gels must be done very carefully, particularly when carried out by the simple evaporation of solvent, in order to avoid the stress build up due to high evaporation rate. The usual method is to keep the atmosphere surrounding the gel saturated with the evaporated liquid, and to allow the vapor to escape very slowly through pinholes in the cover of the mold.[18,22,61,64]

Hench[60,65,66] tried to control the evaporation rate by introducing formamide, which he calls a drying control chemical additive (DCCA), into the precursor solution together with methanol. Formamide has a very low vapor pressure (0.1 Torr at 40°C compared to 100 Torr at 21°C for methanol) and a high boiling temperature (109°C compared to 64.5°C for methanol). The combination of formamide and methanol reduces the rate of evaporation during the initial stages of drying and therefore reduces the stress build up.

The introduction of formamide does, however, cause problems when the gel is heated to sinter it into glass. Apparently the formamide cannot be easily removed from the gel due to its low vapor pressure. If the gel is heated before the complete removal of formamide, the remaining amount turns into a charcoal, which stresses and fractures the gel. Clearly, further investigation of drying control chemical additives is needed.

A particularly efficient method of drying with reduced capillary force consists in evacuating the solvent under hypercritical conditions where the surface tension is equal to zero due to the absence of a liquid-gas interface. The preparation of monolithic gels from silicon alkoxide by the hypercritical solvent evacuation technique was made first by Zarzycki and his colleagues.[14-16,29] In this method which was later modified by the group of Philips Research Laboratories,[67-70] the gel formation from the precursor solution is usually made in an autoclave, though the technique is also applicable to the drying of already gelled samples.

The TMOS- or TEOS-alcohol-water solution contained in a proper mold is placed in an autoclave with some excess alcohol and pressurized with nitrogen gas, followed by heating at a rate of about 100°C/hr to a temperature of about 250°C. The introduction of excess alcohol is to utilize the autogeneous pressure of alcohol vapor to form the temperature and pressure conditions which are such that the critical point of the solvent of the solution (e.g., 240°C, 79.7 bars for the case of methanol) remaining after gelling is exceeded, without the liquid/vapor equilibrium curve of the solvent being broken at any time. The pressurizing with nitrogen gas prior to heating is to prevent the evaporation of the solvent from the gel before the critical condition is reached.

When the hypercritical condition is reached at about 250°C and 200 bars, the solvent is slowly removed under substantially isothermal conditions over a period of several hours. The autoclave is then purged several times with a dry gas (nitrogen, argon or helium). After the autoclave has returned to atmospheric pressure, it is cooled down to room temperature to withdraw the dried gel.

The gels prepared by a technique like hypercritical evacuation of the solvent are called aerogels. The aerogels produced at the absence of surface tension have nearly the same volume as their precursor solution, and have bulk densities of the order of 0.03 to 0.5 g/cm^3 depending on the ratio of alcohol contained in the precursor.

The preparation of aerogels is also possible by Polaron critical point drying (PCD), a technique similar to hypercritical solvent evacuation.[71] In this method, gelled samples are immersed in a large excess of solvent such as amyl acetate and inserted in the PCD autoclave. The sample chamber is then flushed with liquid carbon dioxide at room temperature to replace the solvent. The temperature of the liquid is slowly raised to about 40°C causing the pressure to exceed 80 bars. The carbon dioxide is then slowly vented to ambient.

The monolithic aerogels prepared by this technique are also free of fissures and have low bulk densities. The hypercritical method and this critical point drying are perhaps the most effective way of producing dry monolithic gels.

GEL PROPERTIES

Pore Size Distribution and Specific Surface Area

There are many parameters which affect the densification of an alkoxy-derived gel. These parameters include the properties of the dried gel as well as the heat cycle and the atmosphere in which the heating takes place. If a gel is not properly prepared, stress build up can occur during heating, resulting in fracture or bloating.

Among the gel properties affecting the densification process, the pore size distribution in the gel is perhaps the most important. As will be shown later, almost all the problems that arise during the densification of a gel are closely related to the pore size within the gel. For example, gels containing only small pores tend to fracture or bloat when heated to a high temperature, whereas those containing large pores can be sintered without such problems.[11,18,19,22]

The pore size distribution and bulk density of a gel vary widely depending on conditions during gel formation. In general, gels with low bulk density contain large pores. This is clear from Figures 1 and 2 which show the pore size distributions determined from the nitrogen adsorption isotherms on gels having bulk densities shown in Table 2 and 3. Gels prepared from solutions containing ammonia have a bulk density of 0.82 g/cm^3 and contain pores as large as 150 Å in diameter, whereas gels prepared without ammonia have a bulk density of 1.51 g/cm^3 and contain no pores exceeding 70 Å in diameter. A similar relationship is seen in Figure 2, where the pore size distribution curves for gels prepared at higher temperatures, and thus having lower bulk densities, are shifted toward the larger pores.

It should be noted that the gels shown in Figures 1 and 2 all contain a considerable number of small pores below 20 Å in diameter regardless of the number of larger pores, suggesting that the pores in these gels have many small necks. In other words, although many of the pores in these gels are large, they apparently are connected by fairly small passageways. The presence of the necks in the pores hinder the escape of water generated by dehydration condensation of silanols.

The specific surface area is often used to describe the approximate pore size within a gel,[19,22,36] because it can be determined by a measurement which is relatively simple compared to direct measurement of the pore size distribution using special equipment. The average pore diameter of a material is roughly related to the specific surface area through the bulk density and true density of the material by the following Equations (4) or (5):

$$D = \frac{4}{S_P}\left(\frac{1}{d_B} - \frac{1}{d_T}\right) \tag{4}$$

$$D' = \frac{6}{S_P}\left(\frac{1}{d_B} - \frac{1}{d_T}\right) \tag{5}$$

where, D and D$'$ are the average pore diameters, d_T is the true density, d_B is the bulk density, and S_P is the specific surface area of the material. The Equation (4) expresses the relation between the diameter and the specific surface area when the pores are cylindrical. The Equation (5) is applied to the material containing large spherical pores connected by narrow necks.

The specific surface area of alkoxy-derived silica gels dried under atmospheric pressure varies from about 500 to 900 m^2/g[12,13,18,19] again depending on the gelling temperature and composition of the precursor solution, while the surface area of aerogels is often as high as 1,600 m^2/g.[15,70] In general, gels with lower bulk densities have larger specific surface area.

The true density of an alkoxy-derived silica gel, on the other hand, is about 2.05 g/cm^3, and is nearly independent of the conditions during gel formation.[11] This density is slightly lower than 2.2 g/cm^3, the density of fused silica. The

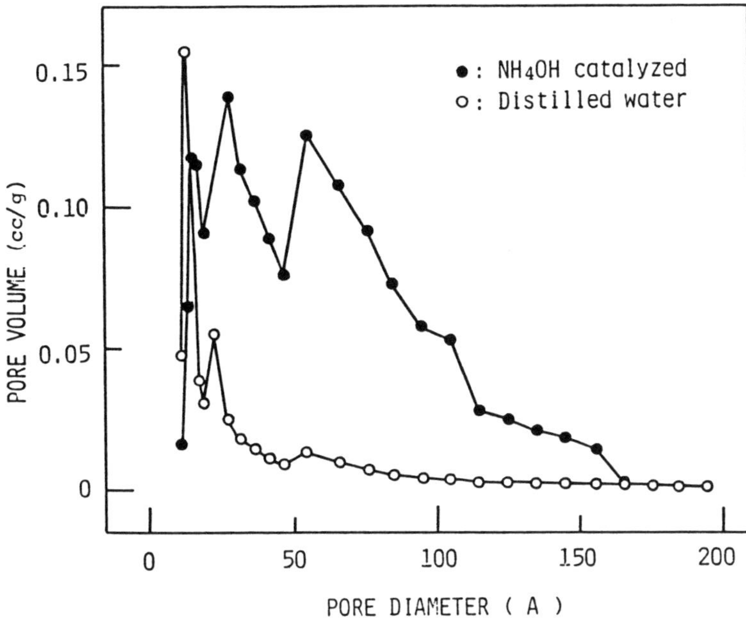

Figure 1: Pore size distribution in silica gels prepared by the hydrolysis of TMOS with distilled water and ammonia water of pH = 10.

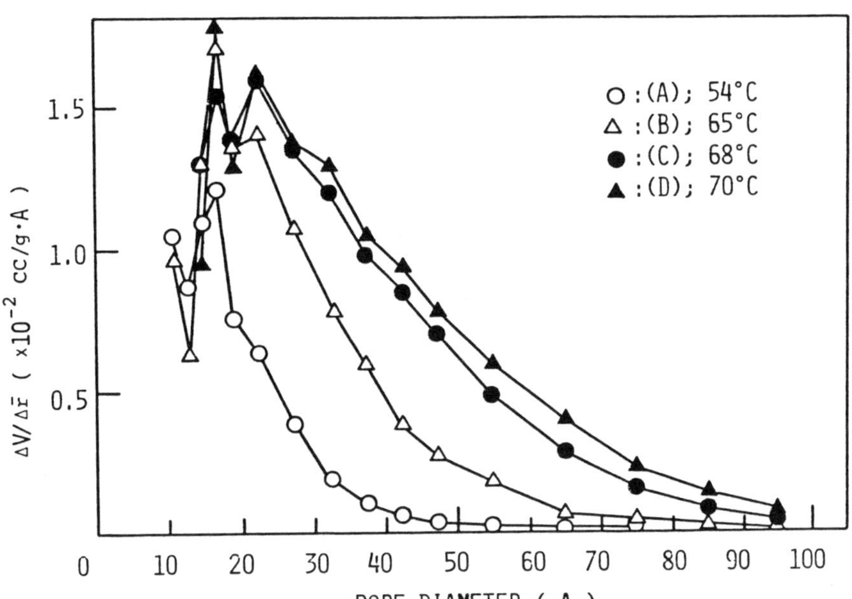

Figure 2: Pore size distribution in silica gels prepared by the hydrolysis of TMOS with distilled water and gelling at: (a), 54°C; (B) 65°C; (C) 68°C; (D) 70°C (Reference 11).

results for the estimation of the average pore diameters using this value of true density along with the measured values of specific surface area and bulk density are shown in Table 4 for various gels.

Table 4: Estimated Average Pore Diameter from Specific Surface Area and Bulk Density

Sample	Surface Area (m^2/g)	Bulk Density (g/cm^3)	Average Pore Diameter (A), Eq.(4)	Eq.(5)
A	560*)	1.46	14	21
B	780*)	1.13	20	30
C	880*)	1.02	22	33
D	870*)	0.98	25	38
I	520	0.15	475	712
II	1080	0.05	722	1084
III	1590	0.03	826	1239

A-D; Ref.(11), I-III; Ref.(70)

*) Unpublished data

It is clear from Table 4 that gels prepared by the hypercritical technique contain pores more than one order of magnitude larger than those obtained at atmospheric pressure. The table also shows that the pore diameters obtained for Samples A through D by using Equation (5) are closer to the average diameters expected from the pore size distribution curves shown in Figure 2 than those obtained by Equation 4. This agrees with the above suggestion of the presence of randomly connected globular pores in these gels, although the presence of such pores is unlikely in gels with extremely low bulk density such as those produced by the hypercritical technique.

The measurement of the specific surface area also provides information about the size and the coordination number of the particles comprising the gel. The specific surface area of a porous material is determined using a BET plot based on nitrogen adsorption and depends on the diameter of the particles. When the particles are small, the apparent value of the specific surface area varies depending on the coordination number of the particles. This is due to the fact that nitrogen molecules cannot be adsorbed onto the gel in regions near the contact area of two adjacent particles. The value of the specific surface area of a gel can be estimated if the density of the spherical particles comprising the gel skeleton and the effective size of a nitrogen molecule are known. Figure 3 shows the result of such an estimation for a hypothetical system as a function of radius, r, and average coordination number, n, of the constituent spherical particles. The particles are assumed to have a density of 2.05 g/cm^3. It is also assumed that a nitrogen molecule is a rigid sphere with a radius of 2.2 Å, which corresponds to a cross sectional area of 16.5 $Å^2$.

Since the coordination number of the particles must be at least 2 in order to develop a continuous skeletal network of silica, the radius of the particles which comprise an aerogel whose specific surface area is 1,600 m^2/g must be smaller than 10 Å. If it is assumed that the xerogels dried at the atmospheric pressure consist of the particles of similar size, the average coordination numbers of the particles in the gels shown in Table 4 are in the range from 7 to 5, decreasing with the increasing gelling temperature. The particle radius of about 10 Å is smaller by one order of magnitude than the radius of silicic acid particles[30] or silica particles obtained by the flame hydrolysis of $SiCl_4$.[36] The small size of the particles in the alkoxy-derived gel is advantageous in preparing multicomponent glasses using aqueous solutions of various salts which may not take part in the formation of network structure but rather stay on the surface of the silica skeleton.

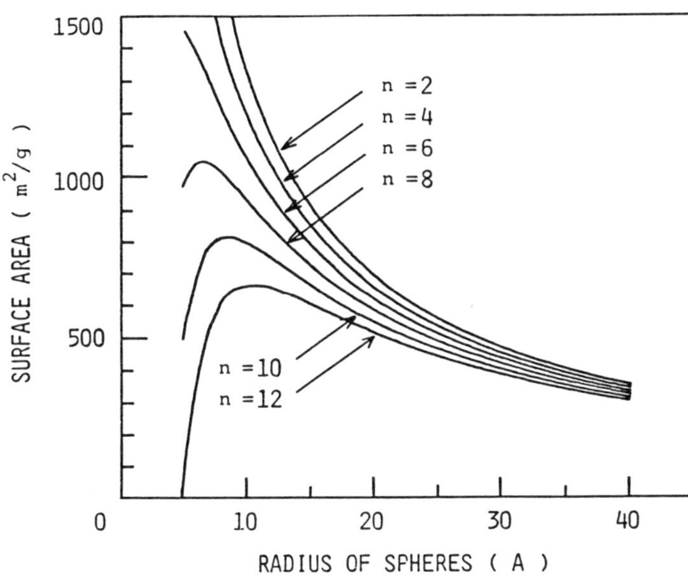

Figure 3: Dependence of apparent BET surface area of gels on the radius, r, and average coordination number, n, of constituent particles.

Hydroxyl Groups and Residual Organic Compounds

Hydroxyl groups and residual organic compounds in a gel are the other parameters which must be taken into consideration in the densification process of a monolithic gel.

Hydroxyl groups generally exist as silanols covering the surface of the silica skeleton of a xerogel.[19,45,72-74] Some of the silanols are hydrogen-bonded (H-bonded) to molecular water remaining in the micropores either as a single molecule or in clusters.[38] The number of silanols, single water molecules and clustered water molecules varies from gel to gel depending on the pore size distribution within the gel.

Figure 4 shows the near IR absorption spectra of ammonia catalyzed and HCl catalyzed gels. In the figure, which gives the absorption due to the coupling of various modes of OH-related vibrations, the sharp band at around

1360 Å is assigned to free SiOH on the surface of the silica skeleton of the gel. The bands at 1390, 1890 and 2200 Å are due to SiOH H-bonded to molecular water. The peaks at 1440, 1930 and 2250 Å are attributed to another type of SiOH H-bonded to molecular water. The absorption in the vicinity of 2300 Å is due to clustered molecular water which is H-bonded to SiOH.[38]

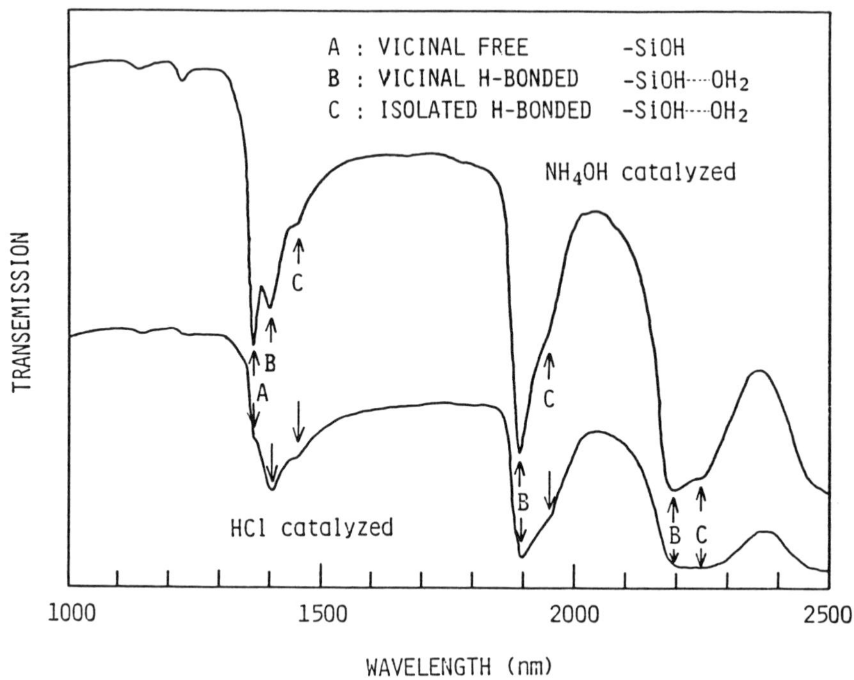

Figure 4: Absorption spectra of ammonia catalyzed and HCl catalyzed gels in near IR region.

It is known from the figure that the HCl catalyzed gel containing only small pores has a considerable amount of molecular water in the pores, whereas the ammonia catalyzed gel containing large pores has only a small amount of molecular water. On heating the gel, this water, as well as that generated by the dehydration condensation of silanols, diffuses toward the outer surface of the gel through the network of micropores, and leaves the system by evaporation.

This diffusion of water vapor through micropores takes a very long time because the pore diameter is roughly the same as the mean free path length of the vapor. Therefore, if a gel contains too many hydroxyl groups, water vapor is gradually accumulated in the gel due to the successive generation of water from silanols as the gel is heated. The accumulated water vapor generates pressure from the inside of the gel, which leads to gel fracture. Thus the preparation of a gel with low hydroxyl content is preferable.

The organic compounds in a gel are mainly attributed to unreacted alkoxy groups.[17-19,61] The unreacted alkoxy groups decompose on heating and turn into gas or charcoal depending on the oxygen partial pressure in the atmosphere surrounding the gel. If the unreacted alkoxy groups turn into charcoal due to insufficient oxygen, local stress is built up in the gel when the gel shrinks by further heating. If the residual organic compounds completely decompose into gas, they leave very fine pores in the gel. Although the formation of fine pores is not as serious as the formation of charcoal, the preparation of a gel with a minimum amount of residual organic compounds is obviously desirable.

The quantity of both hydroxyl groups and organic compounds in a gel is dependent on the mode of gel formation, particularly on the kind of catalyst used. This is shown in Table 5 in terms of the amount of hydrogen and carbon determined by gas chromatography of the released gas from xerogels quickly heated to 1200°C. The ammonia catalyzed gel contains a large amount of carbon attributed to organic compounds, because the polycondensation of silanols occurs so quickly in the sol containing base catalyst, which leaves many alkoxy groups unreacted. The small amount of carbon detected in the HCl catalyzed gel, on the other hand, reveals that an acid catalyst accelerates hydrolysis of alkoxides rather than condensation or coagulation of the formed particles.

Table 5: Effect of pH of Water on the Hydrogen and Carbon Content of a Silica Gel*

Sample	A	B	C
pH of Water	7	3	9.5
Catalyst	—	HCl	NH_4OH
Hydrogen Content (%)	1.51	1.49	0.99
Carbon Content (%)	0.31	0.12	0.53

*) TMOS : MeOH : H_2O = 1 : 4.6 : 4

CHANGE IN STRUCTURE AND PROPERTIES OF A GEL WITH HEAT TREATMENT

Differential Thermal Analysis and Thermogravimetric Analysis

The reactions which occur as a material is heated are accompanied by a gain or loss of heat, and sometimes a loss in weight as well. If there is any reaction which causes a large absorption or release of heat during the densification process of a gel, the heat treatment in the temperature range where such a reaction occurs must be carried out very carefully so that the gel does not undergo stress build up due to a sudden change of its own temperature.

The information on such temperature range is usually obtained by differential thermal analysis (DTA) and thermogravimetric analysis (TGA) of the gel. The DTA and TGA curves of an alkoxy-derived silica gel normally look like those in Figure 5.[17,18] In the DTA curve, a broad endothermic peak, A, an exothermic peak, B, and the gradual deviation of the curve, C, from the base line are observed in the temperature ranges, around 150°, 350° to 600°, and above 650°C, respectively. The reactions corresponding to these peaks are accompanied by gel weight loss as shown by the TGA curve.

The first peak, A, is attributed to the heat of evaporation of water adhering to the micropore walls of the gel.[17] The second peak, B, is due to the combustion of residual organic compounds in the gel.[17-19] This peak becomes sharp if the analysis is made under the flow of oxygen gas.[18] The gradual deviation starting at a temperature around 600°C is caused by the release of water generated by the dehydration condensation of silanols.[17,18]

The shape and the temperatures of the respective peaks are dependent on both the rate of heating during the analysis and the size of the sample. If the rate of heating in the analysis is high compared to the rate of reaction, the peaks become sharp and shift towards the higher temperatures. If the size of the sample is large, the peaks shift slightly towards the higher temperatures and become less sharp due to the long time necessary for material diffusion through micropores.

It should be noted that DTA and TGA are usually carried out at a heating rate of 5° to 10°C/min. This is much quicker than the practical heating rate for the densification of a microporous monolithic gel. A porous gel is a good thermal insulator, and it takes very long for such a gel to attain temperature equilibrium with the ambient atmosphere. Thus, in the vicinity of a reaction peak the temperature should be increased either very slowly or not at all in order to avoid fracture. More rapid temperature increase can be resumed once the reaction is completed.

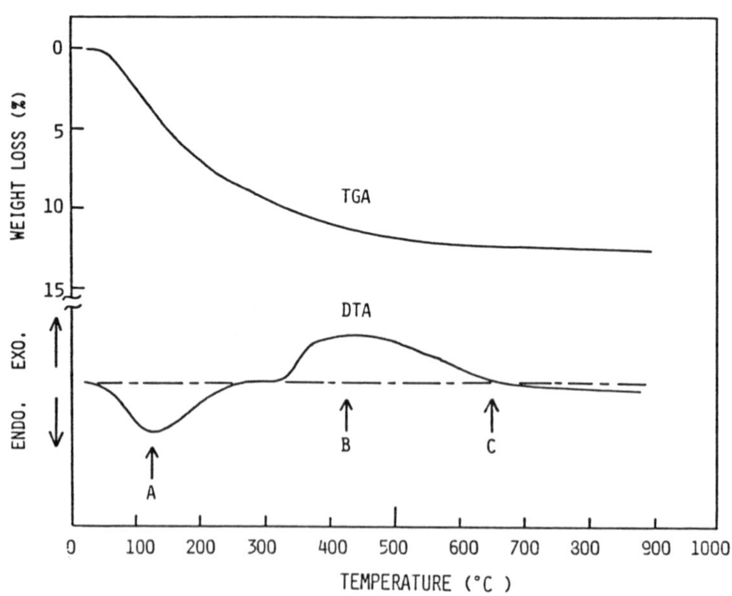

Figure 5: Schematic illustration of DTA-TGA curves of an alkoxy-derived silica gel.

Change in Density and Linear Shrinkage

True Density: When an alkoxy-derived gel is subjected to heat treatment for densification, many properties of the gel change as a result of structural changes in the silica skeleton and the reduction of pore volume. The structural change of the silica skeleton of a gel is attributed to the development of crosslinking of the Si–O bonds due to the dehydration condensation of silanols. This change in pore volume is attributed to the sintering of the particles comprising the gel, which is governed by the viscous flow which dissipates the large surface energy.[75–77] This change is indicated by the change of bulk density or the linear shrinkage of the gel.

The true density of an alkoxy-derived gel, which is almost independent of the mode of gel formation, changes with temperature independently of the size or the volume of the micropores contained in the gel. The true density increases with increasing temperature due to the development of crosslinking of the Si–O bonds and becomes similar to the density of fused silica at a temperature around 700°C. Then it levels-off with further heating to higher temperatures, as shown in Figure 6.[11]

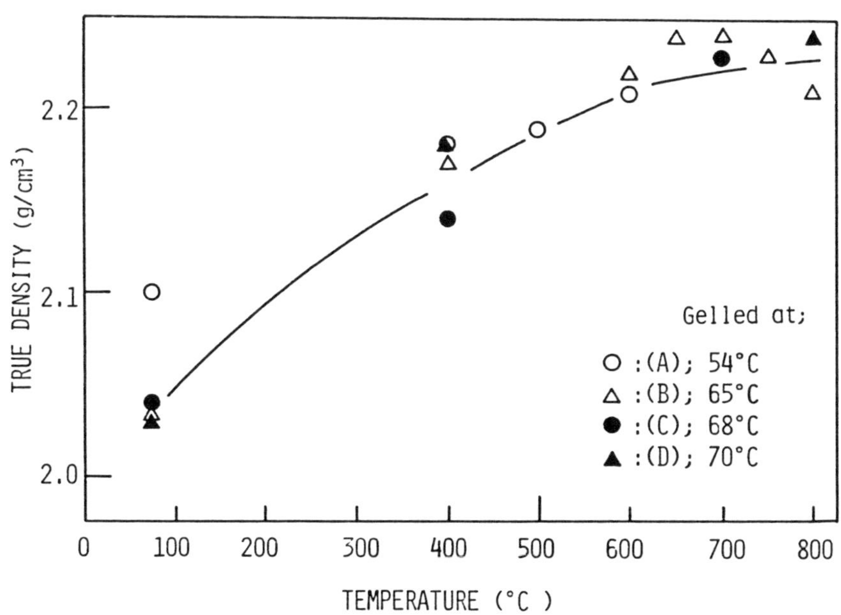

Figure 6: Change in the true density of alkoxy-derived silica gels with temperature (Reference 11).

The leveling-off of the true density of the gel at a value around 2.2 g/cm^3 suggests that the structure of the silica skeleton becomes similar to that of fused silica, regardless of the existence of the remaining pores. This also means that the development of the crosslinking of the Si–O bonds by the dehydration condensation of silanols is mostly completed by the time when the heat treatment reaches 700°C.

Bulk Density: The change in the bulk density of a gel with heat treatment is dependent on the pore structure of the gel. A change in bulk density reflects a change in the porosity of the gel, as well as the development of cross-linking in the solid phase, i.e., the increase in the true density which was discussed above.

The change of the porosity of a gel is not as simple as that of true density. It slightly increases up to about 300°C, due to the evaporation of water adhering to the micropore walls and the formation of fine pores as the result of the decomposition of residual organic compounds.[13,21] Then it decreases gradually with further heating up to around 700°C where the decrease becomes rapid. The gradual decrease in the porosity above 300°C is probably due to the shrinkage of pores accompanied by the dehydration condensation of silanols covering the surface of the silica skeleton.[13,78] The rapid decrease in the porosity at around 700°C is attributed to the micropore collapse due to the sintering governed by viscous flow.[13,78-81]

Since the viscosity of a gel varies with time as well as temperature due to the loss of free volume and the composition change caused by the dehydration condensation of silanols,[72,79,82] the extent of micropore collapse and the resulting bulk density at a particular temperature depends on the thermal history of the gel. The bulk density of a gel containing only small pores begins to increase at a lower temperature than gels containing large pores as long as both types are subjected to the same heat treatment. This is shown in Figure 7 for gels A through D in Table 3 heated up to various temperatures at a rate of 0.5°C/min after being held at 400°C for 8 hours.

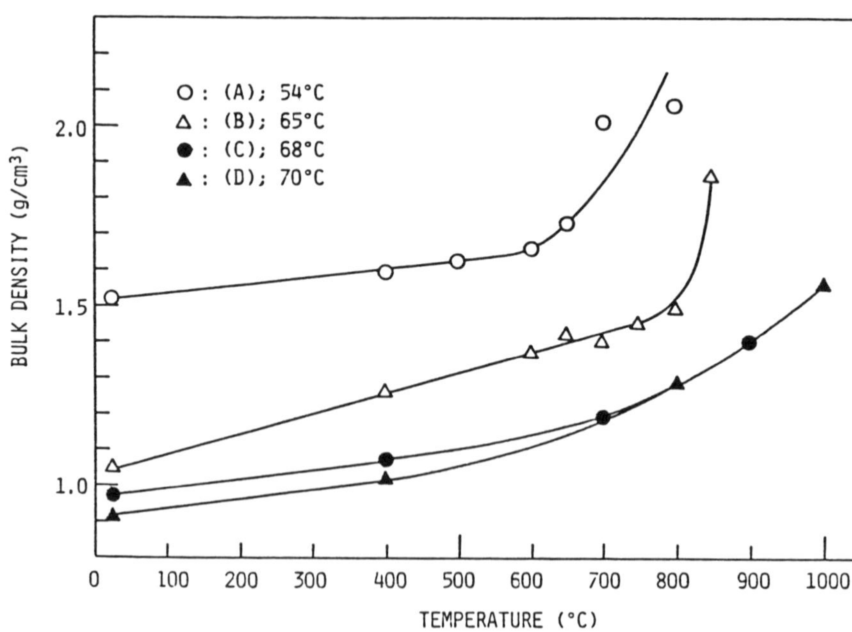

Figure 7: Bulk density of various gels heated up to various temperatures at 0.5°C/min after holding at 400°C for 8 hours: (A) gelled at 54°C; (B) 65°C; (C) 68°C; and (D) 70°C.

Linear Shrinkage: The linear shrinkage of a gel with heat treatment also reflects the phenomena occuring during the densification of a monolithic gel. It is usually determined by measuring the reduction in the length per unit length of a gel sample heated at a constant rate, and therefore, only accounts for the change in the outer dimension of a gel. The measurement of the linear shrinkage is, however, much easier than the measurement of bulk density and provides information on the temperature range where the rapid shrinkage of a gel due to micropore collapse begins.[13,17,18,78,81] The curves in Figure 8 show the linear shrinkage of gels prepared from various precursor solutions. It is again obvious that gels containing only small pores such as HCl catalyzed gels begin to shrink at a lower temperature than those containing large pores like ammonia catalyzed gels.

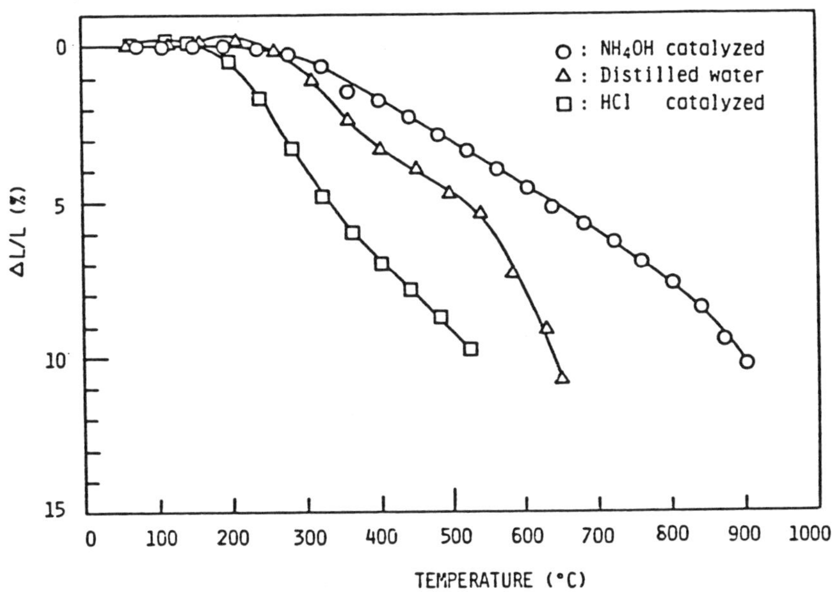

Figure 8: Linear shrinkage of gels prepared from various precursor solutions.

IR and Raman Spectra

The structure of the skeleton in an alkoxy-derived silica gel is very close to that of fused silica as is expected from the similarity of true density.[11,83] It primarily consists of a three dimensional random network of SiO_4 tetrahedra along with a large amount of SiOH covering the surface. The IR spectrum of a silica gel therefore has an absorption band due to the Si–O bonds with non-bridging oxygen, i.e., the bonds in SiOH groups. This absorption appears in practice in the vicinity of 960 cm^{-1} as the shoulder of the intense absorption peak around 1080 cm^{-1} which is assigned to the stretching vibration of the Si–O bonds without non-bridging oxygens.[45,83]

The IR spectra of gels before and after the heat treatment up to various temperatures are shown in Figure 9.[11] In this figure it is shown that the absorption at 960 cm^{-1} decreases as the temperature is raised. At the same time, the absorption peak around 1080 cm^{-1} shifts towards a higher frequency. The decrease in the absorption at 960 cm^{-1} is attributed to the decrease in the amount of SiOH due to the dehydration condensation of silanols. The shift in the absorption peak around 1080 cm^{-1} towards a higher frequency suggests that the Si–O bonds in the tetrahedra are strengthened as the crosslinking develops with temperature.

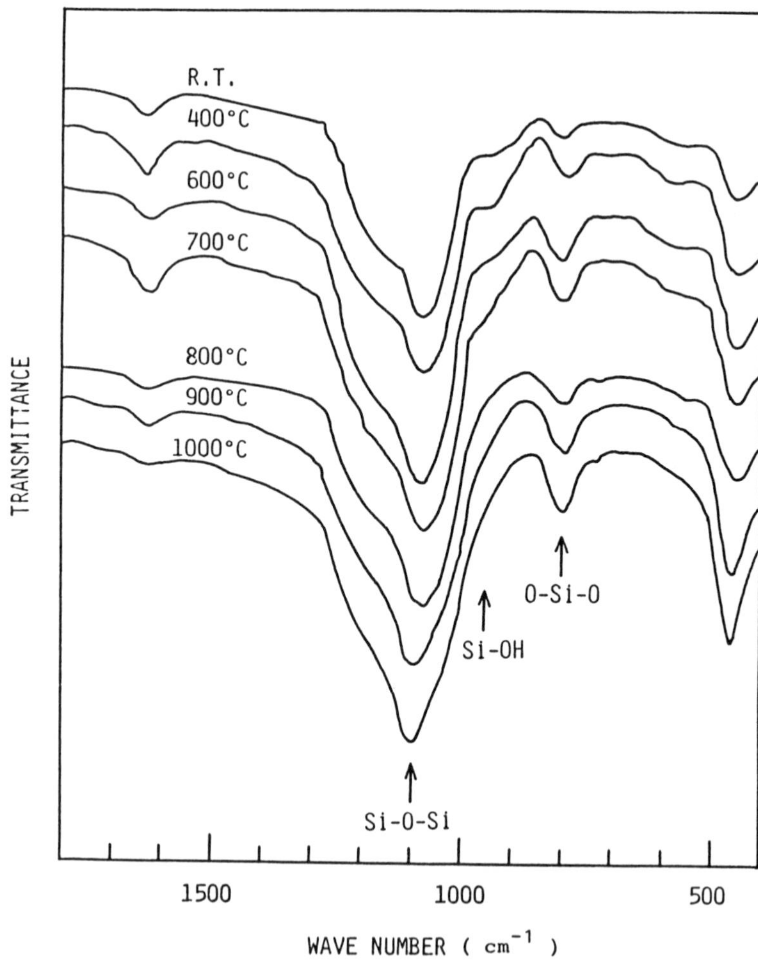

Figure 9: Change in IR spectra of silica gels with heat treatment (Reference 11).

It should be noted that the absorption at 960 cm^{-1} almost disappears at around 800°C, showing that most of the silanols have been spent for dehydration condensation by this temperature. 800°C is slightly higher but very close to the temperature where the true density of the gel levels-off. Thus, the structure of a gel becomes similar to that of fused silica at around 800°C, regardless of the existence of micropores.

According to Gallo et al,[72] the surface hydroxyl coverage in terms of number of hydroxyls per unit surface area of the gels, OH/nm^2, is about 3 at around 300°C in the gels of the borosilicate system. It decreases monotonically to about 0.7 when the gel is heated linearly at 2°C/min up to about 540°C. A similar trend of the decrease in OH coverage with heat treatment is reported by Krol and van Lierop[19] on SiO_2 gels heated at 4°C/min. The OH coverage in terms of the relative intensity of the Raman band at 3750 cm^{-1} to that of 800 cm^{-1} divided by the specific surface area decreases from 16 for a gel desiccated at 120°C and having a surface area of 790 m^2/g to 1.2 for a gel heated up to 700°C and having a surface area of 550 m^2/g. Silanols covering the skeletal surface decreased to less than half of the value at 700°C by further heating to 800°C. It turns out, however, that the SiOH groups in the glass network become appreciable instead of the SiOH groups on the skeletal surface. This suggests that the pores of the gel are already collapsed while the condensation is still proceeding. The water that is still formed in the condensation reaction cannot escape because of this collapse and the gel bloats. Therefore, treatment to allow the escape of the generated water before the extensive collapse of pores is necessary.

Change in Pore Size Distribution

Change with Linear Heating Rate: The effect of temperature on micropore collapse is different from gel to gel depending on the distribution of the pores in the original gel. If a gel contains only small pores, the total collapse of the pores is attained even at 700°C, which is well below the glass transition temperature of fused silica, i.e., about 1200°C. On the contrary, if a gel contains large pores, it remains porous even above 1200°C.

Examples of the influence of pore size distribution on the micropore collapse are shown in Figure 10 A through D.[11] All the gels were pretreated at 250° and 400°C for the purpose of removing physically adsorbed water on the micropore walls and residual organic compounds, respectively, and then were heated at a rate of 1°C/min up to the temperatures shown in the figure.

The curves in Figure 10A illustrate the pore size distributions at successive heating stages of gels which contain mostly small pores with a diameter of less than 50 Å. The pores begin to shrink by heating above 400°C. A drastic change in the pore size distribution occurs at a temperature around 700°C due to micropore collapse. Gels with this pore size distribution are subject to large stress due to the internal pressure caused by water vapor accumulating in the pores which is generated by the dehydration condensation of silanols. Therefore, the gel fractures very easily and disintegrates into small pieces before the temperature of total collapse is reached. Thus, it is very difficult to sinter gels with this pore size distribution into dense monolithic glasses by a simple heating.

Figure 10B shows the change in the pore size distribution in gels having slightly larger pores than those in Figure 10A, i.e., the gels containing the pores of 50 to 70 Å as well. In this case, the main change in the distribution of pore size observed during the heating up to 800°C is the decrease in the

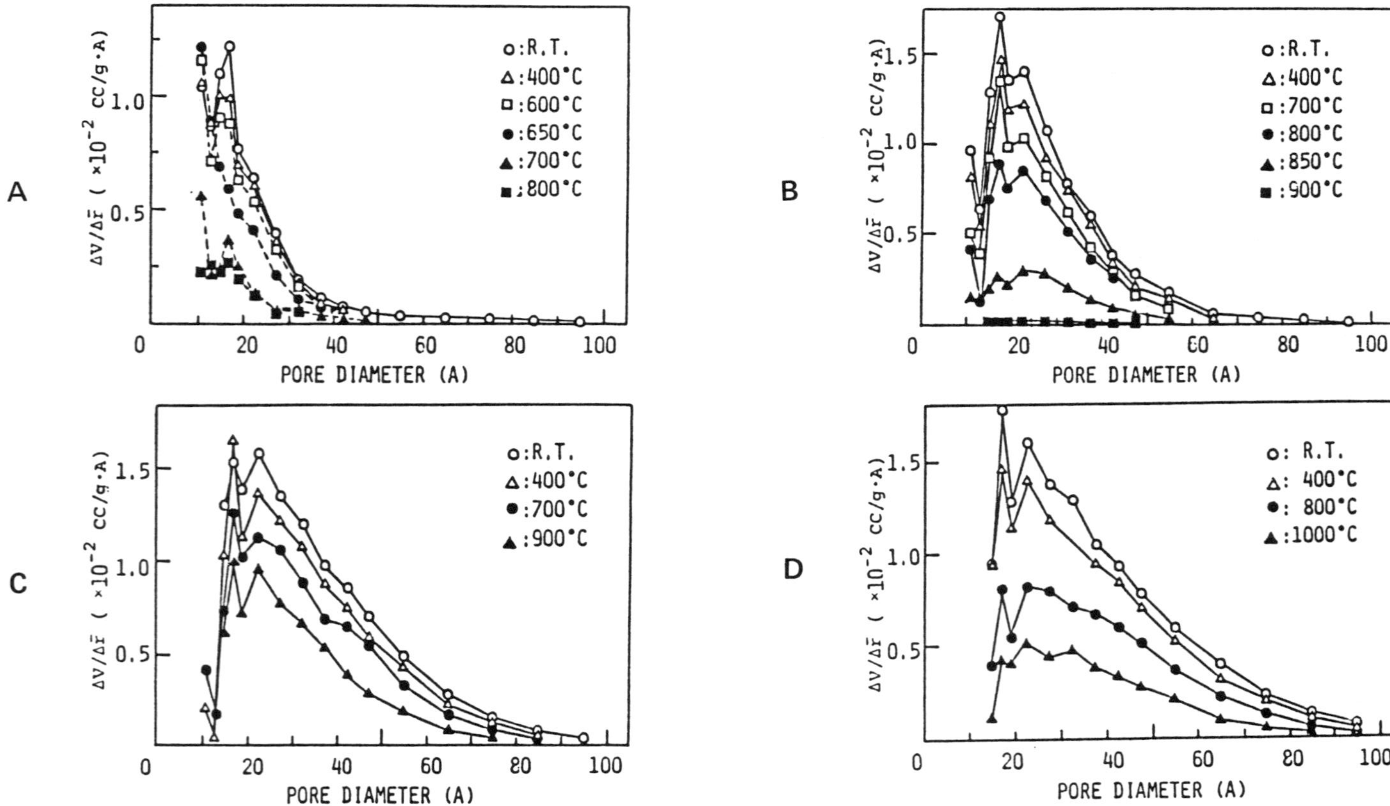

Figure 10: Change in the pore size distribution in various gels with heat treatment: (A) gelled at 54°C; (B) 65°C; (C) 68°C; and (D) 70°C (Reference 11).

number of pores with a diameter of 20 to 30 Å. These gels contain continuous open pores at 800°C. They lose a considerable number of open pores with an increase in temperature of another 50°C, and finally sinter into glasses free from open pores when heated up to 900°C.

Gels with this pore size distribution remain unfractured all the way up to 900°C. They do, however, contain a cloudy center. This is due to the formation of small bubbles of trapped water generated by dehydration condensation of silanols which failed to get out of the system before total collapse of the pores had occurred. These gels, therefore, will bloat when heated above 900°C where the viscosity of the gel-derived glass is low enough to allow the bloating by the stress generated at the inside due to the expansion of the trapped water.

The changes during heat treatment in the pore size distribution of gels containing pores as large as 80 to 100 Å in diameter are shown in Figure 10C and Figure 10D. In these cases, the trend of the changes in pore size distributions are similar to that of the gels shown in Figure 10B. But the temperatures where the drastic shrinkage of the pores begins are much higher than 800°C. Most of the pores remain uncollapsed even at 900°C. These gels turn to transparent monolithic glasses free from bubbles when heated isothermally at a temperature around 1000°C.

Change Under Isothermal Treatment: As is shown by the examples above, pores with a diameter around 80 Å or above remain uncollapsed up to about 800°C where the dehydration condensation of silanols is nearly completed. Though the open pores serve as the path for the generated water to diffuse towards the outer surface, the accumulation of water in the gel is hard to avoid if a gel is simply heated at a constant rate. The accumulation of water causes the gel to densify non-uniformly due to the presence of a gradient in water content throughout the gel. This leads to bloating as the gel is heated further.

One of the ways of preventing the accumulation of water in monoliths is to hold the gel isothermally at a temperature slightly below the temperature of total collapse of the pores, so that there is plenty of time for the water to leave the system. Isothermal treatment at a high temperature also promotes the structural relaxation of a gel. Reduction of water content and structural relaxation both increase the viscosity which, in turn, governs the micropore collapse.

According to Scherer,[75] the sintering of porous glass can be described by a model in which the glass skeleton is assumed to be a cubic array of intersecting cylinders. The pore diameter, D, is related to the cylinder radius, a, and the length, l, by Equation (6):

$$\frac{\pi D^2}{4} \approx (l - 2a)^2 \tag{6}$$

In the model it is assumed that the energy dissipated by viscous flow is equal to the energy change resulting from the reduction of surface area, which is in agreement with Frenkel theory. The rate of energy dissipation by viscous flow, $\dot{E}_f$, as a cylinder decreases in height is given by Equation (7):

$$\dot{E}_f = \frac{3\pi\eta r^2}{h}\left(\frac{dh}{dt}\right)^2 \tag{7}$$

where r and h are the radius and height of the cylinder, respectively, and η is the viscosity. The energy supplied by the reduction of surface area is given by Equation (8):

$$\mathring{E}_s = \gamma\left(\frac{dS_c}{dt}\right) \tag{8}$$

where γ is surface tension and S_c is the surface area of a single full cylinder. Finally, the change in the ratio of radius to length of a cylinder, x = a/l, with time is given by the Equation (9):

$$\frac{dx}{dt} = \left(\frac{\gamma}{2\eta}\right)\left(\frac{1}{l}\right) \tag{9}$$

Thus, the rate of increase in a/l, which is directly related to the rate of pore diameter decrease, is inversely proportional to the viscosity. Therefore, the increase in viscosity during isothermal treatment due to structural relaxation and the reduction of water content reduces the rate of decrease in pore diameter, which allows the gel to remain uncollapsed for a relatively long time.

Figure 11 shows the very slow change in the pore size distribution in gels prepared under the same conditions as the gels in Figure 10B during isothermal treatment at 750°C.[11] The gels which originally contained pores with diameters up to 70 Å remain porous even after holding at 750°C for 24 hours. These samples would become almost pore free within 100 minutes if heated at a rate of 1°C/min from 750°C. Thus, the isothermal treatment at around 750°C is desirable, considering the change in Raman spectra. Also, large pore gels again seem more desirable than small pore gels.

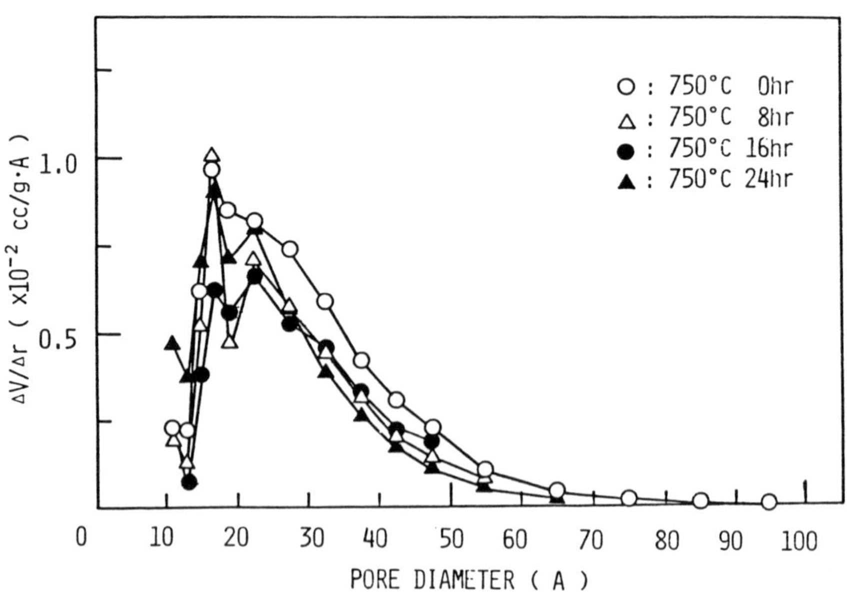

Figure 11: Change in the pore size distribution in a gel by isothermal treatment at 750°C (Gel B in Figure 2) (Reference 11).

HEAT CYCLE FOR THE DENSIFICATION OF AN ALKOXY-DERIVED MONOLITHIC GEL

As discussed above, there are several stages where the fracture or bloating of a gel can occur in the densification process of an alkoxy-derived silica gel for monolith formation due to the stress build-up accompanied by various reactions. Most of the problems are related to the pore size in the gel. Gels containing extremely large pores such as those prepared by a hypercritical solvent evacuation technique are probably immune to serious stress build up. Gels prepared without such a special technique, however, do not contain extremely large pores, and are always subject to fracture problems.

In order to obtain fracture-free monolithic glass from relatively small pore gels it is necessary to take special care in the temperature ranges where stress build-up in the gels is possible. The heating of a gel at a very slow rate is one way of avoiding fracture. A more effective technique is to hold the gel isothermally at the appropriate temperature to allow the reactions to complete before going ahead to the next step. One of the examples of heat cycles proposed for this purpose is shown in Figure 12.[11] The scale of the time axis is obviously dependent on the pore size distribution in the gel and the size of the gel itself, which determines the diffusion path length for the evolved gas during the heat treatment. The smaller the pore size and the larger the gel, the longer is the time necessary for each successive stage.

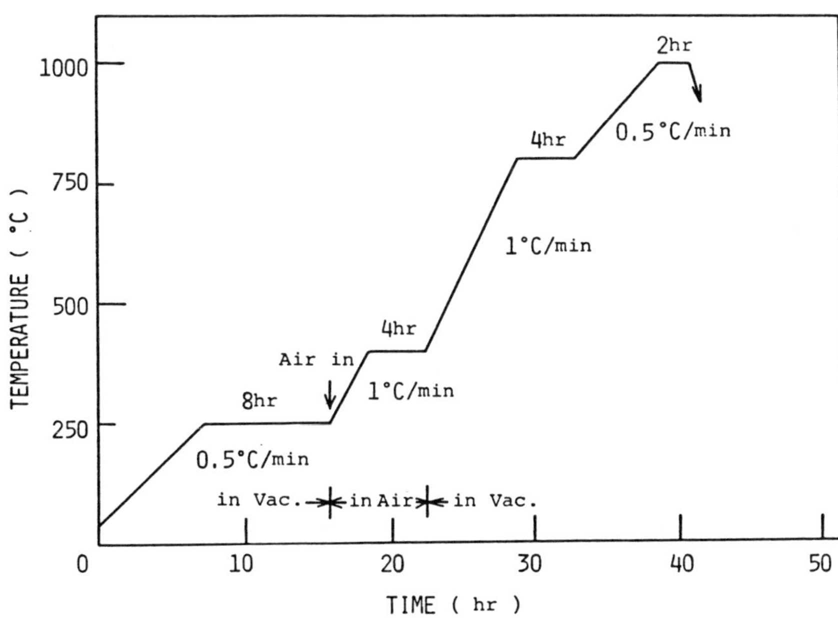

Figure 12: Heat cycle used for the densification of alkoxy-derived silica gels (Reference 11).

The first plateau at a temperature around 200°C is to ensure that the physically adsorbed water on the micropore walls leaves the system. It will be more efficient if the atmospheric pressure in the reaction tube is reduced by pulling a vacuum or if dry air is passed through the reaction tube during the isothermal treatment.

The second plateau at around 400°C is to allow the complete decomposition of the residual organic compounds before the micropore collapse begins. The introduction of oxygen gas or air at this temperature range is obviously advantageous.

The purpose of the isothermal treatment at a temperature around 700°C is to allow the water vapor generated by the dehydration condensation of silanols to diffuse out of the pores as was mentioned in the last two sections. Again the pulling of a vacuum or the introduction of dry helium in the reaction tube is effective. This may be started after the complete decomposition of organic compounds and be continued all the way through the final isothermal treatment for sintering at a temperature at or above 1000°C.

Monolithic silica glasses prepared using this heat cycle have properties similar to commercial fused silica. The glass, however, contains about 1,000 ppm of water, which is an order of magnitude more water than is contained in fused silica. This value is acceptable for some purposes. But in many cases, a glass of much lower water content is required.

For the purpose of reducing silanols, chlorine gas can be used together with helium while the temperature is between 400°C and the pore collapse point. Oxygen is then fed in to remove the chlorine. The water content in a glass prepared by this method is of the order of several ppm, and such a glass can be used as fiber for optical communications since it has a loss of only about 5 dB/km.[12]

REFERENCES

1. Mackenzie, J.D., *J. Non-Cryst. Solids* 48: 1-10 (1982).
2. Mackenzie, J.D., in: *Ultrastructure Process of Ceramics, Glasses, and Composits* (L.L. Hench and D.R. Ulrich, ed.), pp 15–26, John Wiley and Sons, New York (1984).
3. Dislich, H., *J. Non-Cryst. Solids* 57: 371–388 (1983).
4. Sakka, S., and Kamiya, K., *J. Non-Cryst. Solids* 42: 403–422 (1980).
5. Sakka, S., in: *Treatise on Materials Science and Technology* (M. Tomozawa and R.H. Doremus ed.), Vol. 22, Glass III, pp 129-167, Academic Press (1982).
6. Mukherjee, S.P., *J. Non-Cryst. Solids* 42: 477–488 (1980).
7. Dislich, H., *Angew. Chem. Int. Ed, Engle.* 10: 363–70 (1971).
8. Decottignies, M., Phalippou, J., and Zarzycki, J., *J. Mater. Sci.* 13: 2605–18 (1978).
9. Mukherjee, S.P., and Zarzycki, J., *J. Am. Ceram. Soc.* 62: 1–4 (1979).
10. Jabra, R., Phalippou, J., and Zarzycki, J., *J. Non-Cryst. Solids* 42: 489–498 (1980).
11. Yamane, M., and Okano, S., *Yogyo-Kyokai-Shi* 87: 434-438 (1979).
12. Susa, K., Matsuyama, I., Satoh, S., and Suganuma, T., *Elect. Lett.* 18: 499–500 (1982).

13. Klein, L.C., Gallo, T.A., and Gavey, G.J., *J. Non-Cryst. Solids* 63: 23-33 (1984).
14. Zarzycki, J., Prassas, M., and Phalippou, J., *J. Mater. Sci.* 17: 3371-3379 (1982).
15. Prassas, M., Phalippou, J., and Zarzycki, J., *J. Mater. Sci.* 19: 1656-1665 (1984).
16. Zarzycki, J., Prassas, M., and Philippou, J., U.S. Patent 4,432,956; Feb. 21, 1984; assigned to Corning France, Avon France.
17. Klein, L.C., *Ann. Rev. Mater. Sci.* 15: 227-48 (1985).
18. Kawaguchi, T., Hishikura, H., Iura, J., and Kokubu, Y., *J. Non-Cryst. Solids* 63: 61-69 (1984).
19. Krol, D.M., and VanLierop, J.G., *J. Non-Cryst. Solids* 63: 131-144 (1984).
20. Nogami, M., and Moriya, Y., *J. Non-Cryst. Solids* 37: 191-201 (1980).
21. Yamane, M., Aso, S., Okano, S., and Sakaino, T., *J. Mater. Sci.* 14: 607-611 (1979).
22. Matsuyama, I., Susa, K., Satoh, S., and Suganuma, T., *Ceramic Bull.* 63: 1408-1411 (1984).
23. Kawaguchi, T., Iura, J., Taneda, N., Hishikura, H., and Kokubu, Y., Structural Change of Monolithic Silica Gel During the Gel-to-Glass Transition, to be published in *J. Non-Cryst. Solids.*
24. Woignier, T., Phalippou, J., and Zarzycki, J., *J. Non-Cryst. Solids* 63: 117-130 (1984).
25. Kamiya, K., and Sakka, S., *J. Mater. Sci.* 15: 2937-39 (1980).
26. Yoldas, B.E., *J. Mater. Sci.* 12: 1203-1208 (1977).
27. Hayashi, T., and Saito, H., *J. Mater. Sci.* 15: 1971-77 (1980).
28. Yoldas, B.E., *J. Mater. Sci.* 14: 1843-1849 (1979).
29. Brinker, C.J., and Mukherjee, S.P., *J. Mater. Sci.* 16: 1980-88 (1981).
30. Nogami, M., and Moriya, Y., *J. Non-Cryst. Solids* 48: 359-366 (1982).
31. Gonzalez-Oliver, C.J.R., Alkoxide Low Temperature Synthesis of Complex Silicate Glasses in Monolithic Form, to be published in *J. Non-Cryst. Solids.*
32. Gonzalez-Oliver, C.J.R., James, P.F., and Rawson, H., *J. Non-Cryst. Solids* 48: 129-152 (1982).
33. Scherer, C.P., and Pantano, G.C., Titania-Silica Glasses Using Colloidal Sol-Gel Process, to be published in *J. Non-Cryst. Solids.*
34. Soup, R.D., in: *Colloid and Interface Science* (Milton Kerker ed.), Vol. 3, pp 63-69, Academic Press (1976).
35. Rabinovich, E.M., Johnson, D.W., Jr., MacChesney, J.B., and Vogel, E.M., *J. Non-Cryst. Solids* 47: 435-439 (1982).
36. Rabinovich, E.M., Johnson, D.W., Jr., MacChesney, J.B., and Vogel, E.M., *J. Am. Ceram. Soc.* 66: 683-688 (1983).
37. Johnson, D.W., Jr. et al, *J. Am. Ceram. Soc.* 66: 688-693 (1983).
38. Wood, D.L., et al, *J. Am. Ceram. Soc.* 66: 693-699 (1983).
39. Rabinovich, E.M., MacChesney, J.B., Johnson, D.W., Jr., Simpson, J.R., Meagher, B.W., Dimarcello, F.V., Wood, D.L., and Sigety, E.A., *J. Non-Cryst. Solids* 63: 155-161 (1984).
40. Scherer, G.W., and Loung, J.C., *J. Non-Cryst. Solids* 63: 163-172 (1984).
41. Scherer, G.W., *Better Ceramics Through Chemistry* (C.J. Brinker, D.E. Clark and D.R. Ulrich ed.), pp 205-212, North-Holland (1984).
42. Rabinovich, E.M., *J. Non-Cryst. Solids* 71: 187-193 (1985).

43. Aelion, R., Loebel, A., and Eirich, F., *J. Am. Chem. Soc.* 72: 5705-5712 (1950).
44. Strawbridge, I., Craievich, A.F., and James, P.F., *J. Non-Cryst. Solids* 72: 139-157 (1985).
45. Bertoluzza, A., Fagnano, C., Morelli, M.A., Gottardi, V., and Guglielmi, M., *J. Non-Cryst. Solids* 48: 117-128 (1982).
46. Gottardi, V., Guglielmi, M., Bertoluzza, A., Fagnano, C., and Morelli, M.A., *J. Non-Cryst Solids* 63: 71-81 (1984).
47. Klein, L.C., and Garvey, G.J., Effect of Water on Acid and Base Catalyzed Hydrolysis of Tetraethylorthosilicate (TEOS) in *Better Ceramics Through Chemistry* (C.J. Brinker, D.E. Clark and D.R. Ulrich ed.), pp 33-40, North-Holland (1984).
48. Yoldas, B.E., *J. Non-Cryst Solids* 51: 105-121 (1982).
49. Miyashita, S., Kanbe, S., and Toki, M., Japanese Patent Sho-58-181735; October 24, 1983; assigned to Suwaseikosha Co. Ltd.
50. Klein, L.C., and Garvey, G.J., *J. Non-Cryst. Solids* 48: 97-104 (1982).
51. Miyashita, S., Kanbe, S., and Toki, M., Japanese Patent Sho-59-8625; Jan. 17, 1984; assigned to Suwaseikosha Co. Ltd.
52. Klein, L.C., and Garvey, G.J., *J. Non-Cryst. Solids* 38 & 39: 49-50 (1980).
53. Sakka, S., and Kamiya, K., *J. Non-Cryst. Solids* 48: 31-46 (1982).
54. Brinker, C.J., st al, *J. Non-Cryst. Solids* 48: 47-64 (1982).
55. Brinker, C.J., et al, *J. Non-Cryst. Solids* 63: 45-69 (1984).
56. Yamane, M., Inoue, S., and Yasumori, A., *J. Non-Cryst. Solids* 63: 13-21 (1984).
57. Keefer, K.D., The Effect of Hydrolysis Conditions on the Structure and Growth of Silicate Polymer in *Better Ceramics Through Chemistry* (C.J. Brinker, D.E. Clark and D.R. Ulrich ed.), pp 15-24, North-Holland (1984).
58. Schmidt, H., Scholze, H., and Kaiser, A., *J. Non-Cryst. Solids* 63: 1-11 (1984).
59. Brinker, C.J., and Scherer, G.W., *J. Non-Cryst. Solids* 70: 301-322 (1985).
60. Wallace, S., and Hench, L.L., The Processing and Characterization of DCCA Modified Gel-Derived Silica in *Better Ceramics Through Chemistry* (C.J. Brinker, D.E. Clark and D.R. Ulrich ed.), pp 47-52, North-Holland (1984).
61. Yamane, M., Aso, S., and Sakaino, T., *J. Mater. Sci.* 13: 865-870 (1978).
62. Yamane, M., *Yogyo-Kyokai-Shi* 88: 589-594 (1980).
63. Colby, M., Osaka, A., and Mackenzie, J.D., Effects of Temperature on Formation of Silica Gel to be published in *J. Non-Cryst. Solids.*
64. Kanbe, S., Toki, M., and Miyashita, S., Japanese Patent Sho-58-199735; November 21, 1983; assigned to Suwaseikosha Co. Ltd.
65. Wang, S.H. and Hench, L.L., Processing and Properties of Sol-Gel Derived 20 mol % Na_2O–80 mol % SiO_2 (2ON) Materials in *Better Ceramics Through Chemistry* (C.J. Brinker, D.E. Clark and D.R. Ulrich ed.), pp 71-77, North-Holland (1984).
66. Orcel, G., and Hench, L.L., Physical-Chemical Variables in Processing Na_2O–B_2O_3–SiO_2 Gel Monoliths in *Better Ceramics Through Chemistry* (C.J. Brinker, D.E. Clark and D.R. Ulrich ed.), pp 79-84, North-Holland (1984).

67. Van Lierop, J.G., Huizing, A., Meerman, W.C.P.M., and Mulder, C.A.M., Preparation of Dried Monolithic SiO_2 Gel Bodies by an Autoclave Process to be published in *J. Non-Cryst. Solids.*
68. Mulder, C.A.M., van Leeuwen-Stienstra, G., van Lierop, J.G., and Woerdman, J.P., Chain-Like Structure of Ultra-Low Density SiO_2 Sol-Gel Glass Observed by TEM to be published in *J. Non-Cryst. Solids.*
69. Mulder, C.A.M., van Lierop, J.G., and Frens, G., Densification of SiO_2-Xerogels to Glass by Ostwald Ripening to be published in *J. Non-Cryst. Solids.*
70. Mulder, C.A.M., and van Lierop, J.G., Preparation, Densification and Characterization of Autoclave Dried SiO_2 Gels Proc. 1st Internat. Symp. on Aerogel, September 23-25, 1985, Wurzburg/West Germany.
71. Brinker, C.J., Ward, K.J., Keefer, K.D., Holupka, E., Bray, P.J., and Pearson, R.K., Synthesis and Structure of Borate Based Aerogels. 1st Internat. Symp. on Aerogel, September 23-25, 1985, Wurzburg/West Germany.
72. Gallo, T.A., Brinker, C.J., Klein, L.C., and Scherer, G.W., The Role of Water in Densification of Gels in *Better Ceramics Through Chemistry* (C.J. Brinker, D.E. Clark and D.R. Ulrich ed.), pp 85-90, North-Holland (1984).
73. Brinker, C.J., Tallant, D.R., and Roth, E.P., Sol-Gel Transition in Simple Silicates, III to be published in *J. Non-Cryst. Solids.*
74. Krol, D.M., van Lierop, J.G., *J. Non-Cryst. Solids* 68: 163-166 (1984).
75. Scherer, G.W., *J. Am. Ceram. Soc.* 60: 236-239 (1977).
76. Zarzycki, J., *J. Non-Cryst. Solids* 48: 105-116 (1982).
77. Phalippou, J., Woignier, T., and Zarzycki, J., in *Urtrastructure Process of Ceramics, Glasses, and Composits* (L.L. Hench and D.R. Ulrich, ed), pp 70-87, John Wiley and Sons, New York (1984).
78. Brinker, C.J., Scherer, G.W., and Roth, E.P., *J. Non-Cryst. Solids* 72: 345-368 (1985).
79. Scherer, G.W., Brinker, C.J., and Roth, E.P., *J. Non-Cryst. Solids* 72: 369-389 (1985).
80. Brinker, C.J., Roth, E.P., Scherer, G.W., and Tallant, D.R., *J. Non-Cryst. Solids* 71: 171-185 (1985).
81. Brinker, C.J., Droting, W.D., and Scherer, G.W., A Comparison Between the Densification Kinetics of Colloidal and Polymeric Silica Gels in *Better Ceramics Through Chemistry* (C.J. Brinker, D.E. Clark and D.R. Ulrich ed), pp 25-32, North-Holland (1984).
82. Gallo, T., and Klein, L.C., Apparent Viscosity of Sol-Gel Processed Silica to be published in *J. Non-Cryst. Solids.*
83. Tohge, N., Moore, G.S., and Mackenzie, J.D., *J. Non-Cryst. Solids* 63: 95-103 (1984).

11

Thermal Insulation Materials from the Sol-Gel Process

Jochen Fricke
Physikalisches Institut der Universität Am Hubland
Würzburg, West Germany

INTRODUCTION

Shortly after Kistler succeeded in preparing noncollapsed silica aerogel[1] by supercritical drying of an alcogel he started to investigate the thermal transport within the fascinating porous material. Over a period of about ten years he studied the thermal conductivity of evacuated and air-filled granular and monolithic aerogel samples.[2,3,4] Kistler's pioneer work showed for example that:

- aerogel systems have thermal conductivities around 0.02 W/(m·K) at normal air pressure and around 0.01 W/(m·K) if evacuated;
- monolithic aerogel layers have smaller conductivities than granular fillings at ambient pressures, while at low pressures the opposite is true;
- a pore size diameter in the order of 50 nm results from the dependence of thermal conductivity on internal gas pressure.

Aerogel research later on centered around the shortening of Kistler's time-consuming preparation technique[5] and improvement of the optical transparency. The crucial idea which promoted the development of high-quality transparent aerogel tiles was to use this low-density material in Cerencov detectors in high energy physics.[6] Two groups, one at DESY/Hamburg[7,8] and the other at the University of Lund/Sweden[9] succeeded in producing in the order of 10^3 liters of clear crack-free aerogel tiles for the TASSO detector and a CERN experiment, respectively. Both teams diluted tetramethoxysilane (TMOS) with alcohol in order to get tiles with a low enough density and index of refraction. With ammonia as a catalyst and by adding a controlled amount of water the TMOS was hydrolyzed

$$Si(OCH_3)_4 + 4\ H_2O \rightarrow Si(OH)_4 + 4\ CH_3OH.$$

The water produced in the condensation reaction

$$n\ Si(OH)_4 \rightarrow \{SiO_2\}_n + 2n\ H_2O$$

was washed out with methanol. Subsequent supercritical drying with respect to methanol (T_k = 240°C and p_k = 78.5 bar) allowed the small scale production of clear 20 x 20 cm^2 aerogel tiles.

It was only then–about 50 years after Kistler's initial success–that the use of aerogel as a transparent superinsulating spacer in solar collectors and window systems was proposed in a patent specification.[10] A large scale production facility with a 300 liter autoclave was set up at Sjöbo/Sweden in which even 60 x 60 cm^2 tiles could be made.[11] However, in August 1984 the main gasket of the autoclave failed. More than 1,000 liters of methanol were ejected into the building. The explosion which followed destroyed the whole facility. After one year of police investigation, doubt and lack of money the fascination about aerogel obviously won: The Swedish Team is pursuing a new start.[11]

Other groups[12,13] are following a less dangerous production route: Instead of supercritically drying an alcogel they employ a method commonly used by microbiologists, when, for microscopy purposes, water in biological systems is to be replaced by supercritical CO_2 (T_k = 31.1°C and p_k = 73.9 bar). As water or alcohol are immiscible with CO_2, an intermediate solvent has to be used, e.g. amyl acetate. The amyl acetate-gel is placed into an autoclave, where liquid CO_2 replaces the amyl acetate. After several flushings, supercritical drying with respect to CO_2 is performed. At Lawrence Berkeley Laboratory, for example, the process is used to produce base catalyzed monolithic aerogel from tetraethoxysilane (TEOS) in the presence of fluoride ions with excellent optical transparency.[13]

Another production method[14] is pursued by BASF/Ludwigshafen (West Germany). A pilot production facility allows the production of aerogel granules with diameters of about 1 to 8 mm. Sodium silicate (water glass) is reacted with sulfuric acid. The starting materials are sprayed from a mixing jet and are transformed into spherical hydrogels in a receiver flask. The alkali salts formed as by-products in the sol-gel process are removed by after-treatment and washing. By varying the after-treatment the pore structure can be changed. Hypocritically dried silicic acid hydrogels are also used for the preparation of catalysts and of drying agents.

Both forms, the translucent granular as well as the transparent monolithic aerogels open up new possibilities for the improvement of window systems and of covers for passive solar energy usage. If evacuated, aerogel layers of 15 to 20 mm have loss coefficients as low as 0.5 W/$m^2 \cdot$K). Thermal losses thus would be down by a factor of 6 compared to normal double pane windows and a factor of 3 compared to low emissivity double pane windows. Aerogel superinsulations may therefore contribute towards:

- the reduction of heat losses through windows - see Figure 1;[15]

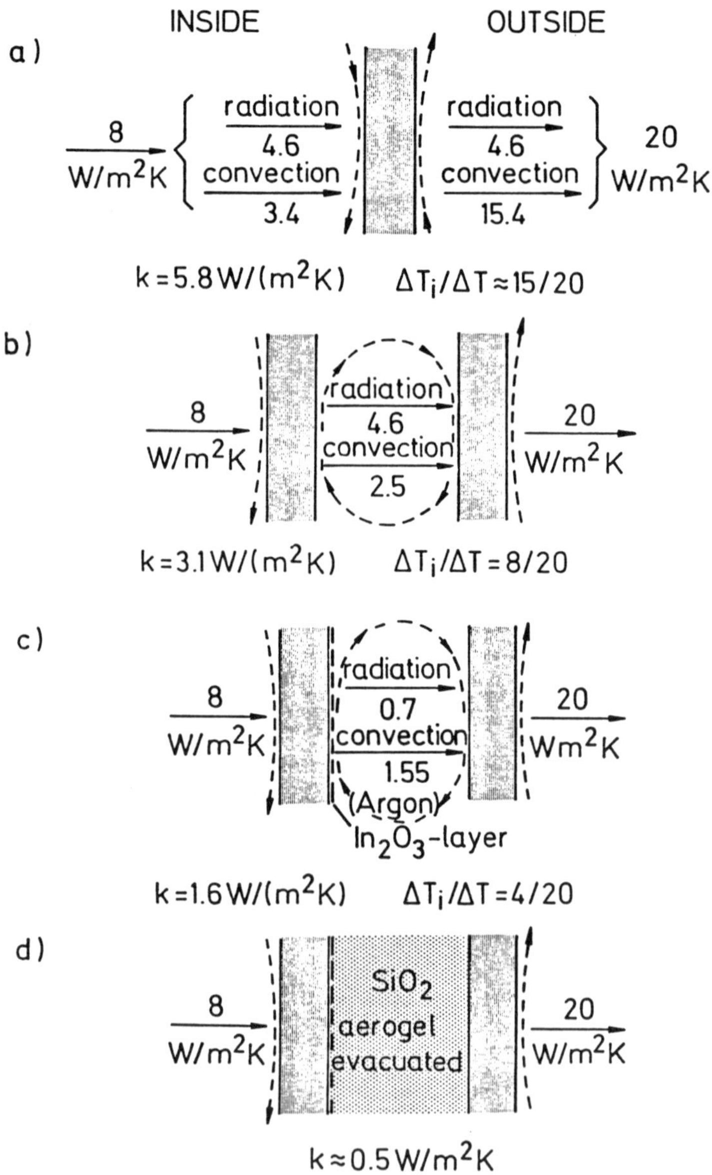

Figure 1: Improvement in window systems: (a) Single pane window, (b) double pane window, (c) double pane window with argon and low emissivity layer, (d) double pane window with argon and low emissivity layer evacuated with aerogel spacer.[15]

- direct passive use of solar energy by installing east-, south- or west-oriented glass facades without the large thermal losses normally coupled with an increase of window area;
- the energy-effective operation of greenhouses;
- the construction of translucent insulations for house walls or modified Trombe wall systems.[16] In such systems the massive wall serves as a solar collector, which is covered by a superinsulating translucent layer (Figure 2). The better the aerogel insulation is compared to the wall insulation, the more heat will be available for heating purposes;

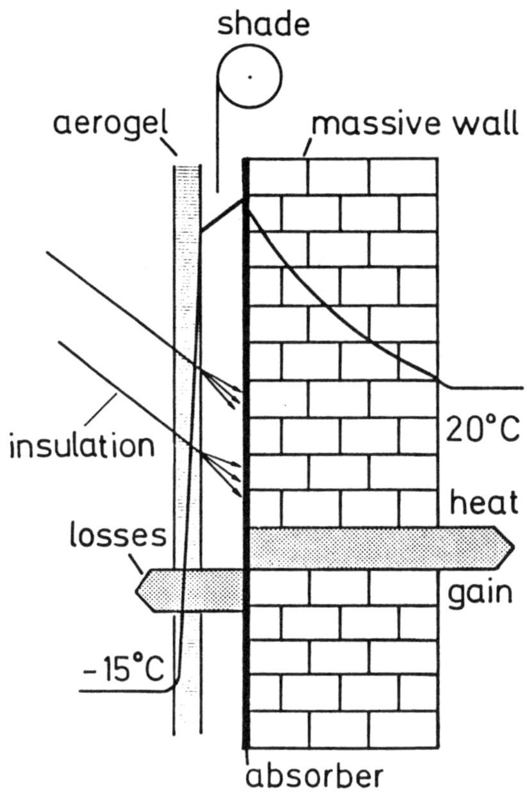

Figure 2: Wall with translucent thermal insulation. The absorbed solar radiation is partly available for room heating.[16]

- the installation of aerogel-superinsulated solar ponds for collection and long-term storage of solar energy (Figure 3). The usual burdensome use of salt gradients in solar ponds thus could be avoided.

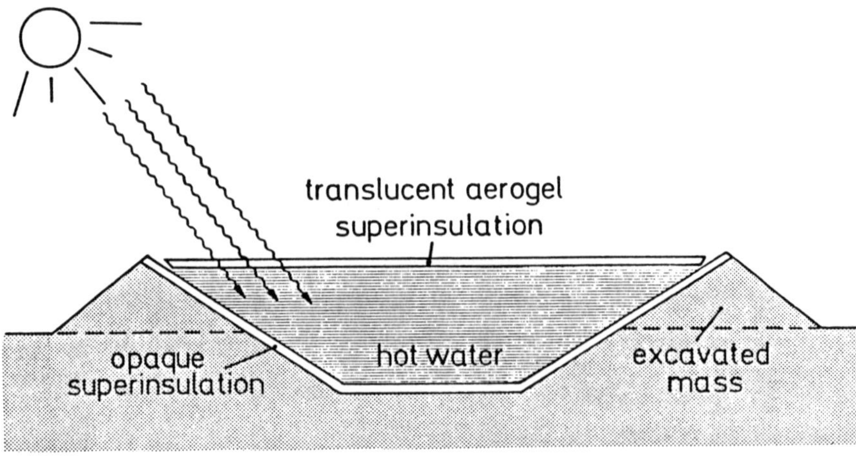

Figure 3: Solar pond with superinsulating translucent cover (no salt has to be used) for collection and long-term storage of solar energy.[15]

THERMAL TRANSPORT IN EVACUATED POROUS SUPERINSULATIONS

Radiative Transport

The radiative heat flux q_r between two parallel walls (spacing evacuated, distance d) of emissivity ϵ and temperatures T_1 and T_2, respectively, is given by

$$q_r = \{\sigma / (2/\varepsilon - 1)\} (T_1^4 - T_2^4) \tag{1}$$

$$= 4\,\sigma' (1/4) (T_1^2 + T_2^2) (T_1 + T_2) (T_1 - T_2) = k_r (T_1 - T_2) ,$$

with

$$k_r = 4\,\sigma' T_r^3 , \tag{2}$$

$$\sigma' = \sigma / (2/\varepsilon - 1) , \tag{3}$$

$$T_r^3 = (T_1^2 + T_2^2) (T_1 + T_2) / 4 . \tag{4}$$

The radiative loss coefficient k_r is proportional to the third power of the radiative temperature T_r, given by Equation 4. Instead of the loss coefficient often the pseudoconductivity $\lambda = k{\cdot}d$ ("pseudo," as it varies with d and c) is used:

$$\lambda_r = 4\,d\,\sigma' T_r^3 , \tag{5}$$

which is dependent on d. In order to reduce the radiative losses, ϵ has to be made small. In addition, in cryogenic insulations with heavy pressure supporting steel walls a large number N of foils is inserted into the spacing (Figure 4). This reduces the radiative flux by a factor (N + 1). The temperature profile in this case is given by

$$\lim_{N\to\infty} T(x) = T_1 \{ 1 - (x/d) [1 - (T_2/T_1)^4] \}^{1/4} , \tag{6}$$

where x is the coordinate vertical to the walls. The temperature is only defined on the foils themselves, not within the spacings.

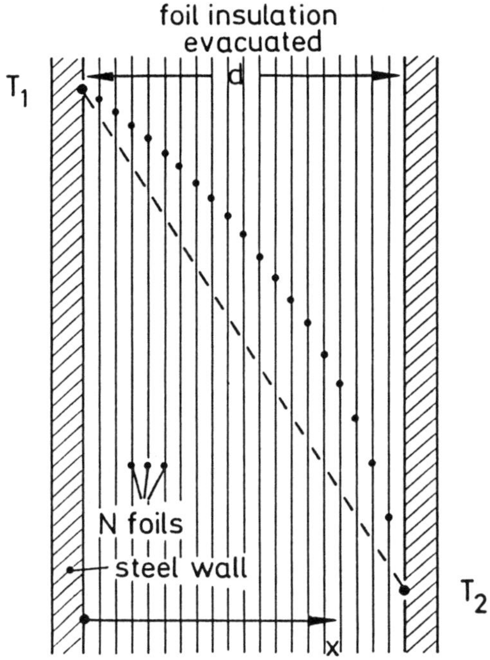

Figure 4: Evacuated foil insulations (N foils) with temperature profile. The radiative losses are reduced by a factor of (N + 1). The heavy walls have to sustain the outer atmospheric load.

For many applications, evacuated porous insulations are sealed into a thin steel or plastic envelope and thus have to sustain the outer atmospheric load. This would also apply for aerogel-filled evacuated window systems. The heat then is transported from the hot to the cold side by infrared radiation and by solid conduction. If the spacer has a large optical thickness $\tau_0 = E{\cdot}d \gg 1$, where E is the extinction coefficient and d the geometrical thickness of the insulation, the radiation diffuses and the radiative conductivity λ_r and the solid conductivity λ_{sc} simply can be summed up:

$$\lambda = \lambda_r + \lambda_{sc} \,. \tag{7}$$

The temperature distribution in those cases is neither a straight line as for pure solid conduction, nor given by Equation 6 as for pure radiative transport, but is located between these two curves. The diffusion of radiation can be treated according to neutron diffusion in a moderator. If the boundary emissivity ϵ is also taken into account, the radiative pseudoconductivity is given by Reference 17

$$\lambda_r = 4\, n^2\, \sigma\, d\, T_r^3 \,/\, (2/\varepsilon - 1 + 3\, E\cdot d/4) \,, \tag{8}$$

where n is the index of refraction of the spacer. For optimized opacified superinsulations we get values for $E{\cdot}d = \tau_o$ in the order of 500, which is very large compared to $1/\epsilon$. The influence of ϵ on λ_r then can be neglected:

$$\lambda_r = (16/3)\, n^2\, \sigma\, T_r^3/E \,. \tag{9}$$

In properly designed powder or fiber insulations, E often is only weakly dependent on T_r. The radiative conductivity then approximately is a linear function of T_r^3 as in Equation 2. According to Equation 9 for a given temperature T_r the radiative transport is reduced by a factor of several hundred compared to λ_r according to Equation 2 with $\epsilon \approx 1$.

One of the main tasks for the design of thermal superinsulations thus is to find materials which effectively absorb or scatter thermal radiation in a wide wavelengths range ($\Lambda \approx 1$ to 100 μm) and therefore have a large extinction coefficient E. If the chosen basis material is not blocking radiation effectively throughout the IR spectrum, a suitable opacifier has to be added. For example, fumed silica (SiO_2) shows high transmission (low absorption) below $\Lambda \approx 8$ μm. If it is combined with iron oxide (Fe_3O_4), which strongly absorbs in this region, a large total extinction coefficient E results: A spacer of density $\rho \approx 0.3$ g/cm^3 made of fumed silica with 10% Fe_3O_4 provides an extinction coefficient E $\approx$ 100 cm^{-1}.[18] If the spacer is 3 cm thick an optical thickness $\tau_o = E{\cdot}d$ in the IR of about 300 results.

The calculation of E from IR transmission data is relatively simple for purely absorbing materials. Let $E(\Lambda)$ denote the extinction (absorption) coefficient at wavelength Λ. $E(\Lambda)$ has to be properly averaged over the IR spectrum (Rosseland mean). As a weight function the net diffusing radiative spectrum is used (and not the Planck spectrum emitted by either of the two boundaries). If the temperature differences are small, the weight function $f(\Lambda{\cdot}T_r)$ is given by the difference of the two Planck spectra, otherwise

$$f(\Lambda\cdot T_r) = \partial I_\Lambda/\partial I|_\Lambda \,, \tag{10}$$

with I_Λ being the Planck distribution and $I = \sigma{\cdot}T_r^4$ the integral intensity. $f(\Lambda{\cdot}T_r)$ always peaks at shorter wavelengths as I_Λ and falls off more rapidly at long wavelengths. As averaged extinction coefficient we get the "Rosseland mean:"[17]

$$\frac{1}{E(T_r)} = \int_0^\infty \frac{1}{E(\Lambda)} \cdot f(\Lambda \cdot T_r)\, d\Lambda \ . \tag{11}$$

E generally depends on the radiative temperature T_r. If scattering is the major cause for radiation extinction, the derivation of E is more complicated.[19] For example then the angular scattering probability has to be taken into account. For particles or fibers with diameters in the order of the diffusing IR wavelengths Mie scattering occurs, which is predominantly forward peaked. It is obvious that forward scattering is less efficient in terms of radiation extinction than isotropic scattering.[19]

Solid Thermal Conduction

The solid conduction λ_{sc} in porous insulations is mainly determined by the size of the point-like contacts between the grains or fibers, which can be calculated from Hertz's formula. For powder insulations, for example, one gets[20]

$$\lambda_{sc} = 3.4 \cdot \lambda_s \cdot \{(\rho/\rho_s)^4 (1-\mu_s^2) p_0 / EM_s\}^{1/3} \ . \tag{12}$$

λ_s is the solid conductivity of the solid material the grains are made of, EM_s its Young's modulus, μ_s its Poisson's ratio and ρ_s its density. ρ is the apparent density of the evacuated porous insulation under the load p_o per area. According to Equation 12 we have to choose a grain or fiber material with a solid conductivity λ_s as low as possible. A high Young's modulus is important as well, i.e. the material should resist deformation. The apparent density ρ of the spacer has to be low, i.e. the porosity high - even under atmospheric load.

For $\rho/\rho_s \approx 0.1$, $\mu_s \approx 0.2$, $p_o = 10^5$ N/m^2 and $EM_s = 6 \cdot 10^{10}$ N/m^2 we get $\lambda_{sc}/\lambda_s \approx 1.8 \cdot 10^{-3}$ and with $\lambda_s \approx 1$ W/(m·K) finally $\lambda_{sc} \approx 1.8 \cdot 10^{-3}$ W/(m·K). The solid conduction in porous media can thus be reduced by a factor of about 500 compared to nonporous systems. Especially via λ_s also λ_{sc} is weakly temperature dependent. If this dependence is neglected the total conductivity $\lambda(T_r^3)$ can be extrapolated towards $T_r = 0$ in order to get an estimate on λ_{sc} (see Figure 5).

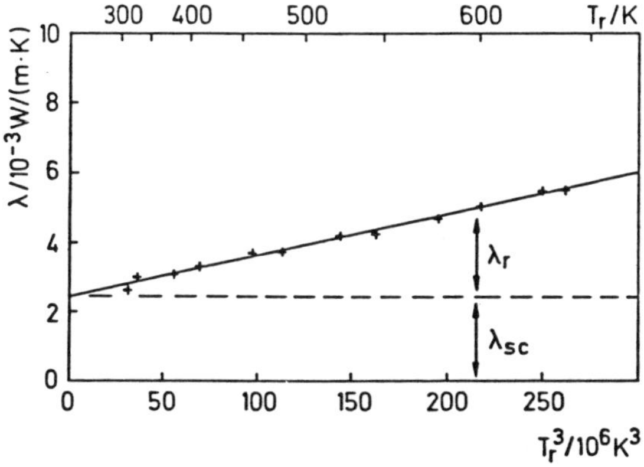

Figure 5: Thermal conductivity λ of a load bearing fiber insulation which is IR-optically thick; λ is a linear superposition of λ_r and λ_{sc}.[19]

It should be clear by now that the design and further development of evacuated porous superinsulation require a detailed knowledge of absorption and scattering mechanisms of granules, powders, fibers and of opacifying agents. Such information can be made available by measuring IR transmission (reflection) as a function of wavelengths between $\Lambda \approx 1\ \mu m$ and $100\ \mu m$ and of angle. Furthermore, the angular scattering distribution must be known. Monte Carlo calculations ought to be performed in order to simulate the radiative thermal transport on a computer.

As far as the solid conduction is concerned the structural parameters, such as contact areas, orientation of fibers or detrimental mechanical effects from opacifiers are to be studied.

THERMAL TRANSPORT IN AEROGEL TILES

General Considerations

Though silica aerogel tiles can be made transparent, and granular aerogel is translucent in the visible, there is appreciable IR extinction, which is caused by absorption. Scattering in the IR ought to be extremely weak, as the structural inhomogeneities have dimensions in the order of 100 nm and below. The absorption in aerogel is strong above wavelengths $\Lambda \approx 7\ \mu m$, however especially low between 3 and 5 μm (Figure 6). Therefore at low temperatures (280 K) the thermal IR spectrum (peaking around 10 μm) is effectively attenuated, the radiative transport through aerogel is weak. For increasing temperatures however more and more radiation can penetrate the spacer within the wavelength range $\Lambda \approx 3$ to 5 μm. Silica aerogel then is no longer a good IR radiative barrier. If a suitable opacifier could be found (which must not absorb in the visible!), its application even as a high temperature superinsulator could be possible. As can be seen from Figure 6 adsorbed water increases the absorption in the 3 to 5 μm range. Therefore one could seal evacuated aerogel tiles together with some moisture into the glass envelope in order to reduce the radiative transport. The adsorbed water might, however, increase the solid conductivity.

The solid conduction in silica aerogel is strongly reduced compared to the case of nonporous silica glass: The aerogel porosity can be made very high (up to 98%) and the "necks" which interconnect the primary SiO_2 agglomerates (diameter roughly 3 nm) formed in the sol-gel process are very narrow. Solid conductivities of only a few times 10^{-3} W/(m·K) are the consequence.

An important question is the dependence of the solid conductivity λ_s on the density ρ_s of the gel. At first glance one would assume a proportionality $\lambda_s \propto \rho_s$. This, however, is not true. Silica aerogels have been shown to display features characteristic for percolating systems:[22-24] Mechanical properties, like the Young's modulus EM_s scale with density[24] according to

$$EM_s \propto \rho_s^{\alpha} , \text{ with } \alpha \approx 3.8 . \tag{13}$$

This was derived from propagation of ultrasound in aerogel.[25,26] Scaling with ρ_s was also observed for the thermal conductivity.[23] Compression experiments showed, that

$$\lambda_s \propto \rho_s^{\beta}\ , \text{ with } \beta \approx 1.6 \text{ or } 2.6\ . \qquad (14)$$

The larger value for β applies to strongly compressed aerogel, while $\beta = 1.6$ is valid for aerogels typically employed in thermal insulations.

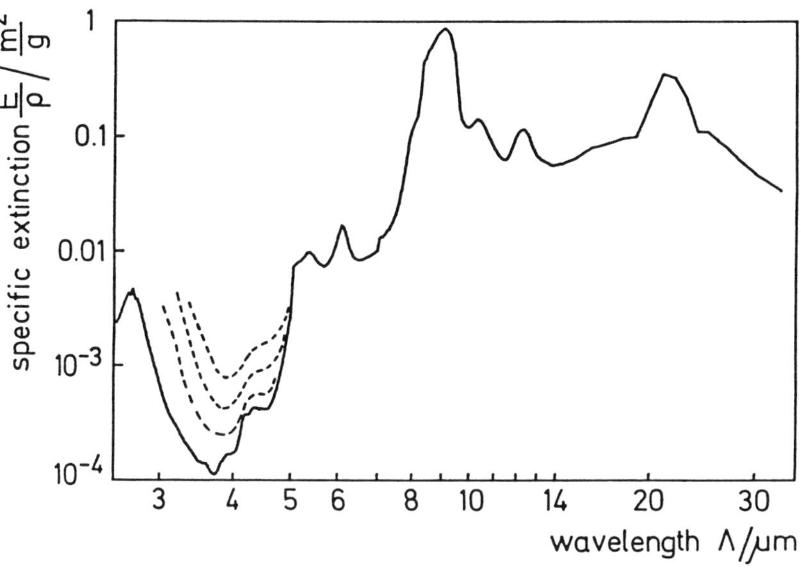

Figure 6: Specific extinction coefficient E/ρ for aerogel. The influence of adsorbed water can especially be seen within the low absorption region below $\Lambda = 5\ \mu m$: Solid line is for samples heated to 400°C under vacuum, dashed lines are for aerogel exposed to air for 0.5 hour, 12 hours and 2 days.[21]

Now we have to look at the combined action of the discussed two routes for thermal transport. Due to the IR transmission window between $\Lambda = 3$ and 5 μm with values of τ_0 in the order of 1 for a 20 mm tile the simple additive superposition according to Equation 7 does not hold in silica aerogel. In this wavelength region, radiative transport is not a local phenomenon anymore. Some IR photons may completely penetrate the tiles and allow for direct radiative "communication" between the boundaries. Consequently, the λ-values strongly vary with the sample thickness and the emissivity of the boundaries. In this case, the coupling between the radiation field and the heat flux caused by solid conduction has to be taken into account. If one considers low emissivity walls ($\epsilon \approx 0.1$) the radiation flux close to the walls is weak. Further inside the aerogel increasingly more radiation is produced by IR emission. In order to conserve the total thermal flux, the temperature gradient thus has to change across the aerogel tile: A more or less steep gradient is expected close to the boundaries, which causes energy transport via solid conduction, while deep within the aerogel the gradient is small (Figure 7). Here the relative contribution of the solid conduction is reduced.

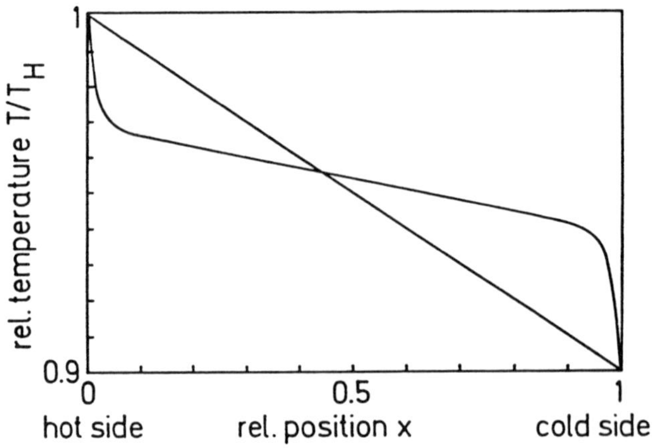

Figure 7: Temperature profile for pure conduction (straight line) and temperature profile for a sample with small optical thickness and small emissivity boundary.[21]

The heat transfer for combined radiation and conduction is in an exact treatment described by a complicated integro-differential equation. It is in general dependent on the wall temperatures T_1 and T_2, the emissivity ϵ of the walls, the amount of solid conduction compared to radiative contributions and the optical thickness as a function of wavelength $\tau_0(\Lambda)$. It is very tedious to solve this integro-differential equation numerically for a nongray medium such as aerogel.

A semiempirical procedure to calculate λ is given in Reference 27. Here the concept of an effective emissivity ϵ^* is introduced. ϵ^* is dependent on the optical thickness τ_0, the conduction-radiation parameter $N_1 = \lambda_s \cdot E/(4\sigma T_h^3)$ (with T_h being the temperature of the hot boundaries) and the emissivity ϵ of the boundaries. ϵ^* is inserted into Equation 8 instead of ϵ and the resulting total heat flux q of the additive approximation is compared with numerical results of exact calculations for a gray medium.[21] By adjusting free parameters one gets the following expression for the effective emissivity ϵ^*:

$$\varepsilon^* = 1 - (1-\varepsilon)\cdot\exp\{-N_1/(2N_1+0.04)\cdot\arctan(\tau_0\cdot(1+0.02/N_1))\} \quad . \tag{15}$$

If N_1 is larger than 0.1 this expression can be simplified to:

$$\varepsilon^* = 1 - (1-\varepsilon)\cdot\exp\{-\arctan(\tau_0)/2\} \quad . \tag{16}$$

As the optical thickness in aerogel is strongly wavelengths dependent Equations 8 and 15 have to be generalized for a nongray medium. The total emissive power $I = \sigma T_r^4$ is replaced by the spectral emissive power $I_\Lambda(T_r)$. Assuming small temperature differences ΔT the total heat flux $q = q_{sc} + q_r$ can be approximated by

$$q/\Delta T = \lambda_s/d + 4\sigma T_r^3 \int_0^\infty f(\Lambda\cdot T_r)/\{2/\varepsilon^*(\Lambda) - 1 + 3\tau_0(\Lambda)/4\}\, d\Lambda \quad , \tag{17}$$

where $\epsilon^*(\Lambda)$ is calculated according to Equations 15 or 16, however, using spectral optical thicknesses $\tau_0(\Lambda)$ and emissivities $\epsilon(\Lambda)$.

Calorimetric Measurements

The low thermal conductivities of evacuated highly porous superinsulating materials can only be determined with accurate guarded hot plate systems.[28] The machines used in our laboratory, LOLA I and LOLA II,[29] are especially designed to investigate the relative and absolute contribution of the solid conduction and the radiative heat transfer in load-bearing powder and fiber insulations.[30] A main question in such studies is the change in solid conduction by varying the external load onto the porous specimen. Together with IR optical measurements from which the extinction coefficient for the thermal radiation can be determined, these investigations are indispensable for a systematic improvement of superinsulations. Accurate measurements of the total conductivity λ and the derivation of the contributions λ_r and λ_{sc} as a function of temperature and external load are rather time consuming. Therefore they are performed with complete computer control.

For properly opacified insulations the IR optical thickness τ_0 is in the order of a few hundred and often only weakly temperature dependent. In a $\lambda(T_r^3)$ plot (see Figure 5) data then are located on a straight line with correspondingly small slope. For silica aerogel with its small extinction coefficient below wavelengths $\Lambda \approx 7$ μm the loss coefficient increases stronger than with the third power of T_r, i.e. the optical thickness τ_0 decreases with rising temperatures. Due to this effect it is necessary to use small temperature differences (<40 K) between hot and cold plates. In addition the emissivity ϵ of the hot plates influences the loss coefficient. Despite guarding, IR radiation with Λ <7 μm originating from the inner parts of the sample escapes through the rims of the aerogel tiles. These losses are reduced as much as possible by installing reflecting boundaries around the rims of the tiles and by using thin samples. Furthermore the (reduced) losses are determined and subtracted from the total amount of heat fed into the hot plate.

Another experimental problem is the contact resistance between the aerogel tile and the plates. As the tile surfaces are only flat to within fractions of a mm such a resistance will be noticeable, especially if a small mechanical load is applied and if low emissivity walls are used. On the other hand, for practical applications such a resistance further reduces the loss coefficient of aerogel window systems.

Three aerogel samples with densities ρ_1 = 270 g/l, ρ_2 = 105 g/l and ρ_3 = 75 g/l and thicknesses d_1 = 11 mm, d_2 = 22 mm and d_3 = 9 mm were investigated. The measurements were performed in our evacuable guarded hot plate device LOLA II (Figure 8). For all evacuated and baked samples the thermal loss coefficient k was determined as a functon of temperature and of boundary emissivity. As an example the data for the tile with 270 g/l are shown in Figure 9. As the IR extinction coefficient of aerogel decreases with rising T_r the concave curvature is expected. With small emissivity boundaries lower loss coefficients are obtained. As these data are taken at high vacuum, the contact resistances between the tiles and the boundaries are included.

In Figure 10 the loss coefficients for the three investigated tiles with different thicknesses and densities are deployed. Though no direct comparison is possible, it is obvious that for low-density tiles a relatively smaller solid conductivity appears. The rise with temperature depends on the optical thickness, i.e. the mass m" per area of the tiles.

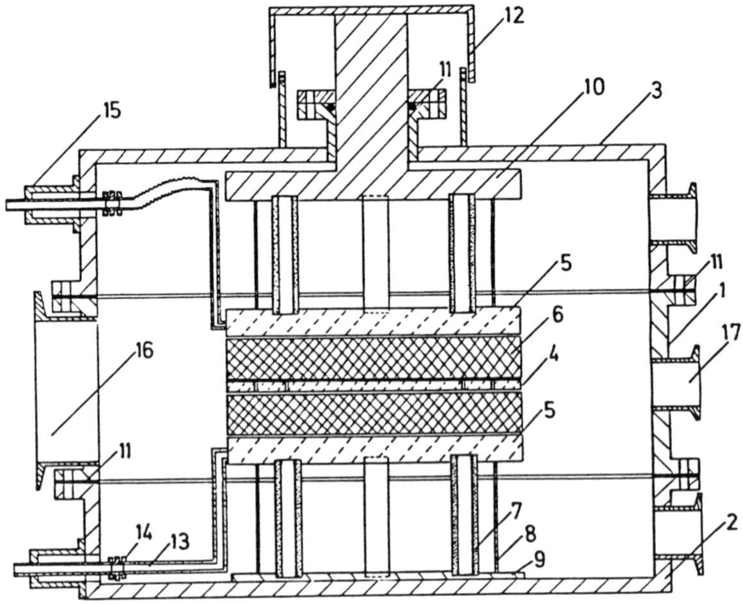

Figure 8: Evacuable guarded hot plate device LOLA II. The numbers refer to the following items: 1 to 3 = vacuum chamber, 4 = hot plate with two guard rings, 5 = reference plates, 6 = samples, 7 = load bearing thermally insulating tubes, 8 = thin metal connectors, 9 = bottom plate, 10 = piston, 11 = vacuum seal, 12 = holder for piston, 13 to 15 = parts for cooling, 16 and 17 = vacuum flanges.

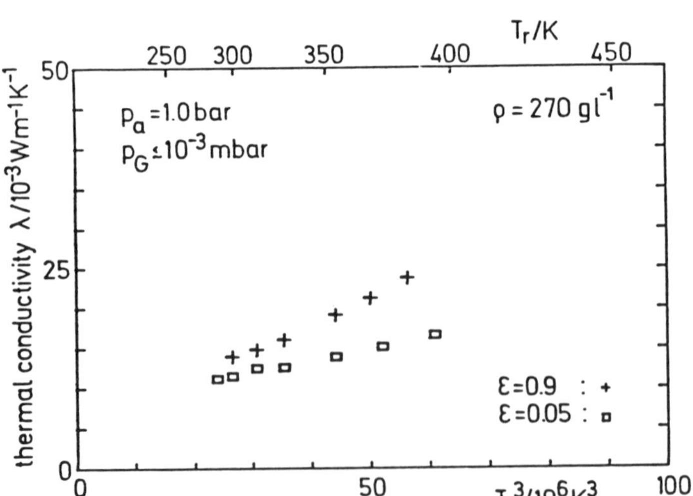

Figure 9: Thermal loss coefficient k of an evacuated aerogel tile with density ρ = 270 g/l, thickness d = 11 mm over third power of mean radiative temperature T_r. Parameter is the boundary emissivity ϵ.[31]

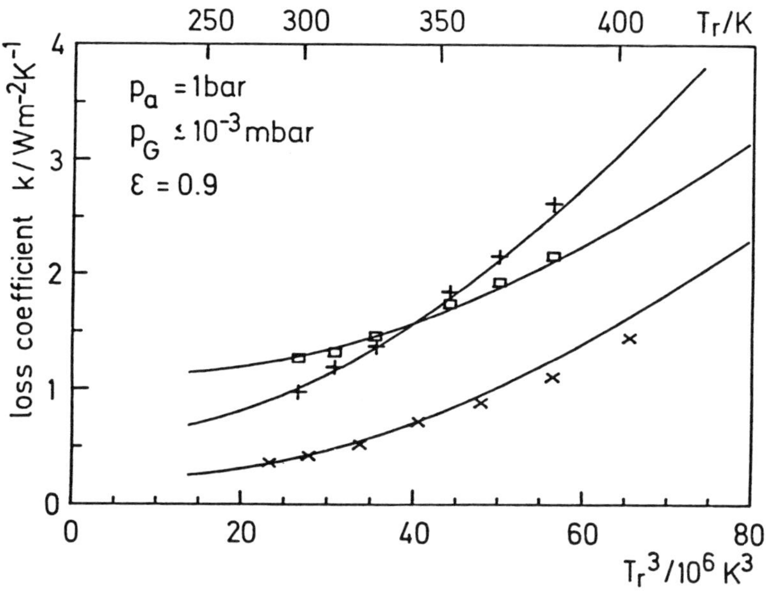

Figure 10: Thermal loss coefficient k for three different evacuated tiles over third power of mean radiative temperature T_r;[31] + : ρ = 75 g/l, d = 9 mm, m'' = 0.68 kg/m^2 ; x : ρ = 105 g/l, d = 22 mm, m'' = 2.3 kg/m^2 ; □ : ρ = 270 g/l, d = 11 mm, m'' = 3.0 kg/m^2. The fit curves shown are for λ_{sc} = 0.004, 0.004 and 0.013 W/(m·K), according to the ϵ^*-method.[27]

The reported results show that with evacuated low density tiles (i.e. $\rho \approx 100$ g/l) of about 20 mm thickness and at ambient temperatures a loss coefficient as low as 0.5 W/(m^2·K) can be reached. For rising temperatures, however, the IR transmission window between $\Lambda \approx 3$ and 5 μm allows an increasingly larger fraction of radiation to pass through the aerogel. This underlines the necessity to develop IR-opacified aerogels, if applications at elevated temperatures are considered.

Effects of Gas Pressure

An important question is how much the conductivity of aerogel tiles increase if air leakage occurs. It is well known that the thermal conductivity of a gas varies according to the Knudsen formula[20]

$$\lambda_G\,(Kn) = \lambda_{Go}/(1 + 2\,\beta\,Kn)\ , \tag{18}$$

where Kn is the Knudsen number

$$Kn = \bar{l}/d\ , \tag{19}$$

with $\bar{l}$ being the mean free path of the gas particles and d a typical dimension of the envelope around the gas. For air $\beta \approx 1.5$ holds. λ_{Go} is the thermal conductivity for air and Kn = 0:

$$\lambda_{Go} = 1.9 \cdot c_V \cdot \bar{v} \cdot \rho \cdot T/3 \ . \tag{20}$$

Here c_V is the heat capacity, $\bar{v}$ the average molecular speed and ρ the density of the gas. Measurements of the loss coefficient of aerogel tiles as a function of internal gas pressure p_G are depicted in Figure 11. Two distinct rises can be observed under variation of p_G

- One occurs between 10^{-2} and 10^{-1} mbar and is caused by the change in contact resistance between tile and the boundaries. For low boundary emissivities and at low gas pressures the thermal coupling between tile and walls is only weak - despite an external load. When ρ_G increases and the Knudsen number for the small gaps between tile and walls approaches Kn $\approx$ 1, heat conduction into the tile is enhanced. This shows as an increase in the thermal loss coefficient of about 20%. For walls with emissivities $\epsilon \approx 0.9$ a stronger radiative coupling between tile and walls occurs for all p_G, thus the rise in k between $p_G \approx 10^{-2}$ and 10^{-1} mbar is much weaker (10%).
- The second rise is observed at pressures $p_G > 10^2$ mbar. It is caused by the onset of gas conduction within the larger pores of the tiles, which therefore must have diameters in the order of 200 nm.

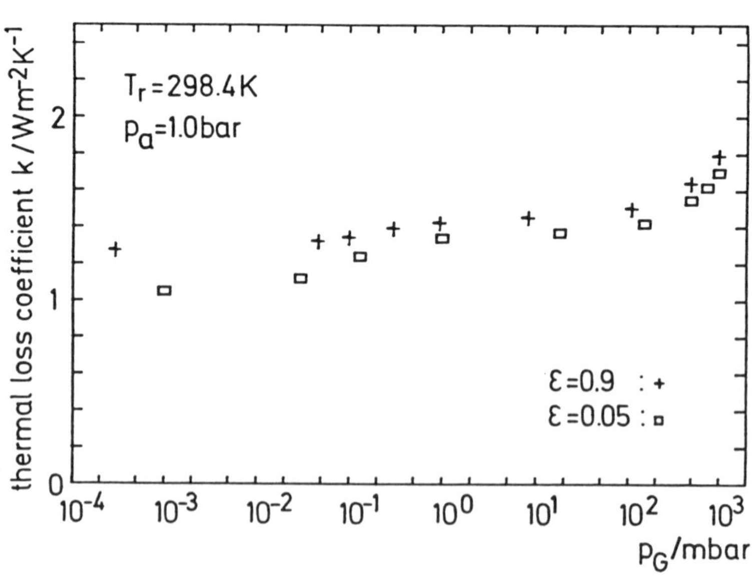

Figure 11: Thermal loss coefficient k of an aerogel tile with density ρ = 270 g/l, thickness d = 11 mm and external load p_a = 1 bar over internal gas pressure p_G. The parameter is the boundary emissivity ϵ.[31]

We recognize that the total thermal loss coefficient of aerogel tiles under external atmospheric load is especially low, if the boundary emissivity ϵ is small, if the tile surface is somehow rugged and if the internal gas pressure p_G is below 10^{-2} mbar.

THERMAL TRANSPORT IN GRANULAR AEROGEL

General Aspects

Kaganer[20] has shown that for a granular filling consisting of massive spheres (radius r_s, density ρ_s and conductivity λ_s) the solid thermal conductivity λ_{sc} mainly depends on the thermal contacts between adjacent spheres. λ_{sc} for a "mono-layer" with thickness $2r_s$ is given by

$$\lambda_{sc} = Q \cdot 2r_s/(A_s \cdot \Delta T) , \qquad (21)$$

where Q is the amount of heat transferred per second through the area A_s for a temperature difference ΔT. Suppose, A_s is the area occupied by one grain on the average, then Q also represents the heat flux through that grain, i.e. through its contact zone (radius r_c) characterized by the thermal resistance

$$R_c \approx 1/(2 \cdot r_c \cdot \lambda_s) . \qquad (22)$$

With N being the coordination number for the spheres then N/3 contacts along any axis in an orthogonal system and N/6 contacts in either direction on each axis exist, therefore

$$Q = \Delta T \cdot (N/6)/R_c . \qquad (23)$$

For two touching spheres the contact radius r_c is given by Hertz's formula

$$r_c = \{(3/4)\ (1-\mu_s^2)\ F \cdot r_s/EM_s\}^{1/3} , \qquad (24)$$

where μ_s is the Poisson ratio of the material the spheres are made of, EM_s its Young's modulus, and F the force the spheres are pressed together with. From Equation 24 we learn that the harder the spheres are, i.e. the larger the modulus is, the smaller r_c will be. The average force F on one contact is

$$F = p_a \cdot A_s/(N/6) . \qquad (25)$$

p_a is the pressure exerted onto the insulation. The area A_s occupied by one grain can be expressed by the macroporosity Π of the pellet filling

$$\Pi = 1 - \rho/\rho_s = 1 - (4r_s^3\pi/3)/(2r_s \cdot A_s) , \qquad (26)$$

$$A_s = (2/3)\ r_s^2\pi/(1-\Pi) . \qquad (27)$$

The coordination number N has been derived empirically[32]

$$N = 2 \cdot e^{2.4(1-\Pi)} . \tag{28}$$

The contact radius r_c can now be calculated by combining the above equations

$$r_c = r_s\{(3\pi/2)(1-\mu_s^2)p_a/[(1-\Pi)e^{2.4(1-\Pi)}EM_s]\}^{1/3} . \tag{29}$$

Equation 21 then can be rewritten as

$$\begin{aligned} \lambda_{sc} &= \lambda_s(2/\pi)(r_c/r_s)(1-\Pi)\ e^{2.4(1-\Pi)} \\ &= \lambda_s\{(12/\pi^2)(1-\mu_s^2)(1-\Pi)^2\ e^{4.8(1-\Pi)}p_a/EM_s\}^{1/3} . \end{aligned} \tag{30}$$

Equation 30 allows an estimate on the ratio λ_{sc}/λ_s. For aerogel pellets with ρ_s = 232 g/l and ρ = 154 g/l (corresponding to $\Pi = 0.34$), $EM_s = 10^7$ N/m^2,[24] and $\mu \approx 0.2$, $p_a = 10^5$ N/m^2 (= 1 bar) one gets

$$r_c/r_s \approx 0.25 \text{ and} \tag{31}$$

$$\lambda_{sc}/\lambda_s \approx 0.5 . \tag{32}$$

Granular layers of aerogel thus are predicted to have smaller solid conductivities than monolithic aerogel layers. Further λ_{sc} should be independent of sphere radius r_s, however, increase proportional to $p_a^{1/3}$ and to $EM_s^{-1/3}$. With Equations 13 and 14 and $1\text{-}\Pi = \rho/\rho_s = 0.55 \ldots 0.7$ for irregularly close-packed systems of equal spheres we get the dependence

$$\lambda_{sc} \propto \rho_s^{1.6}(\rho_s^{-3.8})^{1/3} \approx \rho_s^{0.3} . \tag{33}$$

The ρ_s-dependence of μ_s^2, which is small compared to one is neglected.

We recognize that the solid conductivity λ_{sc} of granular aerogel fillings increases only very weakly with the density ρ_s of the granules. This is in contrast to the case of aerogel tiles, where the solid conductivity λ_s grows much stronger ($\propto \rho_s^{1.6}$). The consequence for practical applications is the use of not too porous granules if one wants to minimize total conductivity, i.e. sufficiently reduce not only solid conduction but also radiative heat transport. For tiles on the other hand the lowest conductivities at room temperature can be expected for the most porous aerogels (see Figure 10).

Calorimetric Measurements

A layer of aerogel pellets was packed into our guarded hot plate device LOLA II. The evacuated layer had a thickness of 15 mm when loaded with an external pressure of 1 bar. The pellets had diameters of about 3 mm. Their density ρ_s was 232 g/l. At the lowest temperatures ($T_r \approx 315$ K) and for low emissivity walls ($\epsilon \approx 0.05$) the loss coefficient for the 15 mm layer was shown to be below $k \approx 0.5$ W/($m^2 \cdot$K), corresponding to a pseudoconductivity $\lambda < 7.5$ mW/(m·K) (see Figure 12). For the same temperature, however with black walls ($\epsilon \approx 0.9$) the loss coefficient was 80% higher. This demonstrates the importance of radiatively decoupling the walls and the adjacent granular layer. Figure 12 also shows the expected steep rise of k with T_r for walls with $\epsilon \approx 0.9$ and the reduced increase with $\epsilon \approx 0.05$. Of interest is also the dependence of k on the external load p_a for the cases $\epsilon \approx 0.9$ and $\epsilon \approx 0.05$, respectively. For black boundaries the walls are radiatively coupled to the aerogel layer. Therefore k changes only by about 10% when the external load p_a is increased from 0.1 to 1.3 bar, i.e. when the mechanical contact between the pellets and the wall is intensified. For $\epsilon \approx 0.05$ on the other hand the radiative flux into the aerogel is suppressed and consequently the mechanical contacts become more important, resulting in a 40% increase of k within the same load range.

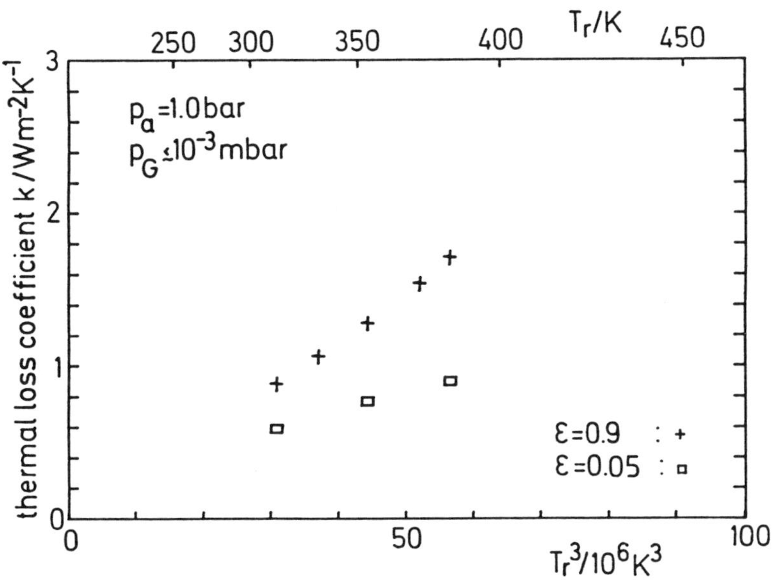

Figure 12: Loss coefficient k of an evacuated 15 mm layer of aerogel pellets ($\rho_s \approx 232$ g/l, diameter 3 mm) vs. third power of radiative temperature for boundary with emissivity $\epsilon \approx 0.05$ and $\epsilon \approx 0.9$.[33]

Effects of Gas Pressure

The variation of the loss coefficient with the gas pressure p_G in the pellet system is shown in Figure 13. The first steep rise of k with p_G occurs around 0.1 mbar and is caused by the onset of air conduction between the pellets. A second weaker rise can be seen around a few hundred mbar. It corresponds to the onset of air conduction within the pellets. This demonstrates that the pellets have pores up to the order of 0.1 μm. Boundaries with $\epsilon \approx 0.05$ provide the smaller loss coefficients. For increasing p_G the differences decrease, because the thermal coupling to the walls is taken over by gas molecules. For practical applications the window system has to be kept at gas pressures $p_G \leqslant 10^{-2}$ mbar. The granular fillings thus have to be degassed at elevated temperatures and air leakage has to be prevented by use of glass-metal seals.

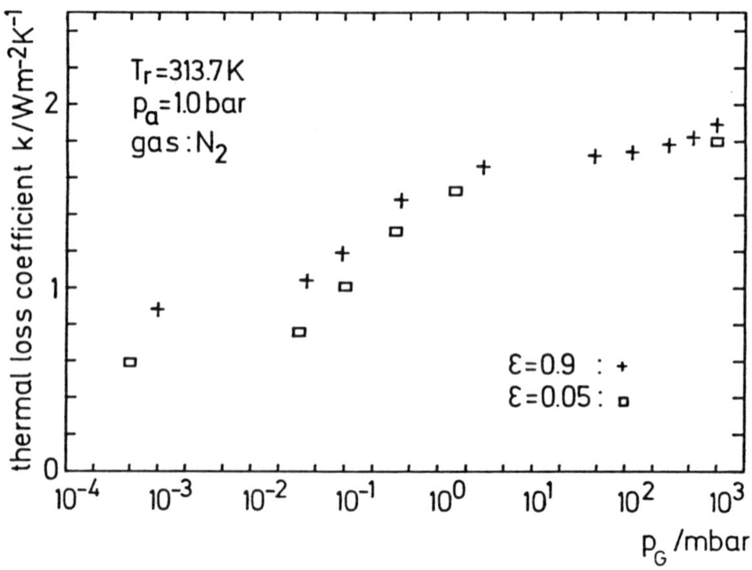

Figure 13: Loss coefficient k of a 15 mm layer of aerogel pellets ($\rho_s \approx 232$ g/l, diameter 3 mm) vs. internal gas pressure p_G; the rise of k at 0.1 mbar is caused by the onset of air conduction between the pellets, the weaker rise at 1,000 mbar by the onset of air conduction within the porous pellets.[33]

OPTICAL TRANSPARENCY

The use of aerogel systems in windows or covers for passive use of solar energy implies that transparency or at least sufficient transmission for visible light can be provided.

Aerogel tiles are transparent, showing only some Rayleigh scattering. The tiles thus look yellowish if viewed against the sunlight or bluish in front of a black background. As only 20 x 20 cm^2 tiles are available, and the problem of surface bending does not seem to be solved yet, window applications are not feasible right now. In addition the tiles are made in a discontinuous batch process and thus are still too expensive for mass applications. Granular aerogels have a far better potential as superinsulating spacers in window-like systems. They are, however, only translucent. Scattering occurs not only within the pellets but also from surface inhomogeneities. The transmittances for direct solar radiation are about 55% and 64% for granular layers of 2 cm and 1 cm, respectively. For diffuse sunlight the transmittance is lower by about 10%.[34] These values seem to be rather promising for passive solar applications.

CONCLUSIONS AND OUTLOOK

From the data presented above it ought to have become clear that evacuated silica aerogel either in tile or in pellet form indeed is a suitable material for transparent or translucent superinsulations with room temperature loss coefficients around 0.5 $W/(m^2 \cdot K)$ for layers 15 to 20 mm thick. Several of the important parameters for window design have been investigated and thus are known to some extent. In general, the monolithic systems are better understood than the granular fillings. The latter still need to be optimized with respect to density and diameter of granules. The question whether the total conductivity of an evacuated granular filling depends on pellet diameter already has been investigated by Kistler.[2,3,4] While in his work small pellets ($\phi \approx 0.1$ mm) have a lower conductivity than larger pellets ($\phi \approx 3$ mm), Kaganer's model (Equations 12 and 30) does not predict such a dependence.

An important task for future sol-gel activities is the (partial) closure of the IR transmission window between $\Lambda = 3$ and 5 μm. With less radiative heat transfer aerogel fillings with lower density and correspondingly lower solid thermal conduction could be used. This would further improve thermal insulation or allow the construction of even thinner window systems. Sufficiently opacified aerogel layers could also be used as superinsulating covers in window systems with elevated temperatures, e.g. solar collectors.

An important task from a theoretical point of view is to solve the multiband energy transport equation for silica aerogel. In such a calculation the coupling between the solid conduction and the large number of interdependent radiative wavelengths - channels has to be solved, quite certainly an ambitious endeavor!

Acknowledgements

I am grateful to the Lund group, G.V. Dardel and S. Henning, to G. Poelz/-DESY, Hamburg as well as to G. Seybold, M. Mielke and A. Stange/BASF, Ludwigshafen for generously providing aerogel tiles and granules, respectively.

REFERENCES

1. Kistler, S.S., *Nature 127,* 741 (1931).
2. Kistler, S.S. and Caldwell, A.G., *Industr. Eng. Chem. 26,* 658 (1934).
3. Kistler, S.S., *J. Chem. 39,* 79 (1935).
4. Kistler, S.S., *J. Phys. Chem. 46,* 19 (1942).
5. Nicolaon, G.A. and Teichner, S.J., *Bull. Soc. Chim. Fr.* 1906 (1968) and U.S. Patent 3,672,833.
6. Cantin, M., Casse, M., Koch, L., Jouan, R., Mestran, P., Roussel, D., Bonnin, F., Moutel, J. and Teichner, S.J., *Nucl. Instrum. Methods 118,* 177 (1974).
7. Poelz, G. and Riethmüller, R., Preparation of Silica Aerogel for Cerenkov Counters, *DESY* 81-055 (1981).
8. Poelz, G., Aerogel Cerenkov Counters at DESY, *DESY* 84-110 (1984).
9. Henning, S. and Svensson, L., *Physica Scripta 23,* 697 (1981).
10. v. Dardel, G., Henning, S. and Svensson, L., European Patent Specification 0018 955 B 1, Date of Filing 17.4.1980, Date of Publication 1.12.1982.

11. Henning, S., Large Scale Production of Airglass, in: *Aerogels, Springer Proceedings in Physics 6,* J. Fricke (Ed.), Springer Verlag, Heidelberg, Berlin, New York, Tokyo (1986).
12. Brinker, C.J., Ward, K.J., Keefer, K.D., Holupka, E., Bray, P.J. and Pearson, R.K., Synthesis and Structure of Borate Based Aerogels, ibid.
13. Tewari, P.H., Hunt, A.J. and Lofftus, K.D., Advances in Production of Transparent Silica Aerogels, ibid.
14. Broecker, F.J., Heckmann, W., Fischer, F., Mielke, M., Schroeder, J. and Stange, A., Structural Analysis of Granular Silica Aerogels, ibid.
15. Fricke, J., Thermal Transport in Porous Superinsulations, ibid.
16. Goetzberger, A., Schmid, J. and Wittwer, V., *Internat. J. Sol. Energy 2,* 289 (1984).
17. Siegel, R. and Howell, J.R., *Thermal Radiation Heat Transfer,* Hemisphere Publ. Corp., Washington (1984).
18. Caps, R., Fricke, J. and Reiss, H., *High Temp. - High Press. 15,* 225 (1983).
19. Caps, R., Trunzer, A., Büttner, D., Fricke, J. and Reiss, H., *Int. J. Heat Mass Transfer 27,* 1865 (1984).
20. Kaganer, M.G., *Thermal Insulation in Cryogenic Engineering,* Israel Progr. for Scient. Translations, Jerusalem (1969).
21. Scheuerpflug, P., Caps, R., Büttner, D. and Fricke, J., *Int. J. Heat Mass Transfer 28,* 2299 (1985).
22. Fricke, J., Aerogels - a Fascinating Class of High Performance Porous Solids, in: *Aerogels, Springer Proceedings in Physics 6,* J. Fricke (Ed.), Springer Verlag Heidelberg, Berlin, New York, Tokyo (1986).
23. Nilsson, O., Fransson, Å. and Sandberg, O., Thermal Properties of Silica Aerogels, ibid.
24. Gronauer, M., Kadur, A. and Fricke, J., Mechanical and Acoustic Properties of Silica Aerogels, ibid.
25. Gronauer, M., Büttner, D. and Fricke, J., Sound Velocities in Microporous SiO_2-Aerogel Tiles, Report E12-1284-2 (1984), Phys. Inst. Universität, D-8700 Würzburg, W. Germany.
26. Gronauer, M. and Fricke, J., Acoustic Properties of Microporous SiO_2-Aerogels (in print in Acustica).
27. Caps, R. and Fricke, J., Radiative Heat Transfer in Silica Aerogel, in: *Aerogels, Springer Proceedings in Physics 6,* J. Fricke (Ed.), Springer Verlag Heidelberg, Berlin, New York, Tokyo (1986).
28. Bode, K.H., *Int. J. Heat Mass Transfer 23,* 961 (1980).
29. Büttner, D., Fricke, J. and Reiss, H., *Vakuum-Technik 34,* 3 (1985).
30. Büttner, D., Fricke, J. and Reiss, H., *High Temp. - High Press. 17,* 333 (1985).
31. Büttner, D., Caps, R., Heinemann, U., Hümmer, E., Kadur, A., Scheuerpflug, P. and Fricke, J., Thermal Conductivity of SiO_2-Aerogel Tiles, in: *Aerogels, Springer Proceedings in Physics 6,* J. Fricke (Ed.), Springer Verlag Heidelberg, Berlin, New York, Tokyo (1986).
32. Meissner, H.P., Michaels, A.S. and Kaiser, R., *I & EC Process Design and Development 3,* 203 (1964).
33. Büttner, D., Hümmer, E. and Fricke, J., Thermal Conductivity of Granular SiO_2 Aerogel, in: *Aerogels, Springer Proceedings in Physics 6,* J. Fricke (Ed.), Springer Verlag, Heidelberg, Berlin, New York, Tokyo (1986).
34. Platzer, W. and Wittwer, V., Solar Transmission of Aerogel Pellets, ibid.

12

Ultrapure Glasses from Sol-Gel Processes

Shyama P. Mukherjee
IBM Corporation
Endicott, New York

INTRODUCTION

In recent years, the demand for ultrapure glasses has greatly increased because ultrapurity is a necessary requirement for making highly transparent glasses that are of great demand in optical waveguide and high power laser system technologies. Moreover, there is a need for the use of ultrapure glasses for verifying scientific theories related to the optical, electronic properties, and nucleation behavior of glasses which are strongly influenced by the presence of trace impurities.

High transparency in glass can be achieved by eliminating optical attenuation or loss. Optical loss in glass arises from: (a) absorption loss, (b) scattering loss. The absorption loss can be of three categories: (i) intrinsic, (ii) extrinsic, (iii) structural defect centers. Intrinsic absorption occurs when materials exist in a perfect state; the contribution due to impurities and defects is zero. Thus, the measurement of intrinsic absorption in glass is impeded by the extreme difficulty of eliminating all impurity absorption. Consequently, the establishment and verification of theories related to intrinsic absorption requires the preparation of ultrapure glass free from extrinsic impurities and defects. Extrinsic absorption is primarily due to the presence of transition metal impurities and hydroxyl groups in glass.[1-3] Many of the transition metal ions have electronic transitions with energies corresponding to the wavelength of light being transmitted. Moreover, structural hydroxyl groups in glass have fundamental stretching vibrations between 2.7 and 4.2 μm depending on the position of hydroxyl groups in the glass network.[4] The absorption bands due to OH groups in the visible and near infrared are associated with overtones of fundamental vibrations. Structural defect centers such as vacancies, interstitials, dangling bonds, overcoordinated atoms, undercoordinated atoms, can cause extrinsic absorption on its own or after radiation with X-rays or γ-rays. The formation of the structural defects is also susceptible to the presence of impurities.[5] Scattering loss[1] in glass has two sources (a) intrinsic scattering due to the fundamental fluctuations in glass density and composition (Rayleigh Scattering), (b) extrinsic scattering due to inclusions, bubbles, chemical inhomogeneities, created due to (a) improper mxing, (b) contamination from environment or containers.

Hence it may be noted that the preparation of ultrapure glasses containing no impurities causing extrinsic absorption and containing no particulates or inhomogeneities causing extrinsic scattering is an essential step to the preparation of highly transparent glasses that are of great demand in optical communication and high power laser system technologies.

The major problems associated with the preparation of ultrapure highly transparent glasses in large pieces are: (a) contamination due to impurities associated with conventional starting materials, e.g. oxides, carbonates, etc., (b) contamination from the environment during crushing, grinding, mixing and melting of conventional raw materials, (c) contamination from containers during melting at high temperatures for a considerable time for achieving homogenity. The lack of container materials that can resist the corrosion by the glass melts at high temperature eliminates the preparation of ultrapure glasses which are corrosive in the molten state. Moreover, the trace impurities incorporated in the glass-melts from the dissolution of containers such as Pt, Al_2O_3, or SiO_2 can produce inhomogenities leading to extrinsic scattering centers or crystallization centers and absorption centers.

METHODS OF MAKING ULTRAPURE GLASSES

The following techniques have been used for making ultrapure glasses: (A) vapor deposition techniques[6,7] (B) conventional glass melting technique[2] (C) sol-gel processes.[8-11] However, each approach has its limitations and advantages. The vapor deposition techniques are based on the production of a porous glassy preform by depositing glassy layers formed by the vapor phase oxidation of volatile metal halides in a closed system.[6,7] The porous preform is subsequently sintered to form glass rods from which fibers are drawn. The prime advantage of this technique is ultrahigh purity that can be achieved by using volatile liquids such as $SiCl_4$, $GeCl_4$, BBr_3, $TiCl_4$, $POCl_3$ which can be distilled to reduce the contamination of most transition metal impurities to below 1 part in 10^9. The vapor pressures of transition metal halides are generally several orders of magnitude lower than those of the metal halides (e.g. $SiCl_3$, BCl_3 etc.). Moreover, since the deposition is done in a closed system, contamination from the environment is minimized. However, the compositions that can be prepared by the vapor deposition techniques are limited to those elements whose halides can be volatilized. Moreover, the shape and size of the glass bodies that can be made are very limited.

The conventional glass melting techniques[2] are based on the melting of mixtures of ultrapure powdered raw materials (e.g. oxides, carbonates, nitrates) in ultrapure silica glass crucibles. This method has several limitations: (a) it is limited to compositions which are not corrosive to silica crucibles; platinum crucibles can be used for high melting compositions but transition metal impurities and Pt metal itself dissolve out from containers, and, thus contaminate the glass; (b) it is extremely difficult to prevent contamination from the environment during grinding and mixing of raw materials; (c) the long melting times at high temperatures needed for homogenization increase the contamination from containers and the environment.

The sol-gel processes of making glasses have several flexibilities and unique features that are of significant importance in preparing ultrapure homogeneous glasses.

SOL-GEL PROCESSES

There are two methods for the preparation of gels which can be used for making glasses.[8-18]

Glasses From Colloids[16-18]

This method is based on the preparation of stable colloidal sols of metal oxides in an aqueous or nonaqueous liquid medium using noncrystalline particles of colloidal sizes. The sol is transformed into monolithic gel bodies which are subsequently sintered to produce glass bodies. The particles for making colloidal dispersions can be prepared either by the vapor phase oxidation of metal halides/organometallics[16-18] or by the sol-gel technique using metal alkoxides.[10,11,19] This method has been used by some workers in making ultrapure high silica glasses using ultrapure glassy powders produced by the vapor phase oxidation of metal halides.[16-18] Published works on the preparation of multicomponent glasses by this method are limited.

Glasses From Gels Prepared by the Hydrolytic Polycondensation of Metal Alkoxides/Metal Organics[8-15]

This method is based on (a) the preparation of homogeneous gels by the hydrolytic polycondensation of alkoxysilanes, metal alkoxides and reactive metal salts in nonaqueous solvents; (b) and subsequent conversion of the gels or gel-monoliths to glasses either by sintering or by melting. The physiochemical principles of the hydrolytic polycondensation of alkoxysilane, and metal alkoxides are reported by several workers.[20] The key technical and experimental aspects in preparing ultrapure and homogeneous glasses will be discussed here in the context of maintaining or improving homogeneity and purity. Following are the steps that should be critically analyzed for preparing ultrapure glasses by this method.

(1) Selection of starting metal alkoxides.

(2) Synthesis of homogeneous gel-powders/gel-monoliths.

(3) Drying of gels/gel-monoliths.

(4) Removal of organics and hydroxyl groups.

(5) Conversion of gel to glass: Either by melting or by sintering.

It should be pointed out that the preparation of ultrapure starting chemicals, and subsequent maintenance of purity at each processing step, are of great concern in actual practice of making ultrapure glass by this method. Another important technical issue, besides the purity, is the synthesis of homogeneous noncrystalline gels which transform into glass without crystallization and will produce glass with minimum or zero extrinsic scattering loss.[22] The key technical aspects of the steps that can impair the purity and homogeneity are discussed briefly in the following sections.

SELECTION OF STARTING METAL ALKOXIDES

The selection of starting metal alkoxides will influence both the volatility and homogeneity; the purification of the starting alkoxides will be controlled by their volatilities which will govern the distillation process used for the purification. The homogeneity of the gels will be influenced by the degree of polymerization of individual alkoxides, i.e. molecular complexities, and by the rate of the hydrolytic polycondensation process.[21]

The volatilities and degree of polymerization of metal alkoxides depend on the following factors:[21] (a) molecular size and shape of the alkoxy group; (b) nature of the central metal atom; (c) the nature of metal-oxygen-carbon bond. Table 1 shows the significant variation of volatilities and molecular complexities of Group IV element ethoxides.

It is evident that as the atomic size increases, the molecular complexities increase and the volatility decreases. Table 2 shows the variation of volatility and degree of polymerization of different isomeric alkoxides of Al, Ti, Zr. The data show that the volatility is sensitive to the shape and size of the alkyl group which by virtue of steric hindrance to intermolecular coordination affects the degree of polymerization. The data shown in Table 2 indicate that usually the degree of polymerization and the volatility decrease from primary to secondary to tertiary alkoxide. It should be noted also that the volatility of alkali metals is quite different from that of the alkoxides of Group III and IV elements. Alkali metal alkoxides are usually insoluble in organic solvents, nonvolatile and decompose on heating to higher temperatures even under reduced pressure. The insolubility of alkali metal alkoxides may be ascribed either to their ionic character or to extensive polymerization. The primary alkoxide derivatives of alkaline earth and other metals of Group II are generally insoluble nonvolatile compounds whereas the secondary and tertiary alkoxides tend to be comparatively more volatile and soluble in organic solvents.[21]

The above discussion indicates that the selection of starting alkoxides is a key technical issue which will dictate the purity of starting compounds, solubility of alkoxide in common solvents and the homogeneity of the multicomponent polymeric solution. Unfortunately, systematic studies on the purification of metal alkoxides by distillation are not available in the literature. However, the work by Gossink et al[10] on the purification of $Si(OC_2H_5)_4$ and Al-isopropoxide by distillation in quartz glass apparatus in a laminar flow box indicates that the

Table 1: Some Physical Properties of Group IV Element Ethoxides[21]

(i) Compound	$C(OEt)_4$	$Si(OEt)_4$	$Ge(OEt)_4$	$Sn(OEt)_4$	$Pb(OEt)_4$	$Ti(OEt)_4$	$Zr(OEt)_4$	$Hf(OEt)_4$	$Th(OEt)_4$
(ii) Electronegativity	2.50	1.74	2.02	1.72	1.55	1.32	1.22	1.23	1.11
(iii) Covalent Radii (A)	0.77	1.11	1.22	1.41	1.47	1.32	1.45	1.44	1.55
(iv) Boiling Pt. (°C/mm)	158/760	166/760	86/12.0	----	----	103/0.1	190/0.1	178/0.1	300/0.05
(v) Degree of polymerization	1.0	1.0	1.0	4.0	----	2.4	3.6	3.6	6.0

Table 2: Volatility and Degree of Polymerization of Metal Amyloxides[21]

R in $(M(OR)_x)_n$	$Al(OR)_3$ B.P. °C/mm	n	$Ti(OR)_4$ B.P. °C/mm	n	$Zr(OR)_4$ B.P. °C/mm	n
$CH_3(CH_2)_4$	255/1.0	4.0	175/0.8	1.4	256/0.01	
$(CH_3)_2CH(CH_2)_2$	195/0.1	4.0	148/0.1	1.2	247/0.1	3.3
$(CH_3)(C_2H_5)CHCH_2$	200/0.6	4.1	154/0.5	1.1	238/0.1	3.7
$(CH_3)_3CCH_2$	180/0.8	2.07	105/0.05	1.3	188/0.2	2.4
$(C_2H_5)_2CH$	165/1.0	2.08	112/0.05	1.0	178/0.5	2.0
$(CH_3)(C_3H_7)CH$	162/0.5	2.06	135/1.0	1.0	175/0.05	2.0
$(CH_3)(C_3H_7)CH$	162/0.6	1.98	131/0.5	1.0	156/0.01	2.0
$(CH_3)_2(C_2H_5)C$	154/0.5	1.97	98/0.1	1.0	95/0.1	1.0

purity of commercial grade $Si(OC_2H_5)_4$ can be markedly increased by single distillation. Table 3 shows the effect of distillation on the purity of $Si(OC_2H_5)_4$ as reported by Gossink et al.[10] It should be noted that the ethyl silicate was converted into SiO_2 by hydrolysis with a 1% HCl solution followed by drying at 300°C; the chemical analysis was done with this powdered SiO_2.[10] These workers reported that multiple distillation of $Si(OC_2H_5)_4$ had little effect on the purity of SiO_2 produced from $Si(OC_2H_5)_4$. Hence it was assumed[10] that at a certain stage, it is no longer the purity of the ethyl silicate but rather contamination from the environment during the hydrolysis and drying process which limits the purity of the resulting SiO_2. Thus, the importance of strict control of clean room conditions and clean room handling procedures are very important in the purification and in maintaining purity.

Table 3: Effect of Distillation on the Purity of Ethyl Silicate[10]

ppb	Cr	Mn	Fe	Co	Ni	Cu
starting product	15	10	86	0.74	200	200
once distilled	2.2	0.79	31	0.3	11	20
residue	140	87	1000	1.1	400	1000

SYNTHESIS OF HOMOGENEOUS GELS/GEL-MONOLITHS

It should be noted that the conversion of gel to glass can be achieved either by the melting of noncrystalline homogeneous gels or by sintering of noncrystalline homogeneous gel-monoliths. In both the techniques the achievement and maintenance of the noncrystallinity and homogeneity of gels are important technical requirements for the preparation of ultrapure highly transparent glasses.[22] However, it is absolutely necessary that when the glass bodies are produced by the sintering of gel-monoliths, the gel-monoliths must remain in a perfectly noncrystalline and homogeneous state during the conversion of gel to glass by sintering around T_g. When this condition is satisfied, the sol-gel process is an excellent and unique way of making large ultrapure glass bodies of any desired shape.

However, systematic fundamental studies on the homogeneity, and the nucleation and crystallization kinetics of ultrapure gels are very limited.[22-25]

The starting metal alkoxides which will undergo hydrolytic polycondensation at different rates would also produce inhomogeneities in multicomponent gel systems.[22,25] The processing parameters such as pH and the concentration of water added during hydrolytic polycondensation also influence the homogeneity and crystallization behavior of multicomponent gels.[22]

Previous work of Konijnendijk et al[9] on the preparation of alkali borosilicate glasses by the melting of gels shows that the homogeneity, both macroscopic and submicroscopic, of glasses produced from gels is achieved in much shorter melting time than in the conventional melting process. The requisite melting time is reduced to 1/2 to 1/3 of the time needed in the conventional process and depends only on whether or not the melt can be made bubble free after the gel has been fired at 200°C.[9] Recently, Mukherjee and Mohr[26] compared the homogeneity of alkali borosilicate glasses prepared by the melting of gels and by the melting of conventional glass batches. Results show that the light scattering coefficient of gel-derived glass is 3.3 dB/Km as compared to 510 dB/Km for the conventional glass melted under *identical conditions,* when both gel-melts as well as the conventional batch melts use were *not stirred at all* during melting for 4 hours. The low scattering coefficient (3.3 ± 0.3) dB/Km compares with the value obtained with conventional glass made by melting and stirring for a long period. The results imply that pure and homogeneous glass from gels can be prepared without contamination resulting from the longer melting times and from stirring the glass melts which are essential for preparing glasses by the conventional method.

DRYING OF GEL-MONOLITHS/GEL-POWDERS[27-30]

Usually two methods of drying are used for the removal of solvents from the gel-monoliths:

(a) Drying by slow rate of evaporation of solvents at normal atmospheric pressure.[27,30]

(b) Evaporation of solvents under supercritical conditions in a pressure vessel:[27,28] This technique is an excellent way of maintaining integrity of gel-monoliths having appropriate pore structure and pore morphology suitable for sintering without entrapment of gases or H_2O into the pore structures.

The selection of drying technique for gel-powders will depend on the technique which (sintering or melting) will be used for the conversion of gel to glass. However, the selection of drying techniques for making ultrapure gels should consider the following aspects: (a) applicability of a closed system; (b) drying time; (c) the maintenance of purity in the subsequent processing steps such as the hydroxyl group removal and the conversion of gel to glass.

REMOVAL OF RESIDUAL ORGANICS AND HYDROXYL GROUPS

The residual organics, molecular water and structural hydroxyl groups should be removed before the densification or melting of gels. The removal of residual organics can be achieved relatively easily by thermal treatment in an oxidizing atmosphere in the temperature range 300°C to 500°C. However, the removal of hydroxyl groups to a few ppm level or below ppm level requires special chemical treatments with reactive gases such as CCl_4, Cl_2 which are normally used to prepare ultra low H_2O-content glass from the vapor deposited porous wave guide preforms. The following two approaches have been used to dehydroxylate the gel-monoliths or gel-derived glass to a very low level (few ppm) of H_2O.[30-33]

Approach 1

When the conversion of gel to glass is done by the consolidation of the gel-monoliths by sintering at around T_g, the removal of hydroxyl groups is done by treating the porous gel with CCl_4 or Cl_2 and subsequently consolidating in O_2/He atmosphere at around T_g. A further reduction of hydroxyl groups can be achieved by exchanging hydrogen with deuterium.[33] The following physico-chemical factors of the gel should be considered for the effective removal of hydroxyl groups from the gel-monoliths:

(1) Porosity and pore morphology of the gels during dehydration should be such that the reaction between hydroxyl groups and reactive gases can take place at high rates. In other words, the reactive gas can have access to the structural hydroxyl groups in gel-networks.

(2) The products of dehydroxylation can be removed readily before the collapse of the pores and the densification of the gel monoliths.

(3) The selection of the dehydroxylation temperature should be done after considering (a) the rate of reaction of hydroxyl groups with the reactive gases which will be used, and (b) the pore structure change in relation to temperature.

The dehydroxylation of supercritically dried SiO_2 gel monoliths by chlorination in the temperature range 600°C to 1000°C and subsequently by exchange H by D at around 1000°C is recently reported.[33] A typical schedule for removing hydroxyl groups to less than 1 ppm level from silica gel-monolith is given below.[30,33]

(1) Room temperature to 300°C: heating rate 300°C/hr.

(2) Holding for 5 hours at 300°C.

(3) 300°C to 600°C: heating rate 300°C/hr.

(4) Holding for 4 hours at 600°C.

(5) Reaction with Cl_2/Ar mixture during treating from 600°C to 1000°C; heating rate: 300°C/hr.

(6) OH–OD exchange reaction at 1000°C.

(7) 1000°C to 1300°C: heating rate 300°C/hr.

(8) Holding at 1300°C for 0.3 hr.

Approach 2

In this approach, the porous gel powders particularly in the system where melting is necessary for the conversion of gels to glass are first dehydroxylated by the treatment with reactive gases and subsequently, are melted in dry nitrogen atmosphere in ultrapure silica crucibles. Mukherjee et al[31] reported the removal of hydroxyl groups from gel-derived alkali-borosilicate systems. The thermal treatment schedule used for the removal of organic and hydroxyl groups from 60 SiO_2–25 B_2O_3–25 Na_2O gels was as follows:[31]

(1) Porous gels were packed in a closed vertical column and were

heated to around 300°C, passed CCl_4 gas carried by dry oxygen through the porous gels for 3 to 4 hours;

(2) The powder was then placed in a platinum crucible covered with a high purity alumina envelope. The melting of gel was then done in dry N_2 (−60°C dew point) flowing atmosphere. The melting was done for about 4 to 5 hours. Hydroxyl concentration of the glasses thus prepared was in the range 2 to 6 ppm level.

CONVERSION OF GEL TO GLASS

It is stated before that the conversion of gel to glass can be done by (a) sintering at around T_g and (b) by melting. The sintering of gel-monoliths or pre-formed powder-compact is the most desirable method for avoiding contamination of impurities from containers. However, the conversion to glass by sintering might not be desirable with gel-systems which tend to crystallize during gel to glass transformation at around T_g.

The conversion of gel to glass by melting is the alternative method. If the homogeneity of the gels produced is high, the melting of gels in a container can be done in very short time and, thus, the contamination can be avoided to a great extent, because the melting time is considerably reduced. The contamination can arise during melting from the furnace environment and from the crucible.[34] A platinum crucible is not desirable because of the dissolution of platinum and transition metal impurities in Pt-alloy by glass melts.[35] Alumina crucibles contain very high concentration of Fe (100 parts in 10^6) impurities.[34] Pure silica glass crucibles are suitable for the melting of low melting glasses. However, if the composition is high melting or corrosive, the dissolution of SiO_2 will introduce inhomogeneities into the glass. Another approach for minimizing the contamination during melting in an SiO_2 crucible is to melt the glass directly in a radio frequency induction furnace while cooling the crucible by water cooled tubes.[36]

In the future, when the processing of materials in a microgravity environment is developed, melting ultrapure homogeneous gel-powders in a containerless mode in the microgravity environment of space will offer a unique approach for making ultrapure glasses from gels.[12,37] Eventually, the acoustic levitator for containerless melting of small gel samples on earth might become a laboratory scale technique for melting small ultrapure gel samples which could be used for scientific studies.

GEL PROCESSING IN A CLEAN ROOM FACILITY

It is mentioned earlier that there are several unique advantages (such as purification of starting chemicals, homogeneity and gel to glass transformation by sintering at lower temperatures). However, the laboratory environment, laboratory techniques and procedures that are needed to synthesize and process gels in the ultrapure state are extremely important in controlling the contamination. The contamination control will depend on the following experimental factors:

(1) Cleanliness of work area: environment[38,39]

(2) Laboratory materials and container materials[40]

(3) Cleaning methods of containers[40]

(4) Heating environment

(5) Laboratory techniques/technical skill in maintaining ultrapurity in clean rooms.

Published systematic investigations related to the contamination control in ultrapure gel processing are practically non-existent. Beam and Mukherjee[41] reported a study on the preparation of ultrapure gels in an excellent clean room facility at the Battelle-Columbus Laboratories. The important features of ultrapure SiO_2 processing as reported[41] in their work are summarized below.

The work area, where gel synthesis or any handling of gels is done in an open environment, should be restricted to class 100 laminar flow benches. Class 100 designates a work area which contains no more than 100 particles per cubic foot which are 0.5 micron in size or larger, and no particles greater than or equal to 5.0 microns in size. However, airborne particles are not the only source of contamination in a clean room. To effectively reduce the transition metal content of the clean area, the entire facility is constructed with plastics, anodized aluminum and stainless steel. No mild steel articles are allowed into the area without proper modification to conceal the steel components. The water used should be 18 megohm-cm^{-1} specific resistivity and free from particulates. It should be pointed out that an ordinary laboratory maintains approximately 2 million particles per cubic foot greater than 0.5 micron in size. A clean room facility itself is of class 10,000 designation while the work bench area is considered class 100. An item in the class 100 clean area maintains its cleanliness as long as it is kept in the clean area. The item should be first sealed in a closed system before transferring to any other area to avoid contamination. Container material is important also. Studies show that container materials can be rated in the following sequence for ultrapure processing applications: polyfluorocarbons $>$ polyethylene $>$ vitreous silica $>$ platinum $>$ borosilicate glass.[41] Hence, class 100 labware and containers are made of Teflon, polyethylene, polypropylene or pure silica glass. Thus, the contamination of transition metal impurities caused by the leaching of containers is reduced. The purification of reactant liquids and solvents by distillation and by filtration through microfilters to remove particulates in the class 100 work area is an important factor in improving purity. If the composition permits, the leaching of impurities from the gel after thermal treatment by treating with appropriate ultrapure acids is an important processing step in improving purity. An analysis of SiO_2 gel samples taken before and after leaching with ultrex acid indicated a significant reduction of certain metallic impurities in the gel.[41] The highest increase in impurity level was observed during heat treatment of the gel performed in a class 10,000 area. Once the gel is removed from clean 100 area, the chances for contamination are greatly increased.[41]

CONCLUSION

In conclusion, it may be stated that we are in the early stages of developing sol-gel processes for preparing ultrapure glasses. Eventually, the sol-gel process will become a useful approach to making ultrapure glasses of various compositions and shapes. While the purity might not be as high as that obtained by vapor deposition techniques, the sol-gel process offers technical as well as economic flexibility, primarily because of generic advantages.

The following chart gives the important processing steps, contamination sources and the steps to be taken for purification:

Processing Steps	Causes of Contamination	Steps for Prevention of Contamination
1. Starting Chemicals Purification	Environment, Containers, Cleaning Procedures	Selection of starting chemical Chemically durable container, cleaning methods, class 100 Clean bench
2. Gel Preparation	Laboratory Environment, Labwares, Lab techniques	Contamination free and non-leachable Labwares; Cleaning techniques, Filtration of solutions, Class 100 Clean Bench facility
3. Heat Treatment of Gels	Environment, Container Flowing gases, (if any):	Closed System, Class 100 Clean Bench facility, Handling, Removal of impurities by leaching, if applicable, purification of gases by microfilters

4. Final Densification or melting	Furnace Environment Flowing Gas and impurity, Container Materials, Period of Experiment	Closed System, Purified and gas filtered Gases. Shorter Period, Choice of Container Materials, Containerless Melting in Microgravity environment, or use of Levitation Systems

REFERENCES

1. Maurer, R.D., *J. Non-Crystalline Solids 47,* 135-146 (1982); Proc. IEEE. 61 (4) 452-462 (1973).
2. Beales, K.J. and Day, C.R., *Phys. Chem. Glasses 21* (1), 5-21 (1980).
3. Schultz, P., *J. Am. Ceram. Soc. 57,* 309 (1974).
4. Keck, D.B., Maurer, R.D. and Schultz, P.C., *Appl. Phys. Lett. 22,* 307 (1973).
5. Griscom, D.C., *J. Non-Crystalline Solids 40,* 211 (1980).
6. Pal, B.P., *Fiber and Integrated Optics 2,* (2) 195 (1979).
7. Keck, D.B. and Schultz, P.C., U.S. Patent 3,711,262 (1973).
8. Dislich, H., *Angew. Chem. Int. Ed. (Eng.) 10* (6) 363-370 (1971).
9. Konijnendijk, W.L., van Duuren, M. and Groenendijk, H., Preparation of homogeneous borosilicate glasses by wet-chemical techniques, *Verres Refract. 27* No. 1, 11-13, Jan.-Feb. 1973.
10. Gossink, R.G., Coenen, H.A.M., Engelfriet, A.R.C., Verheigke, M.L. and Verplanke, J.C., Ultrapure SiO_2 and Al_2O_3 For the Preparation of Low-Loss Compound Glasses, *Mat. Res. Bull. 10,* 35-40 (1975).
11. Mukherjee, S.P., *J. Non-Crystalline Solids 42,* 477 (1980).
12. Mukherjee, S.P., *Materials Processing in the Reduced Gravity Environment of Space.* C.E. Rindone (Ed.), Elsevier, Amsterdam (1982).
13. Susa, K. et al, *Electron. Lett. 18,* 499 (1982).
14. Puyane, R., Harmer, A.L. and Gonzalez-Oliver, C.J.R., 8th Europe Conference. Comman Cannes, France, 623-628 (1982).
15. Zarzycki, J., in: *Glass --- Current Issues,* 203-231, A.F. Wright, J. Dupuy (Eds.), Martinus Nijhoff Publishers, Boston (1985).
16. Rabinovich, E.M., et al, *J. Non-Crystalline Solids 47,* 435-439 (1982).
17. Rabinovich, E.M., Johnson, D.W., MacChesney, J.B. and Vogel, E.M., *J. Amer. Ceram. Soc. 66* (10) 683-688 (1983).
18. Scherer, G.W. and Luong, J.C., *J. Non-Crystalline Solids 63,* 163 (1984).

19. Stober, W., Fink, A. and Bohn, E., *J. Colloid Interface Sci. 26,* 62 (1968).
20. Second Inter. Workshop on "Glasses and Glass-Ceramics from Gels;" *J. Non-Crystalline Solids 63,* 1-2 (1984.
21. Bradley, D.C., Mehrotra, R.C. and Gaur, D.P., *Metal Alkoxides,* Academic Press, N.Y. (1970).
22. Mukherjee, S.P., *J. Non-Crystalline Solids 63,* 35-43 (1984).
23. Mukherjee, S.P., Zarzycki, J. and Traverse, J.P., *J. Mater. Sci. 11,* 341 (1976).
24. Yamane, M., Inoue, S. and Yasumori, A., *J. Non-Crystalline Solids 63,* 95 (1984).
25. Mukherjee, S.P. and Sharma, S.K., *J. Non-Crystalline Solids 71,* 317-325 (1985).
26. Mukherjee, S.P. and Mohr, R.K., *J. Non-Crystalline Solids 66,* 523-527 (1984).
27. Zarzycki, J., Prassas, M. and Phalippou, J., *J. Mater. Sci. 17,* 3371-3379 (1982).
28. Mukherjee, S.P. and Debsikdar, J.C., Influence of Gel Preparation Procedures and Drying Techniques on the Porestructures of Dried Silica Gel-Monoliths, *Ceram. Bull. 62* (1983).
29. Hench, L.L., in Reference 15, 259-262.
30. Mukherjee, S.P. and Cordaro, J.F., Sintering Behavior of Silica Gel Monoliths Having Different Pore Structures. Presented at the 85th Annual Meeting: Amer. Ceram. Soc., Chicago, April 25-27 (1983).
31. Mukherjee, S.P., Beam, T.L. and Holman, R.A., Removal of Hydroxyl Groups from Gels and Gel-Derived Glasses in the $Na_2O-B_2O_3-SiO_2$ System, presented at the 84th Annual Meeting: Amer. Ceram. Soc., Cincinnati, May 2-5 (1982).
32. Phalippou, J., Woigner, T. and Zarzycki, J., in: *Ultrastructure Processing of Ceramics, Glasses and Composites,* L.L. Hench and D.R. Ulrich (Eds.), Wiley (1984).
33. Debsikdar, J.C., Pascucci, M.R. and Wills, Roger, Studies on Dehydration of Silica Gel-Monolith For Producing Low-Hydroxyl-Content Monolithic Silica Glass, presented at the Annual Meeting: Amer. Ceram. Soc. (1985).
34. Scott, B. and Rawson, H., *Glass Tech. 14,* 115 (1973).
35. Ginther, R., *J. Non-Cryst. Solids 6,* 294 (1971).
36. Scott, B. and Rawson, H., *Opto-Electronics 5,* 285 (1973).
37. Barmatz, M., in Reference 12, pp. 25-37.
38. Zief, M. and Nesher, A., Clean Environment for Ultrapure Analysis, *Environ. Sci. Technol. 8* (7), G77 (1974).
39. Zief, M. and Nesher, A.G., Laboratory For Preparation of High-Purity Clinical Chemicals. *Clin. Chem. 18* (5) (1972).
40. Zief, M. and Speights, R., *Ultrapurity–Methods and Techniques,* Marcel Dekker, New York (1972).
41. Beam, T.L. and Mukherjee, S.P., "Ultrapure Gel Processing in a Clean Room Facility," presented at the 85th Annual Meeting: Amer. Ceram. Soc., Chicago, April 25-27 (1983).

13

Particulate Silica Gels and Glasses from the Sol-Gel Process

Eliezer M. Rabinovich
AT&T Bell Laboratories
Murray Hill, New Jersey

INTRODUCTION

When amorphous silica powder of high surface area is dispersed in a liquid (water, alcohols) it can form a gel. This was previously called "colloidal" gel to distinguish it from alkoxide-derived gels.[1-5] However, this may result in some misunderstanding, since "sol" and "gel" are terms of colloidal chemistry, and "non-colloidal" sols and gels are not possible. Following Flory,[6] I prefer, from now on, to refer to the former "colloidal" gels as "particulate" gels, since they are formed from powders or aggregates which appear to be composed of particles.

A classification of the gels partially based on Flory[6] may be as follows when the consideration is limited to silica and high-silica (or silicic acid) sols and gels:

(1) Covalent polymeric networks, completely disordered;

(2) Particulate disordered structures

(3) Particulate-aggregate structures, predominantly disordered but with regions of local order;

(4) Porous particulate-aggregate structures.

The schematics for these 4 cases are shown in Figure 1.

Gelation of the silicic acid formed by hydrolysis of tetraethyl orthosilicate (TEOS) at low pH (1 to 2) is a result of a chemical reaction, growth, and interlocking of a three-dimensional (3-D) network of the first type.[5] A water-alcohol mixture is located in spaces of this network leaving high porosity upon drying. This network of composition $SiO_{2-n}(OH)_{2n}$ is similar to a 3-D Zachariasen network of glass.[5] in fact, a flask with an alkoxide silica gel formed at low pH contains only one giant gel molecule but 2 phases. This "molecule" can be easily broken into smaller ones using mechanical means. Both processes–gelation as a result of a polymerization reaction and mechanical destruction of the gel–are irreversible. Gelation can occur if an original sol is at rest, or if it is under shear stress, although in the latter case the gel will appear to be broken into pieces.

Gelation of particles and aggregates (gels of the 2nd, 3rd and 4th types) is an entirely different process, because no covalent or ionic bonds are responsible for this process. The main feature of this gelation is its reversibility, called thixotropy, and the sols exhibit clearly non-Newtonian behavior. Gelation of a particulate sol occurs only when it is at rest, and fluidity can be restored by shear stress.

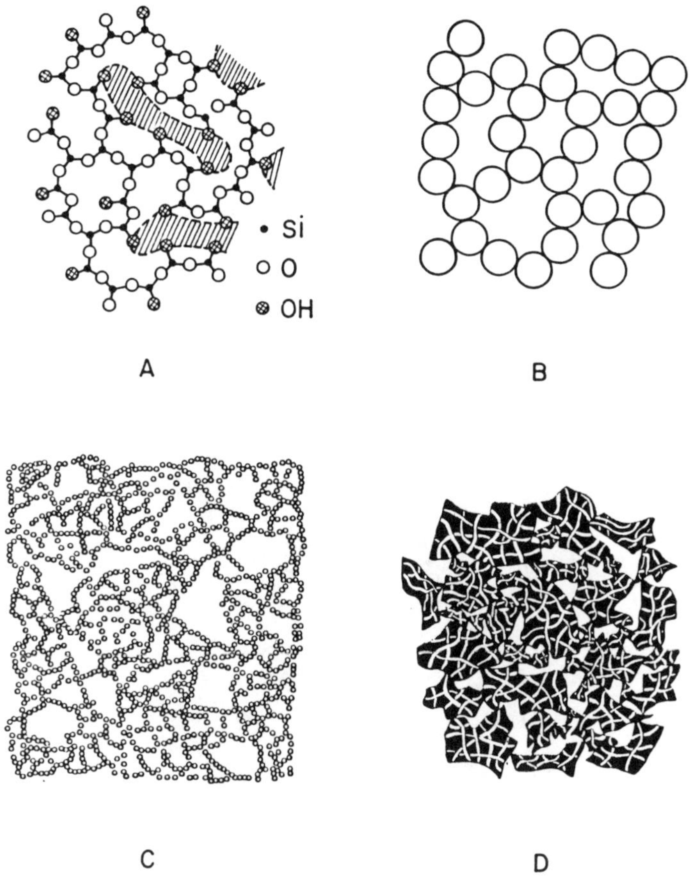

Figure 1: Four types of microstructures in silica gels: (A) disordered covalent polymeric networks (shaded areas represent pores); (B) particulate disordered structures; (C) particulate-aggregate structures with regions of local order inside the aggregates; (D) porous particulate-aggregate structures.

We previously described[1-4] preparation of high-silica glasses from particulate gels by sintering at 1300°C to 1500°C, and other works[8,9] in this area have appeared since. This paper reviews state-of-the-art preparation of particulate sols and gels and glasses made from them.

SOURCES OF SILICA POWDERS AND PARTICLES

There are many techniques for preparation of particles of silica and other oxides with narrow size distribution. A significant part of this paper is devoted to preparation of glass from commercial fumed silica. Two main sources of this silica are Cabot Corp., which sells it under the trade name Cab-O-Sil®,[9] and Degussa, which produces Aerosil®.[10] Both are essentially similar, completely amorphous products made by hydrolysis of $SiCl_4$ in an oxygen-hydrogen flame according to the reaction:

$$SiCl_4 + 2H_2 + O_2 \rightarrow SiO_2\downarrow + 4HCl\uparrow$$

The temperatures of the reaction are significantly different: 1800°C for Cab-O-Sil[9] and 1000°C for Aerosil.[10] The material resembles smoke when it is formed in the flame, hence its name "fumed silica." Cab-O-Sil is produced with a BET surface area ranging from 90 to 380 m^2/g (its average primary particle diameter runs from 24 to 7 nm),[9] while Aerosil's surface area ranges from 50 to 400 m^2/g.[10] Both materials are very pure (e.g., Cab-O-Sil contains less than 40 ppm of metallic contaminants). The properties of Cab-O-Sils are described in Tables 1 and 2;[9] properties of Aerosils are similar.[10] As seen from Table 2, fumed silica is structurally identical to silica glass: it has the same density and refractive index. The particles of fumed silica may be single or clustered (Figure 2).[7]

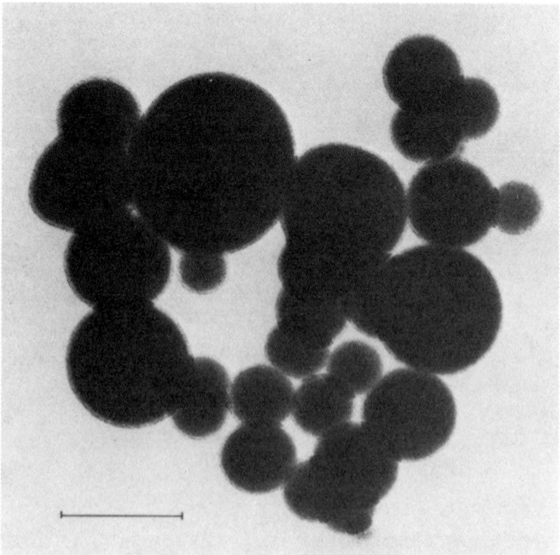

Figure 2: Electron micrograph of Aerosil OX-50 with BET surface area of ~50 m^2/g; the bar = 100 nm (after Scherer and Luong[7]).

Table 1: Characteristics of Different Grades of Cab-O-Sil®[9]

Grade								
Standard	L-90	LM-130	LM-5	M-5	PTG	MS-55	HS-5	EH-5
Densified	L-90D	LM-130D	LM-7	MS-7	—	MS-75	—	S-17
BET Surface Area (m^2/g)	90 ± 15	130 ± 15	160 ± 15	200 ± 20	200 ± 20	250 ± 20	325 ± 25	380 ± 30
Primary Particle Diameter (microns)	.024	.018	.016	.014	.014	.011	.008	.007
Bulk Density (kg/m^3)								
Standard	48.2	48.2	48.2	32.1	32.1	32.1	32.1	32.1
Densified	80.3	80.3	80.3	80.3	—	80.3	—	80.3
Loss on Heating (Max. % at 105°C)	0.5	1	1	1.5	1.5	1.5	1.5	1.5
Loss on ignition (Max. % at 1000°C)	1	1	1	2	2	2	2	2.5

Table 2: Typical Properties of Cab-O-Sil® of All Grades[10]

Property	Typical Values
pH (4% Aqueous Slurry)	3.6 – 4.3
325 Mesh Residue (%)	0.02 Max
Density, kg/m^3	2200
Refractive Index	1.46
Assay (% SiO_2)	>99.8

In his recent review of sol-gel processing of glass powders,[11] Fleming describes preparation of silica by hydrolysis of $SiCl_4$ and silicon alkoxides. Hydrolysis of alkoxides is a method of preparation of amorphous powders and aggregates in the combined alkoxide-particulate method[12-14] described below. Processes of preparation of monodispersed powders of silica and other oxides from alkoxides are described by many researchers.[15-18] These processes are of interest in particulate sol-gel glass technology because they allow for the preparation of doped and multicomponent powders.

GELS FROM ALKALI SILICATES

Shoup[19] has detailed a method for preparing silica bodies with controlled porosity from mixtures of potassium and sodium silicate solutions and colloidal silica sol in the presence of formamide $HCONH_2$. He describes using a 40 wt. % dispersion of colloidal silica (Ludox HS-40 of DuPont), but makes no mention of surface area and particle size. He found that formamide helps to build coherent silica structures (gels) as a result of reducing pH of the sols which are initially strongly alkaline. This process involves the following reaction:

$$HCONH_2 + NaOH + H_2O \rightarrow HCOONa + NH_4OH$$

Eventually a gel is formed which is distinctly separated into silica-rich and alkali-rich phases. The latter phase can be easily leached in weak acidic solution to produce porous silica bodies with <200 ppm of alkali ions. The Ludox gels after the addition of only 20 wt. % of sodium silicate, while preparations with potassium silicate are more stable. The average pore diameter strongly depends on a colloidal silica to silicate solution ratio and could be as low as 30 nm or as high as 600 nm. Figure 3 shows the average pore diameter as a function of the amount of the colloidal silica. In every gel the pore size distribution was narrow: at least 80% of the pores were within the limits of ±30% and, as a rule, ±10% from the average value. The sol in the course of gelation exhibited 13% to 25% linear shrinkage.

Shoup[19] explains the formation of bodies with narrow pore size distribution by a process involving nucleation. He points out that the colloidal silica remains as a stable dispersed phase, and serves as the nuclei for the formation of the polymer network from alkali silicates. When the amount of these silica particles is small, they grow extensively and produce large pores. On the other hand, at high concentrations of nuclei less growth is possible and the pores are

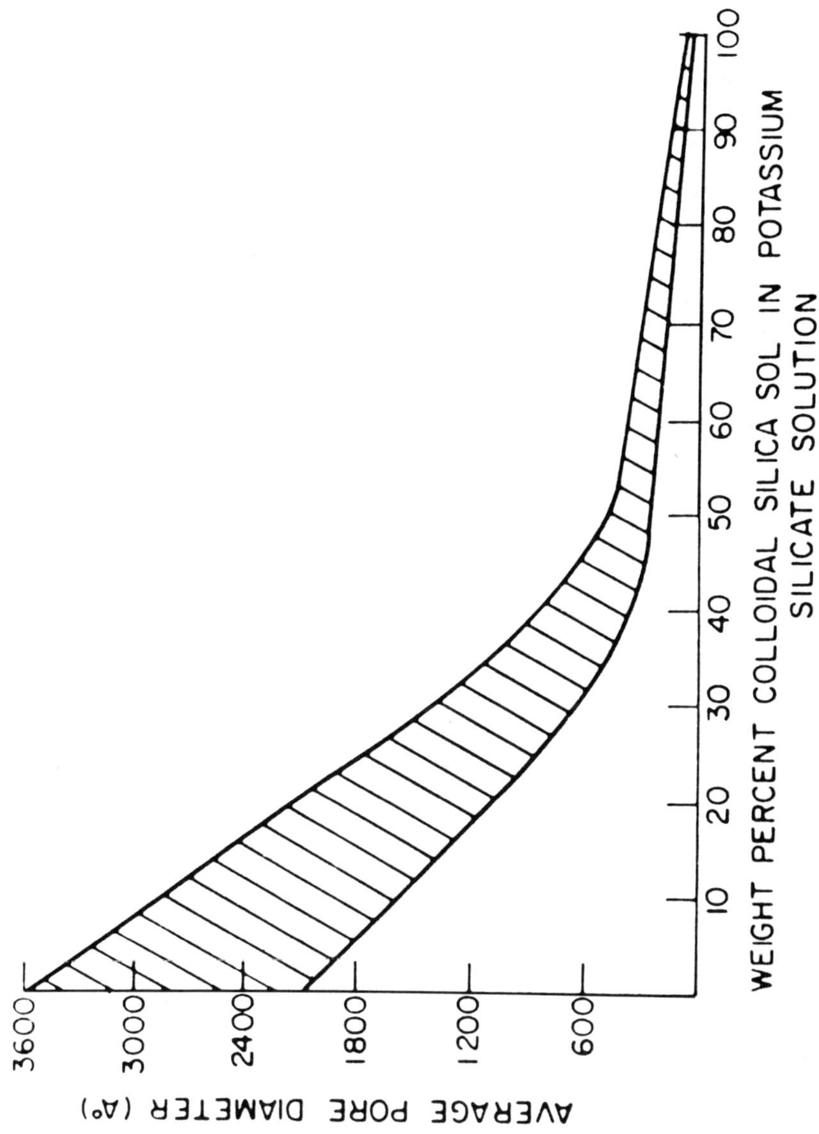

Figure 3: Dependence of the average pore diameter on the amount of a colloidal silica sol in potassium silicate solution (after Shoup[19]).

smaller. This explanation has been supported by electron micrographs. The ability of Shoup's gel bodies to be dried without mechanical fracture was greatly increased with the increase in pore sizes above 60 nm. This is because structures with larger pores are better able to withstand capillary forces during drying. In this way, relatively large bodies were prepared.

There is no description of the sintering of gel bodies in Shoup's paper,[19] but the patent filed by Shoup and Wein[20] claims that these bodies can be sintered to fused silica at temperatures of above 1350°C in a He atmosphere. In this patent, the leaching of originally formed gels was conducted in 1 M NH_4NO_3 solutions.

Apparently, Shoup and Wein's process can produce glass bodies similar to those in the Vycor® (Corning Glass Works) process[21] without a preliminary melting of glass. However, neither the paper[19] nor the patent[20] make clear how long the drying takes or how large the bodies are. A problem can also be caused by traces of alkali ions remaining in the leached material. We know[22,23] that even minor amounts of these ions strongly accelerate crystallization. Because sintering takes place below the liquidus temperature, preparation of transparent glass can be difficult if crystallization is rapid.

GLASSES FROM HIGH-SURFACE AREA PARTICULATE GELS

This chapter mainly describes work with high-silica gels and glasses made at AT&T Bell Laboratories during 1981-1986.[1-5,12-14,24-33]

Fumed Silica Gels

Preparation of glasses from fumed silica Cab-O-Sil M5 (see Table 1) was first described by Rabinovich et al;[1,2] Johnson et al;[3] and Wood et al.[4] Its main steps are considered below.

Dispersion. Dispersion of fumed silica in water is a process strongly affected by pH and various additions. If we take Cab-O-Sil (with a BET surface area near 200 m^2/g) and gradually add it to water in a blender, a castable sol at a solid-water weight ratio of 68:100 can be prepared; its pH will be ~2.7. Adding reagents which increase the pH will transform this flowable sol to an almost dry-looking material. Acidic additions, with the exception of HF and other F-containing compounds, strongly increase fluidity. It is amazing to observe how quickly, sometimes even with the sound of minor explosion, a mixture which looks like wetted powder can be converted to a sol with milky consistency. When citric acid is added, fluidity can be achieved, under continued blending, for the 100:100 ratio, however, this sol cannot be cast: it gels in several seconds when blending stops.

Gelation. The sol (normally without additions, 68:100 ratio) gels after pouring into a stoppered mold and allowing it to rest. Gelation at room temperature normally takes 1 to 2 hours, and can be strongly accelerated by heating to 60°C. Less concentrated sols require more time for gelation.

Unlike the irreversible gelation of alkoxide sols, gelation of particulate sols can be reversed. Mixing again will restore the ability to flow, but when mixing is stopped, the rate of viscosity increase will be higher than the first time. The gel is able to maintain its shape without a container and can be pushed out of molds for drying.

The gelation mechanism was discussed in detail by Iler[34] and by Wood et al.[4] who explained initial contact between particles in terms of the formation of hydrogen bonds. The surface of every silica particle, even dried, when equilibrated with ambient air, is covered by silanol bonds. Belyakova et al.[35] found that the concentration of the silanol groups does not depend on the nature of underlying silica and is equal to about 11 ± 1 μmol/m^2, or 6.6 OH groups per nm^2. The schematic representation of structures on the silica surface, as discovered from infrared spectra by Wood et al.,[4] is shown in Figure 4. The figure shows the formation of clusters of $Si(OH)_m \cdot (H_2O)_n$. These clusters eventually bridge the silica particles, forming a three-dimensional network as shown in Figure 5. However, this is only the first step in the gelation process, during which the particles are still loosely bonded. The next step is a strengthening of the initial bonds as a result of neck growth between the particles, as described by Iler.[34] The solubility of silica in water is strongly dependent on the curvature of the surface: it is higher for the convex external surfaces of a spherical particle pair than for the concave surface in the point of contact of two particles (Figure 6). This difference is the cause of neck growth and strengthens the network. It is more probable that in aqueous media the deposited silica precipitates as a system of particles rather than as a monolith. These particles, in turn, are attached to the original particles and to previously deposited layers through hydrogen bonds and water molecules providing a porous structure of the neck as opposed to the non-porous nature of the original particle. This structure is schematically shown in Figure 7. When this process is complete, it results in the formation of 3-D network and a gel with type B (Figure 1) structure.

The porous necks are not mechanically strong and can be easily broken with high-shear stress. This will restore fluidity although not all the necks will be broken and new gelation will take less time, depending, of course, on amount of energy used for mixing the gel.

HF and other F-containing materials strongly accelerate gelation of colloidal silica. This effect is not unique for particulate silica, but F^- ion also strongly promotes gelation of silicic acid.[34,36] Rabinovich and Wood[36] discussed possible mechanisms for this effect. They suggest that the reason lies in the asymmetry of the electrical fields of participating ions and particles. It is well known[37] that F^- ion is able to replace O^{2-} in the Si–O network because of the similarity in ionic radii of F^- and O^{2-}. Since OH^- has nearly the same radius as O^{2-}, it also can be replaced by F^-. It is more probable that the substitution will take place on the surface with formation of superficial Si–F bonds instead of Si–OH bonds. The fact that $(NH_4)_2SiF_6$ forms when NH_4F is added to water for dispersion of Cab-O-Sil[36] proves that fluoride ion can be incorporated into the silica structure even at room temperature.

In the absence of F^-, when the surface of the micelles is covered by OH^- -groups, collision of two micelles with the same charge may result in repulsion rather than attraction. The effective initial bond can be formed through water molecules using hydrogen bonding. The OH^- groups have locally positive electric fields in the vicinity of protons, but it is not important when F^- ions are not present. When, however, part of the surface is covered with Si–F groups, containing a strongly negative F^- ion, collision of this group on one micelle with the exposed positive charge of the OH^- group on the surface of another micelle results in attraction and fixation of two micelles more frequently than without

F^-. On the other hand, presence of HF in the liquid may enhance the solubility of silica and thus accelerate strengthening of the necks at the point of contact. All this will result in a more rapid formation of the 3-D gel structure.

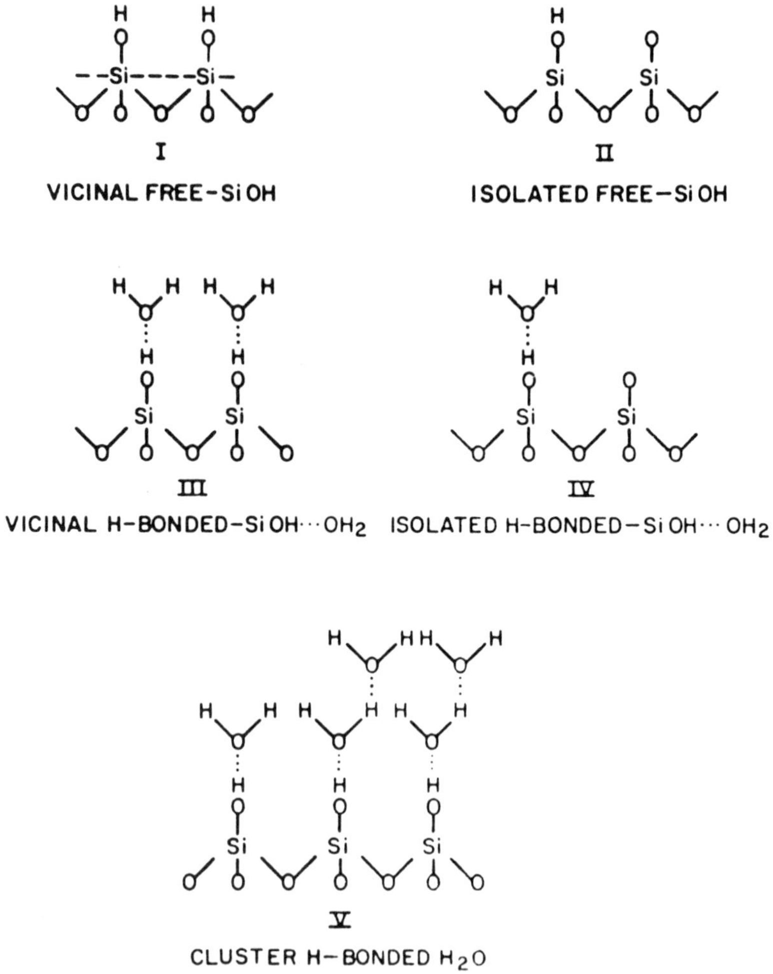

Figure 4: Schematic representation of structures on the silica surface present in gel as revealed from infrared spectra (after Wood et al.[4]). Hydrogen bonds are represented by dotted lines.

Drying and Double Processing. Gel samples prepared from particulates as described and released from molds are difficult to dry without fracture just as in the case of alkoxide gels. The reason lies in their high water content (near 60 wt. %) and small pores narrowly distributed around 14 nm (Figure 8, curve 1). Attempts to solve the problem with freeze drying in a high-humidity chamber proved to be ineffective, but a solution came with introduction of the so-called "double processing."[2]

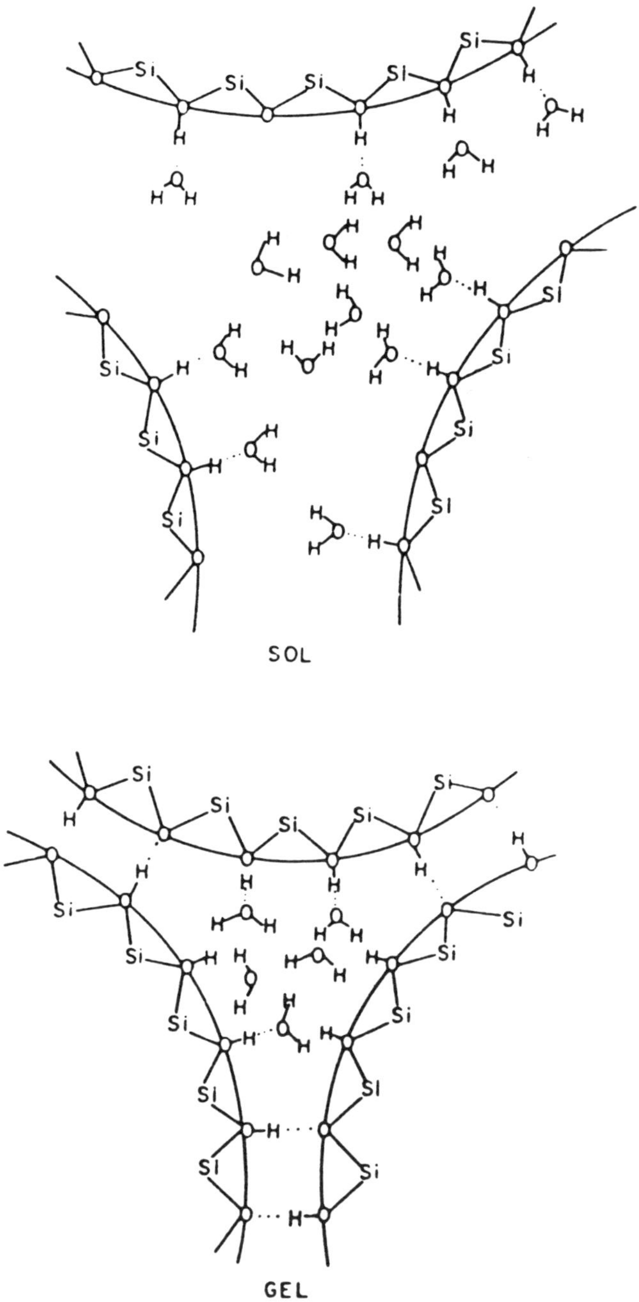

Figure 5: Diagram of changes during the gelation of particulate sols. Dots represent hydrogen bonds; curved lines represent micelle surfaces (after Wood et al.[4]).

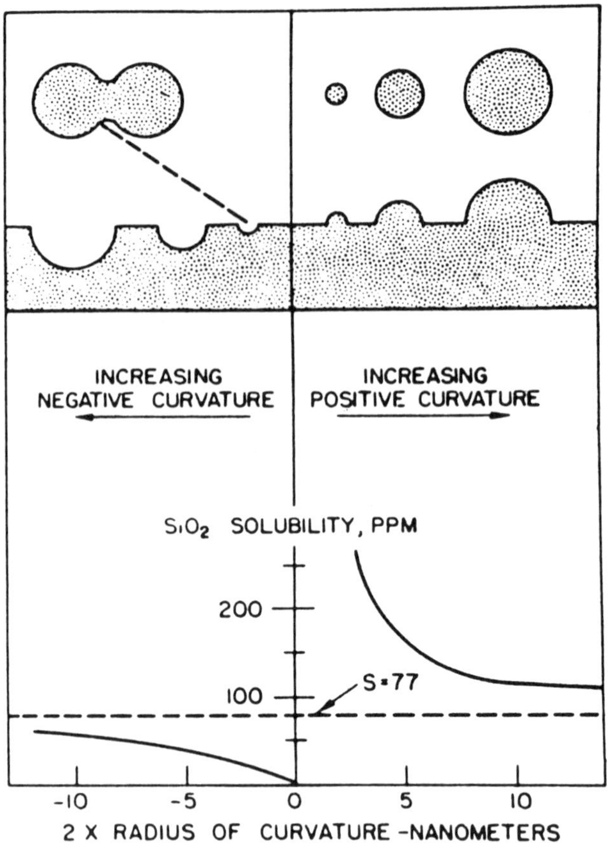

Figure 6: Dependence of solubility of silica on the radius of curvature of surface (after Iler[34]).

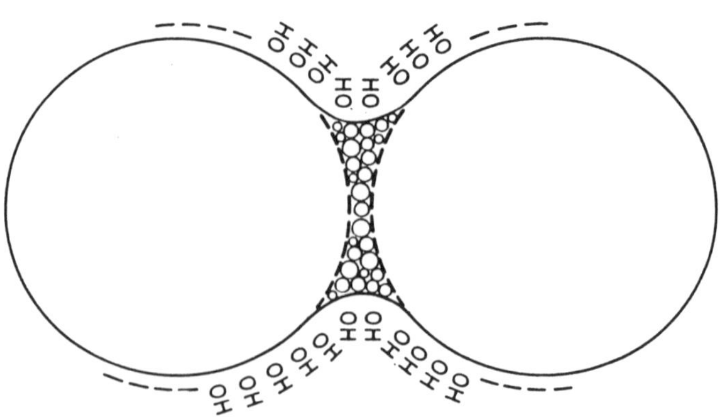

Figure 7: Pair of particles with the neck grown by the mechanism of solution-redeposition.

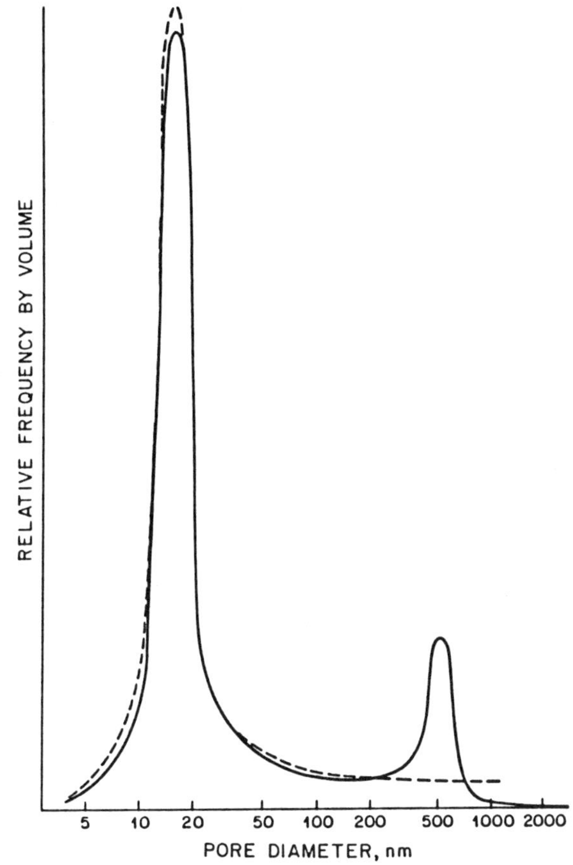

Figure 8: Pore size distribution curves for once- (dashed line) and twice-dispersed (solid line) fumed silica.

In the beginning this process was used for convenience. Dispersion of finely divided Cab-O-Sil in a minimal amount of water required gradual addition of the solids to the liquid, and it is slow. More rapid dispersion of Cab-O-Sil in an excess of water (40 g solids per 100 g H_2O) was achieved in a large blender, and this sol was then dried at 150°C to produce dense fragments with practically unchanged surface area. This denser material (designated as material A) was then redispersed, cast and gelled. When the gelled samples were left on a table for drying, it was found that they did not crack, and rods of dry gel of about 2.4 cm in diameter and up to 120 cm long, or shorter tubes with outside diameter of 3.6 cm and a wall 0.3 cm thick could be easily prepared. After the initial drying in room air the samples could be further dried in an oven.

The effect of easier drying without breaking can be understood in the frame of the gelation mechanism described above. The second dispersion cannot break all the necks and restore original particles. Therefore, the second gel inevitably consists of aggregates of different sizes. While once-dispersed gel contains only small pores of a relatively narrow distribution, the second gel looks as shown schematically in Figure 1C, and this structure was confirmed by electron microscopic pictures.[2] There are 2 kinds of pores as seen from mercury porosimetric curve 2 in Figure 8. In the first place there are intraaggregate pores similar to those in the once-dispersed material, and there also are interaggregate pores with much wider size distribution. The larger interaggregate pores provide convenient paths for removal of water, and the shapeless form of the aggregates results in their interlocking. Thus smaller shrinkage and lower drying stresses occur, and mechanical fracture is avoided.

Because the second dispersion determines the position of the interaggregate pores in the dry gel, it strongly affects the sintering kinetics and quality of the final glass. Therefore, preparation of the material for the second dispersion and the method of this dispersion are very important. Originally the second dispersion was conducted in a blender, and the solid-to-water weight ratio was 68:100. But the sintered glass contained defects, and better glass could be produced only with addition of B_2O_3. Then it was observed that heat treatment of material A to ~800°C before the second dispersion, while not changing noticeably the surface area, improved the quality of the final glass. The systematic study[2] of treatments of material A at different temperatures explained the phenomenon as due to a rearrangement of the pores during heating, and these heat-treated materials were designated as materials E. The peak on the pore-size distribution curve became more symmetric, and the volume of large pores (⩾35 nm in diameter) was reduced from 31% of the total porosity for material A (dried at 150°C) to ~25% for material E treated at 800°C, or 20% for a material treated at 950°C, while the total porosity and the surface area remained unchanged.[2] Besides, infrared spectroscopy showed[4] a progressive decrease in the content of molecular water and hydrogen-bonded Si–OH groups with increased temperature of the heat treatment above 600°C; these bonds were only partially restored during the second dispersion. When material A was dispersed in a blender for the second time, the peak of interaggregate pores of the dried gel was located at 8 μm. The E-material treated at 950°C exhibited this peak at ~2 μm. This reduction in size of interaggregate pores resulted in better sintering of glass with fewer defects.[3] therefore, heat treatment at 800°C to 950°C for 1 to 4 hours between the two dispersions of fumed silica material became an additional step in the process, and that reduced the number and size of defects (bubbles etc.) in the glass even though it did not eliminate them completely.

The next major improvement of the process was the substitution of attrition for blending in the second dispersion.[30,31] The method chosen for this dispersion was essentially classical wet ball milling. However, because even traces of crystalline impurities in the gel would cause inevitable crystallization during sintering to glass, such hard, effective balls as porcelain, ZrO_2, or Al_2O_3 could not be used. Therefore, the dispersion was made using fused silica cylinders of ~12 mm length cut from commercial rod of 12 mm diameter. These cylinders were rotated with sand for several days to round the edges. Milling was done in bottles of borosilicate glass of ~1 liter in volume. It should be realized that although we use words "milling" and "dispersion" as interchangeable, in reality no milling occurs during the process because the surface area is not increased. The purpose of the process is breaking the aggregates and reducing the size of the interaggregate pores.

Either material A (150°C), or material E (900°C/4 hours) can be used for the second dispersion by the "ball milling" process, and glass of good quality can be produced from both materials. The weaker material A is dispersed more rapidly, but the "green" bodies (dried gel) are stronger when made of material E.

Pieces of materials A or E and water could be loaded directly into the bottle for milling, but it was found more practical to blend them initially in the blender for ~2 minutes and after that to load for ball milling. The solid-to-water ratio was 55.5:100–a little lower compared with the second dispersion in the blender alone. The bottles were rotated at ~90 rpm during times varying from 0 to 64 hours. The aggregate size distributions were studied with electronic sensing apparatus (Coulter counter),[31] and these distributions were found to be rather wide. The peak position was used to characterize the aggregate size. The peak position was reduced from 43 to 65 μm for 0 hour to near 3 μm after 16 hours for the weaker A-material (150°C), while the harder E-material (900°C) took 32 to 64 hours for the same reduction. The mercury porosimetric curves, in general, looked like curve 2 in Figure 8, but the interaggregate peak could not be resolved for the E material milled during 32 and 64 hours or for any ball-milled A material. This proves that a more uniform distribution of the porosity occurred in ball-milled double-dispersed gels compared with their blender counterparts. The position of the interaggregate peak is shown in Figure 9: as seen, it rapidly diminishes with increase in the milling time. On the other hand, the linear drying shrinkage of these gels after release from molds went up from 5 to 6% for 0 time to 12 to 14% for 16 to 64 hours of milling.

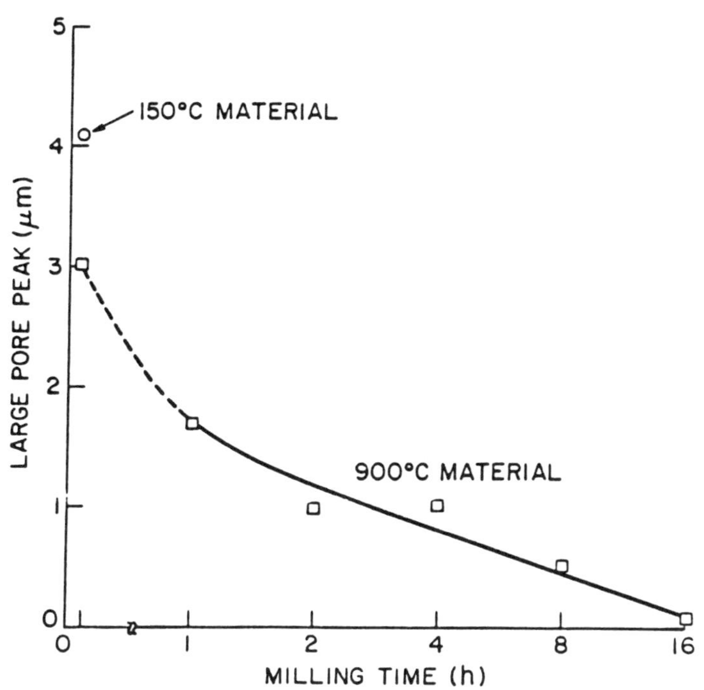

Figure 9: Position of the peaks on the pore size distribution curves for interaggregate pores vs. time of dispersion in ball mills (after Johnson et al.[31]).

The quality of sintered glass produced by the ball milling was much better than in the case of the second dispersion in the blender, and the temperature of sintering reduced. However, the question remains: since the second dispersion was invented to create interaggregate pores and to facilitate drying, will not long dispersion in the ball mills produce sols similar to the once-dispersed sol and restore the pattern of fracture during drying? The compromise depends on the geometry and size of the gel bodies to be dried and the hardness of the dispersed material. In most cases ball milling during 8 to 24 hours is satisfactory, and large bodies, dried without fracture, can be subsequently sintered into essentially defect-free glass.

Combined Alkoxide-Particulate Method

The principle of this method first reported by Rabinovich et al.[12,13] is simple. Powders precipitated from hydrolyzing alkoxides are used for redispersion in water in a process otherwise similar to that for fumed silica. There are two reasons for use of more expensive alkoxides instead of fumed silica. First, although Cab-O-Sil and Aerosil are rather pure products they still contain transition metals on the ppm level unacceptable for such applications as optical fibers. On the other hand, liquid alkoxides can be purified to produce powders of much higher purity. Second, fumed silica is very difficult to dope with foreign ions while this may be easier with alkoxides. This problem is discussed later.

The combined method for pure silica is described in detail elsewhere.[13] TEOS mixed with ethanol is hydrolyzed by water (in the presence of HCl, pH = 1, or NH_4OH, pH $\geqslant$ 11), aged at 60°C for 24 hours, and the mixture is dried by evaporating the liquids above their boiling points. The resulting fragmented gels or powders are redispersed in water using the same ball mills with fused silica cylinders as described above. The sols formed from these powders are further processed similarly to that of fumed silica.

Both acid- and ammonia-catalyzed gels can be used. However, in the case of acid-catalyzed gels (pH $\approx$1), no powder is precipitated but a network gel is formed[5] which can be dispersed in a blender while wet or by fused silica cylinders after drying and remixing with water. However, this dry gel is mechanically strong, and much unpulverized material remains even after long milling times. Although this material has a very high surface area (up to 1,000 m^2/g), this area is mostly inside of relatively large unbroken pieces (type D structure in Figure 1). The whole assemblage has non-uniform particle size and pore size distributions. Hence, each single particle can be easily sintered, but the whole body produces glass with many defects.

In the case of ammonia-catalyzed reactions (pH $\geqslant$11) of TEOS with water, precipitation of powder occurs. The reaction is not exothermic as in the case of low pH, and probably does not involve hydrolysis[38] but rather direct polymerization:

$$Si(OC_2H_5)_4 + 2H_2O \rightarrow SiO_2 + 4C_2H_5OH$$

The particle size and surface area of these powders depend on the conditions of the reaction, such as the water:alcohol:TEOS ratio, the pH, the temperature, or even the order of mixing of the components. Figure 10 shows the dependence of the BET surface area on the concentration of ammonia (pH) and on the amounts of ethanol and water. It can be seen that the surface area, which

a

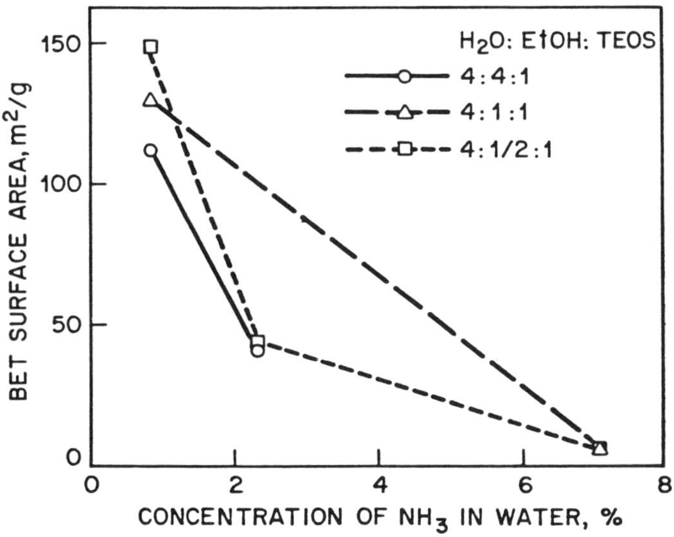

b

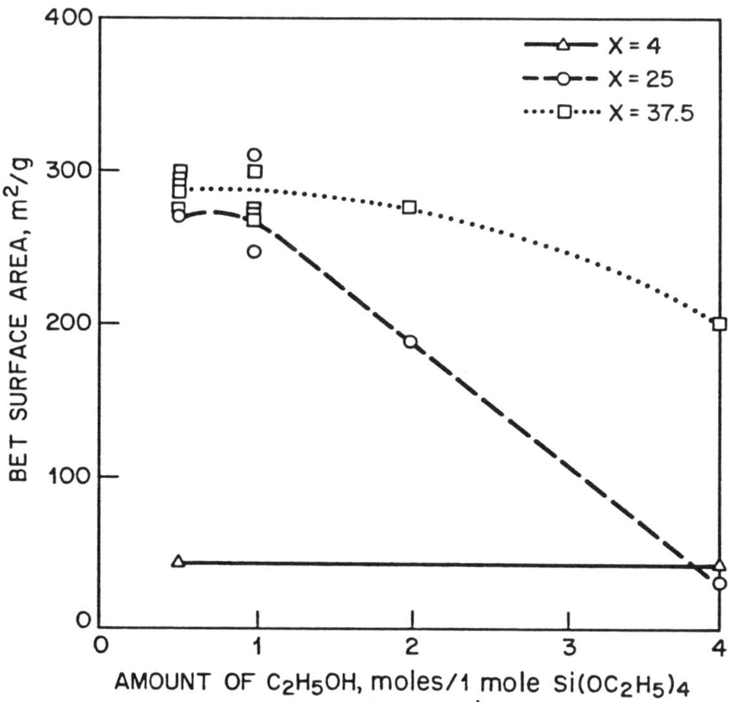

Figure 10: The BET surface area of powder produced by reaction of TEOS with water in the presence of ammonia as a function of: (a) concentration of NH_3; (b) amount of H_2O and ethanol (after Rabinovich et al.[13]).

is generally lower than in the case of acid-catalyzed gels, can still be as high as 300 m^2/g and as low as 6 m^2/g. In the case of low surface area a real powder precipitated. However, material with the area of 200 to 300 m^2/g formed a gel which upon drying produced white or translucent pieces which could be easily crushed with fingers; their strength was much lower than that of the "network" gels. We consider these pieces as aggregates of separate fine particles, although some of these particles still can have internal porosity. Gels formed from them after redispersion can be of types C or D in Figure 1. Those powdered aggregates with surface areas of 200 to 300 m^2/g were selected for further processing.

Figure 11 shows mercury porosimetric curves of one dry material prepared by hydrolysis of TEOS in the presence of ammonia and alcohol, then dried and redispersed in water by ball milling during 12, 24 and 48 hours. These curves, rather similar to curve 2 in Figure 8 for fumed silica, show intraaggregates pores with peaks near 49 nm for all three samples and interaggregates pores with a peak moving from ~1.5 μm (12 hours) to 0.04 μm (48 hours).

Recently a similar method has been described by Shibata et al.[8] who used it for preparation of a pure silica core for an optical fiber preform. The principal difference from the method described by Rabinovich et al.[13] is the absence of the second dispersion. Instead, a particulate sol prepared by hydrolysis of TEOS in the presence of ammonia and alcohol was matured to make larger particles and then directly cast into molds and gelled. The gel rods were dried slowly to prevent cracking and consolidated to transparent glass at 1500°C in an inert gas atmosphere admixed with Cl_2.

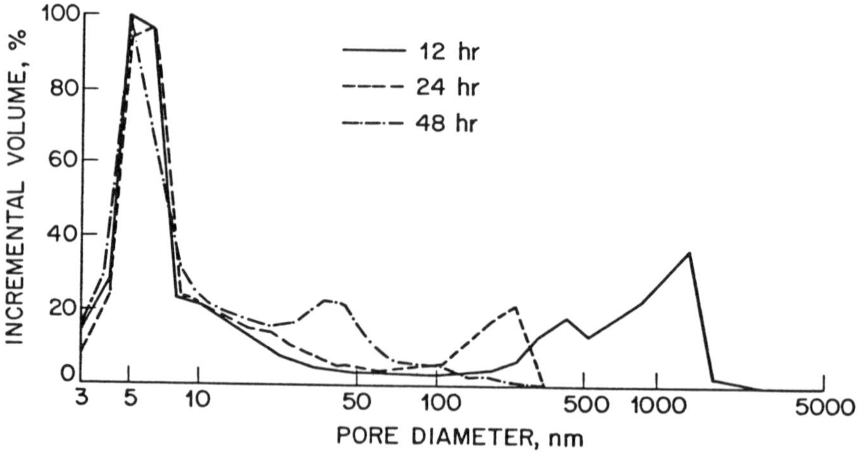

Figure 11: Pore size distribution curves for alkoxide-derived silica aggregate after redispersion in water in ball mills during 12 hours (1), 24 hours (2) and 48 hours (3) (after Sylva and Rabinovich[14]).

Sintering of Gel to Glass

Sintering of various glassy and amorphous powder compacts has been recently reviewed by the present author.[21] The theory of glass sintering is based on theory of viscous flow of Frenkel[39] which was further developed by Mackenzie and Shuttleworth[40] and more recently by Scherer.[41]

Two spherical particles at the point of contact have a small negative radius of curvature compared with the radius of the spheres themselves. On heating, the forces of surface tension cause a transport of material into the pore region by viscous flow. Eventually a large body with closed pores is formed. The annihilation of these closed pores and sintering of glass to the theoretical density and transparency is controlled by diffusion of gases from the pores through the material to the outside. It apparently cannot be achieved in air (except for thin films). Sudo et al.[42] considered the critical pore diameter at constant sintering temperatures for a porous cylinder of silica soot prepared by the vapor-phase axial deposition (VAD) method. This critical diameter (d_c), determines whether the pore will shrink ($d < d_c$) or expand ($d > d_c$) with temperature increase. It was shown that for certain conditions of sintering at 1600K $d_c = 500\ \mu m$ in a He atmosphere but only 0.6 μm in Ar. This is because small He atoms are able to penetrate through the silica glass network. Therefore, sintering in He is essential for preparation of defect-free glass.

This brief consideration shows that conditions for sintering amorphous particulate compacts, such as gels, into glass, should include rather sophisticated temperature-time and atmosphere programs. Four gases may be involved in the firing:

(1) Air or, better, oxygen to ensure that all organics retained from hydrolysis of alkoxides are burnt out. In the case of the sintering of fumed silica-derived gels there would be no need for oxygen.

(2) Helium is required to provide disappearance of pores.

(3) Chlorine is required for dehydration if completely OH^--free glass is needed, as in the case of optical fibres.[3] Recently, MacChesney et al.[33] showed that Cl_2 also effectively removes impurities from particulate silica: up to 1% Fe_2O_3 intentionally admixed to a Cab-O-Sil sol disappeared and gave clear glass after sintering in chlorine.

(4) Fluoride-containing gases or vapors (NH_4F, SiF_4, freon etc.) may be used to eliminate bubble formation in gel glass on reheating[29] and to reduce the refractive index of glass.[43]

Johnson et al.[3] reported sintering of gels prepared of commercial fumed silica (Cab-O-Sil) by double processing with both dispersions carried out in the blender, as described above. The following sintering variables were studied: heating rate, intermediate soak temperature and time, atmosphere (composition and flow rate), temperature and time of the final sintering. As for gases, only He and Cl_2 were utilized in that stage of the research.

Dynamic sintering data (without Cl_2) are shown in Figure 12. The figure shows that the theoretical density was achieved at ~1350°C for B_2O_3-free glass and at ~1300°C for a glass with 3 wt. % B_2O_3, but the glasses only achieved transparency at significantly higher temperatures (near 1570°C for this dynamic schedule without soaks). The B_2O_3 aids densification via lowered viscosity, and it also retards crystallization. The optimal sintering conditions were found to be: (1) heating at the rate of 400°C/hr in He; (2) soaking at 800°C to 1200°C for 8 hours did not improve the optical quality of the sintered glass but was needed to dehydrate the glass with 3% Cl_2 admixtures to the He atmosphere. Chlorine also aided in retarding crystallization by removing any alkali impurities found in the furnace atmosphere; (3) soaking in the range of 1450°C to 1550°C during 1 to 2 hours produced the best glasses since longer soakings could lead to crystallization. The administration of Cl_2 should be stopped before pores start to close.

Fitting of the isothermal sintering data with Frenkel's[39] and Scherer's[41] sintering models allowed the calculation of activation energies of viscous flow for sintered gel-derived glasses, giving 598 kJ/mol for pure SiO_2 and 460 kJ/mol for the glass with 3% B_2O_3. These numbers are close to values for commercial high silica glasses.

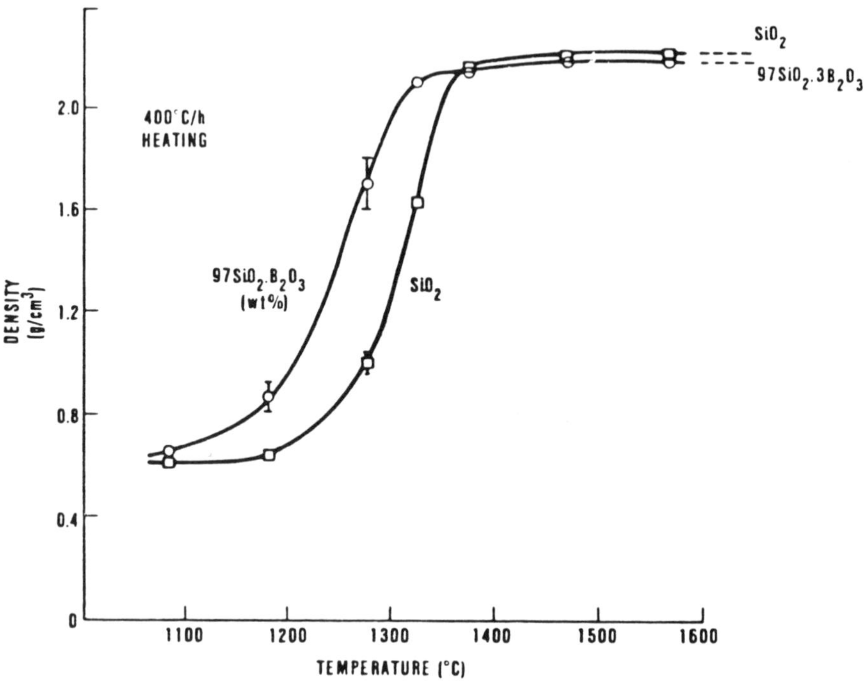

Figure 12: Density of fumed silica gels as a function of temperature for 400°C/hr heating rate. Samples were quenched at temperatures indicated by data points (after Johnson et al.[3]).

Ball milling provided more uniform pore size distributions and, in addition to the elimination of defects, gave significant reduction in the final sintering temperatures, up to 1260°C to 1300°C.[26,31] Figure 13 shows the external appearance of gel specimens milled for different times and sintered at different temperatures.[31] The figure shows that an increase in the time of ball dispersion leads to a reduction in the temperature of preparation of the transparent glass.

Sylva and Rabinovich[14] studied sintering of alkoxide-derived particulate gel rods. All samples were heated in oxygen up to 500°C, and then SiF_4 and Cl_2 were admixed with He before the pores closed. In general, the results are qualitatively similar to those obtained by Johnson et al.[3] for double-dispersed fumed silica. Figures 14 and 15 show the sintering results for the samples whose porosity is given in Figure 11. As seen from Figure 14, small pores almost entirely collapse after 1 hour at 1000°C corresponding with earlier findings for monolithic alkoxide gels.[44] This effect of intraaggregate pores is independent of the dispersion time; the large interaggregate pores do not sinter at this temperature. Figure 15 shows that, due to these pores, complete sintering requires 1400°C (without soak) for a 48 hour dispersion and even higher temperatures for 24 hours. Soaking for 1 to 2 hours at 1300°C to 1400°C permitted preparation of good transparent glasses from these gels if the dispersion time was at least 24 hours.

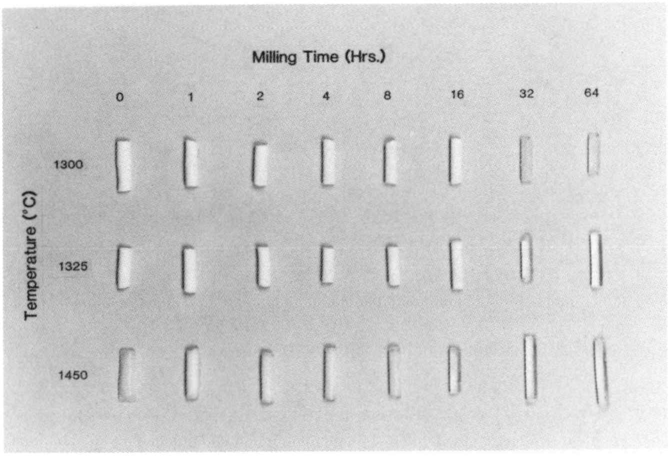

Figure 13: Sintered glasses prepared from material E (heat treated at 900°C) after various milling times and sintering temperatures (after Johnson et al.[31]).

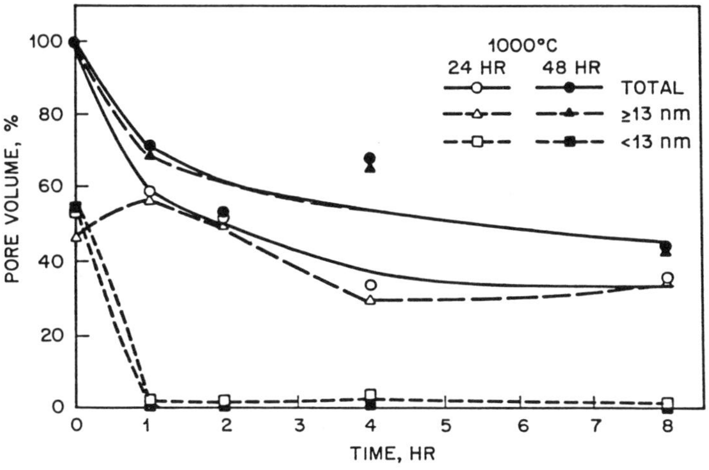

Figure 14: Evolution of porosity in alkoxide derived gels at 1000°C (after Sylva and Rabinovich[14]).

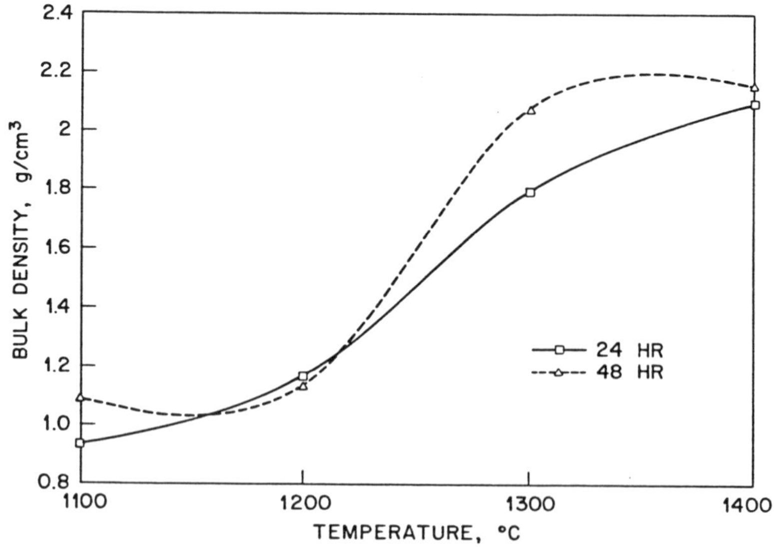

Figure 15: Density of particulate alkoxide-derived gels as a function of temperature for 400°C/hr heating rate (after Sylva and Rabinovich[14]).

Elimination of Bubble Formation on Reheating of Gel Glasses

The described methods produced glass of good optical quality, but these good-looking samples did not endure reheating at higher temperatures, e.g. by torch or during fiber drawing (1700°C to 2200°C). Indeed, they badly bubbled. Figure 16 (left picture) shows a photograph of one such glass rod before and after reheating. Even at the sintering temperatures bloating was observed for very long firings.

Swelling was observed during sintering of alkoxide-derived monolithic gels[45] and was explained by evolution of water from breaking of Si–OH bonds. But this explanation did not work for OH-free sintered silica glasses made of particulate gels. We observed that glasses with as much as 200 ppm OH^- showed minimal bubble formation during reheating, while specimens with more intensive chlorine treatment having <1 ppm OH^- bubbled strongly. Commercial fused silicas may have even 1,000 ppm OH^-, but they do not exhibit bubbles on reheating. Therefore, the reason for bubbling apparently does not lie in the presence of OH^- but in residual Cl^-. Matsuyama et al.[46] also observed bubble formation in alkoxide-derived gel glasses treated with Cl_2. Our study[29] showed that up to 0.4% of Cl^- can be retained in gel glass as a result of reactions with the Si–OH groups on the pore surface and substitution of Cl^- for OH^- during dehydration. Thus significant numbers of Si–Cl bonds are formed. These bonds are not strong enough to survive high temperature reheating and they oxidize and release Cl_2, possibly according to the reaction:

$$\begin{array}{c} O \\ | \\ O-Si-Cl^- \\ | \\ O \end{array} + O \rightarrow \begin{array}{c} O \\ | \\ O-Si-O \\ | \\ O \end{array} + 1/2\ Cl_2$$

Calculations show that even 5% of the amount of Cl^- retained in the glass is sufficient to cause significant bubbling.

Introduction of the F^- ion, practically in any form, prevents formation of Si--Cl bonds by substituting F^- for Cl^- or OH^- and by the formation of Si–F bonds. Up to 3% of F^- can be retained in pure SiO_2 glass[37,43] because of the ability of F^- to substitute for O^{2-} and OH^- as discussed above. If F^- is introduced before dehydration in chlorine it will occupy the sites otherwise available to Cl^-. If F^- is introduced during or after dehydration, it will replace Cl^- in the Si--Cl bonds. The Si–F bonds are much stronger than the Si–Cl bonds and do not release gas on heating.

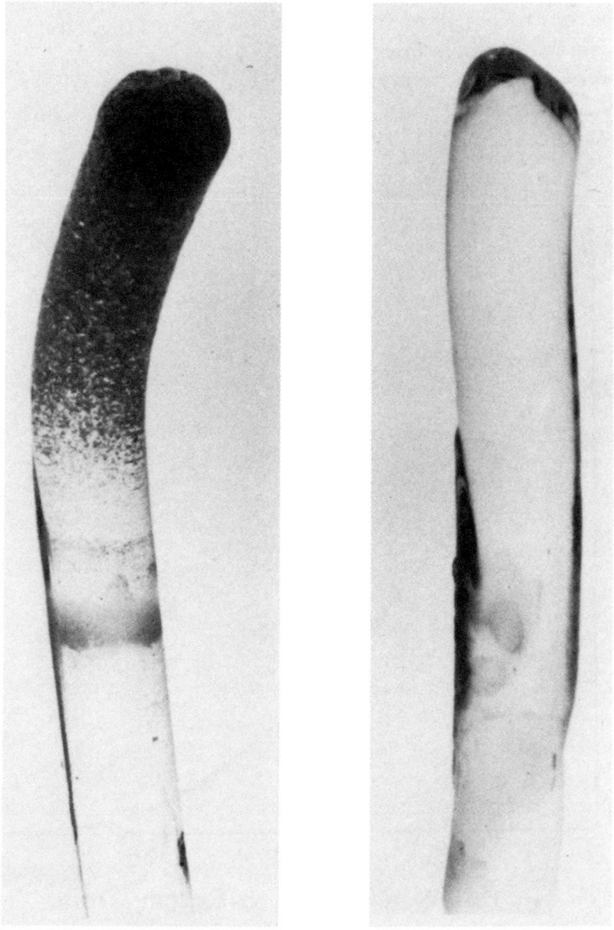

Figure 16: Gel-derived glass rods: lower parts – as sintered at 1450°C; upper parts were reheated by a H_2/O_2 torch (1700°C to 2000°C). Left specimen – sintered without F^- compounds; right specimen – sintered in the presence of NH_4F powder.

The F^- ion could be introduced in different ways: (1) as HF or NH_4F solution during dispersion of fumed silica, but this way is not especially convenient because gelation is strongly accelerated and it is difficult to cast the sol; (2) as HF or NH_4F solution during hydrolysis of TEOS in the combined alkoxide-particulate method;[36,45] (3) as NH_4F powder around the samples during firing. The powder sublimes below 500°C providing F^--containing vapors to react with the silica (Figure 16, right picture);[29] or (4) as SiF_4 gas during or after the dehydration step of firing but prior to closing of the pores.[29] Although all these ways provided removal of Cl^- and prevented or reduced the secondary bubbling, the fourth method using SiF_4 was found to be the most convenient and provided better control. Figure 17 shows the dependence of the amount of Cl^- retained in the glass on the flow rate of SiF_4. Figure 18 shows photographs of some of these samples heated with a torch. As seen from these figures, increasing the flow rate of SiF_4 results in a sharp decrease in the amount of Cl^- retained in the glass and in the degree of bubble formation.[29]

It was also shown that F^- compounds enhance the dehydration of glass by chlorine.[29] They also facilitate sintering because F^- noticeably reduces the glass viscosity.[37]

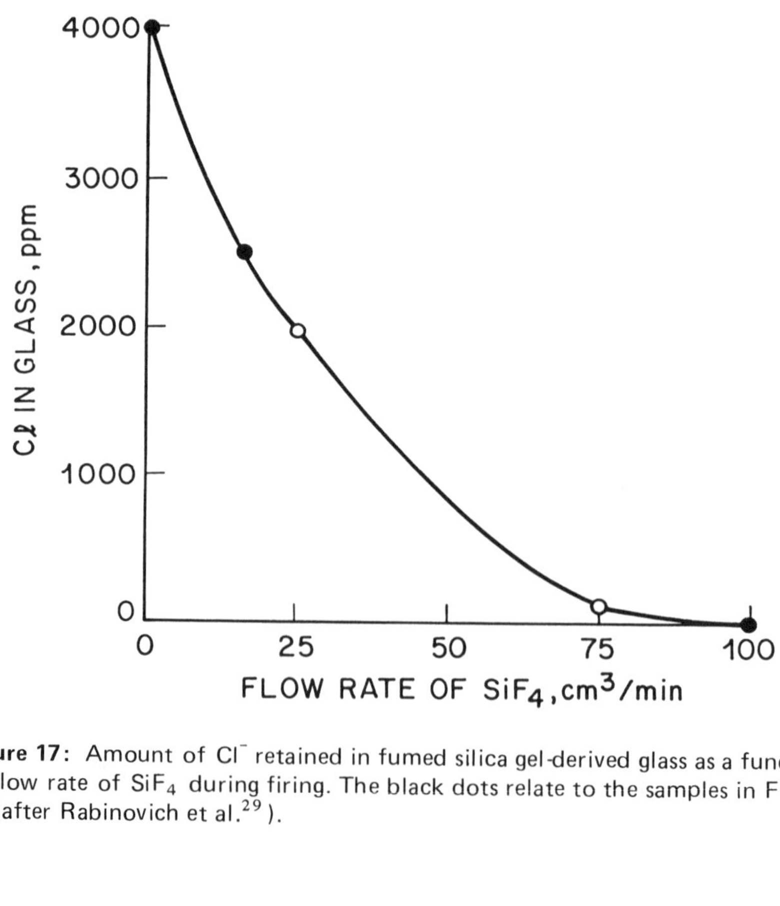

Figure 17: Amount of Cl^- retained in fumed silica gel-derived glass as a function of flow rate of SiF_4 during firing. The black dots relate to the samples in Figure 18 (after Rabinovich et al.[29]).

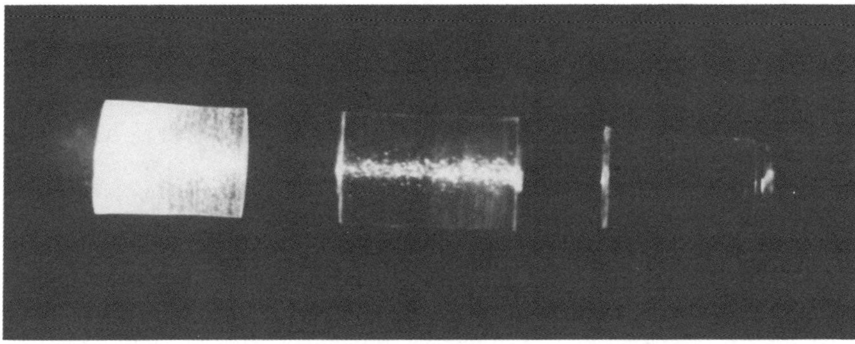

Figure 18: Samples of a gel-derived glass after heating with a torch, photographed in a laser beam entering from the right. The samples correspond to the black points in Figure 17. From left to right: no SiF_4, 15 cm^3/min of SiF_4, 100 cm^3/min of SiF_4 (after Rabinovich et al.[29]).

General Scheme of the Process; Properties of Glasses

A chart of particulate gel processes as described in the works by Rabinovich, Johnson, MacChesney and their co-workers[2,3,12-14,25,26,29-33] is presented in Figure 19. Photographs of unsintered and sintered tubes prepared by the double processing with ball milling are given in Figure 20.[26] The glass bodies were sintered at 1260°C and had a surface haze. When glycerol was applied (left ends of the tubes) to remove this haze, they appeared transparent.

Properties of gel-derived silica glasses prepared by double processing without ball milling and by firing without F compounds[2] are shown in Tables 3 and 4. They are, in general, identical to those of commercial fused silica and high-silica glasses. The mechanical strength of the gel glasses fired at low temperatures is inferior to that of fire-polished commercial silicas although sand-blasting of the samples before testing made the strength values much closer. The samples fired in the presence of fluorine compounds showed depressed refractive indices near 1.450.

GLASSES FROM LOW SURFACE AREA POWDERS

Fumed silica with surface area below ~100 m^2/g cannot be easily gelled when dispersed in water. Even when much higher amounts of solids are dispensed than in the case of high surface area, the sol tends to exhibit almost Newtonian viscosity behavior. If a gel is formed from such a sol (usually after aging at elevated temperatures) it cannot maintain its shape without a container. Recognizing this difficulty Scherer and Loung[7] used organic liquids, such as chloroform and n-decanol with additions of amines or ammonia as a dispersion medium instead of water. Silica powders which they used were produced by flame hydrolysis of $SiCl_4$ (as fumed silica) and had surface areas from 27 to 45 m^2/g with particles of 100 to 60 nm in diameter. Due to the low surface area and larger pores, green bodies could be dried unfractured even though double processing was not used. This seems to be advantageous, but the organic liquids used in the process are significantly more expensive than water, and chloroform is a possible carcinogen. The dried gels were sintered at 1450°C in a He/Cl_2 atmosphere, and a glass rod of 1.3 cm in diameter and 7 cm long was prepared.[7]

Sacks and Tseung-Yuen Tseng[47] studied preparation of silica glass from even larger particles (0.2 to 0.6 μm with surface area near 7 m^2/g) made by hydrolysis of TEOS in the presence of ammonia. The powders dispersed in water did not gel, and green bodies were formed by allowing the particles to settle under gravitational force. The state of particulate dispersion was strongly dependent on pH. At pH = 10, with 20 wt. % solids in the sol, the so-called dispersed structure was formed containing two-dimensional, close-packed hexagonal arrays of particles with some defects which were reminiscent of defects in atomic crystalline structures. Before precipitation this sol exhibited Newtonian behavior of the viscosity.

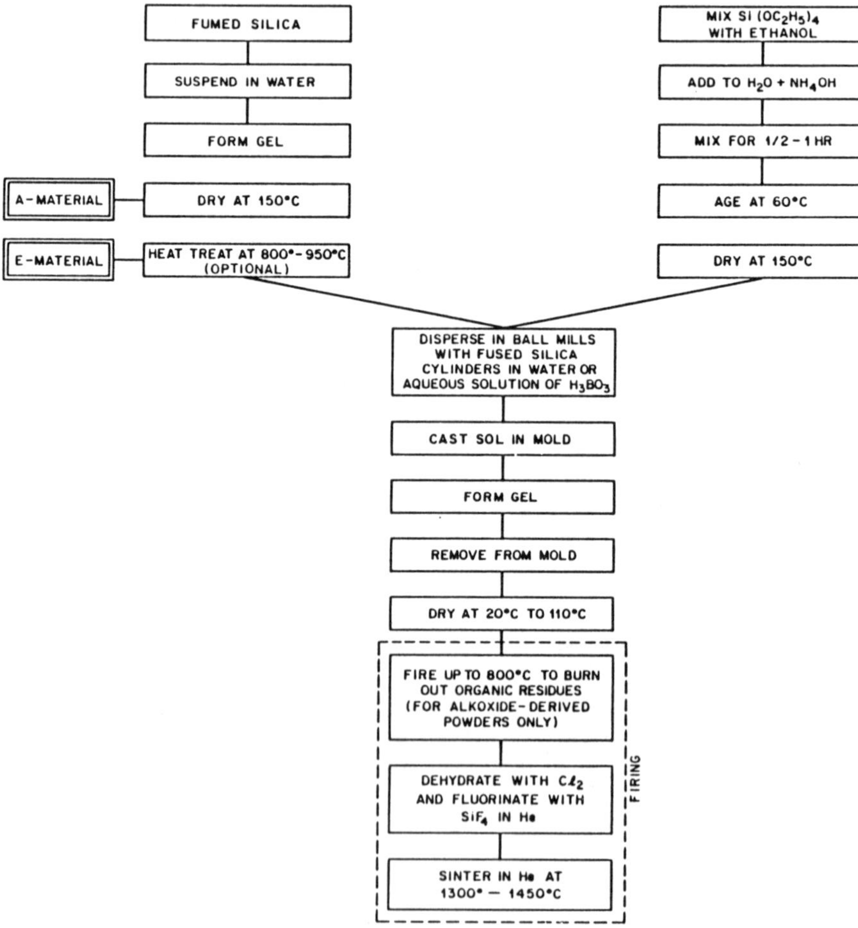

Figure 19: Process chart for preparation of transparent glass from particulate gels.

a

b

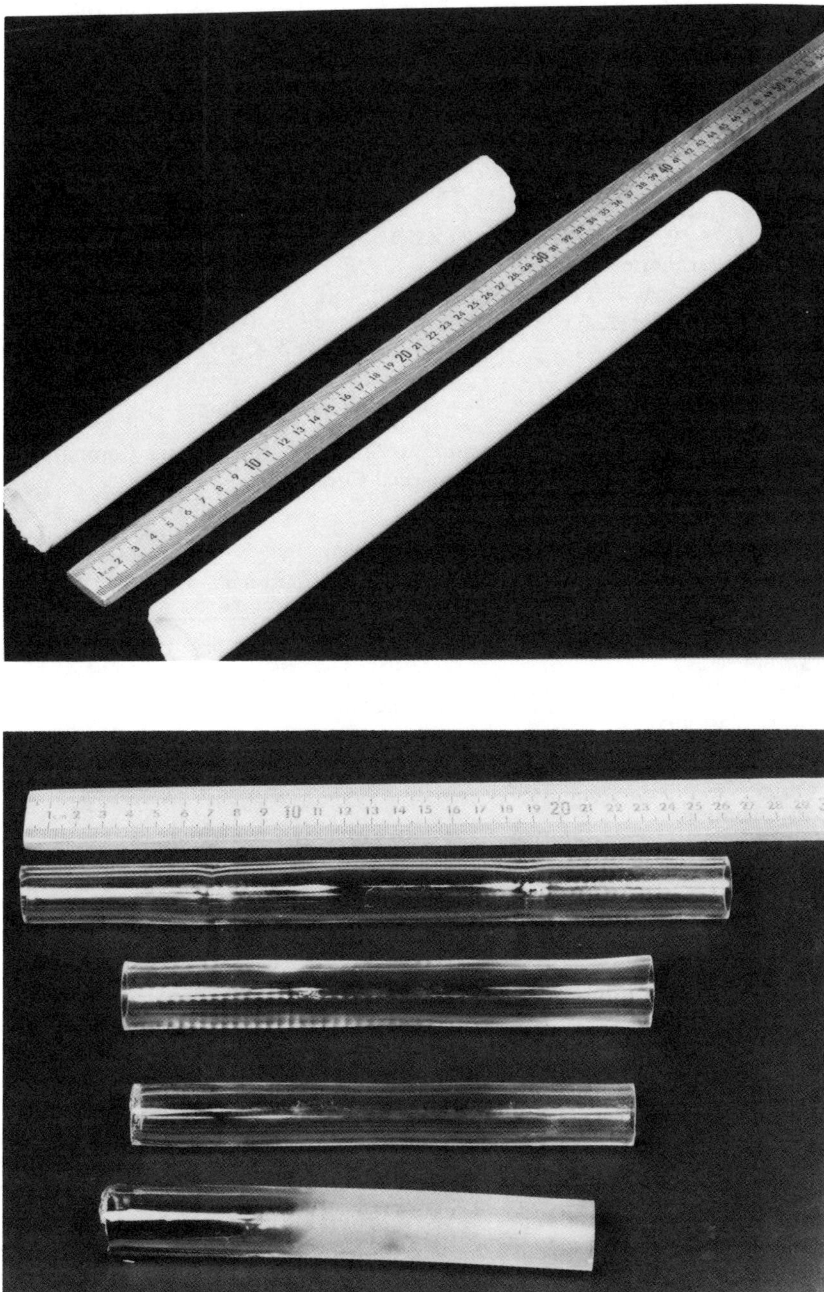

Figure 20: Particulate silica gel tubes before (a) and after (b) sintering (after Rabinovich et al.[26]).

Table 3: Properties of High-Silica Glasses Prepared by the Particulate Silica Gel Method[2]

Property	100% SiO_2 (1460°C)	97 wt % SiO_2/3 wt% B_2O_3 (1450°C)
Density (kg/m^3)	2200	2190
Refractive index (λ = 0.5893 μm)	1.4588	1.4555
Linear thermal expansion coefficient difference, $\alpha_i - \alpha_{fused\ silica}$, ($°C^{-1}$) (at 25° – 700°C)	0	$+3 \times 10^{-7}$

Table 4: Moduli of Rupture* of Sintered "Gel" glasses Compared with Commercial Fused Silica†[2]

Glass	Sintering temp. (°C)	Bulk density (% theor.)	Number of specimens	Modulus of rupture (MPa) (confidence level 95%) As-received or as-fired‡	Sand-blasted
100% SiO_2					
Transparent	1460	≈100	8 and 7	70.1 ± 21.5	65.4 ± 14.1
Nontransparent	1260	98.5	8	22 ± 4	
97% SiO_2-3% B_2O_3, Transparent					
Batch 1	1450	≈100	11	80.0 ± 5.5	
Batch 2	1450		6 and 9	132.3 ± 33.8	100.5 ± 4.5
Nontransparent, Batch 1	1250	99.0	11	72.4 ± 16.3	
Commercial fused silica		100	12 and 10	133.3 ± 21.6	79.9 ± 4.6
Heat-treated	1260§	100	8	230.6 ± 19.4	
Heat-treated	1460§	100	8 and 7	147.9 ± 14.6	92.9 ± 6.5

* Tested with mechanical testing machine (Instron Corp., Canton, MA), 3-point loading, span 2.45 cm, crosshead speed 0.13 cm/min^{-1}, rod diameter 0.6-0.7 cm.

† From Heraeus-Amersil, Inc., Sayreville, NJ.

‡ Firings in bed of sand.

§ Heat-treated with experimental 100% SiO_2 specimens.

On the other hand, a sol with pH = 3 showed viscosity decreasing with increasing shear rate. This was taken as proof that the sol is highly flocculated. The porosimetric curves of both sols after precipitation and drying (Figure 21) are qualitatively similar to those obtained by Sylva and Rabinovich (Figure 11)[14] and show bimodal distributions, but quantitatively they are different even from each other. The peaks due to small, intraaggregate, pores are at the same place for both types of the structures (at a radius of 50 nm). The interaggregate peak, although small, is present even in the "dispersed" structure and it is significantly larger and lies at much higher pore size (>100 nm in radius) for the "flocculated" structure. Sintering in air at 1000°C for 24 hours resulted in almost complete densification and translucency for the dispersed structures, while the flocculated bodies, naturally, remained highly porous after such sintering. Heating to higher temperatures caused crystallization. Only small translucent glass discs (0.8 cm in diameter and 0.08 cm thick) are reported.

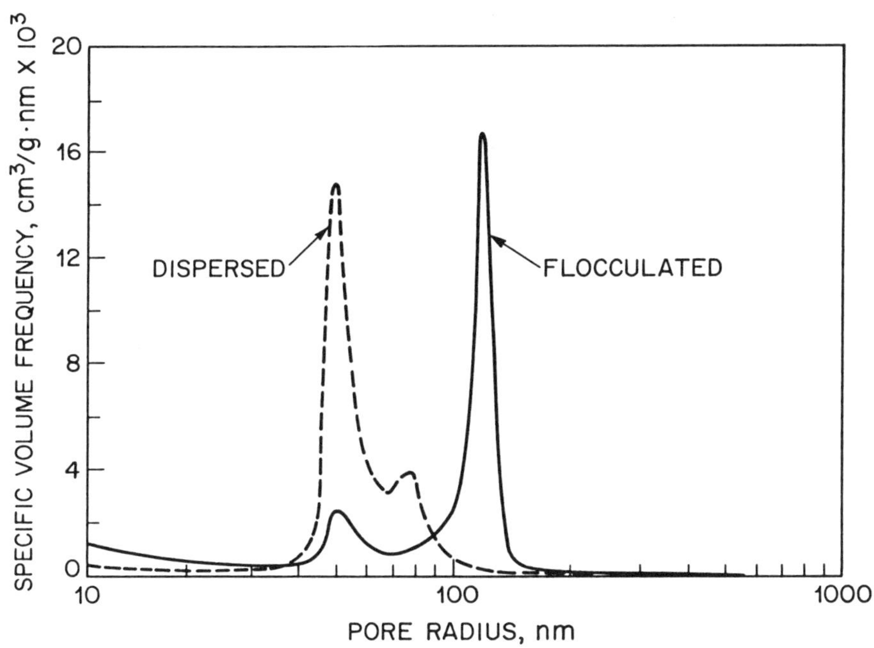

Figure 21: Pore size distribution curves for compacts prepared from dispersed and flocculated suspensions (after Sacks and Tseung-Yuen Tseng[47]).

DOPING PARTICULATE GEL GLASSES

The problem of altering the properties of gel glass can be divided into 2 parts: (1) modification of properties of silica glass by doping with other elements in relatively minor quantities (this is still the field of high-silica glasses), and (2) synthesis of true multicomponent glasses with relatively low SiO_2 content or even non-silicate glasses.

It is easy to review the latter case of multicomponent glasses because, while a lot of such glasses have been made by hydrolysis of alkoxides,[21] as far as we know, there has been no such work with glasses made of particulate sols and gels. Little research has been done with doped high-silica glasses. There are three ways to introduce a cation dopant: (1) as a soluble compound of a second cation, when the solution is used instead of pure water for a dispersion of particulate silica; (2) by mixing silica powder with a dopant powder; or (3) by incorporating a dopant in the amorphous structure of individual silica particles.

The first two ways lead to distribution of a dopant oxide either on the surface of silica particles or between them. Because the sintering takes place at 1300°C to 1500°C, which is below the liquidus for high-silica compositions, apparently only a non-crystalline dopant can be sintered into glass. Therefore, two groups of dopants can be considered: (a) glass-formers or oxide-modifiers with a melting point below 1300°C; and (b) highly reactive oxides which will completely react with fine silica powder forming a structurally incorporated dopant below sintering temperatures. From the first group, glass-formers such as B_2O_3 (introduced as a H_3BO_3 solution), $P_2O_5(H_3PO_4)$, or GeO_2 (through hydrolysis of $GeCl_4$ or Ge alkoxides), and a modifier, such as PbO whose melting point is 886°C,[48] can be considered.

Introduction of B_2O_3 up to 4 wt. % by preliminary dissolution of H_3BO_3 in water was reported by Rabinovich et al.[12] The solution was used either instead of water for dispersion, or as a solution for impregnation of a dried gel body which was then dried again. However, in either case, one of the main advantages of sol-gel processes, high homogeneity of produced materials, was lost here. This is because during drying, the solution was moved out by capillary forces delivering boric acid outside the gel body, and its crystals were seen on the surface after drying. Apparently this will happen with any dopant introduced in a similar way, unless freeze drying is applied. Boron-containing gel-derived silica glass is very similar to Vycor®-glass[21] and may be even better in some respects because it does not contain the alkali ions which are present in Vycor® in amounts of up to 1 wt. %. However, this glass is not suitable for optical fiber applications because the B^{3+} ion cuts transmission in the near infrared region.[49]

Rabinovich et al[24] tried to add GeO_2 and P_2O_5 to fumed silica. Germanium oxide was introduced as a suspension formed by hydrolysis of $GeCl_4$. Binary GeO_2-SiO_2 glasses appeared hazy, but the transparency could have been improved by additions of B_2O_3. However, one of the problems with GeO_2-containing gel glasses fired in the oxygen-poor atmosphere is the reaction:

$$GeO_2 \rightarrow GeO + 1/2\ O_2$$

either during sintering or at reheating. Both GeO and O_2 cause bubbling of the gel glass.

Introduction of low-melting oxides like PbO or a highly reactive oxide like Na_2O or K_2O through solutions could result in their diffusion to fumed silica particles, with the formation of doped amorphous high-silica particles. After that, the problem of sintering these powders without crystallization is similar to that of sintering doped particles such as those formed by cooperative hydrolysis of alkoxides.

It is easy to prepare high-silica amorphous particles which are doped practically with any cation, or to prepare any multicomponent glass particles using hydrolysis of alkoxides. A review of glass compositions prepared in this way is given in Reference 21, and the combined alkoxide-particulate method seems to be suitable for the processing of such particles into bulk glass. However, sintering can be difficult for doped particulate gels. Firing at temperatures below the liquidus means that sintering will compete with crystallization. In the case of alkoxide-derived monolithic gels, sintering of pure silica to the theoretical density can be completed at ~1000°C,[44] which is too low for crystallization. In particulate silica gels, where the pores are significantly larger, the final sintering temperature is at least 1300°C or higher.[3] In this temperature region, the crystallization of silica is more rapid, and its rate strongly depends on impurities. Very pure silica glass has a significant induction period when crystallization does not appear even at 1300°C to 1400°C,[3] so that, because fumed silica or alkoxide hydrolyzed silica are pure materials, sintering can be completed without crystallization. However, almost any glass modifier added even in trace amounts drastically increases the rate of crystallization and makes complete sintering impossible. To illustrate this, compare the linear rates of crystal growth in high silica glasses. Fused silica, depending on purity, crystallizes in air at 1400°C only after the induction period of 30 to 50 hours and, even then, at rates from ~0.35 to ~3 μm/hr (Leko and Komarova, cited in Reference 23). When the glass contains as little as 0.5 wt. % Na_2O, there will be no induction period, and the rate will be as high as 40.8 x 10^3 μm/hr, an increase of 7,000 to 40,000 times! (after Dietzel and Wickert, cited in Reference 23) Although there are insufficient data about crystallization of other high-silica glasses, apparently any cation-modifier will drastically increase the speed of crystallization, thus preventing complete sintering. The possible exception is Al^{3+},[22] but there are no publications describing successful preparation of Al_2O_3–SiO_2 glasses by the particulate gel method. Krol and van Lierop[50] have prepared such glasses by the alkoxide method.

Scherer and Pantano[51] reported preparation of TiO_2–SiO_2 glasses by a version of combined alkoxide-particulate method. They stabilized titanium isopropoxide against hydrolysis by reacting it with ethylene glycol and citric acid at 120°C. Then this liquid was mixed with water and used as the medium for dispersion of fumed silica of surface area of 200 m^2/g to prepare a glass with a composition of 7.5% TiO_2 and 92.5% SiO_2. Analysis of the final glass revealed about 10% TiO_2. The sol was poured into molds (tubes with an inside diameter of 1.28 cm) and gelled at 60°C to 65°C for ~1 day. The firing consisted of heating to 400°C to decompose the organic part and complete densification was achieved at ~1230°C in air. The formed material was translucent.

It could not be expected from the described process that single-phase TiO_2–SiO_2 glass would be prepared. Titania particles formed by decomposition of the alkoxide would preserve their individuality on the surface of the silica particles or between them, and significant diffusion or reaction between TiO_2 and SiO_2 could not be expected at so low temperatures as 1230°C. Actually, the authors[51] discovered TiO_2 in the form of anatase crystals of ~13 nm in size uniformly distributed in a glassy matrix. We believe that if these materials were heated to ~1400°C, rapid crystallization of silica to cristobalite would be observed as well. Therefore, this TiO_2–SiO_2 material cannot be considered as a Ti-doped silica glass.

The above discussion has related to cation doping of silica, but alteration of properties can be achieved by anion doping as well. Two anions are the most promising in this respect: N^{3-} and F^-. Coon et al.[52] incorporated up to 2 wt.% N in multicomponent silicate and borosilicate glasses prepared by melting. They showed that Young's modulus, microhardness and fracture toughness are increased by nitrogen. Several papers[53,54] were devoted to nitrogen-containing thin films of silica and borosilicate glasses prepared by sol-gel methods. It was shown that nitrogen increases the dielectric permittivity of silica.[53]

The fluoride ion reduces the refractive index of silica glass. Up to 3 wt. %

F^- can be incorporated into silica glass structure,[37,43] and this effect is now routinely used for preparation of optical preforms by different chemical vapor deposition methods.[28] One wt. % F^- reduces the refractive index of silica by 0.005,[43] which is quite enough for preparation of single-mode optical fibers. As shown above, fluoride can be easily introduced also into gel glasses.[29,36] Some technological success in this field will be described below. However, it is difficult to introduce F^- into gel glass in such a way that a prescribed refractive index profile will be obtained.

APPLICATIONS OF PARTICULATE GEL GLASSES

Current applications of particulate gel-derived glasses are unknown to the present author. Therefore, the purpose of this paragraph is to discuss possible future applications. These methods have such advantages as: (1) low processing temperatures–1300°C to 1400°C as compared with 1800°C to 2000°C for fused silica; (2) low cost due to relatively simple equipment and low capital investment (no melting kilns); (3) very high homogeneity and purity of the glass; (4) ease of fluorination. The processes have their problems as well, such as: (1) difficulty in doping, or (2) difficulty in producing large bodies, especially thick-wall tubes.

In principle, if large size bodies can be produced commercially by sol-gel methods, any application of fused silica or high-silica glasses becomes available. Probably such materials could replace opaque quartz ceramics, usually produced by sintering of pulverized fused silica cullet.[21] The best bulk density achieved in quartz ceramics is 1,900 kg/m^3 (86% of the theoretical density) and open porosity is 12 to 15%. In comparison, the opaque silica glass prepared from a particulate gel by double processing and described in Table 4 has 98% bulk density. Quartz ceramics are used for production of free-standing insulators in high-temperature nuclear reactors, radomes, nose cones and rocket nozzles.[21] They also can be used for manufacture of crucibles, tubes and plates.

Gel double processing of silica and high-silica B_2O_3-containing glasses seems to be simpler and cheaper than manufacture of a similar glass by leaching Na_2O–B_2O_3–SiO_2 glass and sintering it as in the Vycor® process. The gel glass does not contain alkali and should have better electrical properties, so this gel process may eventually replace the Vycor® process.

One of the most exciting possibilities is the application of sol-gel processes in the production of optical fibers. We recently reviewed[12,21] attempts to prepare such fibers by a variety of sol-gel methods, and new publications have appeared since.[8,13,32] The attempts based on particulate and alkoxide-particulate gels are more promising because until now only these methods allowed the preparation of relatively large glass bodies. On the other hand, the difficulty of doping of particulate gel glass is a serious obstacle.

Rabinovich et al.[26] reported an optical fiber prepared by the Modified Chemical Vapor Deposition (MCVD) process with a double-processed gel-derived silica glass tube 25 cm long, and 2.3/1.7 cm in outside/inside diameter used as a substrate. The tube was fire-polished in an atmosphere of He and Freon 12 and a SiO_2–P_2O_2–F cladding with a SiO_2–GeO_2 core were deposited. The tube was collapsed and the fiber was drawn with 150 μm total diameter and a core of 8.6

μm; the refractive index difference was $\Delta n \approx 0.0045$. This fiber showed a minimum loss of 0.7 dB/km at 1.15 μm.

A more advanced fiber was reported by MacChesney et al.[28] The gel substrate tube was prepared in the same way but it was fluorinated to match its refractive index to that of the down-doped MCVD cladding. Usually fibers based on a decreased cladding index require thick deposits to overcome leaky-mode losses resulting from the higher refractive index of the substrate tube. However such losses have been eliminated by using a fluorinated colloidal gel tube with the index matched to that of the fluorine-doped deposited cladding. In this case the cladding thickness could be reduced to ¼ that normally necessary. One of these tubes was joined with a commercial fused silica tube (not fluorinated), so the MCVD core deposition could be carried out simultaneously in both substrates for better comparison. A deposited cladding/core diameter ratio of 4.5 gave a minimal loss of 0.28 dB/km at 1.55 μm for the gel tube substrate, and as much as ~3.3 dB/km at 0.95 μm (minimal loss) for the commercial tube with a non-depressed index. The same low loss with the commercial tube would have been achieved if a cladding four times thicker had been deposited. The mechanism of this phenomenon is discussed in detail elsewhere.[28]

In these two cases gel glass did not serve as the light guiding part of the fiber. To estimate the quality of gel glass as an optically active part of the fiber, a core rod prepared from fumed silica was drawn into fiber and coated with silicone cladding. This fiber demonstrated a minimal loss of 16 dB/km at 0.83 μm.[32] A fiber prepared in a similar manner using an alkoxide-particulate method had a lower minimal loss of only 7 dB/km,[13] and apparently the TEOS-derived powder was purer than commercial fumed silica. It is possible that the boundary between the silica and the silicone cladding could have increased the loss. Shibata et al.[8] prepared a silica core rod by the alkoxide-particulate method, and the F-doped cladding was deposited by the VAD method. This fiber (2.3 km long) had a minimal loss of 1.8 dB/km at 1.6 μm. Scherer and Luong[7] also reported preparation of an optical fiber from gel silica but they did not describe its properties.

We should conclude that gel methods of glass preparation have started to take their place in optical fiber technology, but they have not yet advanced beyond the stages of feasibility studies.

SUMMARY

Particulate gel methods of preparation of high-silica glasses allow the preparation of significantly larger glass bodies compared with those produced by alkoxide monolithic gel methods. The temperature of their sintering is 1300°C to 1450°C compared with 1800°C to 2000°C for conventional techniques of fused silica manufacture. The simplicity of processing and the high purity of the glass produced suggests a significant future for the particulate gel technology, even though some problems, such as those involved in doping need to be overcome.

Acknowledgments

The author is thankful to D.L. Wood for reading the manuscript and his valuable comments and to Norma J. Kopylov for assistance in its preparation.

REFERENCES

1. Rabinovich, E.M., et al., *J. Non-Cryst. Solids* 47: 435-39 (1982).
2. Rabinovich, E.M., et al., *J. Amer. Ceram. Soc.* 66: 683-688 (1983).
3. Johnson, D.W. Jr., et al., *ibid.,* 66: 688-693 (1983).
4. Wood, D.L., et al., *ibid.,* 66: 693-699 (1983).
5. Rabinovich, E.M., *J. Non-Cryst. Solids* 71, 187-193 (1985).
6. Flory, P.J., *Faraday Disc. of Chem. Soc.* 57: 7-18 (1974).
7. Scherer, G.W. and Luong, J.C., *J. Non-Cryst. Solids* 63: 163-172 (1984).
8. Shibata, S., Hanawa, F. and Nakahara, M., *Electronics Letters* 21 (24): 1145-46 (Nov. 21, 1985).
9. Cab-O-Sil® Properties and Functions, Published by Cabot Corp., Tuscola, IL (1983).
10. Bode, R., Ferch, H. and Fratzscher, H., Basic Characteristics and Applications of Aerosil, 2nd edition, *Technical Bulletin Pigments,* No. 11, Degussa, Frankfurt (Main) (1978).
11. Fleming, J.W., Sol-Gel Processing of Glass Powders, presented at the meeting on *Ceramics Powder Science and Technology,* Aug. 3-6, 1986, Boston, MA; to be published.
12. Rabinovich, E.M., MacChesney, J.B. and Johnson, D.W. Jr., Gel-Derived Glasses for Optical Fibers Prepared from Alkoxides and Fumed Silica, 208-216 in *Sci. of Ceramic Chem. Processing,* L.L. Hench and D.R. Ulrich (Eds.), Wiley, N.Y. (1986).
13. Rabinovich, E.M., et al., Combined Alkoxide-Colloidal Method for Silica Glass, Ref. 24-G-86 in Amer. Ceram. Soc. 88th Annual Meeting Abstracts, 331, April 27-May 1, 1986, Chicago, IL (full text in preparation).
14. Sylva, L.-A. and Rabinovich, E.M., Sintering of Alkoxide-Derived Colloidal Gels, Ref. 26-G-86, *ibid.,* 331 (full text in preparation).
15. Stober, W., Fink, A. and Bohn, E., *J. Colloid. a. Interface Sci.* 26: 62-69 (1968).
16. Shimohira, T., et al., Sintering of Monodispersed Amorphous Silica Particles, in *Proceedings of the Internation. Symposium on Factors in Densification and Sintering of Oxide and Non-Oxide Ceramics:* 119-127, Hakone, Japan, October 1978, S. Sömiya and S. Saito (Eds.) (Gakujutsu Bunken Fukuy-Kai, Association for Science Documents Information, c/o Tokyo Institute of Technology, Tokyo).
17. Barringer, E.A. and Bowen, H.K., *J. Amer. Ceram. Soc.* 65 (12): C-199-201 (1982).
18. Mazdiyasni, K.S., Chemical Synthesis of Single and Mixed Phase Oxide Ceramics, 175-186, in *Better Ceramics Through Chemistry,* Mat. Res. Soc. Symp. Proc. 32, C.J. Brinker, D.E. Clark and D.R. Ulrich (Eds.), North-Holland, Amsterdam (1984).
19. Shoup, R.D., in *Colloidal Interface Sci.* 3, 53-63, M. Kerker (Ed.), Acad. Press, N.Y. (1976).
20. Shoup, R.D. and Wein, W.J., Low-Temperature Production of High-Purity Fused Silica, U.S. Pat. 4,059,658, Nov. 22, 1977.
21. Rabinovich, E.M., *J. Mater. Sci.* 20: 4259-4297 (1985).
22. Rabinovich, E.M., Crystallization of Glass, Chapter IV in *Spravuchnik po Proizvodstvu Stekla* (The Handbook of Glass Manufacture, in Russian), Vol. 1, 79, Stzoiizdat, Moscow (1963).

23. Mazurin, O.V., Streltsina, M.V. and Shvaiko-Shvaikovskaya, T.P., Physical Sciences Data 15, *Handbook of Glass Data,* Part A, Silica Glass and Binary Silicate Glasses, 13,14,190, Elsevier, Amsterdam (1983).
24. Rabinovich, E.M., MacChesney, J.B., Johnson, D.W. Jr., Fleming, J.W. and Chadwick, D.L., Effect of Admixtures on Sintering of High Silica Colloidal Gels, Ref. in *The Impact of Energy and Environment on the Science and Technology of Glass,* 8-9, Intern. Comm. on Glass Annual Meeting, Oct. 3-6, 1982, Toronto, Canada.
25. Johnson, D.W. Jr., MacChesney, J.B. and Rabinovich, E.M., Fabrication of Sintered High-Silica Glasses, U.S. Pat. 4,419,115, Dec. 6, 1983.
26. Rabinovich, E.M., et al., *J. Non-Cryst. Solids* 63: 155-161 (1984).
27. Rabinovich, E.M. and Poulsen, S.D., Mercury Porosimetry of Dry Silica Gels, 329-338, in *Advances in Material Characterization II,* R.L. Snyder, R. A. Condrate, Sr. and P.F. Johnson (Eds.), Plenum, N.Y. (1985).
28. MacChesney, J.B., Johnson, D.W. Jr., Lemaire, P.J., Cohen, L.G. and Rabinovich, E.M., *J. Lightwave Technol.* LT-3(5): 942-945 (985).
29. Rabinovich, E.M., Wood, D.L., Johnson, D.W. Jr., Fleming, D.A., Vincent, S.M. and MacChesney, J.B., *J. Non-Cryst. Solids* 82: 42-49 (1986).
30. Johnson, D.W. Jr., MacChesney, J.B. and Rabinovich, E.M., Sintered High-Silica Glass and Articles Comprising Same, U.S. Pat. 4,605,428, Aug. 12, 1986.
31. Johnson, D.W. Jr., Rabinovich, E.M., Fleming, D.A. and MacChesney, J.B., Improvement in Colloidal Sol-Gel SiO_2 Glass by Attrition of Sols (to be published).
32. Fleming, D.A., Johnson, D.W. Jr. and MacChesney, J.B., Preparation of High Optical Quality Gel Derived Glasses, Ref. 111-G-86 in Amer. Ceram. Soc. 88th Annual Meeting Abstracts, 357, April 27-May 1, 1986, Chicago, IL.
33. MacChesney, J.B., Johnson, D.W. Jr., Fleming, D.A., Walz, F.W. and Kometani, T.Y., Influence of Dehydration/Sintering Conditions on the Distribution of Impurities in Sol-Gel Derived Silica Glass (to be published).
34. Iler, R.K., *The Chemistry of Silica,* Wiley, N.Y. (1979).
35. Belyakova, L.D., et al., *Russ. J. Phys. Chem.* (English translation) 33: 551 553 (1959).
36. Rabinovich, E.M. and Wood, D.L., Fluorine in Silica Gels, presented at the 1986 Spring Meeting of the Mater. Res. Soc., April 15-18, Palo Alto, CA; to be published in MRS Symposia Proceedings, Vol. 73.
37. Rabinovich, E.M., *Phys. Chem. Glasses* 24(2): 54-56 (1983).
38. Wood, D.L. and Rabinovich, E.M., Studies of Alkoxide Gel Formation, presented at the Fall Meeting of Glass Div. of Amer. Ceram. Soc., Nov. 6-8, 1985, Corning, N.Y.; Ref. 49-G-85F in *Ceram. Bull.* 64(10): 1342 (1985) (the full text to be published).
39. Frenkel, Ya. I., *J. Phys. USSR* 9(5): 385-391 (1945).
40. Mackenzie, J.K. and Shuttleworth, R., *Proc. Phys. Soc. (London)* B62: 833-852 (1949).
41. Scherer, G.W., *J. Amer. Ceram. Soc.* 60 (5-6): 236-239 (1977).
42. Sudo, S., Edahiro, T. and Kawachi, M., *Trans. IECE Jpn.* E63: 731-737 (1980).
43. Fleming, J.W. and Wood, D.L., *Appl. Optics* 22: 3102-3104 (1983).
44. Nogami, M. and Moriya, Y., *J. Non-Cryst. Solids* 37: 191-201 (1980).

45. Nassau, K., et al., *J. Non-Cryst. Solids* 82: 78-85 (1986).
46. Matsuyama, I., et al., *Ceram. Bull.* 63(11): 1408-1411 (1984).
47. Sacks, M.D. and Tseng, Tseung-Yuen, *J. Amer. Ceram. Soc.* 67(8): 526-532 (Part I), 532-537 (Part II) (1984).
48. *Handbook of Chemistry and Physics,* 63rd ed., B-112, CRC Press, Boca Raton, FLA (1982-1983).
49. MacChesney, J.B., *Proc. IEEE 68* (10): 1181-1184 (1980).
50. Krol, D.M. and Van Lierop, J.G., *J. Non-Cryst. Solids* 63: 131-144 (1984).
51. Scherer, CP. and Pantano, C.G., *J. Non-Cryst. Solids* 82: 246-255 (1986).
52. Coon, D.N., et al., *J. Non-Cryst. Solids* 56: 161-166 (1983).
53. Carman, L.A. and Pantano, C.G., Dielectric Properties of Silicon Dioxide and Silicon Oxynitride Sol-Gel Films, 187-200, in *Science of Ceramic Chemical Processing,* L.L. Hench and D.R. Ulrich (Eds.), Wiley, N.Y. (1986).
54. Haaland, D.M. and Brinker, C.J., In Situ FT-IR Studies of Oxide and Oxynitride Sol-Gel Derived Thin Films, 257-273, in *Better Ceramics Through Chemistry,* Mater. Res. Soc. Proc. 32, C.J. Brinker, D.E. Clark and D.R. Ulrich (Eds.), North-Holland, Amsterdam (1984).

Part V

Special Applications

14

Electronic Ceramics Made by the Sol-Gel Process

John B. Blum

Norton Company
Northboro, Massachusetts

INTRODUCTION

The use of ceramics in electronic applications is widespread and expanding rapidly. Ceramic materials are used in electrical circuit elements, as substrates for circuits, as sensors and transducers, and in a wide variety of magnetic applications.[1] For many of these increasingly demanding applications, improved materials and methods of fabrication are becoming necessary. One method of fabrication which may be advantageous is the sol-gel process.

In a book of this sort it is not necessary to define sol-gel processing but it is useful to specify in which sense I will use the term. Yoldas[2] has pointed out that in ceramics there are three types of solutions which can lead to gel formation and could therefore be referred to as sol-gel processing. Two types of these solutions lead to gels which then form particles. The third type of solution, soluble polymer solutions or polymerizable species, leads through chemical polymerization to a gel with a continuous network. It is in only this last sense, in which a sol is used to produce a monolithic gel rather than powder, that I will use the term "sol-gel processing." Furthermore, I will restrict my discussion to sol-gel processing of oxide ceramics based on metal-organic precursors.

The field of electronic ceramics is quite diverse, and even with the restricted definition of sol-gel processing described above, a review of all of the work on electronic ceramics made by the sol-gel process would be too lengthy. In this chapter, a few materials and applications will be discussed. It is intended that this chapter offer some insights to the possibilities of using sol-gel processing for electronic ceramics.

ADVANTAGES AND DISADVANTAGES

Electronic ceramics are typically polycrystalline, multicomponent oxides. It has become increasingly apparent that improved materials require higher purity and more control over micro- and macrostructures than were previously achieved.

Sol-gel processing as defined offers advantages over standard ceramic or glass processing for some electronic applications. The process is a series of wet chemical reactions occurring at relatively low temperatures. The chemical reactants are available in solution form which allows for distillation of the materials. Thus, high purity "raw materials" for electronic ceramics can be prepared in the sol-gel process.

Multicomponent oxides can be formed via sol-gel processing. The individual constituents are combined as liquids resulting in atomic level mixing. The final material can be more homogeneous with respect to chemical species than could be expected from processing in which solid-state or melt-phase diffusion is necessary to get compound formation.

It should be noted that purity and homogeneity could also result from other types of solution preparation of electronic ceramics. These advantages result from the liquid nature of the reactants regardless of whether a powder or a monolithic gel is formed. These advantages, therefore, are not limited to sol-gel processing. The three dimensional network (rather than particulate) structure does allow for some additional advantages.

Before gelation is complete, the material is still fluid. Films or coatings of desired compositions can be easily formed on a substrate. The method of application of the film could be dipping, spraying, or spinning among others. This allows for tailoring the process to specific requirements.

After drying, the gel is porous. The pores in dried gels are often extremely small (approximately 10 nm). This allows lower firing temperatures to be used for sol-gel derived pieces than for standardly processed pieces. The low firing temperature of sol-gel derived material can also be important in compositions which undergo undesired phase transitions or have volatile components. It is possible in certain cases to process at sufficiently low temperatures, e.g. below the phase transition, thus avoiding any deleterious effects.

The properties of an electronic ceramic device result from the composition of materials used as well as the micro- and macrostructure of the piece. Because of the options available in each of the steps during sol-gel processing there can be somewhat independent control over porosity, crystal structure, and grain size. Sol-gel processing thus allows for a variety of compositions and structures to be formed. The resultant piece can, therefore, possess a unique combination of properties.

In review, due to the liquid nature of the constituents and the diversity of the processing variables possible, sol-gel processing can offer many advantages for use in electronic ceramics. They are: (1) purity, (2) homogeneity, (3) ease of forming a variety of structures, (4) low temperature processing, and (5) unique combination of properties.

There are, of course, also disadvantages to the sol-gel process for some applications. As discussed above, the technique allows a variety of shapes to be made. Due to the nature of the drying process, however, a problem arises when thick or complex shapes are fabricated. The drying times required become too long to merit using the sol-gel process. In some instances this has been avoided by using supercritical drying but this process leads to a very porous piece. The prime application of sol-gel processing is currently in forming thin films and coatings on substrates.

Another potential problem with sol-gel derived materials is the large amounts of solvents and water used or formed during this process. These trapped liquids in the gels, however, can be removed by heat treatment.

POTENTIAL USES IN ELECTRONIC APPLICATIONS

As described above, the sol-gel process offers advantages in certain applications. In this section some specific applications of the process to electronic ceramics are discussed. As examples, I will use some of the work from my laboratory and from the literature. This is not intended as a comprehensive list of applications but rather to suggest some of the potential applications for which sol-gel processing could be used for electronic ceramics. Included are also some applications which, to the author's knowledge, have not yet been implemented but are included to show some additional possibilities.

In each of the following sections, one class of electronic ceramics is discussed. A representative material will be detailed and some of the work on sol-gel processing of the material is discussed. Superionic conductors are the subject of a separate chapter in this book and so are not discussed here.

Piezoelectrics

Piezoelectric materials convert mechanical energy to electrical energy and vice versa. Piezoelectric ceramics are used in many applications as sensors and transducers. The materials most commonly used are referred to generically as $Pb(Zr,Ti)O_3$ or PZT. PZT ceramics are solid solutions of $PbZrO_3$ and $PbTiO_3$. PZT,[3] $PbZrO_3$,[4] and $PbTiO_3$,[5] have each been made by the sol-gel process.

Pure lead titanate ($PbTiO_3$) has been shown to have potentially useful piezoelectric properties but has not proven to be an important technological material to date because of the destructive phase transformation it undergoes upon cooling below 490°C. This leads to cracked or stressed polycrystalline pieces. As with other members of the PZT family, lead volatility is often a problem during firing. Sol-gel processing of lead titanate may allow processing of samples at low enough temperatures to avoid these problems and is also of interest as a study of forming a multicomponent piezoelectric material by the process.

The details of the sol-gel processing of $PbTiO_3$ have been previously reported.[6] Briefly, a "complex alkoxide" containing lead and titanium was synthesized from lead acetate and isopropyl titanate mixed in a 1:1 molar ratio. Gels of the complex alkoxide were prepared by dissolving the complex in a solvent and then adding water for hydrolysis. The solutions were then poured into petri dishes in which gelation occurred. The gels were dried at 35°C for approximately 10 days.

Gels prepared from the complex alkoxide remained transparent after drying.

The gels shrank approximately 60 to 70 percent during drying and cracking of the samples did occur. Monolithic pieces up to 1.5 cm in diameter have been obtained from the cracked gels.

A systematic study was then made to examine the effects of process parameters such as amounts of water, acid, and alkoxide molarity on the gelation time as well as on the structure of the gels obtained.[7] The gel structure was characterized by the elastic modulus which was determined via sound wave propagation through the gels. It was found that an increase in both the amount of water and alkoxide molarity caused rapid gelation. Acid additions not only suppressed gelation but also reduced the elastic moduli of the resultant gels. Gels having a low elastic modulus were found to be desirable for the suppression of cracking during aging and drying. Addition of an excessive amount of acid, however, prevented the formation of a large scale polymer network and resulted in a powder-like gel.

Samples of the dried gel were analyzed using x-ray diffraction and differential thermal analysis.[8] Crystallization of a metastable intermediate phase was observed to begin upon annealing at 400°C for approximately 6 hours. Tetragonal $PbTiO_3$ appeared after longer annealing times. The crystallization temperature at a constant heating rate of 80°C/min was approximately 470°C.

The Raman spectra were obtained for the sol-gel derived lead titanate powder as-dried and heated to 600°C for one hour.[9] The spectrum of the as-dried material, which was x-ray amorphous, yielded little information. The spectrum of the annealed powder corresponded to that of tetragonal $PbTiO_3$ in agreement with the x-ray diffraction results. The five highest-energy modes observed were consistent with the sample being under a hydrostatic pressure of 1 GPa. This pressure was ascribed to the strain developed from the phase transformation during cooling.

As mentioned above, however, because of the destructive phase transformation at 490°C, $PbTiO_3$ has found few applications. In those applications, though, sol-gel processed material could be advantageous. Banno has found pure $PbTiO_3$ to be useful in piezoelectric-polymer composites.[10] Sol-gel derived $PbTiO_3$ has been tried in this application with moderately successful results.[11] It is thought that due to the wide range of chemical and physical processing variables inherent in the technique, sol-gel derived materials could be engineered for improved performance in this application.

Another application in which $PbTiO_3$ has found use is as a thin-film pyroelectric infrared sensor.[12] Because of the liquid nature of the sol, sol-gel materials are useful for thin-film formation, as described in another chapter of this book. By using a process similar to that described above, thin-film $PbTiO_3$ has been made by the sol-gel method.[13] (PZT films have also been prepared via sol-gel processing.[14]) These films may prove to be useful IR sensors.

Sensors

Ceramic based sensors are used quite extensively to monitor oxygen, combustible gases, and humidity among other parameters.[15,16] As an example, in order to reduce exhaust pollution, automobile manufacturers use devices which are sensitive to the oxygen fugacity in the exhaust stream. The currrently used sensors are based on ZrO_2. There are efforts underway to try to use TiO_2 based resistive sensor instead.[16] Recently, oxygen sensors based on TiO_2 films have been reported.[17] Fast response in these devices resulted from the thin dimension of the film and its high porosity which allowed for rapid gas permeability.

As discussed above, sol-gel processing can easily be used to form thin layers on a substrate. Furthermore, it leads directly to a high porosity product. Thus, sol-gel processing seems to be a promising way to form these titania films. Sol-gel derived TiO_2 monoliths[18] and thin-films[19] have been produced.

In forming TiO_2 films, isopropyl titanate was mixed with ethanol and water.[19] Glass substrates were dipped into the hydrolyzed sol. The coated substrates were heated to 500°C for 10 minutes and the process repeated. Films of up to 2 microns were built up in this manner. The resultant films had high surface areas and were found to be useful as a photochemical electrode, an application related to oxygen sensing.

Microelectronic Packaging

An important use of ceramics is as substrates in packaging microelectronic circuitry. This use requires a material with a high electrical resistivity, low dielectric constant, high thermal conductivity, high mechanical strength, and a thermal expansion that matches that of the integrated circuit material, which is usually silicon. Al_2O_3 is a good compromise material for this application. Sol-gel derived Al_2O_3 has been studied for several years.[20]

Recently, "ultra-thin" Al_2O_3 substrates have been made using a sol-gel process.[21] In this process, after hydrolysis and peptization a "viscosity control" step is used to allow the sol to be tape cast. The green tape is dried, cut, and fired. Substrates 60, 80, and 100 microns in thickness are available. Due to the fine grain size resulting from the sol-gel processing these substrates have a surface finish which will allow them to be used, as is, for thin film microelectronic applications.

As discussed above, the sol-gel process is well suited to forming thin films or coatings. Other packaging applications which might be suitable for this process are: (1) sealing, (2) encapsulating, (3) insulating layers, and (4) passivation layers.

The prime advantage of a sol-gel derived seal would be due to the lower temperatures necessary for processing. By lowering their processing temperatures many more materials could be used as seal materials if they were made by a sol-gel process. In other words, for a given sealing temperature (e.g. 450°C) many more compositions could be processed at this temperature using the sol-gel process than are now possible. Thus a wider choice of sealing materials could be available. Analogously, for a given seal material, the processing temperature could be lowered if the sol-gel technique is utilized.

Some of the potential problems of a sol-gel derived seal are cracking of the coating, densification without harming the circuitry, and the water involved in the process. Cracking of the coating is due to differential drying and therefore shrinkage of the layer. Careful control of the drying process can avoid this problem. The water associated with the process can also be removed. It has been shown by Nogami and Moriya[22] that SiO_2 glass formed by the sol-gel process initially had a higher water content than standard SiO_2 glass. After suitable heat treatment, however, the gel-derived glass had a lower water content. The other problems will require more detailed investigations.

Magnetics

Ceramic materials are used in a wide range of magnetic applications. Ceramic based magnets are used in motors and transformer cores among others. The magnetic properties are critically dependent upon the chemical and physical characteristics of the materials used. The somewhat independent control over these characteristics which can be obtained from the sol-gel process make it extremely interesting for magnetic applications.

There have been reports on magnetic ceramic powders made by the sol-gel process, e.g. $LiFe_5O_8$,[23] as well as a preliminary report on making magnetic thin films.[24] In the thin film process iron oxide was dip-coated onto glass substrates. By controlling the heat treatment of the resultant films, distinctive magnetic properties were obtained.

Ferroelectrics

For many electronic circuits improved capacitors are required. This may mean higher values of capacitance or smaller sized capacitors. These can be accomplished in two ways, by increasing the dielectric constant of the material used in the capacitor or decreasing the thickness of the dielectric layer. In some applications it would be useful to be able to apply the capacitor directly to the integrated circuit as a thin film.

Ferroelectric materials are useful as capacitor dielectrics because of their high dielectric constants. Ceramic based ferroelectrics are increasingly being used in this application. These materials are often composed of several components. Because of the homogeneity and ability to form thin films associated with the sol-gel process it has been considered for use in making ferroelectric ceramics.

While the piezoelectric materials described above are ferroelectrics, they are not very useful in this application. Other ferroelectric materials which would be more useful have also been made by the sol-gel process. Films of $BaTiO_3$[25] and $K(Ta,Nb)O_3$[26] have been produced. These materials have shown promise as dielectric films.

SUMMARY

The sol-gel process offers many potential advantages for electronic ceramics. Because of the liquid nature of the sol, thin films of multicomponent compounds can be synthesized homogeneously. With the number of options available during each of the stages of processing, somewhat independent control over materials properties, such as porosity, crystal structure, and grain size can be achieved. This can and has led to materials being made with a unique set of properties.

Acknowledgements

The work described in this article was done while the author was at Rutgers University. The financial support of the G.N. Howatt Foundation was greatly appreciated. I would also like to thank my colleagues Lisa Klein, Steve Gurkovich, Yoshi Hayashi, and Rouz Haghighat for their useful discussions and interesting work.

REFERENCES

1. See for example, Buchanan, R.C. (ed.), *Ceramic Materials for Electronics,* (Marcel Dekker, New York (1986).
2. Yoldas, B.E., Colloidal vs. Polymerized Gel Systems, presented at the Glass Through Chemical Processing Conference, Piscataway, NJ (1980).
3. Dosch, R.G., *Mater. Res. Soc. Symp. Proceed.* 32, 199-204 (1984).
4. Ibrahim, D.M. and Hennicke, H.W., *Brit. Ceram. Trans. & J.* 80, 18-26 (1981).
5. Blum, J.B. and Gurkovich, S.R., *J. Mater. Sci.* 20, 4479-4483 (1985).
6. Gurkovich, S.R. and Blum, J.B., "Preparation of Monolithic Lead Titanate by a Sol-Gel Process," in *Ultrastructure Processing of Ceramics, Glasses, and Composites,* L.L. Hench and D.R. Ulrich (eds.), 152-160, John Wiley, New York (1984).
7. Hayashi, Y. and Blum, J.B., "Sol-Gel-Derived $PbTiO_3$-II: Structural Control of Monolithic Gels" (unpublished).
8. Gurkovich, S.R. and Blum, J.B., *Ferroelectrics* 62, 189-194 (1985).
9. Blum, J.B., *Mater. Lett.* 3, 360-362 (1985).
10. Banno, H., *Ferroelectrics* 50, 3-12 (1983).
11. Monroe, D.L., Blum, J.B. and Safari, A., *Ferroelectric Lett.* 5, 39-46 (1986).
12. Okuyama, M., Ohtani, K., Ueda, T. and Hamakawa, Y., *Int. J. Infrared & Millimeter Waves* 6, 71-78 (1985).
13. Ariizumi, A., et al., *Japan J. Appl. Phys.* 24, Supplement 24-3, 7-9 (1985).
14. Fukushima, J., Kodaira, K. and Matsushita, T., *J. Mater. Sci.* 19, 595-598 (1984).
15. Kulwicki, B.M., *J. Phys. Chem. Solids* 45, 1015-1031 (1984).
16. Sheppard, L.M., *Adv. Mater. & Process.,* 19-25 (July, 1986).
17. Kaiser, W.J. and Logothetis, E.M., "Exhaust Gas Oxygen Sensors Based on TiO_2 Films," Paper No. 830167, SAE International Congress, Detroit, MI (1983).
18. Yoldas, B.E., *J. Mater. Sci.* 21, 1087-1092 (1986).
19. Yoko, T., Kamiya, K. and Sakka, S., *Denki Kagaku* 54, 284-285 (1986).
20. Yoldas, B.E., *Ceram. Bull.* 54, 286-288 (1975).
21. Shimura, M., Fukuda, Y. and Ono, M., Ultrathin Alumina Substrates, Paper No. 51-E-86, American Ceramic Society, Chicago (1986).
22. Nogami, M. and Moriya, Y., *Yogyo-Kyokai-Shi* 89, 675-676 (1981).
23. Oda, K. and Yoshio, T., *J. Mater. Sci. Lett.* 5, 545-548 (1986).
24. Kordas, G., Weeks, R.A. and Arfsten, M., *J. Appl. Phys.* 57, 3812-3813 (1985).
25. Dosch, R.G., *Mater. Res. Soc. Symp. Proceed.* 32, 157-161 (1984).
26. Wu, E., Chen, K.C. and Mackenzie, J.D., *Mater. Res. Soc. Symp. Proceed.* 32, 169-174 (1984).

15

Superionic Conductors from the Sol-Gel Process

Jean Pierre Boilot and Philippe Colomban

Groupe de Chimie du Solide
Laboratoire de Physique de la Matiere Condensee
Ecole Polytechnique
Palaiseau, France

INTRODUCTION: FAST ION CONDUCTION

In conventional ionic solids (particularly alkali halides), the defects are thermally generated and the conductivity (σ) varies according to

$$\sigma \sim T^{-1} \exp(-H_f/2kT) \times \exp(-H_m/kT) \tag{1}$$

where T is the absolute temperature, H_f the enthalpy of formation of a defect pair (for instance of Schottky type), H_m the activation energy of migration of mobile species and k the Boltzmann constant. In these solids, the number of charge carriers which insure electrical conductivity is very weak and the activation energy for conduction ($H_f/2 + H_m$) is high (a few eV).

In contrast a few types of ionic solids, generally referred to as superionic conductors,[1] have values of conductivity (σ) in the solid state comparable with that found in the melt and, as a consequence have practical application e.g. solid electrolytes in electrochemical systems. In these materials, the skeleton of the structure offers a great number of vacant sites which may be occupied by the "mobile" ions (Na^+, Li^+, Ag^+, H_3O^+, etc.) with easy migration along preferred pathways between neighboring sites. Therefore, the activation energy for conduction involves only the migration energy ($H_f = 0$) and is rather low (a fraction of an eV). The conductivity varies according to

$$\sigma \sim T^{-1} \exp(-H_m/kT) \tag{2}$$

Up to now the most important application of superionic materials has been their use for energy storage by advanced batteries. In particular, this explains why so many publications are dedicated to sodium electrolytes used in sodium/sulfur batteries.

The most widely studied among the sodium-ion conductors is β-alumina,[2] for which, the structure (hexagonal stacking) is currently described in terms of spinel-like blocks, $Al_{11}O_{16}$, separated by loosely packed layers of oxygen and sodium ions. These layers are linked to the blocks by a bridging oxygen. In these planes are found the vacant sites available for migration of the sodium ions, giving rise to a high two-dimensional conduction ($\sigma \sim 10^{-2}\,\Omega^{-1}\,cm^{-1}$ at room temperature).

Another extensively studied material is NASICON.[3] Hong[4] found a new phase in the system $Na_{1+x}Zr_2Si_xP_{3-x}O_{12}$ ($0 \leqslant x \leqslant 3$) possessing a three-dimensional network of SiO_4 and PO_4 tetrahedra sharing corners with ZrO_6 octahedra. This electrolyte was named NASICON (sodium superionic conductor). Many properties of NASICON are very attractive: three-dimensional conductivity, sintering at a substantially lower temperature than β-alumina, the conductivity of ceramics being similar ($\sigma \cong 5 \times 10^{-2}\,\Omega^{-1}\,cm^{-1}$ at R.T.). Unfortunately, many structural transitions have been observed in all NASICON-type compounds. Variations of electrical conductivity and dilatometric anomalies which result from these transitions can induce some degradation of ceramics. Furthermore the properties are very dependent on the history and on the ceramic processing of the NASICON-type compounds.[3]

Solid oxide electrolytes[1] with high oxygen ion conductivity have, during the last two decades, attracted much interest because of their technological applications in the fields of energy conversion and storage. As examples of such applications can be mentioned: high temperature fuel cells, membranes for high temperature water electrolysis and oxygen sensors for measurement in automobile exhaust and molten steel. The best known and most thoroughly investigated oxides exhibiting high oxygen conductivity are those with the fluorite and fluorite related structures, in which high concentrations of oxygen vacancies are introduced by substitutional dissolution of lower valent cations (doping); in particular, defect stabilized zirconia ceramics such as $CaO{:}ZrO_2$ and $Y_2O_3{:}ZrO_2$. In fact, these compounds only exhibit a poor ionic conductivity at room temperature, but their refractory properties make it possible to use them for applications at high temperature ($>1000°C$) in the superionic regime.

On the one hand, it is clear that most solid electrolytes used in electrochemical cells are crystalline compounds. But, as in NASICON ceramics, dilatometric anomalies and microstructure effects can affect mechanical and electrical properties of ceramics and decrease the lifetime of the electrochemical system. Consequently, amorphous superionic conductors are intensively researched. On the other hand, non-power applications (gas monitors, electrochemical display devices, switches, etc.) generally implying microionic systems appear now to be a new opportunity for superionic conductors.[5] Therefore, it is necessary to prepare solid electrolytes as thin or thick films.

All these problems are at the origin of the association between fast-ion conduction and sol-gel technology which leads to:

- Better superionic ceramics (controlled microstructure, homogeneity, low sintering temperature)

- new amorphous superionic materials (outside the usual range of glass formation)
- thin or thick superionic films (by dip coating and silk-screen printing techniques respectively).

NEW AMORPHOUS SUPERIONIC CONDUCTORS

It is more than thirty years since reports of ionic exchange behavior of zirconium phosphate gels first appeared in the chemical literature.[6] Phosphate gels were obtained by addition of phosphate ion containing solutions (generally a phosphoric solution) to $Zr^{(IV)}$ salt solutions, and can be formulated as $Zr(HPO_4)_2 \cdot nH_2O$. A great variety of ion exchange studies and ion separations have been carried out with amorphous zirconium phosphate which exhibit a layered structure type, as observed in $\alpha ZrP[Zr(HPO_4)_2 \cdot H_2O]$ single crystals.[7] Similarly, numerous other phosphatic gels such as uranyl phosphate are already well known for showing a protonic mobility.[8] With recent advances in sol-gel technology, new amorphous compositions with ion conduction have been reported:

- Transition metal oxide gels (V_2O_5, WO_3 etc.)
- Sodium and lithium zircono-silicophosphates
- Lithium salts containing organically modified silicates.

In this part of the chapter, special attention is given to new sodium and lithium gels and glasses in the $Na_2O-ZrO_2-SiO_2-P_2O_5$ system. Superionic properties of these materials clearly show the interest of the sol-gel technique for fast ion conduction.

Transition Metal Oxide Gels

In an alkaline aqueous medium, numerous transition metal compounds lead to complex ions such as VO_4^{3-}, WO_4^{2-}, CrO_4^{2-}. Addition of an acid solution can give gels through an inorganic polymerization process

protonation $VO_4^{3-} + H^+ \rightarrow HVO_4^{2-}$

condensation

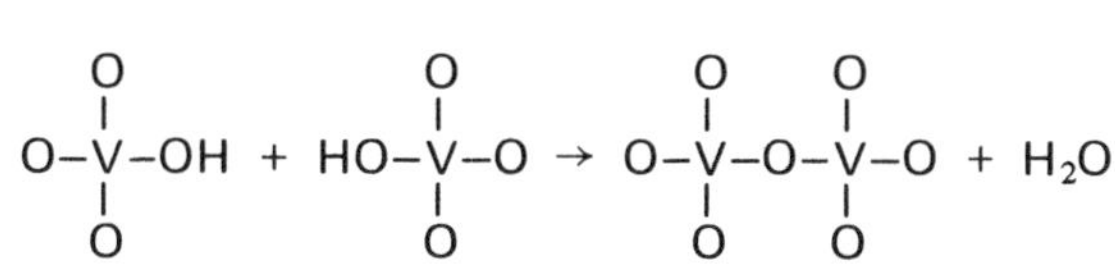

Livage and Lemerle,[9] for instance, obtained V_2O_5 gels by acidification of a metavanadate aqueous solution through a protonic exchanger resin. A clear yellow solution was obtained which spontaneously polymerized to give rise to a dark-red viscous gel. Semiconducting properties of V_2O_5 gels have been intensively studied and the main interest of such gels is that they can be easily deposited onto a glass or polymeric substrate giving semiconducting layers of large area.[9] V_2O_5 gels still contain some water molecules and can be described as a polyvanadic acid hydrate $V_2O_5 \cdot nH_2O$ in which lamellar oxide particles are held together by water layers. In the formula, the number of water molecules depends on the drying process. Therefore, these xerogels exhibit electronic properties arising from the vanadium oxide network [small-polaron hopping process between $V^{(IV)}$ and $V^{(V)}$] and ionic properties from the solvent water molecules. The ionic conductivity of $V_2O_5 \cdot nH_2O$ gels strongly depends on the amount of water: $\sigma \cong 10^{-2}\ \Omega^{-1}\ cm^{-1}$ at 300K for $V_2O_5 \cdot 1.8H_2O$ and $\sigma \cong 10^{-5}\ \Omega^{-1}\ cm^{-1}$ at 300K for $V_2O_5 \cdot 0.5H_2O$.[10]

Because of their layer local structure, gels provide a nice host lattice for intercalation of various species (metallocene, acetone, alkylamine ions, etc.).[8-9] The intercalation of guest species involves both electronic exchange with the oxide lattice and ionic exchange with the intercalated solvent. For instance, the reversible electrochemical Li insertion has been realized in V_2O_5 gels by Araki et al.,[11] following the half-cell reaction:

$$V^{(V)} + e + Li^{+} \rightarrow Li^{+}_{solid} + V^{(IV)}$$

At the beginning of the reaction, the organic solvent (propylene carbonate) is intercalated between the layers. Then the solvent is disintercalated owing to Li insertion leading to a $Li_{0.4}V_2O_5$ formula. As the reversible topochemical reaction of various transition metal crystalline oxides (MoO_3, TiO_2, V_2O_5, etc.) with lithium, this Li insertion in gels may be used as the cathodic reaction in high energy density batteries.

A very similar insertion has been realized by A. Chemseddine et al.[12] in WO_3 gels, obtained by deposition from tungsten oxide colloids, for applications in electrochromic display devices. (Electrochromism is the property of a material or system to change color reversibly in response to an applied potential.)

In fact, it is already well known[3] that thin films of a number of transition metal oxides when employed as electrodes in the appropriate electrochemical cells exhibit electrochromism by a reversible ion-insertion mechanism. Consequently, for these electrochemical applications (cathodic material, electrochromism, etc.) the one and only advantage of the sol-gel process is that thin layer films can be easily made at low temperature.

Sodium and Lithium Superionic Gels and Glasses

Recently, a large variety of new sodium and lithium superionic conductors have been prepared thanks to the low temperature polymerization technique.[14] These materials have been called NASIGEL or LISIGEL (sodium and lithium superionic gels) and NASIGLAS or LISIGLAS (sodium and lithium superionic glasses)[15] and are potentially well adapted for use in low power electrochemical devices such as sensors, microbatteries or displays.

Optically Clear Monolithic Gels. The sol-gel technique, from metal or nonmetal alkoxide hydrolysis can lead to non-crystalline solids outside the usual range of glass formation. Figure 1 shows the domain of clear monolithic gels in some ternary diagrams belonging to the $Na_2O–ZrO_2–SiO_2–P_2O_5$ system:[16]

(a) $Na_2O–ZrO_2–SiO_2$

(b) $ZrO_2 \cdot Na_2O–SiO_2–ZrO_2 \cdot P_2O_5$

(c) $4ZrO_2 \cdot 6SiO_2 - Na_2O - 4ZrO_2 \cdot 3P_2O_5$

(d) $Na_2O - ZrO_2 \cdot 4SiO_2 - P_2O_5$

In the second ternary diagram, compositions of $xZrO_2 \cdot Na_2O - ySiO_2 - zZrO_2 \cdot P_2O_5$ type are represented by points with x, y, z coordinates. The last three diagrams (b, c, d) correspond to sections passing respectively through the (1.1.-,- -1.), (1.- -,-1. 1.5) and (1.- -, -1.4.) lines of the first diagram (1a). Besides, on the 1b diagram, the usual NASICON solid solution $Na_{1+x}Zr_2Si_xP_{3-x}O_{12}$ with $0 \leqslant x \leqslant 3$ is symbolized by a straight line from the (2.3.-) point (x = 3) to the (1.-3.) point (x = 0). Finally, in all these diagrams, the gel domain is limited by a dashed line and the darker the area is, the easier the preparation of gels becomes. For all these compositions which easily lead to non-crystalline materials, partial substitution of Zr atoms by Al, Mg, Ta or Nb or total substitution by Ti , partial or total substitution of Na atoms by Li (Figure 2) or K do not drastically decrease the domain or formation of gels.[17-18] Nevertheless the substitution of Si by Ge atoms significantly decreases the amorphous range (Figure 3).

Therefore, the first important point is that the low temperature polymerization technique leads to numerous new amorphous solids in the $Na_2O - ZrO_2 - SiO_2 - P_2O_5$ systems, whereas the conventional melting method generally gives glass-ceramics in particular with ZrO_2 crystallization. Another preliminary conclusion concerning the fast ion conduction, is that some usual crystalline superionic conductors (NASICON and also β-eucryptite) can be easily prepared in an amorphous state by the sol-gel technique.

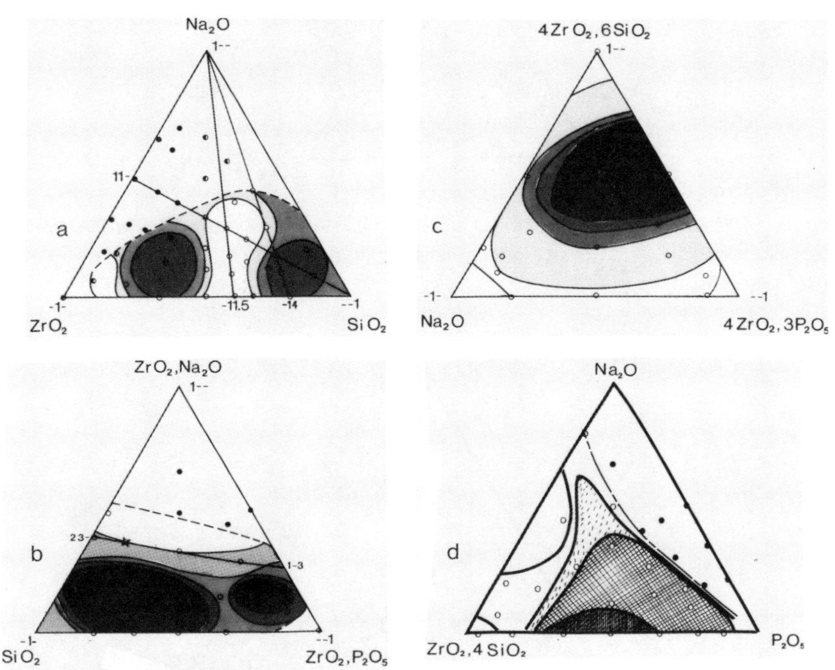

Figure 1: Optically clear monolithic gels in the $Na_2O - ZrO_2 - SiO_2 - P_2O_5$ system. The dashed line limits the gel domain. The volume increases when compositions (circles) are localized in the darker area.[16]

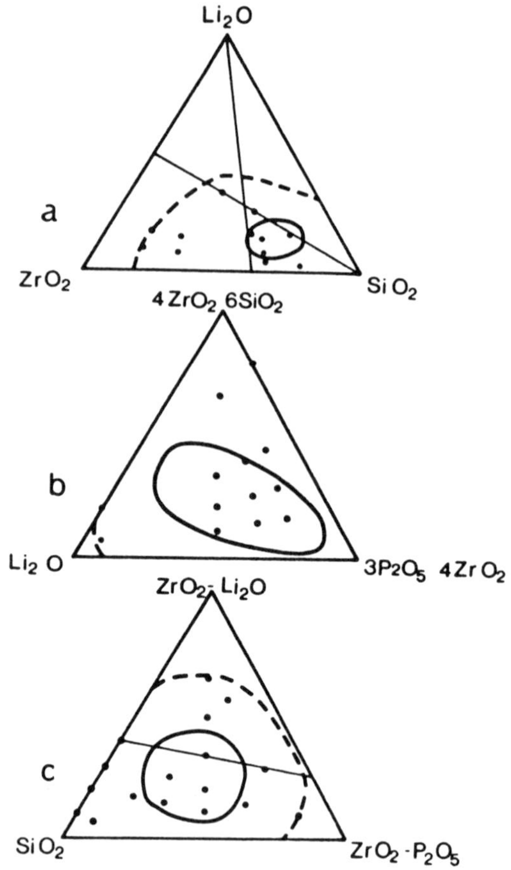

Figure 2: Monolithic gels in the $Li_2O-ZrO_2-SiO_2-P_2O_5$ phase diagram: Domain of easy elaboration.[18] The straight line of diagram c corresponds to the NASICON solid solution.

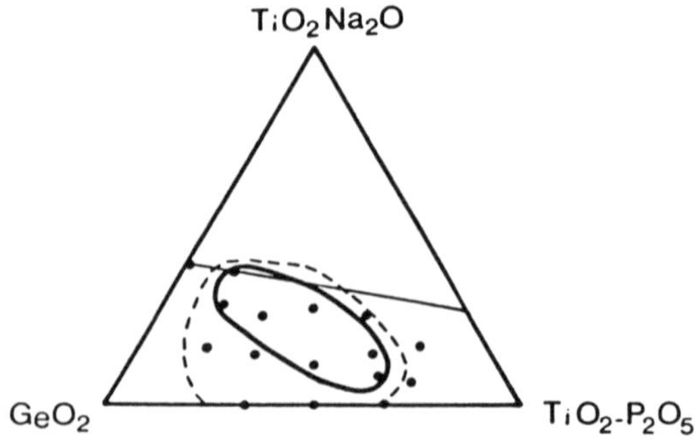

Figure 3: Monolithic gels in the $Na_2O-TiO_2-GeO_2$ phase diagram: Domain of easy elaboration.[18] The straight line corresponds to the NASICON solid solution.

Homogeneity and Densification of Gels. For multicomponent oxides, it is generally admitted that the sol-gel process leads to great homogeneity. This is rather unexpected at a microscopic scale, because alkoxides exhibit different rates of hydrolysis. Therefore one can expect that self-polymerizing species are initially formed from the fast hydrolyzing alkoxides, implying microscopic inhomogeneities in the gel. It is also important to know the capacity of gels for leading to dense glasses by thermal treatment below the crystallization temperature.

Structural investigations by small angle x-ray scattering (S.A.X.S.) of the gelling process (hydrolysis by atmospheric water vapor) and of the gel-glass conversion for compositions belonging to the $Na_2O-ZrO_2-SiO_2-P_2O_5$ have been reported by A. Dauger et al.[19] Two main behaviors have been observed in these gels.

Low Alkaline Compositions (e.g., $0.25Na_2O-ZrO_2-P_2O_5-4SiO_2$). The structural advancement can be summarized as follows:

(i) Gelation: formation of the network and of scattering particles (Figure 4). In the liquid phase, i.e. before the gel point, ^{29}Si-NMR and power law analysis of S.A.X.S. curves show the existence of a distribution of weakly branched (Si–O–Si) oligomers. Subsequent hydrolysis steps lead to a weakly ramified polymeric structure and, at the gel point, the network is less crosslinked than in a corresponding melted silica glass. Moreover, near the gel point, a liquid phase remains in the interstices of the silica network structure. This phase contains unhydrolyzed (Si–O–Si) oligomers, as shown by NMR and hydrolysis products of fast hydrolyzing alkoxides (Zr, Na). Therefore, active polymerizing species, such as $Zr(OH)_x(OR)_{4-x}$, give by self-polymerization, small particles of zirconia.

Assuming that zirconia clusters contribute in a large part to the scattered intensity, the increasing of this and the shift of the maximum towards higher angles, during the gelation, can be explained by the growth of complex scattering particles. Each particle is comprised of a spherical nucleus of high electronic density (zirconia clusters where surface oxygens are associated with protons and alkyl groups), surrounded by a depleted shell (H_2O, alcohol and ramified Si–O–Si oligomers) of density that is less than the average electronic density of the structure "frozen in" at the gel point.

(ii) Conversion from gel to glass (Figure 5). Between 20°C and 300°C, in the ramified Si–O–Si structure, continued cross-linking occurs, resulting from condensation reactions. As shown in Figure 5, the progressive densification of the gel is associated with a modification in the low angle part of the S.A.X.S. curve and with the shift of the maximum towards higher angles. Probably, these changes only correspond to the decreasing size of the low electronic density areas by removal of the volatiles and by condensation between Si–O–Si oligomers. Between 300°C and 700°C, the gel evolves towards fully cross-linked glass via condensation reactions. The decreasing of the scattered intensity demonstrates that complex particles are progressively dissolved in the network.

High Alkaline Compositions (e.g., $Na_2O-2ZrO_2-P_2O_5-4SiO_2$). Concerning these compositions (Figure 6), X-ray scattering curves from xerogel, and power law analysis suggest that zirconium atoms are homogeneously distributed in the gel. In fact, for this high alkaline composition, the rapid hydrolyzing sodium alkoxide leads to a high OH concentration which probably activates the slow hydrolyzing P and Si alkoxides (in agreement with the short gelation time for this sample). Active polymerizing species, such as $Si(OR)_x((OH)_{4-x}$, then react

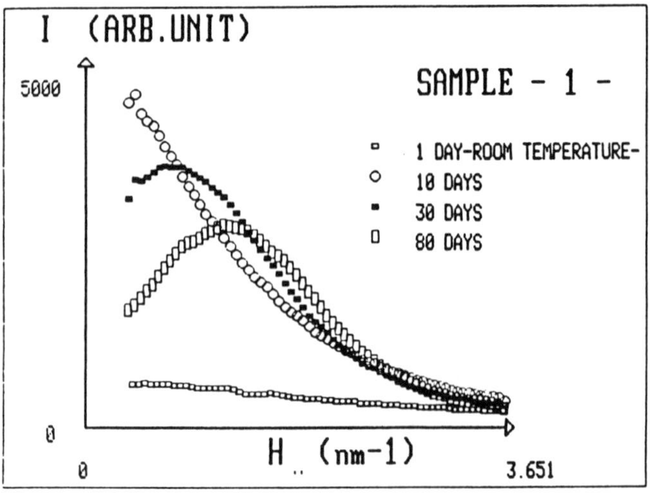

Figure 4: Gelling process (low alkaline compositions): aging in ambient atmosphere after the gel point. Small angle X-ray scattering curves versus time (CuKα). One can note that near the gel point, the scattered intensity does not decrease continuously and that, after a few days, curves exhibit a well pronounced maximum. With increasing time, the maximum occurs at larger angles ($H = 4\pi \sin \theta/\lambda$). The analysis of the intermediate angle region displays the cross-over from a polymeric state with randomly branched chains to a state formed by particles with well defined boundaries.[19]

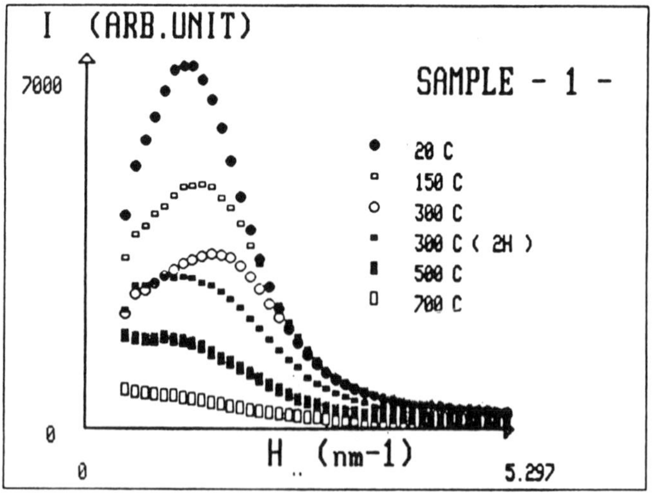

Figure 5: Gel to glass conversion (low alkaline compositions). S.A.X.S. curves at different temperatures (CuKα). Between 20°C and 300°C, the scattered intensity decreases in the lower angle part. Above 300°C, the maximum occurs at smaller angles and the scattered intensity decreases corresponding to the progressive disappearing of the scattering particles.[19]

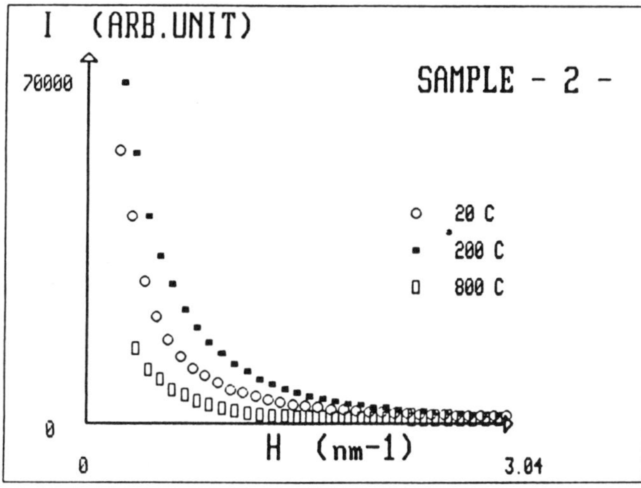

Figure 6: Gel to glass conversion (high alkaline compositions). Evolution of the scattering curves (CuKα) during the heating of the gel. For this sample, at all temperatures, scattering decreases continuously from a maximum at zero angle and the low angle Guinier region leads to a radius of gyration of 36 Å.[19]

with unaltered rapid hydrolyzing Zr alkoxide, and dissimilar constituents tend to become neighbors, i.e. M_1–O–M_2 bonds rather than M_1–O–M_1, and thus greater homogeneity is attained. By heating, the scattered intensity decreases, corresponding to the crossover towards a dense homogeneous amorphous state.

The results of S.A.X.S. experiments show that the formation of multicomponent oxide gels is generally not straightforward, because alkoxides exhibit different rates of hydrolysis. Therefore, self-polymerization of active polymerizing species, can lead to microinhomogeneities in gels. Nevertheless, inhomogeneities can be progressively dissolved during the conversion from gel to glass, as a consequence of the high reactivity of the ramified silica structure which is more open than in melt-prepared glass.

Crystallization. Figure 7 shows isotherms of crystallization (generally partial) for the different diagrams of the Na_2O–ZrO_2–P_2O_5–SiO_2 system and Figure 8 presents the first phase observed after crystallization. These phases are zirconia (cubic, tetragonal or monoclinic), zirconium phosphate, NASICON solid solution, cristobalite or vlasovite (Na_2ZrSiO_5). Concerning high phosphorus compositions, a large range is glassy up to 900°C and sometimes remains amorphous up to 1200°C. The substitution of sodium atoms by lithium ones or of silicium atoms by germanium ones, decreases the temperature of devitrification by about 300°C.[16,18]

Ionic Mobility and Conductivity. Amorphous compositions of the Na_2O–ZrO_2–SiO_2–P_2O_5 system generally exhibit exchange properties in aqueous solution and in molten salts. The ionic exchange in $LiNO_3$, $AgNO_3$ and KNO_3 molten salts, has been studied for glassy sodium zirconosilicates.[16] The starting compositions, localized on the line joining SiO_2 (0.1.0. point) to $ZrO_2 \cdot Na_2O$ (1.0.0. point) in Figure 1b, correspond to different associations of tetrahedra: isolated (number of bridging oxygens: n = 0), linear (n = 1), triangular (n = 2), tridi-

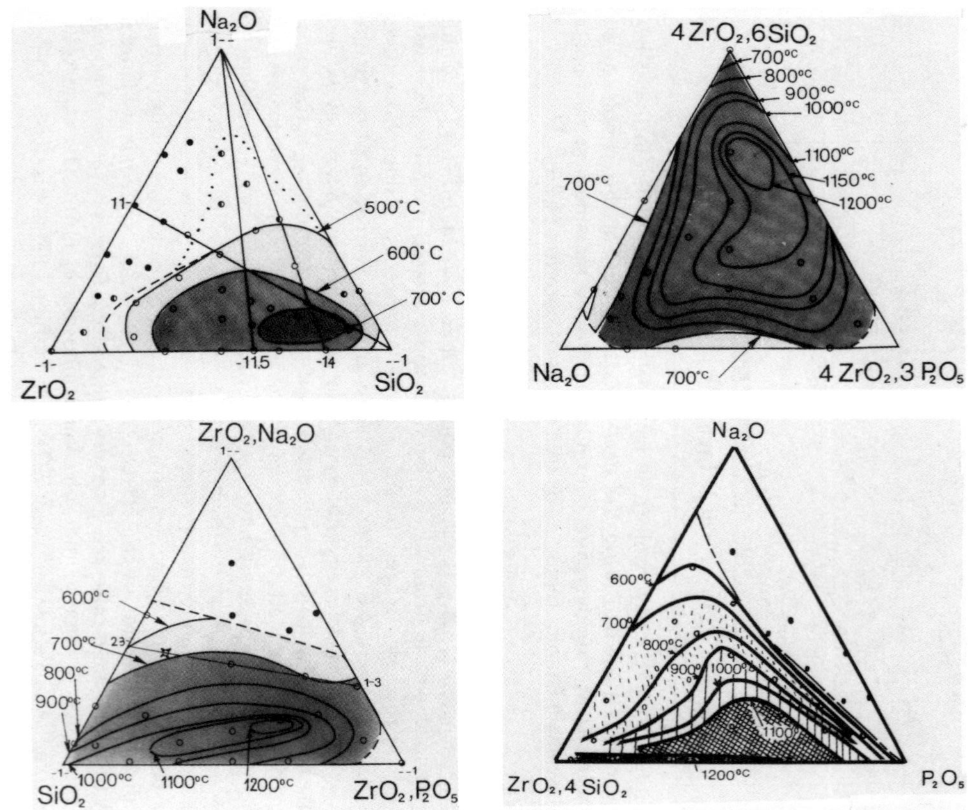

Figure 7: Isotherms of crystallization in the Na_2O–ZrO_2–SiO_2–P_2O_5 system, heating rate 8°C/hour. Firing step time 10 hours. Cooling rate 50°C/hour.[16]

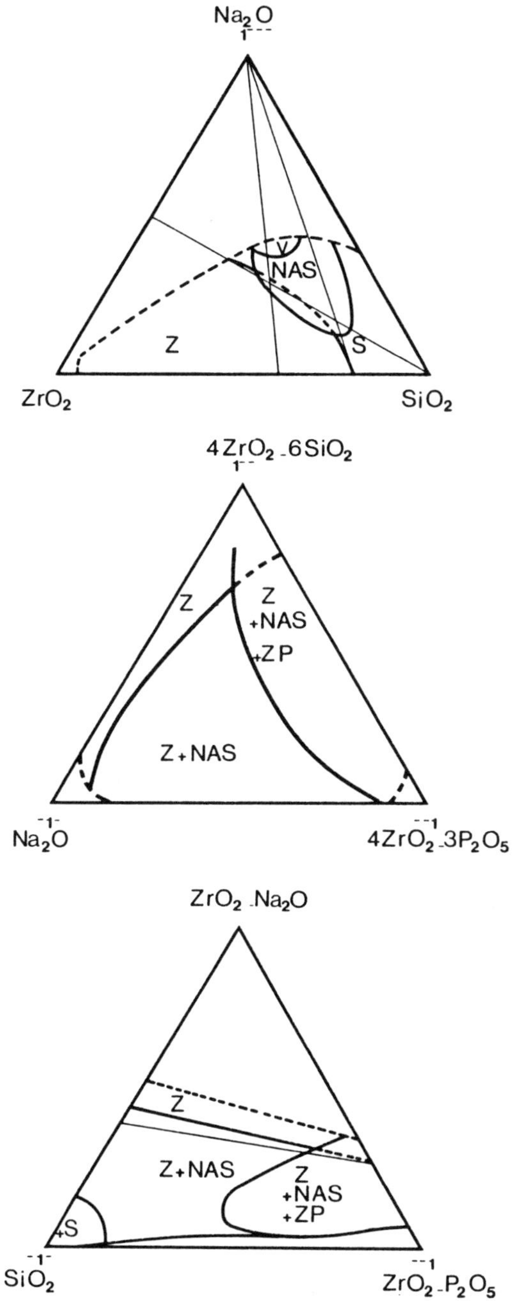

Figure 8: Main phases formed after glass-crystallization in the system Na_2O–ZrO_2–SiO_2–P_2O_5; V: vlasovite $NaZrSi_2O_5$; NAS: NASICON solid solution; Z: ZrO_2; ZP: ZrP_2O_7; S: cristobalite.[16]

mensional (n = 4). As expected, the exchange reaction is easier for a structure with n = 0 (NASICON type), whereas chemical durability is maximum for a structure with n = 4 (SiO_2). The substitution of Na^+ by other ions changes the devitrification temperature, which decreases with Li^+ and increases with K^+ ions (Table 1).

A.C. conductivity measurements have been achieved at room temperature and at variable frequencies (10^{-1} to 10^6 Hz) by means of the complex impedance method with evaporated Au electrodes for glass samples or with silver electrodes for gel samples. As usually observed for solid state ionic conductors, the electrical impedance plotted in the complex plane exhibits an arc of a circle and the bulk conductivity can be derived from the high frequency intercept of the impedance diagram with the real axis (Figure 9). At room temperature, conductivity of xerogels (Table 2), in the $Na_2O–ZrO_2–SiO_2–P_2O_5$ system, changes from 4×10^{-5} $(\Omega cm)^{-1}$ for high alkaline compositions to 10^{-6} $(\Omega cm)^{-1}$ for low alkaline ones. This suggests that sodium ions are more mobile than protonic species and largely contribute to the conductivity. The conductivity of glasses (Table 2) varies from $4\times10^{-4}\Omega^{-1}cm^{-1}$ for high alkaline compositions to $10^{-5}\Omega^{-1}cm^{-1}$ for low alkaline ones. In fact, these values are similar to those observed on equivalent ceramic pellets,[20,21] where activation energies are higher than the ones observed for NASICON ceramics.

In all cases, the values of ionic conductivity are high in comparison with those previously observed in other amorphous alkaline ion conductors [generally lower than $10^{-6}(\Omega cm)^{-1}$ at R.T.].[22,23]

Lithium Salt Containing ORMOSILS Gels

It is now well known that the mixture of polyethylene oxide, PEO, or polypropylene oxide, PPO and alkaline salts leads to complexes which exhibit a fast ion conduction.[24] In the past few years, practical emphasis has been placed on lithium conductors and the polymer electrolyte concept has generated a widespread interest among both polymer specialists and electrochemists, with the perspective of high energy density batteries in a thin film configuration.

Recently, D. Ravaine et al.[25] prepared and studied a new family of organically modified silicates (ORMOSILS) with PEO radicals, by mixing TEOS (tetraethoxysilane) and PEG (polyethylene glycol). Transparent dried gels were prepared and doped gels were easily obtainable with high Li^+ conductivities by dissolution of various lithium salts. At room temperature, values of the ionic conductivity appear to be stable and reproducible: they increase with the concentration of lithium (Table 3) up to values which are more than four orders of magnitude higher than for the base gels. This suggests that the major contribution to the conductivity comes from the lithium ions. Consequently, ORMOSILS gels appear as a very attractive host lattice for alkali metal salts-polyethers interactions and the discovery of other superionic gels is expected in the future.

SOL-GEL ROUTES LEADING TO CERAMICS AND THICK FILMS

A solid electrolyte, especially as used in all-solid devices, requires some specific properties:

(1) high chemical purity to lower the electronic conductivity.

Table 1: Exchange Properties of Glasses in Molten Salts[16]

Composition	n	Phases and crystallization temperature as a function of the ion: starting compound	Li⁺	Ag⁺	K⁺	Note
$Na_{0.75}Zr_{2.5}Si_{2.2}P_{0.8}O_{10}$	0	700° C ZrO_2,C 800° C NAS+ZrO_2 *m*	ZrO_2,C 700° C NAS+$ZrSiO_4$	ZrO_2,C 900° C NAS	ZrO_2,C 800° C ZrO_2 *t*+NAS	Devitrification during exchange
$Na_3Zr_2Si_2PO_{12}$	0	700° C ZrO_2,C 800° C NAS	ZrO_2,C 700° C ZrO_2 *t* 800° C +(Li_3PO_4)	700° C Ag+Ag_3PO_4+ZrO_2 *m* 800° C Ag+Ag_3PO_4 +SiO_2 *t*+ZrO_2 *m*	800° C ZrO_2,C 900° C $K_2ZrSi_3O_9$	crumbling of samples
$Na_2Zr_{1.25}Si_2P_{0.25}O_{8.75}$	1	750° C ZrO_2,C 900° C NAS	ZrO_2,C 700° C ZrO_2 *t*+SiO_2 *t* 900° C $ZrSiO_4$+(Li_3PO_4)	700° C Ag+ZrO_2 *t* 900° C Ag+ZrO_2 *t* +SiO_2 *t*+NAS	800° C $K_2ZrSi_3O_9$	
$Na_2Zr_{1.25}Si_3P_{0.5}O_{10.75}$	2	800° C ZrO_2 *m*	700° C ZrO_2 *t*+SiO_2 *t* 900° C $ZrSiO_4$+SiO_2 *t*	700° C Ag+ZrO_2 *t* 900° C ZrO_2 *m*+Ag+SiO_2 *t*	1 100° C $K_2ZrSi_3O_9$	
$NaZrSi_8PO_{17}$	4	900° C ZrO_2,C 1 000° C SiO_2 *t*+NAS +ZrO_2 *m*	ZrO_2 *t* 800° C $ZrSiO_4$+SiO_2 *h* +$Li_4P_2O_7$	700° C SiO_2 *t*+Ag+Ag_3PO_4	950° C ZrO_2 *t* 1 100° C SiO_2 *t*+$ZrSiO_4$	

n : probable number of bonding oxygens between tetrahedra
t : tetragonal symmetry, m : monoclinic h : hexagonal
glasses have been prepared by firing at 600°C of xerogels.

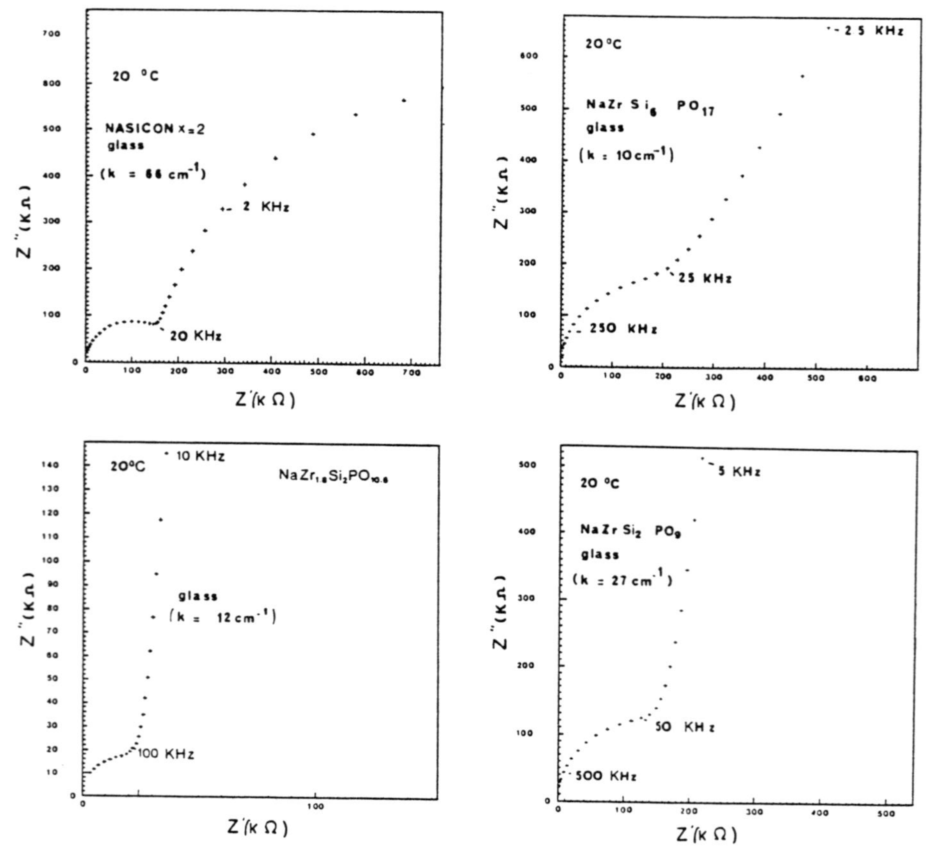

Figure 9: Complex impedance diagrams [Z'' = f(Z)] for different glasses.[16]

Table 2: Conductivity of Some Xerogels and Glasses[14-16]

Composition	σ 27°C (Ω^{-1} cm^{-1})	Ea_{20}^{100} (e. V.)	Note
$Na_4ZrSi_3O_{10}$, $(ROH)_n$	$4\ 10^{-5}$		xerogels (70°C)
$Na_3Zr_2Si_2PO_{12}$, $(ROH)_n$	$3\ 10^{-5}$		
$Na_{0,95}Zr_{1,8}Si_2PO_{12}$, $(ROH)_n$	$1,5\ 10^{-5}$		
$NaZr_{1.8}Si_2PO_{10.6}$, $(ROH)_n$	$4,5\ 10^{-6}$		
$Na_{0.75}Zr_{1.5}Si_{2.2}P_{0.8}O_{12}$, $(ROH)_n$	$3\ 10^{-6}$		
$NaZrSi_2PO_9$, $(ROH)_n$	$2,8\ 10^{-6}$		
$Na_2Zr_5SiP_8O_{33}$, $(ROH)_n$	$2,1\ 10^{-6}$		
$NaZr_4Si_3P_3O_{22}$, $(ROH)_n$	$2,1\ 10^{-6}$		
$Na_{0.9}Zr_2Si_{2.5}P_{0.5}O_{10.7}$, $(ROH)_n$	$1,2\ 10^{-6}$		
$Na_{0.75}Ti_2Si_2PO_{10.9}$, $(ROH)_n$	$1\ 10^{-6}$		
$NaZr_{1.8}Si_2PO_{10.6}$ (*a*)	$5\ 10^{-4}$	0,46	glass (c)
$Na_{1.5}ZrSiP_{0.5}O_6$ (*a*)	$4,6\ 10^{-4}$	0,35-0,40	glass-ceramic
$NaZrSi_2PO_9$ (*b*)	$1,8\ 10^{-4}$		glass (c)
$Na_{0.8}Zr_2Si_2PO_{10.9}$ (*b*)	$1,2\ 10^{-4}$	0,41	glass (c)
$NaZrSi_6PO_{17}$ (*a*)	$6,6\ 10^{-5}$		glass (c)
$NaZr_{2.5}Si_2P_4O_{19.5}$ (*b*)	$4,4\ 10^{-5}$		glass (c)
$NaZrSi_6PO_{16.5}$ (*b*)	$2\ 10^{-5}$		glass (c)
$NaZr_{10}Si_{12}P_3O_{52}$ (*b*)	$2\ 10^{-5}$		glass (c)
$NaZr_3Si_{1.5}P_3O_{17}$ (*b*)	10^{-5}		glass (c)
$NaZr_4Si_3P_3O_{22}$ (*b*)	10^{-5}		glass (c)
$LiZr_{1.5}Si_2P_2O_{12.5}$ (*a*)	$8\ 10^{-4}$		glass (c)
$LiZrSi_{0.5}PO_6$ (*a*)	$3,3\ 10^{-4}$		glass (c)
$Na_3Zr_2Si_2PO_{12}$	$7\ 10^{-4}$	~0,35	monoclinic ceramic (d) (20)
$Na_3Zr_2Si_2PO_{12}$	$\sim 5\ 10^{-4}$	~0,37	rhombohedral ceramic (21)
Nasiglass (*a*)	$\sim 10^{-6}$	0,60	glass (usual technique) (22)

(a) the temperature of crystallization is about of 700°C
(b) the temperature of crystallization is higher than 900°C

Ea_{20}^{100} Activation energy between 20 and 100°C
(c) glass has been fired at a temperature which vary from .7 to .9 the one of crystallization
(d) structural symmetry at room temperature
(ROH)n : organic and/or protonic species (30–40wt %)

Table 3: Room Temperature Conductivities of Lithium-Doped ORMOSILS Gels[25]

PEG/TEOS	Solvent	Salt	Li/Si	σ (Ω^{-1}cm^{-1})
2	MeOH		o	$1,4\ 10^{-9}$
2	MeOH	$LiNO_3$	1:16	$2.3\ 10^{-5}$
2	MeOH	$LiNO_3$	1:4	$3.0\ 10^{-5}$
2	MeOH		o	$1.0\ 10^{-9}$
2	MeOH	$LiClO_4$	1:16	$1.0\ 10^{-7}$
2	MeOH	$LiClO_4$	1:8	$1.2\ 10^{-5}$
2	MeOH	$LiClO_4$	1:2	$6.4\ 10^{-5}$
4	THF	$LiClO_4$	1:4	$3.2\ 10^{-5}$
4	THF	$LiClO_4$	1:2	$6.9\ 10^{-5}$
4	THF	$LiClO_4$	1:1	$7.4\ 10^{-5}$

(2) high crystallographic purity to have optimal physical properties (i.e. the ionic conductivity, the mechanical strength, etc.).

(3) no second phases arising from incomplete reaction of the raw materials, leading to a barrier for fast ion diffusion.

(4) homogeneous microstructure to avoid aging under the electric field.

(5) various compositions (electrolyte, electrodes, etc.) must be associated without deterioration of the properties of each material.

In the first part of this chapter, we have pointed out how the synthesis of amorphous electrolytes partly solves the problems. However, the conductivity of amorphous ionic conductors generally remains lower than that of crystalline ones (i.e. β-Al_2O_3, NASICON, etc.) and, particularly the activation energy is higher than is required for numerous applications. Furthermore, the crystallization of the glassy electrolyte often occurs at a low temperature and prevents, for example, cofiring of the various materials constituting the electrochemical device.

For many years, ceramicists have been interested in obtaining pure phase fine powders. Fine powders can be obtained by grinding (but impurities are introduced), or from salts such as oxalates, nitrates, acetates, carbonates, etc., which are thermally decomposed. However, their use appeared to be difficult, often leading to badly densified products. Presence of anion traces and of aggregates inhibit a complete disappearance of porosity.

Colloidal gels are typically formed from metallic salt solutions and oxide or hydroxide sols. These methods have been used e.g., for aluminum and uranium oxides[26] but the presence of parasite cations is an obstacle to good densification and multicomponent materials are difficult to obtain, especially when a precise composition is required.

K.S. Mazdiyasni and C.T. Lynch have prepared zirconium oxide in 1963 by means of the hydrolysis of alkoxides.[27] Translucent yttria-stabilized zirconia was obtained by sintering at 1450°C with a final grain size of 2 to 4 μm.[28] In these works an "all-alkoxide" route was used: all the elements being introduced in the form of metal alkoxides. Multicomponent compositions such as transparent-ferroelectric PLZT Ceramics (a lanthanum doped lead zirconate-lead titanate perovskite) were also prepared.[27] In this last case G.H. Haertling and C.E. Land[29] pointed out that the advantages of the sol-gel route and those of the usual ceramic technology could be added by using "partial" sol-gel routes: only one part of the elements is incorporated via the sol-gel process.

Using of sol-gel routes: (1) allows the preparation of homogeneous submicronic powders which sinter a few hundred degrees lower than the usual oxide powders, (2) enables one to overcome the reactivity of the powder and to displace the sintering temperature in a large range, (3) results in obtaining a homogeneous fine-grained microstructure required in microionic devices. Here we report works concerning both "all" and "partial" sol-gel routes leading to superionic ceramics and thick films. Previous reviews dealing with gel technology in ceramics have been published.[27,30,31]

Pure Phase Powders and Ceramics

In previous works, the very low impurity content has already been shown.[27]

Table 4: Typical Solid Electrolyte Synthesized Using Alkoxides

Electrolyte	Elements added as alkoxide	sintering temperature	remarks	Ref.
ZrO_2 : Y	Zr, Y	1100°C-1400°C	Translucent ceramics	27,28,32
$Na\beta Al_2O_3$	Al,(Na)	1100–1400°C	ceramic, films	33,34
$K\beta Al_2O_3$	Al	1400°C	ceramic	33
NASICON	Si	1200°C	ceramic	35,39
	Zr	800-1200°C	ceramic, film	43 48
	Zr,Si		ceramic, film	43–52
	Zr,Si,P		ceramic, film	11
	Zr,Si,P,Na^+(Li^+)	600-1250°C	ceramic	14-17,21,43-52
β Eucryptite	Si	1000°-1100°C	film	49

Another specific advantage of sol-gel routes is that different elements are mixed together at a very short scale (mainly in the liquid state) which lowers the temperature needed to make the batch more homogenous: in all cases the homogeneity is better than the one arising from the mixing of pure oxide raw materials. Thus, when crystallization occurs, typically in the 500°C to 900°C temperature range, the target phase is obtained and lower stabilizing amounts are needed, e.g. in yttria-zirconia.[28,32] Generally, the resultant phases remain highly disordered and a higher temperature treatment is needed to develop the well crystallized state.

Gel processing has been used on Na and K β-Al_2O_3 powder preparation using citrate, oxalate or alkoxide.[2,33,34] The amorphous phase crystallizes above 1000°C. NASICON solid electrolyte is difficult to prepare free from zirconia traces with usual ceramic processing. Attempts to use more reactive raw materials have been made by various scientists: Quon et al.[35] and Lin et al.[36] have used tetraethylorthosilicate, Boilot et al.,[37] Gordon et al.[38] and J.J. Bentzen et al.[39] have used silicate sol for the elaboration of NASICON and of the parent $Na_5GdSi_4O_{12}$ compounds. The most satisfactory results were obtained by using zirconium propoxide:[40-43] however, due to the high intense Raman spectra of zirconia traces, more advanced sol-gel routes leading to zirconia free NASICON powders and ceramics, have been developed for vibrational spectroscopy analysis. These methods have also been used for various silico-phosphate superionic conductors, e.g. β-eucryptite ($LiAlSiO_4$), wadeite ($K_2ZrSi_3O_9$).[41] Since then, sol-gel routes have been used in numerous works.[44-45]

Low Temperature Sintering and Fine Grained Microstructure

It has already been shown that sol-gel routes allow the synthesis of powders which sinter at very low temperatures: e.g. yttria stabilized zirconia was sintered to 99.5% of the theoretical density in a 1 hour 1100°C cycle[32] whereas zirconia was usually sintered above 1500°C. NASICON was sintered at 700°C instead of 1200°C.[21] In all these cases, the low temperature sintering results from the following points:

- first, the sol-gel powder is very fine (<0.1 μm) and highly reactive, the specific area usually being between 30 and 300 m^2/g.
- second, powders can be compacted with a green density higher than 60% (in respect to the theoretical composition), without binder. This point ensues from the possibility of developing a room temperature pressure-sintering of green powder via the protonic species and the viscous character.[21] Thus it is important to use direct sintering of the xerogel powder, without calcination.
- third, the homogeneity is sufficient to avoid long thermal treatments to develop the target phase.

However, the high densification is only obtained if any agglomerates are suppressed.[21,31,32] Figure 10 compares, for zirconia, the effect of temperature on sintering for agglomerate-free and agglomerate-containing powders. Figure 11 compares green samples of NASICON prepared by different sol-gel processes. One can note that the green densification is not directly related to the optical clearness of the specimens: the most transparent specimen is prepared with the slow hydrolysis of an alcoholic alkoxide solution whereas the one with the highest density is nearly opaque and arises from the rapid hydrolysis of zirconium propoxide in alcoholic solution with two aqueous solutions, a silica sol and a $NH_4H_2PO_4$ solution both containing Na^+ ions.[48]

Ceramics obtained from sol-gel routes exhibit a fine grained microstructure. This was the first interest of these routes.[21,27,29,46] Dense ceramics exhibit a typical 0.5-1 μm grain size where usual ceramic process involves typically 10 μm grain size.

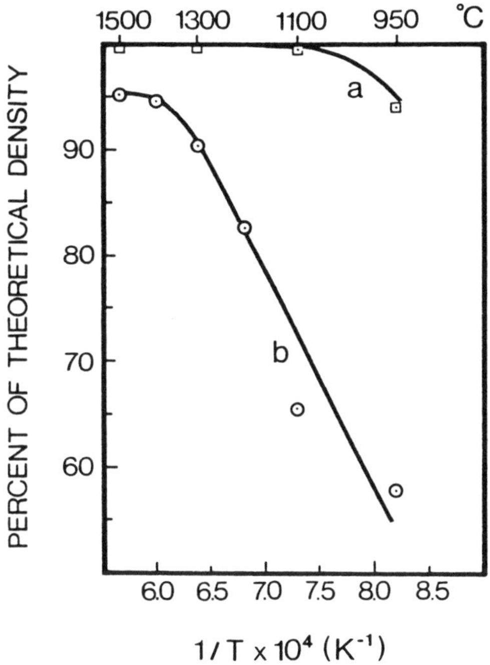

Figure 10: Effect of temperature (1 hour cycle) on yttria-stabilized zirconia sintering for agglomerate and agglomerate-free powders: (a) centrifuge cast fine, (b) as-received powder.[32]

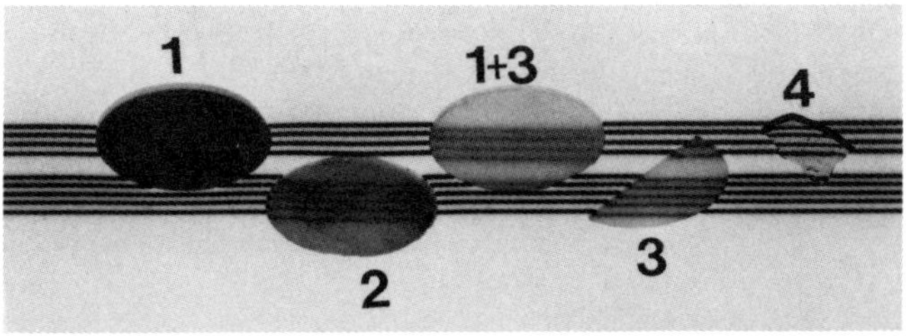

Figure 11: Green samples of NASICON: monolithic optically clear xerogel issued of a total alkoxide route (method 4), translucent room-temperature pressure sintered xerogels (methods 3, 1+3 and 2 respectively), opaque sample (method 1). Diameter 20 mm, thickness ~1.5 mm).[21]

How to Choose the Sintering Temperature and Overcome the Powder Reactivity

Preparation of microionic devices with both electrode and electrolyte film requires a low temperature sintering according to a non-deformation of the substrate and the keeping of the target composition of non-stoichiometric material which can easily evolve during the sintering. Sol-gel routes allow the preparation of powders used to make ink for silk-printing.[47,49] These powders have an adapted reactivity to limit shrinkage after their deposition on the substrates and sintering. Figure 12 shows various sol-gel routes leading to NASICON precursors, in the form of powder or monolithic samples. Figure 13 compares the evolution of the specific area as a function of the temperature with the process—(a) alkali excess, (b) part of elements incorporated in the form of alkoxide.[50] Figure 14 points out the modification of the temperature of the densification.

Using hybrid methods gives a greater possibility of overcoming the reactivity of powders: as in the case of usual ceramic technology (for China, faiency, etc.) the same element can be introduced simultaneously from various raw materials: i.e. silicium can be introduced both from an aqueous colloidal solution or from tetraethoxysilane alcoholic solution.[21,48] Thus it is possible to develop various surface behaviors of the powder and to control the sintering rate.[50,51] For example the presence of sodium silicate[51] or of lead oxide[29] at the grain boundaries favors the sintering process. In the case of NASICON, a method involving sodium silicate develops a liquid phase at the grain surface while methods involving a large part of alkoxide lead to smooth grains. By resorting to hybrid methods, in which for instance, one-third of the elements are brought via method 2 (in order to develop a liquid phase at the grain boundaries) and two-thirds via method 1 (so-called method 1+2) or respectively via methods 1 and 3 (method 1+3), we tend to gain the specific advantages of each method.

How to Choose the State of Ordering

Sol-gel routes lead to amorphous materials which crystallize at a relatively low temperature. However this crystallization is not drastic and the local structure changes "gradually" from the amorphous state to a well crystallized state.

The structure of superionic conductors consists of a loose packed framework in which some ions diffuse rapidly; a high degree of disorder is specific to the structure, especially in the case of nonstoichiometric materials. The thermal history fixes the state of disorder as clearly displayed in the NASICON solid solution. Sol-gel routes allow one to choose the temperature of the thermal treatment and thus the crystallization degree. Figure 15 shows the evolution of the Raman spectra of $Na_3Zr_2Si_2PO_{12}$ samples, synthesized from the sol-gel method.[52] Two reactions occur in the solid state:[14,21,22]

$$\text{amorphous NASICON} \xrightarrow{\cong 950^\circ\text{C}} \text{disordered NASICON I}$$
(mean symmetry—rhombohedral)

$$\text{disordered NASICON I} \xrightarrow{> 1050^\circ\text{C}} \text{ordered NASICON II} + \text{monoclinic } ZrO_2$$
(monoclinic at R.T.)

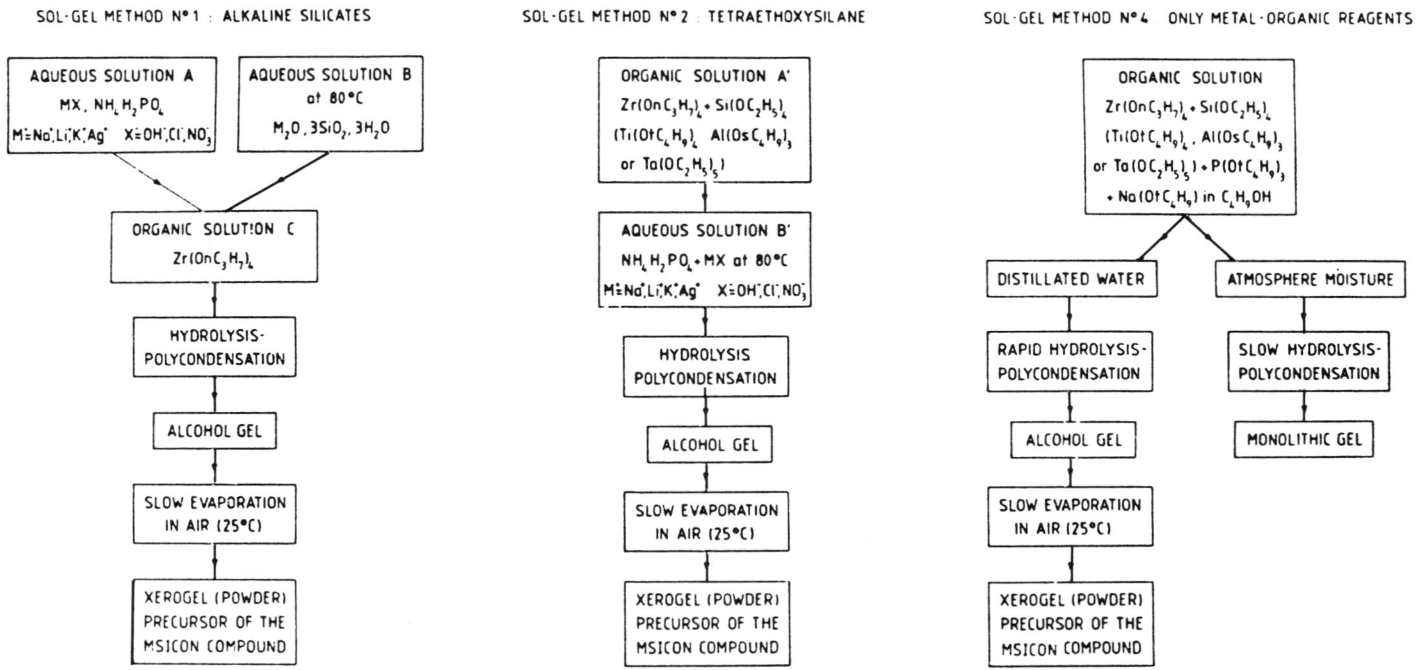

Figure 12: Schematic representation of various sol-gel routes.[18]

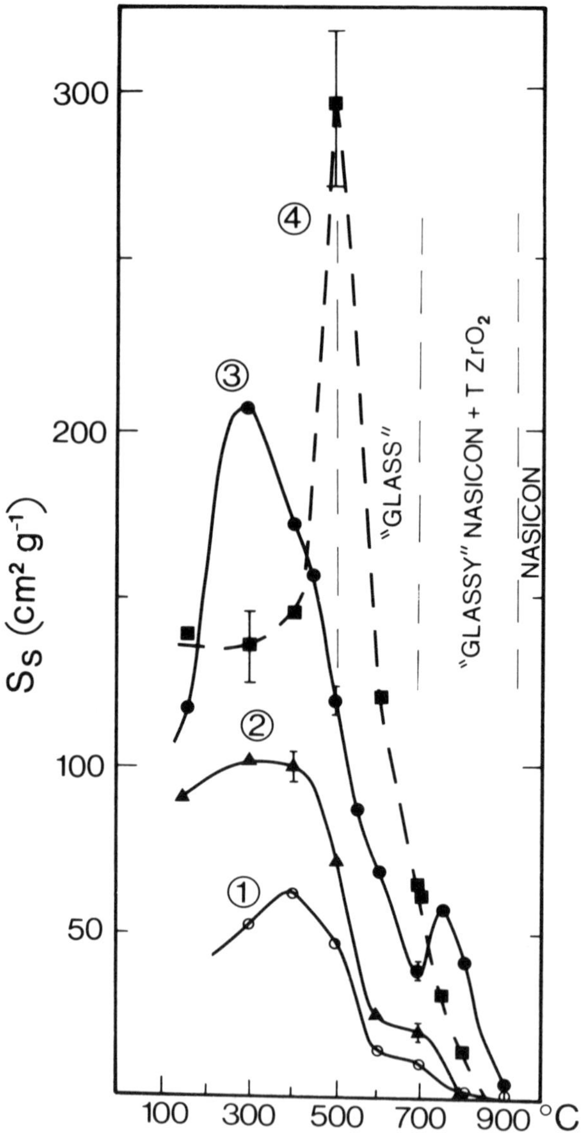

Figure 13: Comparison of the specific area (Ss) plots as a function of the temperature for various sol-gel routes ($Na_3Zr_2Si_2PO_{12}$ composition).[50]

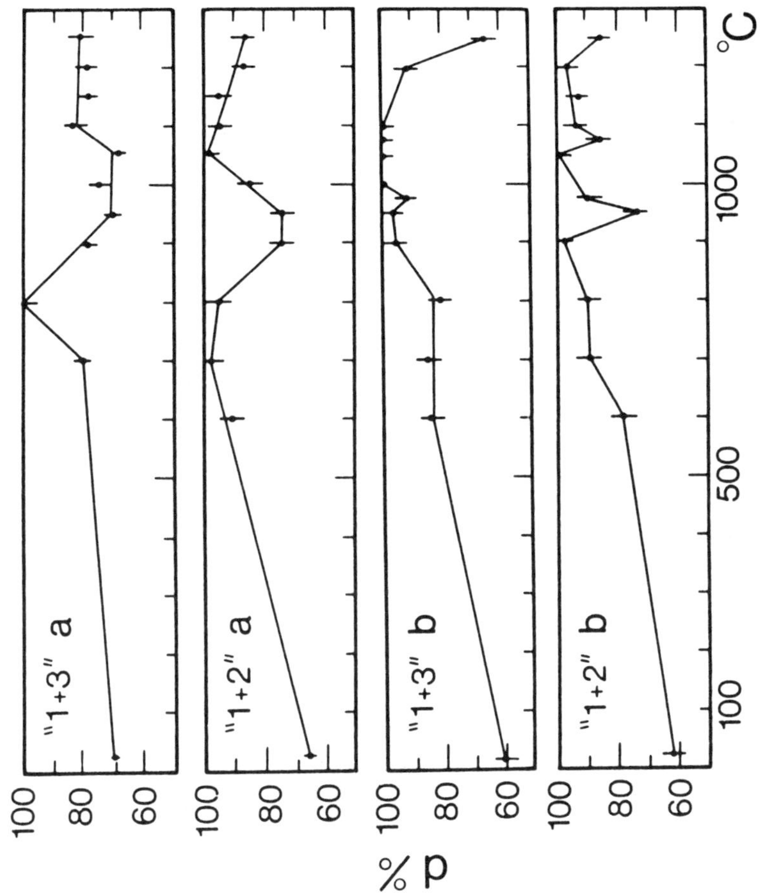

Figure 14: Evolution of the densification ratio d% (apparent/theoretical density of the $Na_3Zr_2Si_2PO_{12}$ composition) as a function of the sintering temperature for hybrid methods 1+2 and 1+3, for a pressing step at 350 MPa (a) 10 hours in duration, (b) 2 minutes.[21]

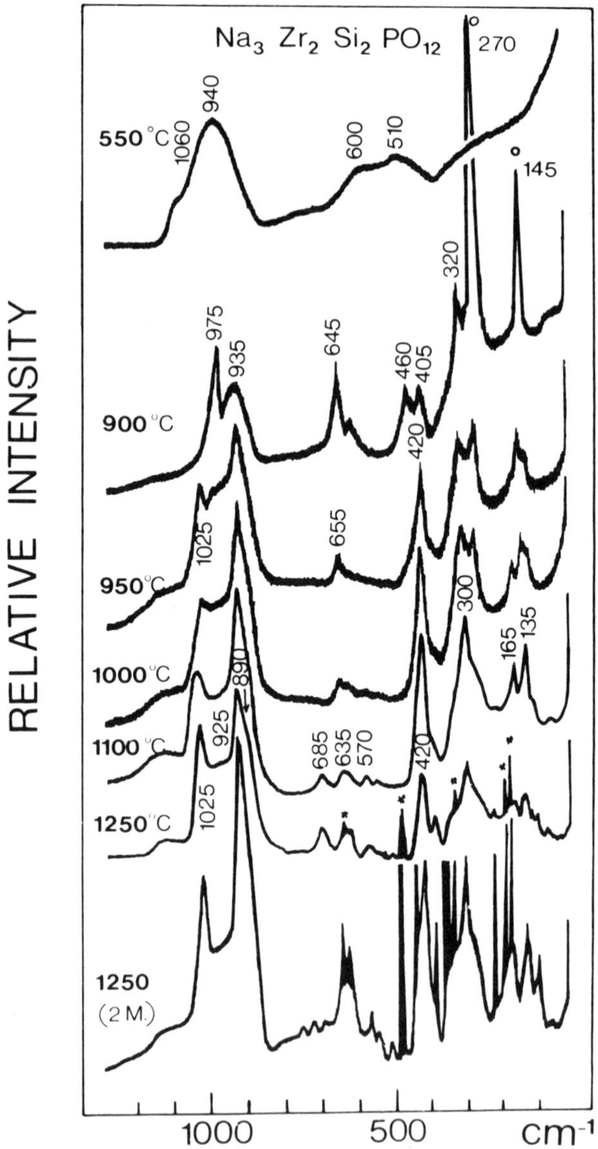

Figure 15: Raman spectra of $Na_{1+x}Zr_2Si_xP_{3-x}O_{12}$ ($x \sim 2$) samples as a function of the temperature of the preparation. Stars and circles correspond to the monoclinic and tetragonal zirconia traces.[52]

CONCLUSION

Sol-gel routes allow the preparation of better electrolytes in the ceramic or thick film form: controlled microstructure, good homogeneity, low impurity content, low and selected sintering temperature. These methods lead to new amorphous superionic materials in the gel or glass state. In conclusion, we will not discuss the general drawbacks of sol-gel methods, e.g. the important and often anisotropic shrinkage, the presence of microcracks, the cost and possible toxicity of raw materials and of intermediates, the difficulties of keeping raw alkoxides from self-polycondensation and the use of oxygen-free water. We only emphasize three problems dealing specifically with ionic conductors: first, the difficulty of achieving exactly the target composition; second, the presence of carbon traces and third, in some cases an abnormal grain growth.

Composition shifting can take place as a consequence of the various rates of hydrolysis of alkoxides and of the evolution of some elements in the atmosphere during gelation. Furthermore, the amount of each element in the organic liquid raw materials must be carefully controlled for all batches. Of course, as in the case of the usual ceramic technology, shifting of the composition can occur during the thermal treatment and it must be precisely taken into account to achieve target compositions.

Presence of carbon traces in materials issued from alkoxides hydrolysis was first established by Lin and Yu[36] by chemical analysis: the amount can vary between ~0 and 0.3% (weight) and is strongly atmosphere-dependent (pure O_2 or air). Presence of carbon traces can be clearly shown in ceramics using Raman spectroscopy,[52] even in high temperature sintered materials (<1100°C).

Another specific behavior has been observed on some monolithic specimens synthesized using the slow hydrolysis of alkoxides. At a temperature superior to the temperature of densification an abnormal grain growth occurs;[31,46] this phenomenon has been related to the absence of well characterized grain boundaries where the concentration of impurities regulates and brakes the grain growth.

REFERENCES

1. Hagenmuller, P. and Van Gool, W., *Solid Electrolytes,* Academic Press, London (1978).
2. Collongues, R., Gourier, D., Kahn, A., Boilot, J.P., Colomban, Ph. and Wicker, A., *J. Phys. Chem. Solids* 45: 981-1013 (1984).
3. Boilot, J.P., Collin, G. and Colomban, Ph., in: *Progress in Solid State Electrolytes* (T.W. Wheat, A. Ahmad and A.K. Kuriakose, eds.), 91-122, Energy, Mines and Resources CANMET, Ottawa, Canada (1983).
4. Hong, H.Y-P, *Mat. Res. Bull.* 11: 173-182 (1976).
5. Velasco, G., *Solid State Ionics* 9/10: 783-792 (1983).
6. Kraus, K.A. and Philipps, H.O., *J. Am. Chem. Soc.* 78: 249, 263 (1956).
7. A. Clearfield (ed.), *Inorganic Ion Exchange Materials,* CRC Press, Inc., Boca Raton, Florida (1980).
8. Pham-Thi, M. and Colomban, Ph., *Journal of the Less-Common Metals* 108: 189-216 (1985).
9. Livage, J. and Lemerle, J., *Ann. Rev. Mater. Sci.,* 12: 103-122 (1982).

10. Barboux, P., Baffier, N., Morineau, R. and Livage J., in: *Solid State Protonic Conductors III* (J.B. Goodenough, J. Jensen, A. Potier, eds.), 173-179, Odense University Press (1985).
11. Araki, B., Mailhe, C., Baffier, N., Livage, J. and Vedel, J., *Solid State Ionics* 9/10: 439-444 (1983).
12. Chemseddine, A., Morineau, R. and Livage, J., *Solid State Ionics* 9/10: 357-362 (1983).
13. Beni, G. and Shay, L. in Adv. in: *Image Pickup and Display,* (B. Kazan ed.), Vol. 5, 83-136, Academic Press, New York (1982).
14. Boilot, J.P., Colomban, Ph. and Blanchard, N., *Solid State Ionics* 9/10: 639-644 (1983).
15. Boilot, J.P. and Colomban, Ph., *J. Mat. Sc. Letters* 4: 22-24 (1985).
16. Colomban, Ph. and Boilot, J.P., *Revue de Chimie Minerale* 22: 235-255 (1985).
17. Boilot, J.P., Colomban, Ph., Gay, A. and Lejene, M., French Patent 8,306,934, 1983.
18. Blanchard, N., Boilot, J.P., Colomban, Ph. and Pouxviel, J.C., New glasses from metal organic precursors: preparation and properties. *J. Non-Cryst. Solids* (to be published).
19. Dauger, A., Chaput, F., Pouxviel, J.C. and Boilot, J.P., *Journal de Physique* C 8: 455-459 (1985).
20. Goodenough, J.B., Hong, H. Y-P. and Kafalas, J.A., *Mat. Res. Bull.* 11: 203-220 (1976).
21. Bouquin, O., Perthuis, H. and Colomban, Ph., *J. Mat. Sci. Lett.* 4: 956-959 (1985).
22. Susman, S., Delbecq, C.J., Mc Millan, J.A. and Roche, M.F., *Solid State Ionics* 9/10: 667-674 (1983).
23. Tuller, H.L., Button, D.P. and Uhlmann, D.R., *J. Non-Cryst. Solids* 40: 93-118 (1980).
24. Armand, M., *Solid State Ionics* 9/10: 745-754 (1983).
25. Ravaine, D., Seminel, A., Charbouillot, Y. and Vincens, M., A new family of organically modified silicates prepared from gels. *J. Non-Cryst. Solids* (to be published).
26. Dell, R.M., *Proc. 7th Int. Symp. Reactivity of Solids,* (J.S. Anderson ed.), 553-566, Chapman and Hall, London (1972).
27. Mazdiyasni, K.S., *Ceramics Int.* 8: 42-56 (1982).
28. Mazdiyasni, K.S., Lynch, C.T. and Smith, J.S., *J. Amer. Ceram. Soc.* 50: 532-537 (1967).
29. Haertling, G.H. and Land, C.E., *Ferroelectrics* 3: 269-280 (1972).
30. Zelinsky, B.J. and Uhlmann, D.R., *J. Phys. Chem. Solids* 45: 1069-1090 (1984).
31. Colomban, Ph., *L'Industrie Ceramique* 792: 186-196 (1985).
32. Rhodes, W.H., *J. Amer. Ceram. Soc.,* 64: 19-22 (1981).
33. Lenfant, P., Plas, D., Ruffo, M., Colomban, Ph. and Boilot, J.P., *Mat. Res. Bull.* 15: 1817-1827 (1980).
34. Yoldas, B.E. and Parlow, D.P., *Am. Ceram. Soc. Bull.* 59: 640-642 (1980).
35. Quon, D.H.H., Wheat, T.A. and Nesbitt, W., *Mat. Res. Bull.* 15: 1533-1539 (1980).

36. Lin, Z. and Yu, H., Preparation of NASICON by liquid drying method. *Ceramics Powders,* (P. Vincenzini ed.), 583-586, Mat. Sci. Monograph 16 (1983).
37. Boilot, J.P., Salanie, J.P., Desplanches, G. and Le Potier, D., *Mat. Res. Bull.* 14: 1469-1477 (1979).
38. Gordon, R.S., Miller, G.R., Mc Entire, B.J., Beck, E.D. and Rasmussen, J.R., *Solid State Ionics* 3/4: 243-248 (1981).
39. Bentzen, J.J. and Nicholson, P.S., *Mat. Res. Bull.* 15: 1737-1745 (1980).
40. Barj, M., Colomban, Ph. and Lucazeau, G., Vibrational Study of NASICON-Like Fast Ionic Conductors. *Proc. of the 8th Int. Conf. Raman Spectry,* Sept. 6-11, 1982, Bordeaux, (J. Lascombe and P.V. Huong, eds.)., 461-462, John Wiley and Sons, New York (1982).
41. Colomban, Ph., Perthuis, H. and Velasco, G., New Protonic Conductors for Hydrogen Sensors, the Thick Film Route. *Proc. European Workshop on "Solid State Protonic Conductors II"* Sept. 6-10, 1982, Hindsgarde Danmark, Odense University Press (J.B. Goodenough, J. Jensen and M. Kleitz, eds.), 375-391 (1983).
42. Barj, M., Perthuis, H. and Colomban, Ph., *Solid State Ionics* 11: 157-177 (1983).
43. Perthuis, H., Colomban, Ph., Boilot, J.P. and Velasco, G., Chemical Preparation of NASICON-Type Compound by Sol-Gel Method. *Proc. of the 5th Int. Meet. Modern Ceramic Technology,* June 14-19, 1982, Lignano-Sabbiadoro, Ceramic Powders, P. Vincenzini ed., Mat. Sci. Monograph 16: 575-582, Elsevier, Amsterdam (1983).
44. Yoldas, B.E. and Lloyd, J.K., *Mat. Res. Bull.* 18: 1171-1177 (1983).
45. Engell, J., Mortensen, S. and Moller, L., *Solid State Ionics* 9/10: 877-884 (1983).
46. Colomban, Ph., Densification and Microstructure of NASICON Ceramics as a Function of Sol-Gel Process. *Proc. of the 10th Int. Symp. on the React. of Solids,* 27 Aout-1er Sept. 1984, Dijon, Reactivity of Solids, Mat. Sci. Monogr. 28A, P. Barret and L.C. Dufour, eds., 485-492, Elsevier, Amsterdam (1985).
47. Perthuis, H., Velasco, G. and Colomban, Ph., *Jap. J. Appl. Phys.* 23: 534-543 (1984).
48. Perthuis, H. and Colomban, Ph., Well Densified NASICON Type Ceramics, *Mat. Res. Bull.* 19: 621-631 (1984).
49. Perthuis, H. and Colomban, Ph., *J. Mat. Sci. Letts.* 4: 344-346 (1984).
50. Colomban, Ph., How to displace to sintering temperature of oxides with sol-gel routes, Ceramic Powder Science and Technology: Synthesis, Processing and Characterization. Boston, Aug. 3-6 1986 (to be published).
51. Perthuis, H. and Colomban, Ph., Sol-Gel Routes Leading to NASICON Ceramics, *Ceramics Int.* (to be published).
52. Colomban, Ph., Raman study of the inorganic polymer→superionic NASICON transformation: dynamic, static orientational disorder and superionic conductivity, *J. Molecular Structure* (to be published).

16

Hollow Glass Microspheres by Sol-Gel Technology

Raymond L. Downs, Matthias A. Ebner and Wayne J. Miller
KMS Fusion, Inc.
Ann Arbor, Michigan

INTRODUCTION

Sol-gel technology, and in particular the metal-organic variations of that technology, has been applied with considerable success in the production of high quality hollow glass microspheres used in inertial confinement fusion (ICF) research. Inertial confinement fusion is one of two routes to controlled thermonuclear fusion being investigated as a potential source of inexhaustible energy to help meet the world's energy demands in the twenty-first century. An international research effort is underway in both inertial confinement fusion and magnetic confinement fusion. For thermonuclear fusion to occur, an intimate mixture of the heavy isotopes of hydrogen, deuterium and tritium must be compressed and heated to conditions found in the interior of the sun or in thermonuclear weapons. In magnetic confinement fusion, very strong magnetic fields generate these conditions and confine the fuel long enough for fusion to occur. Large quantities of fuel are maintained in a plasma state. In inertial confinement fusion the conditions of fuel compression and heating required for fusion are achieved by extremely rapid acceleration of the fuel container. The energy for this acceleration is provided by photons from high energy laser sources, hence the term laser fusion, or from particle beams (electron, light ion, and heavy ion drivers are all under investigation).

Glass starting materials prepared from sol-gels, when combined with careful control of processing conditions, can be converted into shells having a wide range of sizes and wall thicknesses and that meet the stringent specifications imposed by their use as ICF targets. This chapter will describe the preparation and processing of sol-gel materials, especially metal-organic-derived, for the production of hollow glass microspheres (HGS) that have a high degree of uniformity and ultra-smooth surfaces that are resistant to degradation by ambient weathering in a laboratory environment.

Glass Shell Uses

Hollow glass microspheres are used in many commercial products, primarily as fillers. They are produced in tonnage quantities, by a variety of patented processes, for markets requiring hollow spheres with low density and moderate chemical durability and strength.

When added to plastic resins, hollow glass microspheres (or hollow spheres of other materials) produce a composite known as syntactic foam, whose properties are superior in many ways to composites containing conventional fillers. The use of shells as fillers in a variety of commercial products has been reviewed elsewhere.[1,2,3]

With the advent of laser-induced thermonuclear fusion research in the early 1970's[4] a new application was found for hollow glass microspheres. The first laboratory demonstration of neutron production by laser-induced thermonuclear fusion utilized a 70 μm glass shell with a 1 μm wall filled with 15 atmospheres of a deuterium-tritium (DT) fuel gas mixture.[5] (A fuel-filled fiber mounted hollow glass microsphere at the instant of implosion by the 1.06 μm Chroma laser at KMSF is shown in Figure 1.) The first fuel-filled glass shells used by the participants in the U.S. ICF program, KMS Fusion, Inc. (KMSF), Lawrence Livermore National Laboratory (LLNL), Los Alamos National Laboratory (LANL), and the University of Rochester, Laboratory for Laser Energetics (UR/LLE), were obtained from commercial sources, primarily Emerson and Cuming, Inc. and the 3M Company. It was readily apparent that the quality of these shells was inadequate for the exacting demands of ICF research. Computer code calculations of the implosion process indicated that optimum implosion performance could only be achieved if the fuel containers were nearly perfect with respect to sphericity, uniformity, and surface smoothness. Other essential requirements were that the diffusion time, at a temperature below the softening point of the glass, be reasonable and that the shells retain the fuel at room temperature for weeks and preferably longer. Because of the many basic implosion physics questions that had to be explored it was necessary that fuel shell containers be available over a large range of sizes (up to 1,000 μm) and wall thicknesses (0.5 to 30 μm) and generally, be able to retain pressures of 100 atmospheres or more. The commercially available hollow glass microspheres were (and still are) very limited in diameter (10 μm to ~200 μm) and were also very nonuniform, i.e., only one in 10,000 to 100,000 could meet the uniformity requirement.[6,7] This situation prompted Lawrence Livermore National Laboratory and KMS Fusion to initiate programs to fabricate glass shells that could meet both the high quality standards and the dimensions imposed by ICF research. Since only a few experiments a day could be conducted, and at great expense, shell yields and cost were not major considerations.

Review of Glass Shell Production Methods

The production of hollow glassy spheroids appears deceptively simple, as

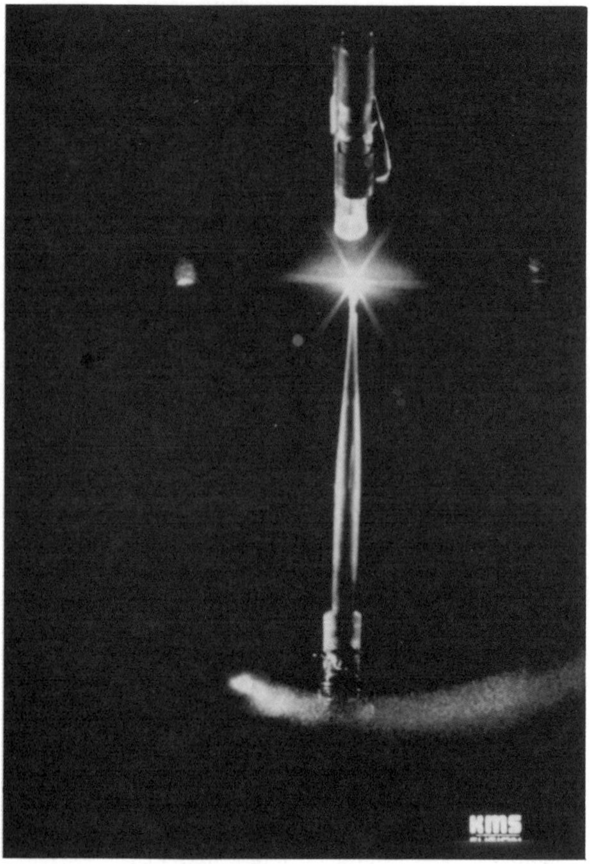

Figure 1: Laser driven implosion of a DT-filled glass shell. The starburst effect is produced by emission from the target.

evidenced by their natural occurrence in the fly ash of coal burning power plants and in volcanic ash. These products, particularly the latter, are very crudely formed hollow microspheres but, nonetheless, identify the basic components of shell formation, namely, the presence of glass forming ingredients, gas generating species and sufficiently high temperatures for glass flow. (We demonstrated the ease of glass shell formation in our laboratory by showing that compacted pellets of crushed soda lime glass powder form into shells when dropped through a vertical tube furnace.[8] The gas trapped between glass powder particles served as the blowing gas.) Although fly ash has commercial value, more controllable shell size, density and strength is generally desired.

Many patented processes for shell production have appeared, some of which are presumed to be used by the three major commercial manufacturers. The processes utilize either liquids or powders as the feed material in shell manufacture.

Liquid Feed Materials. Aqueous sodium silicate solutions are spray dried to produce hollow glass microspheres in the PQ Corporation process.[9] Another spray dry process, using aqueous silica sols to produce hollow shells is described in a patent to Monsanto wherein a sol is fed into an atomizing gas burner of a vertical combustion tower.[10]

Spray drying of viscous hydrogels of silica, alumina and titania has been reported to give hollow porous beads useful as catalyst supports.[11] Using a liquid-liquid extraction process aqueous sols of metal oxides, e.g., TiO_2, have been converted into hollow gel microspheres having porous walls. Controlled heating converts these gelled products to polycrystalline or amorphous nonvitreous microspheres, which may be porous or nonporous, depending on the final firing temperature.[12]

Molten glass can also be a liquid feed material. A method described in the patent literature but apparently not yet the basis for any existing commercial product is the production of glass shells from molten glass using multiple concentric dual orifices.[13] Glass shells with an inner surface coating of metal are claimed to result from heating glass coated metal spheres in a stream of hot (~1000°C) air.[14] The starting material is prepared by a high temperature dual nozzle extrusion process. A dual orifice technique has been utilized by workers at the Jet Propulsion Laboratories to prepare millimeter size glass, metal and plastic hollow spheres from the corresponding pure melts.[15,16]

Powder Feed Materials. Powder feed materials are blown into shells when heated rapidly. The Emerson and Cuming Company has prepared shells using a process patented by Sohio wherein an aqueous slurry of an alkali silicate solution containing boric acid and urea is oven dried to form a solid water-glass product which is ground to a powder, classified and fed into a shell blowing furnace.[17] In a later patent to Emerson and Cuming a homogeneous sodium borosilicate sol is spray dried to form the feed material for a subsequent shell blowing process.[18] In the 3M process, preformed glass particles containing a blowing agent (incorporated by gas diffusion or addition of sulfate to the original glass melt) are blown into shells in a gas flame.[19]

Glass Shell Production for ICF Targets

The approach taken by both KMSF and LLNL was to prepare the glass for the hollow microspheres by a solution process, i.e., the glass oxide precursors used are pure chemical compounds mixed in a solvent to provide compositional homogeneity on a colloidal or, in most cases, a molecular scale. This is important because the total mass of a single shell can be as small as 35 nanograms (70 μm diameter with 1.0 μm wall) and it is highly desirable that shell to shell properties be constant. This constancy is particularly important since several of the critical measurements made on ICF targets, e.g., tests for fuel gas pressure, are based on destructive testing of what is required to be a representative sample of shells from a batch. Also, preparing glass shell precursors from solution results in the homogeneous incorporation of shell blowing agents, i.e., gaseous components or gas generating species that are released or decomposed upon heating of the shell precursor particles or solution droplets. The solution process also provides access to a wide concentration range for the blowing agents.

At LLNL a droplet generator process was developed by C.D. Hendricks et al. that used aqueous alkali silicate solutions, similar to a spray dry feed material described by Veatch et al.,[20] in a liquid drop technique that had been developed earlier at the University of Illinois by Hendricks and coworkers.[21,22,23] Detailed descriptions of this technology and its evolution at LLNL can be found in several texts and patents[24,25,26] and in LLNL reports.[27,28]

At KMSF, work on glass shells began in early 1973, also with an aqueous silicate system similar to that described by Veatch,[20] but instead of forming droplets or using powder produced by simply evaporating an aqueous silicate slurry,[17] a gel was produced that was subsequently dried, crushed, and sieved. In early experiments gel powder was blown into glass shells with a powder-fed, oxy-acetylene torch, but this was soon replaced by a much more controllable electrically heated vertical drop tower furnace. The use of aqueous silicate solutions [actually colloidal silica (Ludox) was the silica source] permitted only a very limited choice of glass compositions. The subsequent use of metal-organic based sol-gels allowed a much wider range of glass compositions to be investigated and also led to glass shells with much more uniform wall thickness than that obtained from the KMSF Ludox-based process.[29,30,31] In the following discussion we will describe how KMS Fusion has utilized sol-gel technology, mostly metal-organic-derived gels, to produce full shell containers that have been used in laser fusion experiments by all the major participants in the national ICF program. Shells manufactured by KMS Fusion have also been used in ICF experiments in France, Italy and Great Britain. Reports from Japan's laser fusion program at Osaka University and France's program at Centre D'Etudes De Limeil describe the preparation of glass shells from metal-organic gels in a process similar to that developed at KMS Fusion.[32,33,34]

Specifications of Glass Shells for ICF Targets

A glass shell used as an ICF fuel container must meet stringent specifications. These specifications cover geometrical characteristics of the shell and the chemical and physical properties of the glass. There are at least seven properties which are critical.

(1) *Size.* Shell size requirements are dictated both by the purpose of a particular experiment and the available laser energy; the more energy the larger the shell that is required. Shell diameters used in ICF experiments have ranged from 50 μm to 1,000 μm and are increasing as lasers become more powerful.

(2) *Wall Thickness.* Shell wall thickness is generally specified as an independent variable and ranges from 0.4 μm to ~20 μm.

(3) *Wall Thickness Uniformity.* Ideally, for optimum compression of the fuel, the wall thickness of ICF shells should be constant over the 4 π surface of the shell. In reality, shells are generally less than perfect. The most difficult defect to eliminate arises when both the inner and outer surfaces of a shell are truly spherical but have centers that are offset, i.e., not concentric. Interferometry[35] or microradiography[36] is used to measure this defect. In selecting shells for use as ICF targets, we make an assessment of the relative number of shells in the batch that have a nonconcentricity of 10% or less in a single plane of view, i.e., shells whose inner and outer

diameters in this view are offset ≤10%. For simplicity of discussion, such shells are designated as Class A shells. Thus, by increasing shell uniformity we mean producing shells that have a more uniform wall thickness. An example of a shell batch that could be used for ICF target selection is given in Figure 2, which shows a photomicrograph of a batch of shells by standard optical transmission microscopy and by interference microscopy.

This property is perhaps the most important for ICF shells and is one of the most difficult to achieve. Some specifications have called for shells to have a $(wall_{max}\text{-}wall_{min})/wall_{av}$ of 2%. Maximizing uniformity has been a major goal in our sol-gel research leading to glass shells, and the principal constraint in shell size and wall thickness limits.

(4) *Surface Finish.* The typical goal is a surface finish of ≤0.1 μm. This property is determined primarily by the glass composition and secondarily by the shell forming conditions.

(5) *Strength.* For some types of ICF experiments the shells must hold internal pressures of 100 atmospheres or more. Achieving this requires making shells uniform, not exceeding some minimum aspect ratio (o.d./w), (determined empirically), and using glass compositions that are resistant to static fatigue.

(6) *Gas Permeability.* Glass shells for ICF targets are first and foremost hydrogen isotope fuel containers. The permeability of glass to gases is strongly dependent on temperature and composition. The temperature dependence allows the shells to be filled by diffusion in a pressure vessel at 300°C to 400°C and at room temperature the gas to be retained for periods greatly exceeding the fill time, e.g., for days to years depending on the composition of the glass. Permeation rates drop rapidly as glass-former oxides are replaced by glass-modifier oxides.

(7) *Glass Composition.* Although composition affects most of the chemical and physical properties of shells, for implosion diagnostic or other purposes, some ICF target experiments require that specific elements be included or excluded from the shells.

Meeting all of these sometimes conflicting specifications has been the driving force of our research. The use of sol-gel technology to make the hollow glass microspheres used in ICF research has been critical to our success to date. Work to improve the properties and extend the range of sizes for ICF targets continues and spin-offs from this technology should lead to many other useful applications of high quality glass shells.

GLASS SHELL STARTING MATERIALS

As just noted, both aqueous silicate and metal-organic reagents, in organic solvents, have been used as starting materials at KMSF for glass shell formation.

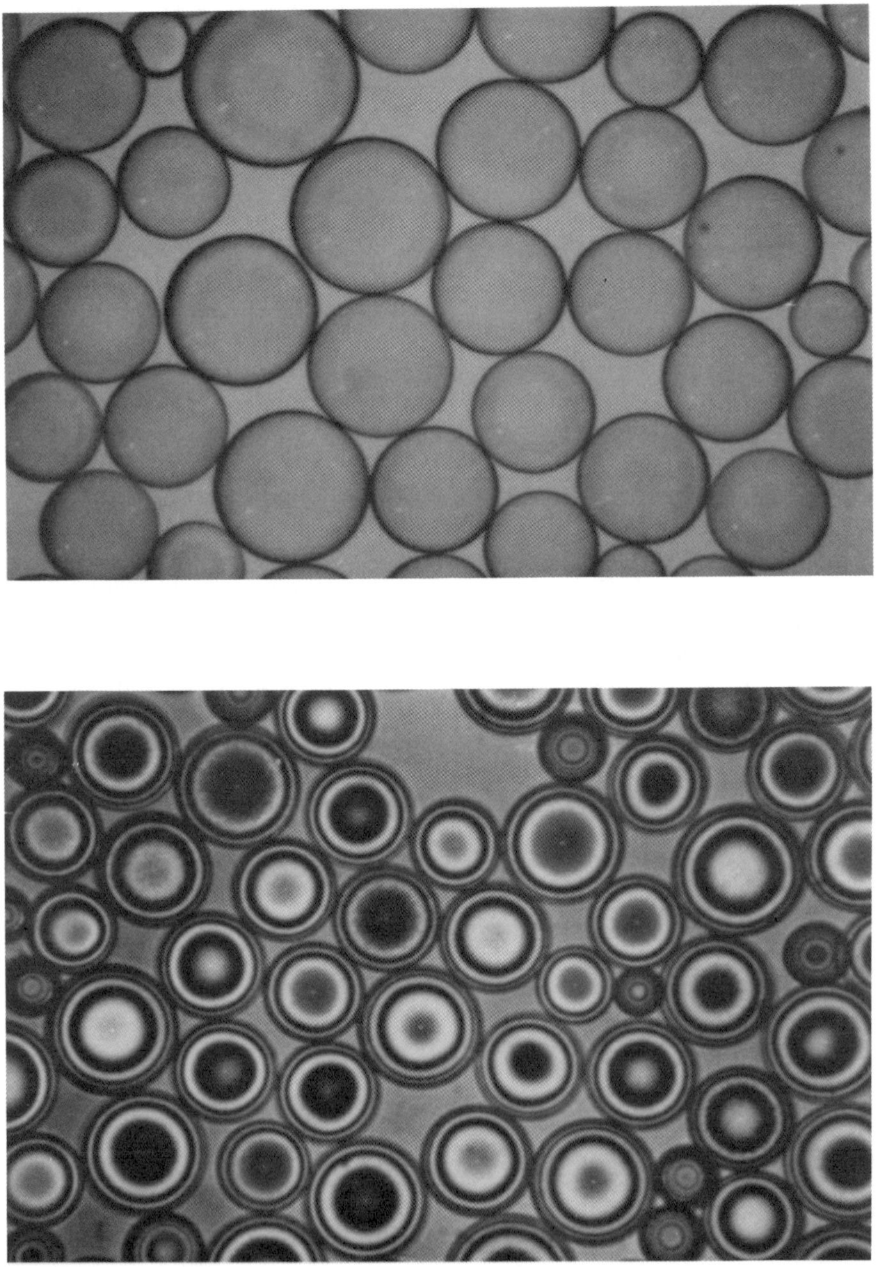

Figure 2: Optical and interferometric photomicrographs of a batch of highly-uniform glass shells having an average diameter of 250 μm and wall thickness of 1 μm.

For the first two years of shell fabrication inorganic sol-gels were the primary feedstock material but by 1975 had been replaced by metal-organic-derived gels.

Inorganic Gels

There are many variations on the method to make a sol-gel based on aqueous water-glass solutions or colloidal silica dispersions, but the following was found to provide a satisfactory feed material for shell blowing. Boric acid is added with stirring to a Ludox® aqueous colloidal silica dispersion (HS-30) in a plastic beaker. Urea (variable amounts, but 10 wt. % often used) is added and followed by incremental additions of sodium methoxide (NaOH and $NaNO_3$ also studied). Stirring is continued throughout the additions. The resulting clear solution is placed in a hot water bath until it gels spontaneously (less than one hour). The gel is then vacuum dried, crushed and sieved.

Metal-Organic-Derived Gels

Glass shells were first prepared from metal-organic-derived gels in 1974. The gels in current use for shell blowing are synthesized by methods that have evolved from the work of Dislich[37] and Thomas[38] and early work by King and Ebner at KMS Fusion.[29] Preparing glass from metal-organic sol-gels has been studied extensively at KMSF since 1975. The metal-organic sol-gel method of preparing glass has several attractive attributes:

(1) Chemical homogeneity of glass components on a molecular level.

(2) Gel-to-glass transformations at temperatures below the melt.

(3) Very low levels of impurities.

(4) Nearly unlimited compositional ranges.

(5) Homogeneous incorporation of shell blowing gases.

The metal-organic sol-gel method of preparing glass also has several unattractive attributes:

(1) Difficulty of preparation. Keeping all components in solution, and then gelling without some components preferentially precipitating is impossible or nearly impossible for some multicomponent systems.

(2) High cost. The starting materials are expensive.

(3) Carbon and/or gas bubble inclusions in the glass. Processing conditions for the gel-to-glass transformation are complicated since most gels contain organic species and/or water which must be removed (generally by pyrolysis under carefully controlled temperature steps).

For glass shells used in ICF research, the choice of the metal-organic derived gel method is based on the advantages of chemical homogeneity and low levels of impurities attainable by this technique. The major disadvantage of its use is the difficulty of preparing the gels; nevertheless, this difficulty has been overcome for a great many compositions (see below).

The gels are prepared by the hydrolysis/condensation reactions of alkoxides which are glass formers. Polymers are formed by these reactions and a gel is produced when the chain lengths become sufficiently long. The condensation reactions allow glass modifier metal cations to become chemically attached to the chain through metal-oxygen-metal bonds. However, not all glass modifiers become attached to the chain, but, under appropriate synthesis conditions will be homogeneously trapped in the gel. Since most of our glasses are silicates, tetraethoxysilane is our usual source of the glass former. For nonsilicate gels, i.e., germanium, tantalum, and aluminum, we have used alkoxides containing up to five carbon atoms.

The metal-organic derived gels are synthesized by two methods, differing by whether the glass modifiers are added as alkoxides (All-Alkoxide Gels) or metal salts (Salt Gels). The gelling agent in both methods is the glass forming alkoxide. All-alkoxide gels are prepared under basic conditions due primarily to the alkalinity of the alkali alkoxides. (The gelation of pure tetraethoxysilane is an exception. Our best results have been under acidic conditions using inorganic acids.) Salt gels are prepared under acidic conditions, usually in the presence of acetic acid. The preferred glass-modifier salts are the acetates, although nitrates and other organic acid salts such as lactates have also been used.

The following are examples of the All-Alkoxide method and the Salt method for preparing a gel with an oxide equivalent weight percentage of 73.0% SiO_2, 17.4% K_2O, 2.6% Na_2O, and 7.0% B_2O_3: (The oxide equivalent weight percentage is the weight percentage composition if the gel was completely converted to oxide glass. Different gels can have much different organic content but still have the same oxide equivalent compositions.)

All-Alkoxide Gel. For 200 g of glass, 3.86 g of sodium metal and 29.98 g of potassium metal are added to a 2 liter resin flask continually purged with argon. A total of 491 ml of absolute ethanol is added carefully (caution must be taken to safely dissipate the heat and hydrogen gas that are produced). Tetraethoxysilane, 506.28 g, is added after the metals are completely reacted. This mixture is allowed to cool to room temperature.

A 50:50 by volume ethanol:water solution is added dropwise to the alkoxide solution while stirring rapidly. After 150 ml has been added the reaction mixture is cooled to below 10°C. An additional 19 ml of the ethanol:water solution is added such that the ratio of water to silicon ethoxide in the reaction is brought within the range of 2.0 to 2.3. This is followed by 58.72 g of triethoxyborane. This final solution is allowed to warm to room temperature. Gelation occurs spontaneously.

Alkoxide solutions of the usual alkali borosilicate compositions thicken and gel rapidly after the addition of 2.0 to 2.3 mols water per mol of silicon tetraethoxide. The addition of larger quantities of water almost always forms precipitates, regardless of the rate of addition of water or the speed of mixing. Attempts to force the hydrolysis of the alkoxides further by the addition of dilute solutions of water in ethanol and by cooling the metal alkoxide solution to temperatures as low as -60°C have always resulted in the formation of precipitates when the reaction was warmed to 0°C.

This conclusion does not apply to an alkoxide solution of our usual alkali-lime gel with an oxide equivalent composition (weight %) of 69.9% SiO_2, 10.0% CaO, 9.1% K_2O, 6.0% Na_2O, and 5.0% B_2O_3. This solution could be successfully gelled by hydrolysis with 2 to 4 mols of water per mol of silicon tetraethoxide.

When 4 mols of water per mol of silicon tetraethoxide were used to prepare the alcogel, the resulting dried xerogel had an exceptionally low density, one third of the density of the usual xerogels, and a specific surface area of 560 m^2/g compared to the usual 200 to 350 m^2/g.

Salt Gel. For 200 g of glass, 506.22 g tetraethoxysilane, dissolved in 314 ml of absolute ethanol, is prehydrolyzed by the addition of 44 ml of water and 1.0 ml of 1 M HCl catalyst. The reaction mixture is heated to 65°C for at least an hour and then allowed to cool to room temperature.

A second solution is prepared containing 22.83 g of sodium acetate trihydrate, 72.52 g potassium acetate, 340 ml water and 170 ml of glacial acetic acid. This latter solution is added slowly to the partially hydrolyzed tetraethoxysilane with rapid stirring. Finally, 41.9 g of trimethoxyborane is added and the solution gels spontaneously.

This acid-catalyzed hydrolysis reaction readily forms alcogels and is not as prone to precipitation as is the base-catalyzed all-alkoxide reaction. Gelation is routinely accomplished by the addition of 2 mols of water per equivalent of ethoxide, or in other words, 8 mols water per mol silicon tetraethoxide *plus* 6 mols water per mol trimethoxyborane.

Parameters Affecting Shell Properties. The number of experimental parameters in gel preparation is large, including:

1. The oxide equivalent composition.
2. The chemical form of each modifier cation; i.e., the particular alkoxide or inorganic or organic acid anion.
3. The H_2O/Si ratio.
4. The amount of solvent.
5. Which solvent. Usually the solvent is alcohol, but many alcohols can be used.
6. The order of addition of each species.
7. The time and temperatures at and between each addition.
8. How the water is added.
9. The addition of other species, such as 2,4-pentanedione.
10. The amount of gel prepared.

Each of the parameters may affect the properties of the resulting glass shells.

Compositions of Glasses From Gels

The composition chosen for the glass directly determines the chemical and physical properties of the glass and in combination with gel synthesis, gel treatment and shell blowing conditions determines the physical (geometrical) properties of the shell itself. Glass composition has a significant effect on diffusion rate of gases through the shell wall and also on weathering characteristics of the shell, i.e., the reaction of the glass surfaces with atmospheric gases (external surface) and gel decomposition gases such as water and CO_2 (internal surface). Glass composition also has an effect on shell strength, particularly static fatigue. Glass composition plays a role in shell wall uniformity since, for a given particle size, the glass or gel-glass shell precursor must have a sufficiently low viscosity, in the temperature range and time interval accessible by the drop-tower furnaces used to allow glass refining and shell formation to occur. An intermediate shell forming stage with a relatively low viscosity appears to be a necessary but not sufficient condition for the production of shells with uniform wall thickness.

Glass choice for shells used in ICF targets often is based upon a complex set of compromises of shell properties and ease of shell forming. Additionally, some laser fusion experiments have occasionally required the specific exclusion (potassium and calcium) or inclusion (aluminum and rubidium) of specific elements in the glass composition.

Silicates. Most ICF target shells are made from silicate glasses. As silica content is increased, generally with a corresponding decrease in alkali, chemical durability and resistance to weathering increases but fuel retention decreases.

While the compositions used to prepare shells from aqueous silicate solutions have been restricted to sodium-potassium-borosilicates, a relatively large number of elements have been incorporated in the metal-organic-derived gels. The bar graph in Figure 3 shows the various elements and maximum concentrations that have been used in gels that produced shells. By including different combinations of the same elements, over 50 different silicate compositions have been prepared. Because of trade-offs in shell properties and forming characteristics, three silicate compositions have found the most applications; an alkali lime glass with a final shell composition of CaO ~10 to 12 mol %, $Na_2O + K_2O$ ~2 to 5 mol % (for shells smaller than about 250 μm); an alkali silicate containing 0 to 7 mol % Rb_2O and 0.5 to 17 mol % of $Na_2O + K_2O$ for shells larger than 250 μm; and >96% to 99% silica for thin (~1 to 2 μm) walled shells 100 to 1200 μm.

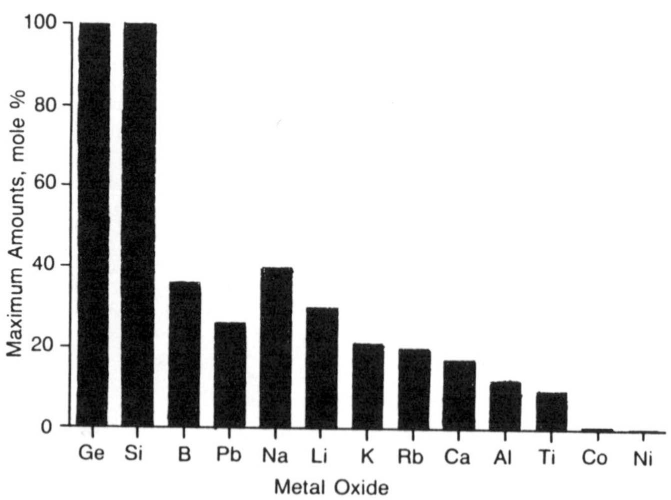

Figure 3: This graph shows the maximum amount of elements, as oxide equivalent mol percentages, which have been incorporated into metal-organic-derived gels for glass shell production. The number of metal oxides per gel composition has varied from one, e.g., pure SiO_2 and GeO_2, to six, e.g., Na_2O, K_2O, CaO, Al_2O_3, SiO_2 and Na_2O, K_2O, CaO, CoO, B_2O_3, SiO_2.

The preparation of ICF target quality shells with an SiO_2 content >99% at a furnace temperature of 1500°C is particularly noteworthy. This is accomplished by using a metal-organic gel with a sodium-potassium-borosilicate composition adjusted to optimum elemental ratios such that substantially all of the alkali and boron are volatilized during shell blowing in a vertical tube furnace.[39] A rubidia borosilicate composition (20 mol % Rb_2O and 7 mol % B_2O_3, oxide equivalent in the gel) can also be used to prepare >99% SiO_2 glass shells.[8] Pure silica gel of identical particle size and prepared in a similar manner, produces only glassy foamed spheroids when processed identically.

Germanates. The uniformity of germania shells prepared from sol-gels has generally been poor, nevertheless, a few ICF experiments have been conducted using these shells. The amount of work with this system has been very limited; additional effort at optimizing the processing conditions should produce more uniform shells.

Shells of germanium dioxide were blown in a drop tower furnace at 1250°C to 1300°C from several germania sol-gel precursors. First attempts at preparing a gel by hydrolysis of $Ge(OC_2H_5)_4$ were unsuccessful. The precipitates that formed instead of a gel contained small amounts of residual alkoxy groups, and formed some GeO_2 shells when dropped through the furnace at 1250°C to 1300°C. However, most of the product was beads. Hydrolysis of $Ge(OC_2H_5)_2(t-OC_4H_9)_2$ also formed precipitates but these were converted to higher quality shells in better yields. Monolithic gels were subsequently formed by carefully controlled hydrolysis of $Ge(OC_2H_5)_4$ and its mixtures with $NaOC_2H_5$ in ethanol at -50°C. Powders from these dried gels were blown into shells at 1300°C.[40] Shells as large as 600 μm were prepared from sodium germanate gels containing 5% Na_2O.

Other Oxides. The ability to form shells from oxide systems not generally considered as glass former was dramatically demonstrated with metal-organic-derived sodium tungstate.[41] Although many hydrolysis schemes were attempted, clear monolithic gels could not be produced from alcohol solutions of $NaOCH_3$ and $W(OCH_3)_6$; only precipitates formed. Nevertheless, when the precipitate was dropped through our 4 μm vertical tube furnace at 1000°C a small fraction of the product was clear, colorless, hollow glass microspheres. Most of the product was in bead form, presumably because of the low concentration of residual alkoxy groups in the precipitate from the hydrolyzed metal alkoxide mixture.

Metal-organic gels have been prepared from metal alkoxides in two other nonsilicate systems, calcium aluminate and calcium tantalate, but shell blowing experiments have not been attempted.[41,42]

GEL PROCESSING AND SHELL BLOWING PROCEDURES

The usual application of sol-gel technology is the formation of continuous films or monoliths of xerogel for subsequent conversion into glass or polycrystalline products. In this application, the volume shrinkage that occurs during the drying of the films or monoliths of alcogel usually results in the development of internal stresses that can flaw or destroy the continuous or monolithic structure. In addition, the complete removal of the residual organic moieties in the metal-organic sol-gel usually requires special processing to prevent the formation of elemental carbon or carbon dioxide gas inclusions in the final glass films or monoliths. The use of metal-organic sol-gels for the production of small hollow glass spheres takes advantage of these "undesirable" alcogel properties.

In the production of microspheres, the required configuration of the sol-gel is a powder or grain of mass appropriate for the final hollow sphere. The reduction of the alcogel to a dry xerogel powder of desired size is a straight-forward procedure, involving controlled drying of the alcogel, comminution, sizing, and supplementary treatment.

Alcogel Drying

The wet alcogel is usually dried in a dynamic vacuum at ≈100°C for 16 hours. In the initial stages of drying, there is often some separation of solvent (ethanol) from the wet alcogel. However this separation causes only negligible leaching of alkali from the wet gel. The concentration of sodium or potassium in the ethanol has been found by atomic absorption analysis to be extremely low. It is therefore unlikely that this separation of solvent causes any significant inhomogeneity in the gel.

The rapid drying of the alcogel invariably causes significant shrinkage and the rapid development of stresses within the bulk of the gel. As the drying and shrinkage of the alcogel progresses and stresses build up, the bulk of the gel becomes increasingly laced with fissures. At the end of the drying cycle, the bulk volume of the gel is usually reduced by 75 to 80%. The corresponding intensification of internal stresses converts the monolithic wet alcogel to a bed of fractured, highly-stressed xerogel nuggets no larger than ≈1 cm in dimension.

Comminution and Sizing

While grains of the xerogel are hard and appear vitreous, their residual internal stresses render them very friable, and they are easily reduced to granules smaller than 1 mm. Comminution is easily accomplished with a mortar and pestle or by rubbing the granules across the surface of a steel wire sieve screen. In practice, kilogram quantities of the dry granular xerogel are easily and rapidly comminuted using an Alpine laboratory mill.

The powder resulting from comminution is classified by size by iterative sieving with U.S. Standard Testing sieves, A.S.T.M. E-11 specifications, using a sieve series ranging from 38 to 600 μm. Sieve cuts less than 125 μm are dedusted in an Alpine air jet sieve, a procedure necessitated by the copious quantities of fines produced in the comminution and in the mechanical sieving. (Dedusting of the gel powder improves the wall uniformity of the resulting shells.) Because of its sensitivity to ambient water vapor, the resulting powder is stored in glass containers (bottles or vials) in a desiccator.

Supplementary Treatment

In this state, after characterization for size distribution, specific surface area, residual organic content, and composition, the powder is suitable as feedstock for the production of hollow glass microspheres. Further processing of the feedstock, such as pyrolysis or controlled hydrolysis, is usually performed just prior to use. These treatments reduce the concentration or alter the composition of residual organic species in the solid gel, thereby reducing the concentration of potential blowing agents, and can result in the production of smaller and thicker-walled shells. (See following section for a more complete discussion.) When pyrolyzed, the gel is heated in air for one hour at temperatures of 500°C to 675°C, depending on the gel and the geometric characteristics required in the shells. These conditions are not sufficient to remove all of the organic carbon and will usually produce enough elemental carbon inclusions in the powder to render it black. Oxidation during shell blowing converts this carbon to CO_2, and a clear colorless glass is produced.

Hydrolysis of the gel in the solid state is accomplished either by autoclaving at 20 psi water vapor pressure for one hour or in a controlled humidity chamber at 80% relative humidity for 24 hours at 30°C. The hydrolyzed powder is subsequently air dried at 100°C for 5 to 10 minutes to restore its free-flowing characteristics.

Gel-To-Glass Transformation

Hollow glass spheres are produced by rapid heating of the gel powder. To accomplish this, we prefer to use 4 m high 1500°C and 1650°C electrically-heated vertical drop tower furnaces. The furnaces are lined with 7.5 cm diameter 99.9% alumina process tubes. The bottom of the tube is hermetically sealed to provide a turbulence-free ambience in the furnace. When a controlled water vapor pressure is desired in the furnace ambience for additional control on the geometric and compositional characteristics of shells, the hermetic seal can be achieved with a water siphon, whose temperature is thermostatically controlled to vary the water vapor pressure up to 0.9 atmospheres. The gel powder is manually injected or mechanically trickled into the open top of the furnace process tube.

While powder-fed oxy-acetylene spray torches are extremely effective for high-volume shell production, the control of the quality of the product and the size range of hollow spheres are very limited by the short residence time of the powder in the flame. Vastly better control of shell wall uniformity and a larger range of shell diameters and aspect ratio (diameter/wall thickness) are possible by firing the powder in an electrically heated vertical drop tower furnace. However, production rate and recovery efficiency are significantly lower than is possible with spray torches. Production rates (throughput rates) are limited by the risk of interparticle collisions, due to the slow sedimentation of the buoyant hollow spheres and by the high loss rates ($\approx$50%) to the walls of the process tube.

REDUCTION OF ORGANIC CONTENT IN METAL-ORGANIC XEROGELS: GEL CHARACTERIZATION

Gaseous blowing agents are the driving force in the formation of a hollow sphere. The sources of these blowing agents are the water, alkali carbonate, and residual organic species in the gel. There are several gel treatments that can affect the relative concentrations of these constituents, and thus, in principle, affect the properties of the resulting shells. How the properties of the metal-organic xerogel change as a result of hydrolysis (by water vapor) and pyrolysis is the subject of the following discussion.

Water Vapor Hydrolysis of Xerogel

Water vapor reacts with residual metal alkoxides in the dried xerogel, producing, as during gel synthesis, silanol moieties, alkali hydroxides and borates, and the corresponding free alcohol. Hydrolysis can be accomplished by exposure of the gel to humid air of the ambient atmosphere (gel aging), the controlled water vapor concentration (humidity) in a humidity chamber, or the controlled steam pressures of an autoclave. (However, the higher water vapor pressures and temperatures produced in an autoclave, and to a lesser extent in a humidity chamber, will cause the rapid deliquescence of high-alkali xerogels.) Because the reaction consumes the residual alkoxides in the gel, hydrolysis should thus reduce the amount of residual organic moieties in the gel that can potentially contribute to the gases that form and expand the hollow sphere. This treatment involves the reaction of alkoxides only; the residual acetates used in acid-catalyzed syntheses of xerogels are inert to hydrolysis.

Xerogels synthesized by the all-alkoxide method are rather sensitive to vapor phase hydrolysis. Relatively stable when stored under anhydrous conditions, such xerogels age perceptibly when exposed to the water vapor (and CO_2) in ambient air. These reactions are markedly accelerated at higher temperatures and water vapor pressures, such as those attainable in a constant humidity chamber or an autoclave. The usual effect of such treatments is the removal of most of the residual organics in the gel and the saturation of the gel with water, as shown by the evolved-gas chromatograms of Figure 4 and in Table 1. These data were obtained by pyrolysis-gas chromatography (PGC) of several sodium silicate xerogels hydrolyzed by exposure to 80% relative humidity air for 5 days. The

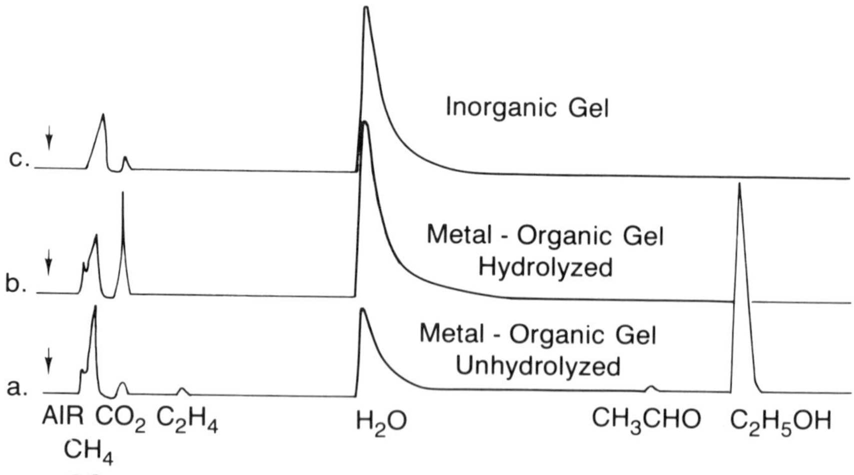

Figure 4: The effect of vapor-phase hydrolysis on the residual alkoxides and carbonates in a metal-organic-derived sodium silicate xerogel. Compared are the PGC chromatograms of gaseous pyrolysates from: (a) pristine xerogel, containing ~20% ethoxides and water; (b) hydrolyzed xerogel, after exposure to 80% RH at 30°C in air; and, (c) gel powder of the same composition, but derived from an aqueous water-glass solution.

Table 1: Effect of Gas-Phase Hydrolysis on the Residual Alkoxides and Carbonates in the Xerogel, as Indicated by the Relative Amounts of Pyrolysates Analyzed by Pyrolysis-Gas Chromatography

Gel Sample[a)]	Gel History[b)]	Gas in Gel	Volatiles[c)] (by TGA)	CO_2	H_2O	C_2H_5OH	CH_3CHO	CO	CH_4	C_2H_2	C_2H_6
1	Unhydrolyzed	Air	17%	4.5	51.4	25.9	6.1	7.3	3.2	1.2	0.3
	Hydrolyzed	Air	13%	9.8	87.9			1.3	1.0		
2	Unhydrolyzed	Air	20%	2.4	45.6	39.6	6.4	2.5	1.7	1.3	0.3
	Hydrolyzed	Air	13%	8.9	88.6			1.5	1.0		
3	Unhydrolyzed	Argon	12%	1.6	17.8	64.0	4.4	6.5	3.3	2.2	0.2
	Hydrolyzed	Argon	15%	1.3	93.9	1.5	1.0	1.4	0.9		

a) Samples 1 and 2 are two separate batches of xerogel synthesized and stored under dry air. Sample 3 was synthesized and stored under dry argon to exclude the CO_2 in air.

b) The metal oxide glasses were prepared from metal-organic xerogels which were either kept dry (unhydrolyzed) or exposed to 80% RH for about 5 days (hydrolyzed).

c) Gels had been dried to constant weight at 100° prior to analysis.

intermediate stages of this treatment with moist air were monitored by a second technique, pyrolysis-mass spectrometry (PMS). Pristine sodium silicate xerogels, which initially contained 15 to 25 wt % residual ethoxides, experienced a sharp decrease (≈90%) in ethoxide content within two days of hydrolysis in 80% relative humidity, and were effectively rendered devoid of organic species within one week, as shown by the PMS data in Figure 5. At the same time, the water and CO_2 concentration in the pyrolysates of the treated gel rose steadily and significantly. (Note that the PMS data for water are significantly lower than the PGC data of Table 1; this is an artifact of the PMS system.) Autoclaving the xerogel with 20 psig steam at 125°C produces similar results in one or two hours, but without the increase in CO_2.

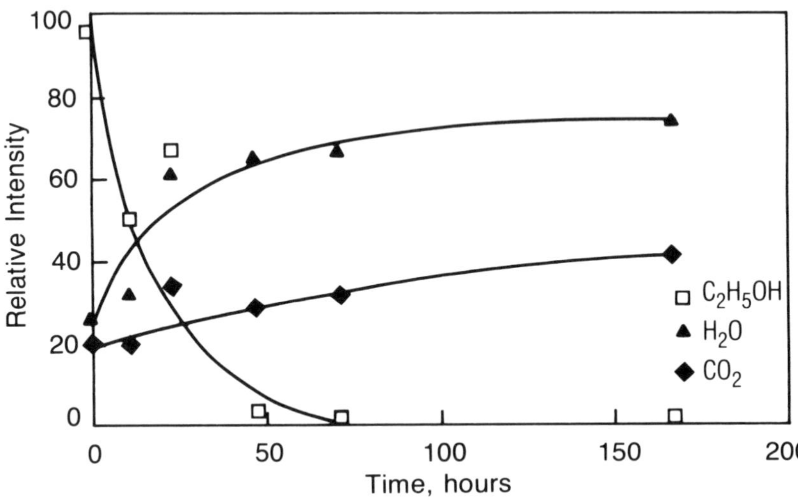

Figure 5: Rate of hydrolysis of pristine xerogel in 80% RH at 30°C, as indicated by the change in the relative intensity of the pyrolysates evolved from the gel during PMS analysis.

In addition to hydrolysis of the residual alkoxides in the gels, water vapor can also cause ion exchange with the alkali in the gel, and promote alkali diffusion to the surface of the xerogel. When the hydrolysis occurs in ambient air, the reaction of the alkali with the ambient CO_2 forms alkali carbonates in the gel, a conclusion supported by the high levels of evolved CO_2 found in PGC and PMS analyses. (See Table 1 and Figures 4 and 5.) [In these analyses the gels were pyrolyzed in an inert helium atmosphere (PGC) or vacuum (PMS), and thus, the evolved CO_2 cannot be the product of oxidation of residual organic species. The source of the evolved CO_2 must be alkali carbonates and bicarbonates produced by the reaction of water vapor and ambient CO_2 with the alkali in the highly porous gel.] The high porosity and large surface area of the gel probably provide an opportunity for deep infiltration of the reactants and a significant extent of carbonate formation within the xerogel; we believe this accounts for the higher levels of CO_2 evolved by the metal-organic-derived xerogel as compared to those from the nonporous inorganic sodium silicate powder of similar composition (Figure 4). Indeed, when one of the metal-organic gel samples was processed under argon to prevent the formation of the carbonates due to exposure to ambient air, the PGC analysis of this sample showed a significant reduction of the CO_2 levels in the gel pyrolysates (Table 1).

As a consequence of ion exchange and alkali diffusion, hydrolysis of the xerogels may also affect the compositional homogeneity of the gel. Hydrolysis has caused significant alkali migration in labile xerogels of sodium silicate (20% Na_2O, 80% SiO_2), although xerogels of soda-lime or mixed alkali borosilicate composition are not as sensitive. Hydrolysis of such xerogels in 80% relative humidity air results in the rapid agglomeration and fusion of contiguous powder particles, a consequence of the formation of significant amounts of surface scale and crystallites on the particles. EDXS analysis of the surfaces of the hydrolyzed gel have indicated that the scale and crystallites are indeed enriched in sodium, while the substrate surface has a lower sodium content than the corresponding unhydrolyzed surface. In fact, as shown in Table 2, the surfaces of powder that has been resieved after hydrolysis have a much lower sodium composition than the unsieved agglomerate or the unhydrolyzed parent stock, due to the removal of the surface scale by the abrasive action of sieving. Bulk analyses were nearly identical within the error of analysis, indicating that the alkali migration is limited to a shallow surface layer for the usual conditions of humid air-induced vapor-phase hydrolysis.

Table 2: Effect of Humid Air on Compositional Homogeneity of Metal-Organic-Derived Gel Powders

Chemical Analysis	Gel History: 0% RH	Gel History: 80% RH for 5 Day
Bulk sample analysis (Wt.% by AA)	SiO_2 80.5% Na_2O 19.5%	SiO_2 81.3 Na_2O 18.7
Surface analysis (EDXS) Na/Si ratio	0.10	0.09 (before sieving) 0.06 (after sieving)

The sequence of the hydrolysis and the CO_2 reactions can have a significant effect on the rate of gel hydrolysis and the morphology of the products. Carbonate formation upon exposure to CO_2 seems to occur in both the pristine and the hydrolyzed xerogels; what differs is the extent of alkali migration and carbonate formation. A pristine carbonate-free sodium silicate xerogel, synthesized and processed under argon to minimize carbonate formation, remained physically unaffected by an extended exposure to an atmosphere of pure dry CO_2. However the PGC analysis of that same gel (sample 2 in Table 3) yielded high levels of CO_2 in the pyrolysates compared to the as-made pristine gel (sample 1), indicative of a considerable extent of carbonate formation as a result of exposure to CO_2. The absence of any physical changes to the gel surface suggests that carbonates have formed internally, on the pore surfaces of the gel, rather than on the external gel surface. (Because of the quantity of evolved CO_2 it is unlikely that the source was the free CO_2 gas in the pore volume; furthermore, in PGC analyses, pore volume gases are purged by the helium carrier gas prior to analysis.) Such a mechanism may account for our observation of gradual aging of stored gel, which manifests itself as a subtle change in shell-blowing performance.

Xerogel that was processed and hydrolyzed under argon did not exhibit any changes in external physical appearance after the hydrolysis. It produced a PGC signature that was almost free of organic species and CO_2 (sample 3, Table 3). Subsequent exposure of this hydrolyzed gel to pure dry CO_2 produced a fluffy loose mat of long crystalline needles of sodium carbonate and bicarbonate that completely enveloped the grains of gel powder. The PGC analysis of this product showed the expected high levels of CO_2 in the pyrolysates of the gel (sample 4, Table 3). The resieved gel powder, stripped of the crystalline mat, produced only modest levels of evolved CO_2 similar to that from gel hydrolyzed in ambient air (sample 5, Table 3). Evidently only a small portion of the reacting CO_2 penetrated into the pore structure before the reaction products plugged the pores; further reaction occurred only at the surface, producing the massive crystalline mat. Hydrolysis of the argon-processed pristine gel in a pure CO_2 ambience produced a thick, dense, carbonate crust on the gel surface that was easily abraded. However, in this case the hydrolysis progressed very slowly, requiring more than three times the usual reaction time to remove the residual ethoxides. This suggests that again the pore structure was rapidly plugged in the initial stages of hydrolysis by the reaction of the excess of CO_2 with surface alkali. It is not known whether the presence of this carbonate crust hindered the diffusion of water into the gel, but PGC and TGA analysis of the intermediate stages of the hydrolyzing gel indicated that the rapid plugging of the pores retarded the escape of the hydrolysate, ethanol. Such an entrapment mechanism can explain the retention of small quantities of residual organic species in hydrolyzed gel, whose carbonization during subsequent pyrolysis invariably renders the hydrolyzed gel coal black. Such a mechanism is also consistent with the dramatic reduction of the specific surface area of the gel, from 400 m^2/g to less than 1 m^2/g, the usual consequence of hydrolysis.

Pyrolysis of Xerogel

Gas Evolution. Gases are evolved by the gel during pyrolysis, and the generation of these gases provides the driving force that forms and expands the hollow shell. The pyrolysates from these gels consist typically of CO_2 and H_2O; those gases evolved from unhydrolyzed all-alkoxide gels also include ethanol and its pyrolysates, including acetaldehyde, ethane, ethylene and methane (Table 1). While the pyrolysates from gels containing acetates or lactates probably also consist of the corresponding acids, in addition to the above hydrocarbons, these gels have not been analyzed by PGC.

Table 3: Alkali Carbonate Formation in Argon-Processed, Metal-Organic-Derived Xerogels that were Exposed to CO_2 and/or H_2O, as Indicated by the Relative Amounts of CO_2 in the Pyrolysates. Gels were Analyzed by Pyrolysis-Gas Chromatography

Gel Sample	Gel Treatment	% Volatiles (by TGA)	Relative Concentration, Volume %: CO_2	H_2O	C_2H_5OH	CH_3CHO	CO	CH_4	C_2H_4	C_2H_6	Comments
1	As-made, dry	12.3	1.6	17.8	64.0	4.4	6.5	3.3	2.2	0.2	
2	1 exposed to 100% CO_2 dry		20.3	42.1	21.2	7.0	7.1	1.7	0.5		
3	1 hydrolyzed under Ar	14.9	1.3	93.9	1.5	1.0	1.4	0.9			
4	3 exposed to 100% CO_2, unsieved	14.4	39.8	39.8	0.4	0.5	← 19.0 →		0.6		Unsieved
5	4 resieved		6.1	78.3	0.8	0.9	2.3	10.4	1.2		Sieved
6	1 exposed to 100% CO_2 and 100% RH simultaneously	19.0	41.6	36.6	0.9	1.2	← 16.7 →		3.0		

The source of the water consists of chemically bound species as well as molecular water adsorbed to the highly active gel surface. Chemically bound water is present in the form of hydrated alkali oxides, silanol groups and alkali bicarbonates. The source of the organic gases is the residual alkoxides and free ethanol entrapped in the gel network:

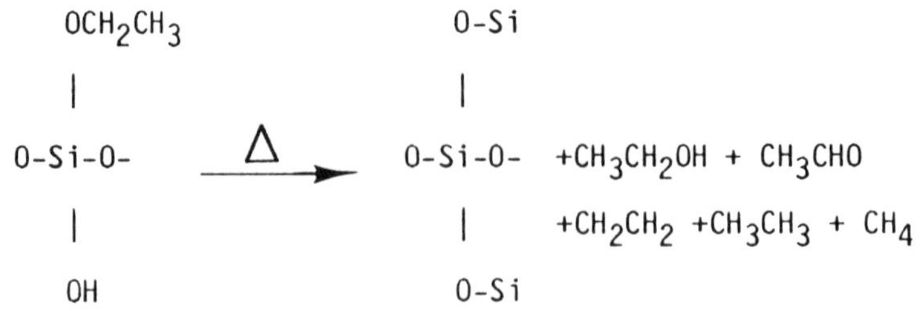

Although these organic gases may serve as initial gel foaming agents, the bulk of these gases do not have a role in the blowing of the shell. The major portion of these gases is evolved by the gel at relatively low temperatures; indeed, it is these evolved pyrolysates that are detected and analyzed in PGC, PMS, TGA, and DTA analyses. Most of these gases are generated and lost almost immediately after the gel powder is injected into the furnace. The organic species often ignite audibly in the mouth of the furnace process tube.

Pyrolysis of metal-organic-derived xerogel powder generates copious quantities of volatile pyrolysates. Thermal gravimetric analyses of metal-organic-derived gels indicate an evolution of gases equivalent to 15 to 25 wt % of the gel; the amount of evolved gases for a given gel depends on the history of that gel, specifically the extent of metal alkoxide hydrolysis in forming the gel and in subsequent aging of the gel or vapor phase hydrolysis of the gel. Almost all of the volatile species are generated and lost before the gel reaches 600°C, except for residual amounts which are carbonized. The TGA of pristine gels display a characteristic broad mass loss inflection between 100°C and 250°C, due to the desorption of water and the evolution of alcohol and carbon dioxide (from decomposition of bicarbonates). A steeper mass loss, characteristically at 400°C to 450°C, is due to the pyrolysis and evolution of the residual hydrocarbons. A barely distinguishable mass loss inflection could be detected at 800°C in the TGA of some gels, probably due to the oxidation of elemental carbon. A correlation of the results of TGA, DTA, and PMS analyses for an unhydrolyzed metal-organic gel sample is shown in Figure 6. Aged gels and hydrolyzed gels do not exhibit mass loss inflections at 450°C and 800°C, and exhibit a stronger, broader inflection between 100°C and 300°C due to dehydration and the decomposition of alkali bicarbonates (see above).

Specific Heat. The specific heat of the gel is markedly affected as the organic and inorganic species are stripped from the gel by pyrolysis. Figure 7 shows the specific heat (Cp) as a function of temperature for an unpyrolyzed sodium silicate metal-organic gel and for the same gel pyrolyzed at 600°C; the response of a sapphire calibration standard is shown for comparison. The unpyrolyzed gel shows considerably higher specific heats than the pyrolyzed gel or the sapphire due to the energy input required for the water desorption, silanol condensa-

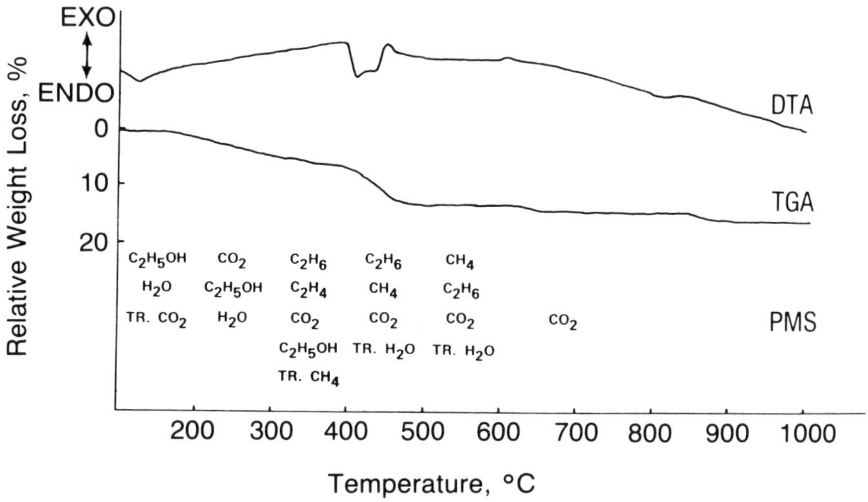

Figure 6: Identification of the gases that are evolved as a function of temperature in the pyrolysis of pristine xerogels, and correlation of these with the characteristic weight-loss and thermal inflections in the TGA and DTA curves. Evolved gases within each temperature bracket in the PMS analysis are listed in order of decreasing quantity.

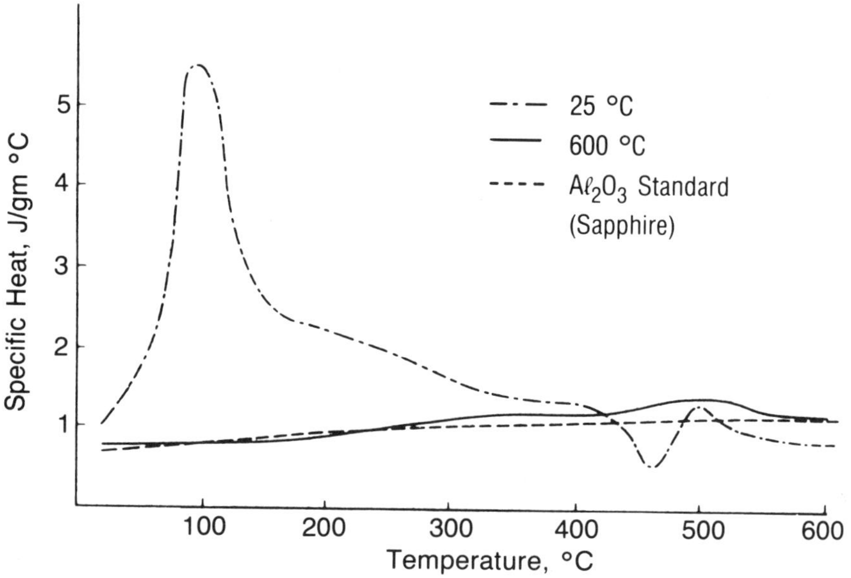

Figure 7: The effect of pyrolysis on the specific heat of metal organic xerogel. Curves compare the specific heat, determined by DSC as a function of temperature, for: (a) pristine xerogel, showing the additional heat capacity due to dissolved water and residual alkoxides; (b) xerogel stripped of water and residual organics by pyrolysis at 600°C for one hour; and, (c) sapphire standard.

tion, sodium bicarbonate decomposition and organic pyrolysis reactions. At 450°C, the gel shows a specific heat lower than the standard, due to the release and oxidization of hydrocarbon pyrolysates. The gel pyrolyzed at 600°C has a flat Cp curve, similar to that of the standard except for a sintering endotherm at 500°C.

Gel Morphology. The pyrolysis of gels has profound effects on gel morphology. Simple alkali-borosilicate gels produced from aqueous inorganic solutions or colloidal suspensions are particularly responsive to pyrolysis, forming low-density foam balls and spheroids at temperatures in the range of 150°C to 300°C. Metal-organic-derived xerogels, because of their polymeric structure, are not as responsive to pyrolysis, and are quite rigid and resistant to flow. Response to the internal stresses generated by the gaseous pyrolysates does not begin until a temperature of 400°C to 600°C is reached, when the gel particles begin to resemble popcorn, forming low-density foams.

A physical consequence of the pyrolysis of pristine xerogels, synthesized with the usual 2 to 2.3 mols of water per mol of silicon ethoxide, is a significant reduction in specific surface area as a function of the pyrolysis temperature. This is shown in Table 4 for an alkali-lime metal-inorganic xerogel (Gel A). Such a reduction is consistent with the formation of a closed-cell foam structure in the gel, and readily accounts for the entrapment and subsequent carbonization of residual organic species in the gel. Pristine soda-lime xerogels, synthesized with 4 mols water per mol ethoxide (designated Gel B in Table 4), were unique in retaining a relatively high specific surface area after pyrolysis. The pyrolysis of these highly crosslinked, rigid gels resulted in a sharp reduction of the initial specific surface area that stabilized at a respectably high value of 100 m^2/g, as shown in Table 4. Because the rigid gel structure maintained a high degree of porosity, the pyrolysates were not retained and carbonized; when pyrolyzed, these gels acquired a beige coloration instead of the coal-black coloration that is characteristic of the usual pyrolyzed gels.

Table 4: Effect of Gel Pyrolysis on Specific Surface Area (m^2/g) of an Alkali Lime Metal-Organic-Derived Gel Prepared with 2:1 (Gel A) and 4:1 (Gel B) Molar Ratio of H_2O to $Si(OC_2H_5)_4$

	Gel Surface Area	
Gel Treatment	Gel A	Gel B
As Prepared	220	560
350°	NA	108
400°	56	NA
525°	NA	99
600°	<4.0	128
675°	<6.0	NA

EXPERIMENTAL PARAMETERS THAT INFLUENCE SHELL PROPERTIES

The geometric characteristics of shells are governed by the gel powder properties and the shell processing parameters. Gel properties are affected by gel synthesis procedures and post treatments of the dried xerogel powder. How gel properties change as a result of gel treatments was discussed in the previous section. How the gel treatments and the procedures used in shell blowing affect the properties of the glass shells is the subject of the following discussion.

Drying of the Alcogel

The method of drying the alcogel to form the xerogel affects the wall uniformity of the resulting shells. Vacuum-dried xerogel powder produced batches of shells of slightly smaller diameter and glass volume, which has as much as twice the number of shells with good wall uniformity as did the air-dried gel of similar composition and powder size. Vacuum-drying the alcogel produced a xerogel powder whose individual grains were highly fissured and of lower density than the air-dried gel. The fissures in the vacuum-dried powder are the result of the greater internal stresses generated by the faster rate of drying. The smaller diameter and glass volume produced by vacuum-dried material is a consequence of the lower density of vacuum-dried material, and not of increased fragmentation of the stressed grains during firing.

We have not identified the chemical reaction mechanisms and/or physical transformations which are responsible for the decreased uniformity of shells produced from air-dried xerogels. However, a probable mechanism is the development of radial compositional inhomogeneities in the air-dried gel powder, caused by exposure to ambient humid air during the extended drying time. Even more so than glass, the xerogel is expected to be susceptible to ion exchange (involving water and the alkali in the gel) during prolonged exposure to ambient water vapor, resulting in diffusion of the liberated alkali to the surface of the powder grains. Such a compositional segregation would be expected to produce radial composition gradients in the grain, and consequently viscosity gradients that would cause uneven flow of glass in the developing, molten shell.

Hydrolysis and Pyrolysis of Xerogels

Even though gel hydrolysis reduces the concentration of potential organic-derived blowing agents in the gel, this treatment has only modest effects on the final diameter and aspect ratio of the shells. Autoclaving the gel accomplishes a modest reduction of the aspect ratio of shells, at best from 90 to 40. This effect is usually obscured by slight variations of the furnace parameters. On the other hand, the hydrolysis of the gel in an ambient-air humidity chamber has no effect on the aspect ratio whatsoever. The inefficiency of these methods in reducing the aspect ratio of shells is probably a consequence of the ubiquitous alkali carbonates that form so readily in the gel during synthesis, processing and subsequent treatment, and to the very small mass of actual gas required to form a shell. In addition, water, which also serves as a blowing agent, is deliberately introduced in the gels in these processes.

An additional effect of hydrolysis of dried gel, especially of autoclaving, is a reduction in uniformity of the resulting shells. This effect has been strongly negative for some gel compositions, and is probably the consequence of accelerated alkali migration and the development of radial compositional inhomogeneity in the gel. An extreme case of this effect was the sodium silicate gel that was hydrolyzed in a CO_2 environment (sample 6, Table 3). This sample produced shells with extremely poor uniformity and gas bubbles in the walls, presumably due to a viscosity gradient in the melt, generated by an initial alkali composition gradient in the gel.

The anticipated effect of gel pyrolysis is control of shell aspect ratio, since in principle the concentration of potential blowing agents, whether water, organic species (alkoxy or acetate groups), or metal carbonates, can be reduced to trace quantities by appropriate heating of the gel. In practice, such control has not been realized for all xerogels. Under the best of circumstances, pristine xerogels synthesized in the usual manner [2.0 to 2.3 mols H_2O per mol $Si(OC_2H_5)_4$] produced shells having a slightly reduced aspect ratio as a result of gel pyrolysis. As shown in Figure 8 for a soda-lime glass with gel designation A, most of that reduction is accomplished by heating the gel to 200°C which removes primarily the adsorbed water. Subsequent pyrolyses at higher temperatures produced scattered data in which no significant reduction in aspect ratio was detectable. This insensitivity is due either to the fact that the actual blowing gases are alkali carbonates, e.g., Na_2CO_3, which decompose above 950°C and are unaffected by pyrolysis at $\leqslant$700°C, or to entrapped elemental carbon, which is oxidized and lost quite slowly from the closed pores of the pyrolyzed gel. However, as shown by the data for Gel A in Figure 9, there was no relationship between the shell aspect ratio and the amount of entrapped potential gas formers in the pyrolyzed gel.

Control of the shell aspect ratio by gel pyrolysis has been realized only with soda-lime xerogel that was synthesized using 4 mols water per mol $Si(OC_2H_5)_4$ instead of the usual 2.0 mols H_2O per mol $Si(OC_2H_5)_4$. When 4 mols of water per mol of silicon tetraethoxide were used to prepare the alcogel, the resulting dried xerogel, designated as Gel B, had an exceptionally low density, one-third of the density of the usual xerogels, and a specific surface area of 560 m^2/g compared to the usual 200 to 350 m^2/g. An unusual feature of this xerogel was the relative insensitivity of its specific surface area to pyrolysis up to 600°C, as shown in Table 4. Though we have no direct evidence, the high degree of solution-phase hydrolysis has apparently resulted in a very extensively crosslinked polymer structure in the gel, one that is not readily affected hydrolytically or thermally. A consequence of this structure is a high degree of porosity in the gel, conducive to the rapid loss of pyrolysates or hydrolysates and the facile removal of residual organic species from the gel. As shown by the Gel B data in Figure 8, the aspect ratio of shells produced from this gel was, like Gel A, independent of the pyrolysis treatment for temperatures up to 500°C. For pyrolysis temperatures from 500°C to 675°C however, the aspect ratio of the shells produced from the pyrolyzed gel was a linear function of pyrolysis temperature. This is more clearly shown in Figure 10. Furthermore, as shown in Figure 9, the shell aspect ratio was linearly dependent on the amount of residual potential gas formers in the gel.

For the "fully" hydrolyzed gel, synergism between autoclaving and pyrolysis results in increased control of the shell aspect ratio. As shown in Table 5, autoclaving alone had a negligible effect on the shell aspect ratio. Pyrolysis at 625°C

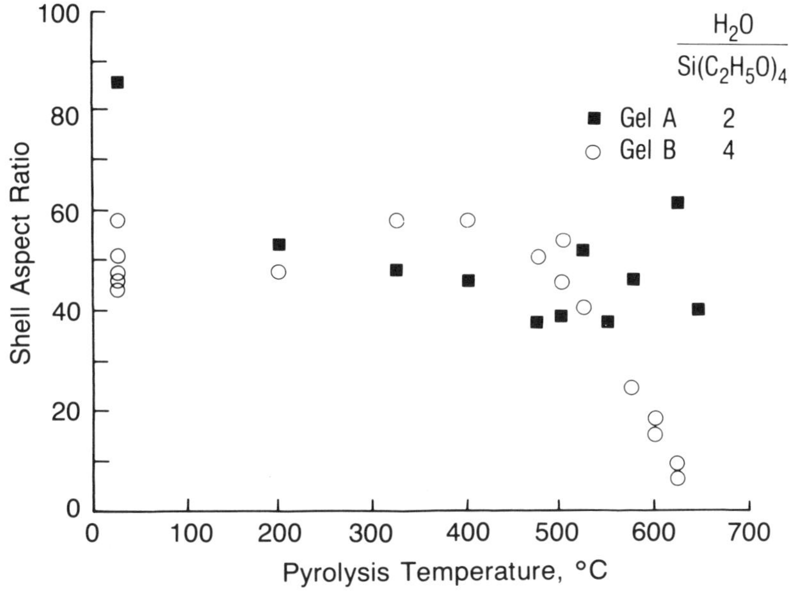

Figure 8: Effect of gel pyrolysis on the shell aspect ratio (o.d./wall), shown for an alkali-lime xerogel that was synthesized by: (a) partial hydrolysis of the alkoxide solution, with 2 mols water per mol silicon tetraethoxide, yielding Gel A, and, (b) hydrolysis with 4 mols water per mol silicon tetraethoxide, yielding Gel B.

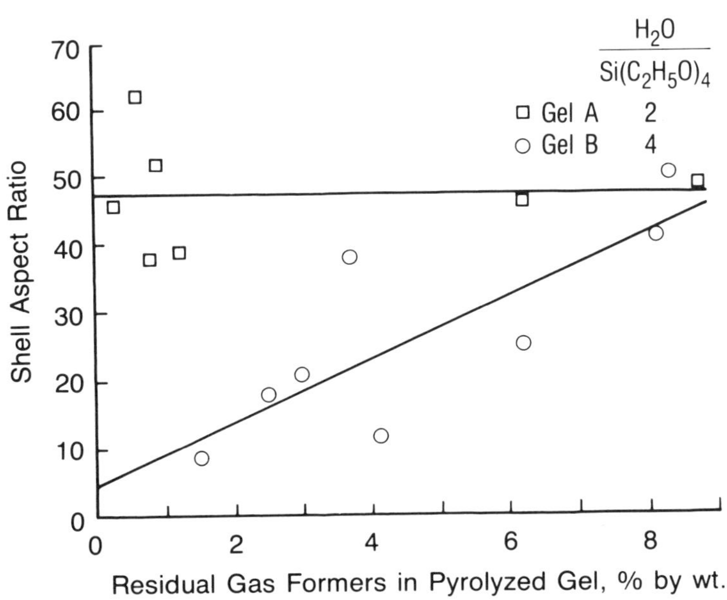

Figure 9: Dependence of shell aspect ratio on the concentration of potential gas formers in the pyrolyzed gel: (a) Gel A, partially hydrolyzed with $H_2O/Si(OC_2H_5)_4$ = 2; and, (b) Gel B, hydrolyzed with $H_2O/Si(OC_2H_5)_4$ = 4.

for one hour had a very strong effect, reducing the shell aspect ratio from 60 to 10. Autoclaving the gel, followed by pyrolysis at 625°C produced aspect ratios less than 4.5; reversing the sequence was not as effective.

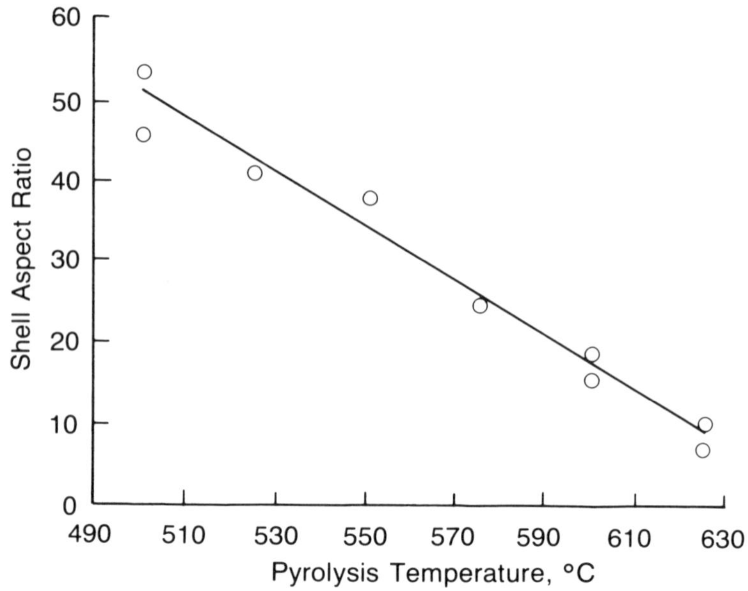

Figure 10: Gel pyrolysis-aspect ratio data for Gel B in greater detail, showing the linear dependence of the shell aspect ratio on gel pyrolysis in a limited temperature range.

Table 5: Synergistic Effect of Autoclaving and Pyrolysis on the Shell Aspect Ratio of an Alkali Lime Xerogel (Gel B, Table 4).

Gel Treatments	Average Shell Aspect Ratio (o.d./W)
Autoclave	61
625°	9
Autoclave then 625°	<4.5
625° then autoclave	8

Shell Processing Parameters

Several shell processing parameters have been identified that broadly influence the properties of the shells. The shell properties influenced are primarily geometrical; uniformity and aspect ratio. The magnitude of the effect of the processing conditions can vary from gel to gel, i.e., from one glass composition to another. Most of our work has been with metal-organic-derived gels but it is not appropriate to ascribe all the correlations observed between shell processing parameters and shell properties exclusively to the use of metal-organic sol-gel shell precursors. Many of the processing-property correlations should hold for shells blown from other shell precursor materials. To a large degree, the upper mass (size) limit and quality of the hollow spheres are determined by the maximum furnace temperature, the residence time of the particle in the hot zone, the temperature profile of the furnace, the amount of gel per injection and the ambient gas in the furnace.

Furnace Parameters. Furnace parameters include the length and temperature of the furnace. These parameters influence the rate of heating, the viscosity, and the residence time of the particle.

Furnace Length. The furnace length influences both the limits of shell sizes that can be produced as well as the wall uniformity that can be achieved. A 1500°C furnace with a one meter long hot zone is sufficient for producing hollow glass spheres $\leqslant$100 μm diameter of acceptable uniformity if wall thicknesses are less than about 2 μm. The quality of larger spheres decreases rapidly with increasing size. In a 1500°C four meter high furnace, shells as large as 800 μm to 1,000 μm with walls up to 4 μm can be produced. An example of the effect of furnace length on uniformity is illustrated qualitatively by comparing the optical interferograms of shell batches produced from an identical gel source, gel size, and furnace temperature (1500°C) but different furnace length, 4 m compared to 12 m (Figure 11). The shells from the 4 m furnace are noticeably more uniform, as indicated by the increased number and quality of interferometric bulls eye images, than those from the 1.2 m furnace.

Furnace Temperature. The furnace temperature influences the limits of diameters, the aspect ratio, and wall uniformity of shells that can be produced. Small shells can be made with furnace temperatures of 1200°C to 1300°C. However, for most glasses, a furnace temperature of 1500°C is required; an upper limit of 1650°C to 1700°C is desirable for greater versatility and flexibility with respect to size, quality, and composition of shells. While furnaces with temperature capabilities greater than 1800°C should increase the range of accessible compositions, the enormously higher costs make these uneconomical. Equivalent performance may be obtained more economically by increasing the residence time of the particle in the hot zone, i.e., by increasing the length of the hot zone in a furnace with a lower temperature limit.

The limits of shell sizes are increased by increasing furnace temperature. For instance, by increasing the temperature of the four meter furnace, from 1500°C to 1650°C, the size limit is increased from 1,000 μm to 1,200 μm, with wall thicknesses of 3 μm. Shells as large as 2,000 μm have been produced; however, beyond 1,200 μm the diameter, the sphericity, the uniformity and the throughput yield of the shells decreases rapidly.

Shell aspect ratio (o.d./wall) can increase or decrease with increasing furnace temperature. As the metal-organic-derived gel particles are heated in the furnace,

a

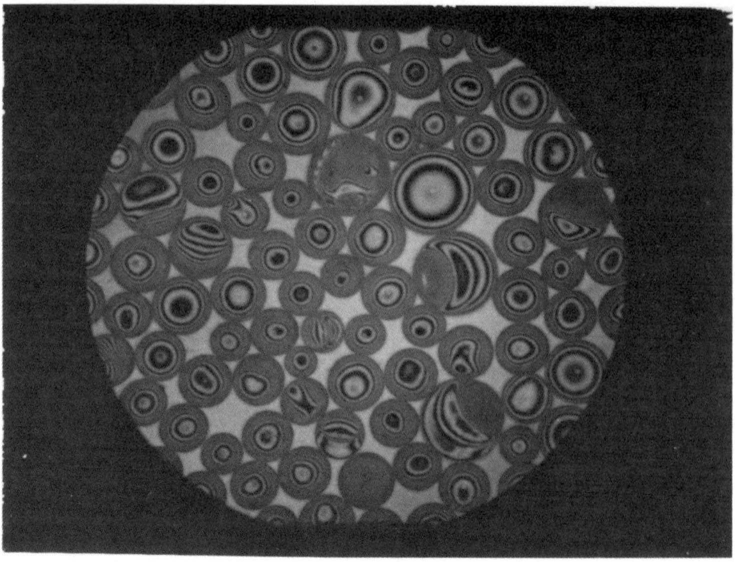

10X

b

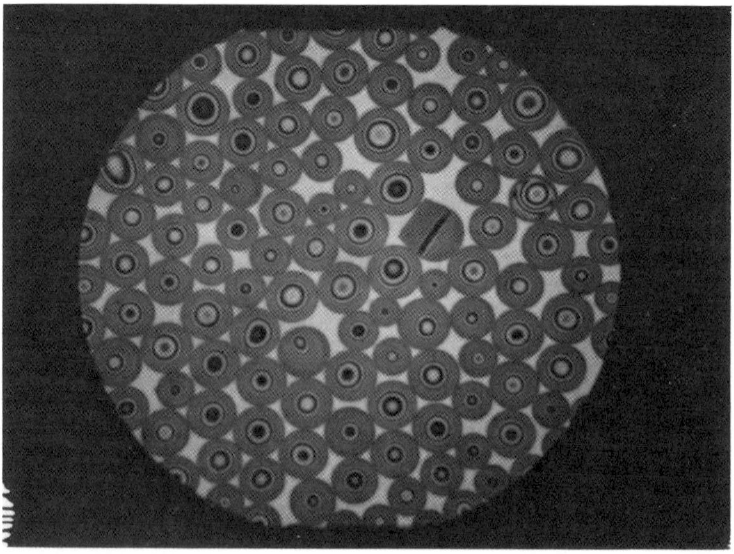

10X

Figure 11: Interferograms of glass shells (~100 μm) produced at 1500°C in: (a) 1.2 m vertical furnace; and, (b) 4 m vertical furnace.

as noted previously, CO_2 and H_2O are generated by the oxidation of residual carbon and decomposition of residual carbonates. Additional water vapor is produced by condensation of metal hydroxy groups and the release of absorbed water. The single gas bubble that is ultimately produced to form the shell expands the shell until an equilibrium state is reached between the external gas pressure in the furnace (one atmosphere) and the sum of the internal gas pressure and the surface tension of the molten glass. If the shell is held indefinitely at high temperature the gases in the shell would eventually permeate out resulting in formation of a bead. [In practice this does not occur since the permeation rate of CO_2 (principal gas) is very slow and the shell residence time in the furnace is limited. When the shell exits the furnace, thermal quenching raises the viscosity and freezes the shell wall.] For a given residence time in the furnace, increasing the furnace temperature will either increase the shell aspect ratio in an amount directly proportional to the excess temperature (Charle's Law), or decrease it by diffusion of the blowing gases through the shell wall and the shrinkage force of glass surface tension. The rates at which these phenomena occur are dependent upon the composition of the glass and the gases inside the shell. One further complicating factor is the effect of furnace temperature on the earliest stages of shell formation at the top of the furnace. If a gel particle is heated too rapidly, i.e., very high furnace temperature, the gel/shell particle may lose an excess of gas before a gas-tight glass skin can be formed. This will result in a shell with a smaller aspect ratio than would otherwise be produced.

Wall uniformity is strongly affected by furnace temperature. An example of the effect of furnace temperature on uniformity is shown in Figure 12 for an alkali-borosilicate composition over the temperature range of 1200°C to 1500°C. This is a computer-generated plot based on an empirical model relating shell properties to processing parameters. Varying the temperature over the furnace length (temperature profiling) also affects uniformity and, for small shells, the aspect ratio. This is illustrated in the results of a series of experiments we conducted using a 53 μm to 63 μm sieve cut of metal-organic-derived gel with an alkali-lime-glass composition.[43] The gel powder was dropped through a vertical tube furnace heated by six independently controlled furnace units. The heated region was 4 m. The results of the experiment are summarized in Figure 13 and Table 6. The best compromise for maximum aspect ratio and acceptable uniformity was a three zone temperature profile. In more recent experiments, we examined the effect of temperature profiling up to 1650°C, using relatively large alkali-borosilicate particles 250 μm to 300 μm. Slight improvements in uniformity were observed but the aspect ratio was not affected.

Gel Addition Parameters. Gel addition parameters include the size (mass) of the gel particles, the amount of gel powder ("shot" size) added per injection, and the time intervals between shots. Each parameter affects wall uniformity. Obviously, gel particle size has a major effect on shell size; more massive gel particles produce more massive shells. Unfortunately, as the gel size (i.e., shell mass) increases, the wall uniformity decreases, as shown in Figure 14, a computer-generated plot based on an empirical model relating shell properties and processing parameters.

The amount of gel powder injected into the furnace per shot can have a dramatic effect on shell uniformity. In a series of experiments with an alkali-lime glass using different gel sizes, the percent of Class A shells in a batch was

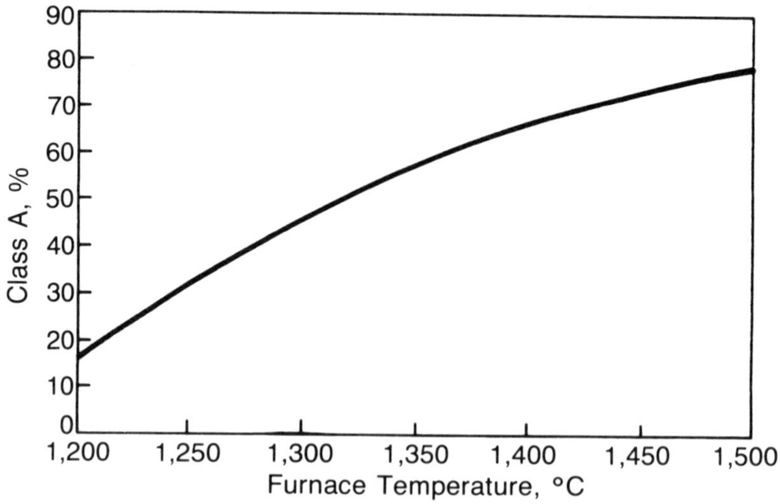

Figure 12: Calculated effect of furnace temperature on shell uniformity of a soda-lime glass for constant gel particle size (106 to 125 μm).

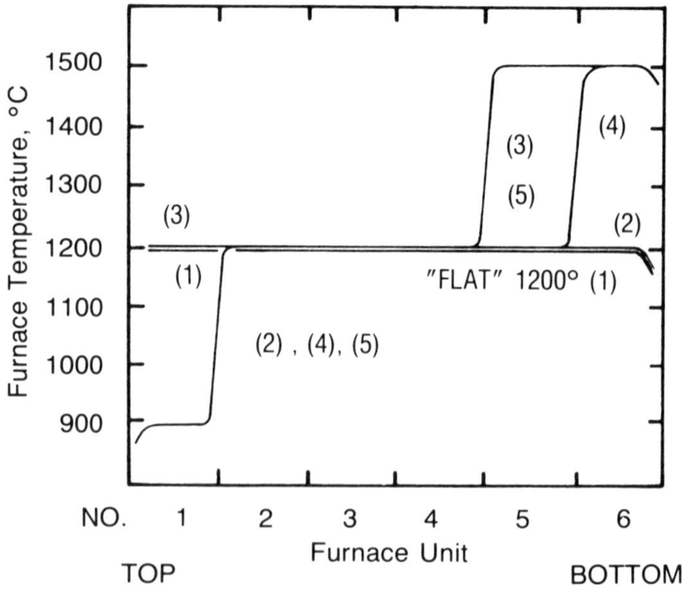

Figure 13: Furnace temperature profiles used in the study of temperature profile effects on shell uniformity. Experiment No. 1 (see Table 1) used profile No. 1, etc.

Table 6: Comparison of Average Shell Batch Properties as a Function of Furnace Temperature Profiles

Experiment and furnace profile No.[a)]	Average shell aspect ratio (OD/W)	Average shell diameter OD	Percent shells with <10% nonconcentricity
1	110	145	50
2	b)	150	40
3	55	125	80
4	160	170	60
5	90	140	70

a) See Fig. 13 for the temperature profile corresponding to each experiment.

b) Could not be measured by interferometry due to poor wall thickness uniformity.

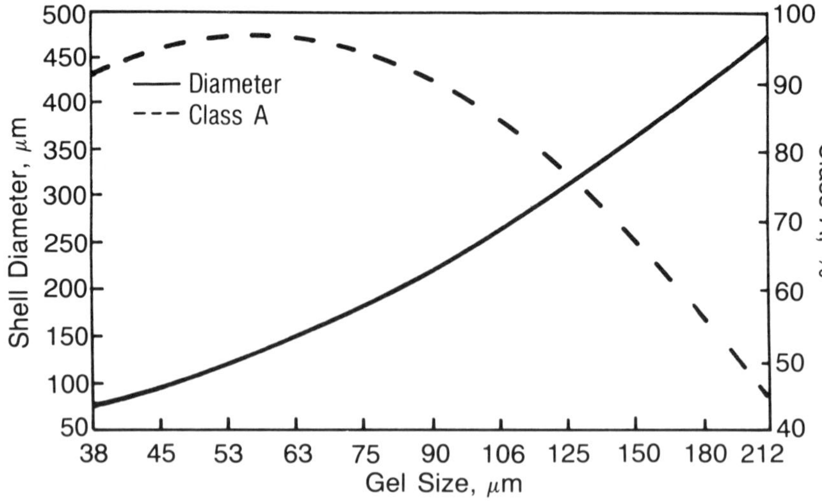

Figure 14: Gel size affects shell diameter and uniformity, calculated for an alkali borosilicate gel.

observed to decrease by a factor of 1.5 to 4.0 by increasing the shot size from 5 mm^3 to 324 mm^3. Perhaps the most noteworthy aspect of these experiments was the observation that relatively high rates of shell production can be achieved if poor shell uniformity can be accepted. The relatively large thin-walled shells in this experiment (300 to 400 μm, o.d. x 3 to 4 μm wall) were produced at rates up to 25 g (200 ml) per hour at a yield (mass of shells produced per mass of gel powder) of ~45%. Even higher shell production rates may be possible if gel injections are increased. The poor wall uniformity for the shell batch (20% Class A) discouraged us from pushing production rates higher since such shell batches are not suitable for ICF target selection.

The time interval between shots can affect the wall uniformity of small shells. For very small gel sizes, shell uniformity is maximized by using small amounts of gel powder per shot and also by extended time intervals between shots. The effects of the time interval between shots is illustrated in Figure 15 which compares the uniformity of three batches of shells made with 30 second and 3 minute intervals between gel powder shots (KMSF, 1980). The production of the most uniform shells occurred in the experiments where the furnace tube was purged with air after five shots. Although we have no explanation for this, the results suggest that the build-up of evolved gases, including alkali and boron oxides, must act in some way to inhibit the formation of uniform wall thickness in the shells.

Furnace Gas Parameters. Gases which can diffuse rapidly through glass shell walls at high temperatures have a pronounced effect on shell size. Experiments to determine the influence of various gases on the shell forming process have involved purging of the furnace tube with the gas of interest before and usually during the injection of the gel powder. Purge gases are introduced from the bottom of the furnace tube. These experiments have been done with the top of

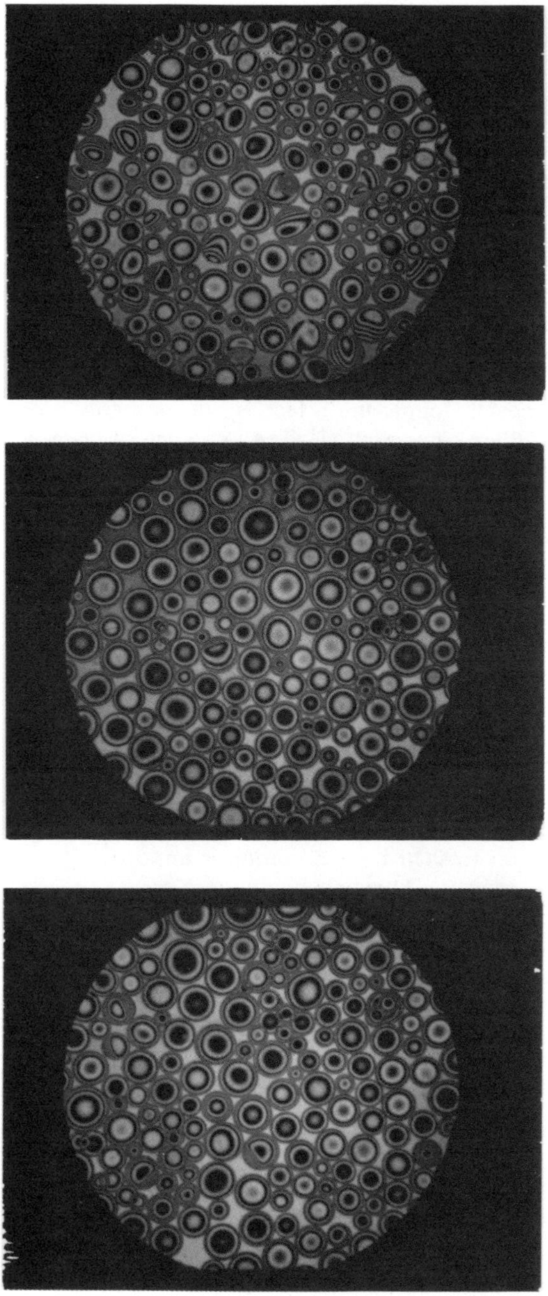

a

b

c

Figure 15: Optical interference photomicrographs showing the shell uniformity of glass shells blown from 45 to 53 μm gel powder injected into the furnace tube at: (a) 30 sec intervals; (b) 3 min intervals; and, (c) 3 min intervals interrupted after every 5 injections by flushing the furnace tube with air.

the furnace tube open so some mixing with air has been unavoidable. Shells have been blown with a variety of gases in the furnace tube, e.g., CO_2, Ar, O_2/N_2 (air), He and $H_2O(g)$, although most experience has been with air and water vapor. The introduction of water vapor into the furnace tube during shell blowing produces a minor but usually positive change in uniformity, and a major increase in shell aspect ratio when compared to experiments with identical gel powder in an air-filled furnace. This phenomenon results from the fact that water vapor permeates through the glass wall of the forming shell at a much faster rate than air so the expansion of the shell, by the gases evolved during shell blowing, is much less inhibited by the pressure of the furnace gas external to the shell.[44] Experiments run in helium, a gas which also readily permeates through the glass, likewise produced shells with much higher aspect ratio than air. Although water vapor and helium differ from air in other respects, i.e., by decreasing glass viscosity (water) and increasing the thermal conductivity of the furnace ambience (helium), the effect is due primarily to higher permeability of the glass to these gases than air. The influence of water vapor is sufficiently large and controllable that it has been used as a means of adjusting shell aspect ratio to desired values for ICF targets. Other effects of water vapor in the furnace tube are promotion of alkali loss and corrosion of the inner surface of the glass shell. This latter effect is dependent on glass composition, i.e., the higher the alkali content in the glass the greater the degree of weathering features on the inner surface of the shell.

MODELS

Physical Transformation From Gel to Shell

Because of the hostile environment within the furnace and the rapidity of the transformation, information concerning the chemical and physical processes comprising the gel to shell transformation is impossible to acquire by real time observation. Instead the chemical component of the process has been deduced from TGA, DTA, DSC, PGC, and PMS analyses as discussed previously. The physical component of this transformation has been deduced from hot-stage microscopy and box furnace studies at relatively low temperatures ($<800^\circ C$) and slow heating rates, and from recovered intermediates formed in the vertical furnace with intermediate temperatures or short hot zones.

Regardless of the thermochemistry involved in the pyrolysis of the gel, the sequence comprising the physical transformation is common to all the xerogels, and typically involves five distinct phases:

(1) Closure of the surface of the porous structure.

(2) Formation of internal cellular void (foam) structure.

(3) Particle spheroidization and/or foam expansion.

(4) Void coalescence and further expansion to form a spheroid.

(5) Fining of the glass to remove carbon inclusions and gas bubbles in the wall.

Each of these stages has associated with it some minimum onset temperature and minimum time of heating. The onset temperatures of each of these phases can vary vastly, depending on the source, composition and thermal and hydrolytic history of the gel. Representative onset temperatures for each of these phases, as observed in box furnace and hot stage pyrolysis studies, are given in Table 7 for several simple sodium silicate powders, both metal-organic-derived xerogels and powders derived from aqueous silicate solution.

Table 7: Morphological Stages in the Physical Transformation of a Xerogel Particle to a Glass Shell

	Temperature for Structure Changes (°C)				
Sodium Silicate Gel	Rigid Gel	Sintering	Foaming and Expansion	Spheroid	Crude Shell with Central Void
Aqueous-Based	25°	-	-	150°	300°
Metal-Organic Hydrolyzed	25°	-	150-200°	240-400°	800-900°
Metal-Organic	25°	400°	500°	800-900°	-

Simple alkali-borosilicate gels produced from aqueous inorganic solutions or colloidal suspensions are particularly responsive to pyrolysis, undergoing foaming and spheroidization at temperatures in the range of 150°C to 300°C. Produced by destablizing an aqueous colloidal sol to form an aqueous gel, or by evaporative precipitation of a water-glass solution, such powder has a specific surface area less than 0.1 m^2/g, indicative of a closed surface, if not of zero porosity, and does not require a separate surface closure or sintering process. Gently heating this powder at 140°C to 150°C causes opalescence in the translucent particles, indicative of the formation of internal microvoids or vesicles. Depending on the extent of hydration, the particles will rapidly expand to form a foam spheroid; further heating rapidly produces a crude, transparent hollow "glass" sphere at temperatures below 300°C. Hollow spheres several millimeters in diameter have been produced in this way, both in the crucible of a hot stage microscope as well as in a vertical oven. Apparently, the dissolved water and the carbon dioxide produced by thermal decomposition of alkali bicarbonates comprise the blowing agents that drive the formation of the sphere; the agglomerated hydrated silicate structure of this gel has a very high viscoelasticity compared to the rigid polymeric backbone of metal-organic-derived gels. However, if these gel particles are desiccated prior to pyrolysis, they behave in a manner identical to the hydrolyzed metal-organic xerogel powder.

Metal-organic-derived xerogels, because of their polymeric structure, are not as responsive to pyrolysis as aqueous silicate gels. Pyrolysis of xerogels that are completely hydrolyzed and saturated by exposure to water vapor causes opalescence in the translucent particle at temperatures as low as 150°C to 200°C. This opalescence, as noted above, is attributed to the formation of microvoids, i.e., a foam structure. Because of the rigidity of the polymeric gel structure, size expansion of the foam occurs gradually at temperatures of 250°C to 400°C. In this temperature range, the viscosity of the surface of simple hydrolyzed metal-organic-derived sodium silicate xerogels becomes low enough for surface tension to cause "flow" and formation of a crude spheroid. For unhydrolyzed pristine xerogels in this temperature regime, the polymer structure remains quite rigid. Significant viscoelastic response to the internal stresses generated by the gaseous pyrolysates does not begin until a temperature of 400°C to 600°C is reached, when the gel particle begins to swell, forming a closed cell foam 75 to 120% larger in volume but with 25% less mass than the original particle. However, the viscosity of the structure does not drop low enough to permit the surface tension to cause viscous flow in the "glass" until temperatures of 600°C to 900°C are reached, whereupon the foam structure either collapses or forms a crude, carbon-loaded spheroid.

Further insight into the high-temperature phases in the gel transformation process was gained from intermediates recovered from shell-forming experiments in a vertical furnace.[40] In these experiments we employed metal-organic-derived xerogel powder pressed into cylindrical pellets. Because of their uniform mass and size, these pellets facilitated the quantitative analysis of the changes that occurred in the transformation of the gel to shells.

The intermediates recovered from these experiments ranged from partly-expanded gel pellets to crude, carbon-laden spheroids. These samples, shown in the series of SEM photomicrographs of Figure 16, clearly illustrate the various morphological phases in the gel transformation. The formation of microvoids within the individual grains of compacted gel powder, shown in Figure 16a, causes the slight expansion of the individual grains comprising the pellet as a consequence of the foaming, and gives the pellet a "popcorn" surface texture. The incipient spheroidization is indicated by the extensive surface flow obvious in Figure 16b; the core of the pellet shows evidence of sintering and foam expansion. Figures 16c and 16d show the progressing spheroidization of the pellet and the coalescence of the internal void structure, while Figure 16e shows the crude shell, the product of the process. The last phase, not pictured here, is the fining and further expansion of the crude shell, driven by the oxidation of the carbon inclusions and the diffusion of the gas bubbles out of the wall.

The limiting step in the gel to shell transformation is the fining of the crude shell. Mathematical modelling of the heat transfer from the furnace to the gel indicates that in a 4-meter vertical furnace the heat transfer rate is not a limiting factor for gel particles of ⩽200 μm. The model predicted, and empirical tests have corroborated, a temperature rise of 1000°C to 3000°C/sec for a gel mass in the range of 20 to 200 μm.[40] The temperature of a 20 μg gel pellet, sufficient to make a 300 to 500 μm shell, is predicted to reach the equilibrium furnace temperature in less than one-half second after injection of the pellet into the

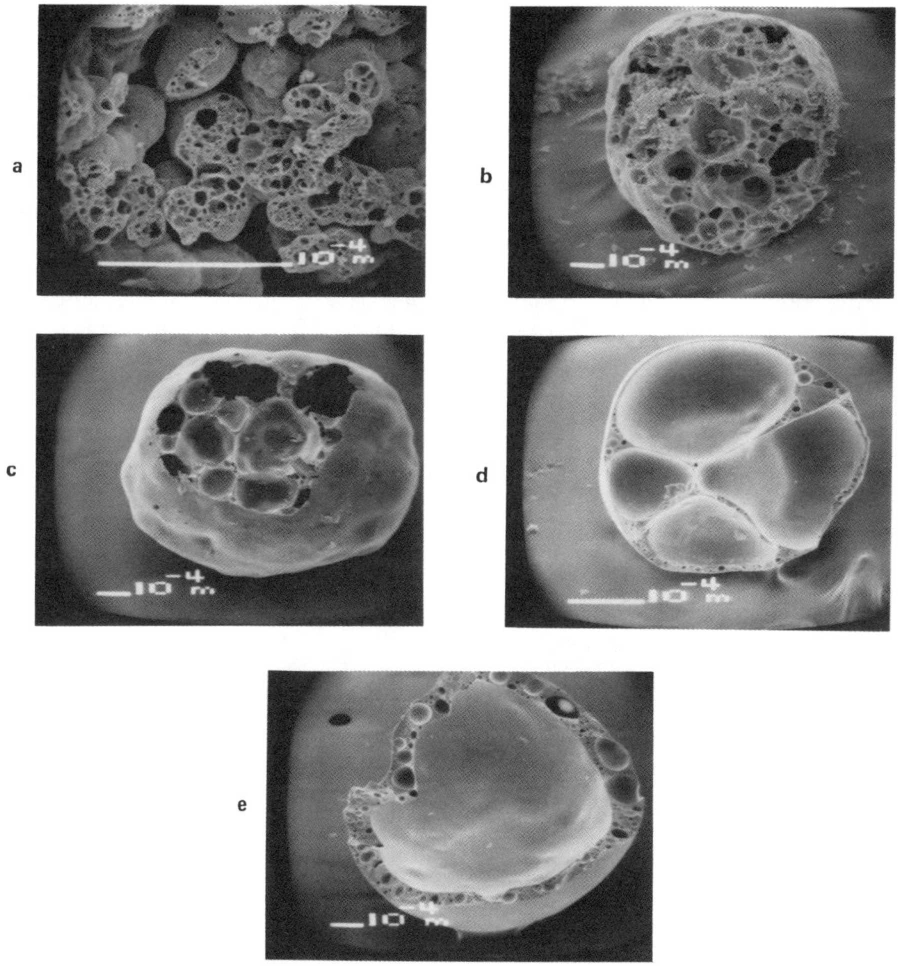

Figure 16: Intermediate stages in the physical transformation of pelletized metal-organic-derived xerogel particles into glass shells. Samples were generated from xerogel powder pressed into a cylindrical pellet, and fired in a vertical tube furnace. (a) Formation of an internal closed-cell foam structure (microvoids). (b) Internal foam expansion and initial phase of spheroidization, as the decomposing gel structure becomes more compliant to stresses. (c) Advanced stage of spheroidization and expansion of the internal foam structure. (d) Coalescence of the internal voids. (e) Crude hollow shell, with intrawall microvoids and carbon inclusions.

furnace. This estimate was corroborated visually by timing the development of equilibrium incandescence by the falling gel pellet. While the heating of the gel is extremely rapid, the chemical and viscous responses within the gel to that temperature rise are somewhat slower. However the limiting factor in shell production is the fining time required to produce a flawless glass shell from a crude, carbon-loaded hollow sphere. This was clearly shown in a set of experiments wherein the number of hot zones (1300°C) in the vertical furnace were incrementally brought on line to increase the time at temperature for a 20 μg gel pellet in a controlled fashion. The calculated rise time to 1000°C was 0.2 second for such pellets in that environment. An additional 1 second at 1000°C to 1200°C, equivalent to the drop-time through the first hot unit of the furnace, was required to complete the physical transformation from a cylindrical pellet to a crude hollow sphere such as shown in Figure 16e. Incrementally increasing the heating time at 1300°C, by bringing more 1300°C units on-line, resulted in shells with steadily increasing diameters, as shown in Figure 17, and steadily decreasing carbon inclusions. Thus, after the formation of a crude hollow sphere within 1 second after injection of the gel pellet into the furnace, a fining time of 5 seconds at 1300°C was required to produce an acceptable shell.

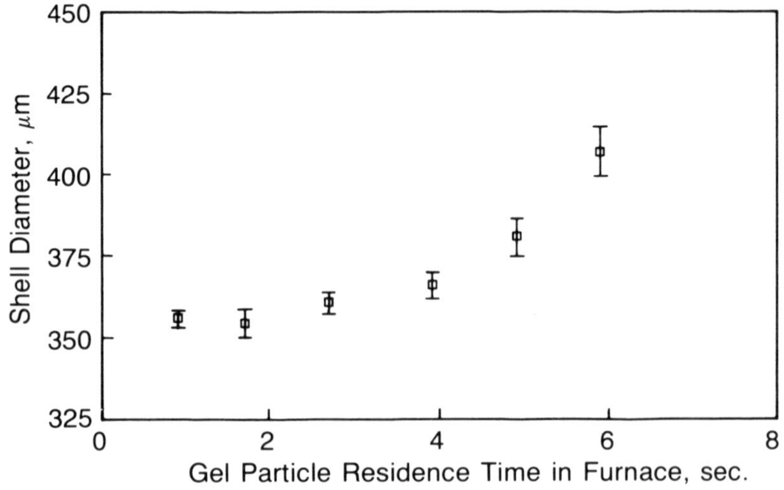

Figure 17: The effect of residence time of the xerogel in the vertical furnace at 1300°C on the final shell diameter.

Analyses of the residual gases in shells have indicated that the primary blowing gas involved in the final stages of shell formation is CO_2. This gas is produced both by oxidation of the residual carbon and by the decomposition of the alkali carbonates in the metal-organic-derived xerogels. Water, which is highly permeable in glass at 1200°C to 1500°C, is a significant component of the residual gases only in shells made in a furnace ambience saturated with water vapor. Molecular oxygen and nitrogen are secondary but significant constituents, together comprising 10 to 40% of the residual gases. Other possible gases, such as carbon monoxide or the nitrogen oxides, were not detected by gas chromatography or mass spectrometry.[45]

The total pressure of the residual gases in shells made in a dry furnace ambience is typically 0.17 to 0.20 atmosphere at 25°C (Table 8). On the basis of gas law and surface tension considerations, this is the expected pressure range for shells that had equilibrated at one atmosphere at 1200°C to 1500°C, and were subsequently cooled to 25°C. The residual pressures in soda-lime shells, however, are consistently and inexplicably somewhat high, i.e., 0.23 atmosphere.

Residual water vapor is detected only in those shells made in a furnace ambience saturated with water vapor. The water vapor in such shells usually comprises 30% of the residual gases, a concentration in equilibrium with an undetermined amount of water adsorbed to the internal surface and dissolved in the glass. (In very large shells, typically with diameters greater than 800 μm, microscopic pools of liquid water have been detected.) As shown in Table 9, the total residual gas pressure and the CO_2 content are considerably less than that found in shells made in ambient air. The reduction in total CO_2 content is caused by the reaction of the CO_2 with the alkali liberated from the glass matrix by ion exchange with the encapsulated water (after shell manufacture), and not from an increased rate of CO_2 diffusion out of the glass shell during its formation. In fact, rapid corrosion of the internal surfaces of the shells, resulting in the formation of alkali and calcium carbonates, is an inevitable result when alkaliborosilicate and alkali-lime glass shells are made in a furnace ambience with a high water vapor pressure. It does not occur in shells made in a dry furnace ambience.

Further evidence of the ease with which water vapor permeates the glass wall at high temperature was provided by repassing shells made in a water vapor saturated environment through a furnace tube containing ambient air. Analysis of the residual gases in the repassed shells gave a total pressure of 0.23 atmosphere and a concentration of CO_2 comparable to that found in shells made in ambient air. No water was found in the residual gases. This facile transport of water vapor has been used as a method of controlling shell aspect ratio.

Empirical Predictive Model for Shell Production

A predictive model for shell production is highly desirable, for each set of ICF experiments requires shells with unique and diverse geometric specifications. Ideally, the predictive process model would be a physical model with well-defined heat transfer and hydrodynamic components, one that quantitatively describes the chemical and physical changes that occur in the gel to shell transformation as a function of gel and furnace parameters. Because such a model has not yet been developed we constructed an empirical process model that successfully predicts the gel and furnace parameters required for the desired shell properties.[46] The model consists of a set of polynomial equations that describe each measured shell characteristic as a function of gel and process parameters. This approach assumes that each empirically-determined polynomial function adequately approximates the curvature of the real but unknown response function (i.e., the physical model). Consequently, the model is quantitative only over narrow, carefully determined ranges of values for the process parameters.

The empirical process model is developed using two experimental statistical methods, screening experiments and response surface experiments. Screening experiments are used to explore qualitatively but statistically a large number of potential experimental parameters and to identify those parameters that significantly affect the products' properties. The method will also identify those

Table 8: Residual Gases in Shells of Several Compositions

Shell glass system[a]	Relative concentration volume percent, by GC CO_2	 O_2	 N_2	Residual Gas Pressure (atm)
Na, Si (unhydrolyzed)	71	19	10	0.19
Na, Si (hydrolyzed)	80	13	7	0.175
Na, K, B, Ca, Si (unhydrolyzed)	63	16	21	0.23
Na, K, B, Si (unhydrolyzed)	88	8	4	0.20

a) The metal oxide glasses were prepared from dried metal-organic gels which were either subsequently kept dry (unhydrolyzed) or exposed to 80% RH for about 5 days (hydrolyzed). Shells were blown in a dry air environment.

Table 9: Residual Gases in Shell Exposed to Water Vapor During Shell Blowing

Shell glass system	Shell blowing environment	Relative concentration volume percent, by GC				Pressure (atm)
		CO_2	O_2	N_2	H_2O	
Na, K, B, Ca, Si	Dry air	63	16	21	0	0.23
Na, K, B, Ca, Si	H_2O (g) and air	21	15	31	33	0.096
Na, K, B, Si	Dry air	88	8	4	0	0.20
Na, K, B, Si	H_2O (g) and air	36	17	16	31	0.031

parameters that act synergistically. Response surface experiments are used subsequently to determine quantitatively the effects of these significant parameters on the final shells.

Screening Experiments. The motivation for using screening experiments is to test a large number of parameters with a minimum number of statistically designed experiments. We have chosen to use the Plackett-Burman experimental screening design, a carefully balanced subset of a two level factorial design[47] to identify efficiently and reliably the significant process parameters. In a typical application of this method for a given gel, only 20 experiments were required to screen 14 possible parameters; six of the parameters were found with significant effects on the shells (Table 10).

The sol-gel process for glass shell production has many potentially significant parameters that must be considered in the screening. These include those related to gel synthesis (gel composition and preparative techniques), gel treatments (gel hydrolysis and pyrolysis), and shell blowing (furnace temperature, temperature profile, gel size, shot size, water vapor in the furnace ambience, etc.). The gel synthesis parameters are too numerous to screen comprehensively, for each composition may be prepared by several different techniques. Therefore, our practice is to model the process with gel synthesis as a fixed constant, i.e., to identify empirically and to control for each specific gel those parameters related only to gel treatment and furnace conditions that are critical to shell production.

Response Surface Experiments. Response surface experiments are used to establish with a minimum of experiments a quantitative, statistically-valid relationship between the significant process parameters and final shell properties. This relationship is a polynomial function that approximates the real response of the system over a limited range of values for the independent variables, and is generated by a least squares fit to the empirically determined data:

$$y = b_0 + \sum_{j=1}^{P} b_j x_j + \sum_{\substack{j=1 \\ j' > j}}^{P} b_{jj'} x_j x_{j'} + \sum_{j=1}^{P} b_{jj} x_j$$

where

y = the shell property

x = a selected experimental parameter

b = a constant determined from the least squares fit

p = number of selected parameters

The empirical equations are used to:

(1) Predict optimum levels of the experimental parameters to produce glass shells.

Table 10: Screening Experiment Parameters

Gel Treatment Parameters	Shell Blowing Parameters
Exposure to 80% relative humidity	*Tower furnace temperature
*Exposure to 20psi H_2O at 109°C	*Gel size
Vacuum drying at 100°C	*Shot size
*Pyrolysis	Number of shots
	Shot interval
	Collector seal to process tube
	*Water vapor pressure in the furnace tube
	Stir bar type in beaker of water
	Stirring rate of the water

* Determined to be significant parameters

(2) Predict the range of shell properties resulting from varying the experimental parameters over their entire range.

(3) Examine the effect of each experimental parameter on the resulting shell properties.

For optimum results, a complete three (or more) level factorial experimental design should be performed. However the number of experiments needed, L^n, can be enormous (where L is the number of levels in the factorial and n is the number of selected parameters). The Box-Behnken[48] and the face-centered cube designs are well suited for glass shell response surface experiments. For instance a Box-Behnken design for five parameters requires 46 experiments. A full three-level factorial design requires 243 experiments.

In principle, all of the significant parameters identified by the screening experiments should be evaluated quantitatively by the response surface experiments. Although the response surface experiments are statistically designed, the number of required experiments increases rapidly with the number of parameters to be examined. The six significant parameters identified in the screening experiments (Table 10) require a prohibitively large number of experiments, even with a Box-Behnken design. In practice, however, not all significant parameters need be examined as part of the response surface experiments. Those parameters that have a pronounced negative effect do not need to be examined further, and are held constant at an optimum value. Since good shell uniformity is one of the most important and most difficult specifications to achieve, those parameters with detrimental effects on uniformity, i.e., gel autoclaving, gel pyrolysis and furnace temperature, were fixed. (Gels were not autoclaved or pyrolyzed; maximum furnace temperatures were used.)

In our usual procedure for developing a process model for each individual gel batch, only the three parameters of gel size, shot size and water vapor pressure in the furnace ambience are tested, using a Box-Behnken design with 15 experiments. For each shell characteristic, a polynomial function is developed, with appropriate adjustments of the constants to give a least squares fit to the data, to describe the shell characteristic as a function of the three independent process variables. The three functions can be solved simultaneously to generate predictive contour plots such as shown in Figure 18, which indicate the appropriate conditions required for making shell batches with characteristics shown. These functions also yield performance maps, i.e., contour plots that show (predict) the possible combinations of critical shell characteristics such as diameter, wall thickness and uniformity, that are attainable within the limits of our operating parameters. An example of this is shown in Figure 19. Such a contour plot can be used to determine whether a specific set of shell characteristics is attainable at all with a given gel, without regard to specific parameters. Thus, although these functions are not exact theoretical models of the physical process, they do provide a quantitative insight into the empirical relationship between the process parameters and the shell properties; each of these relationships, i.e., a set of solutions for each polynomial function, is shown in Figure 20.

The utility of these process models rests in the validity of their predictions. The accuracy of the predictions that can be achieved, in responding to requests of ICF glass fuel containers, is shown in Table 11. This indicates ex-

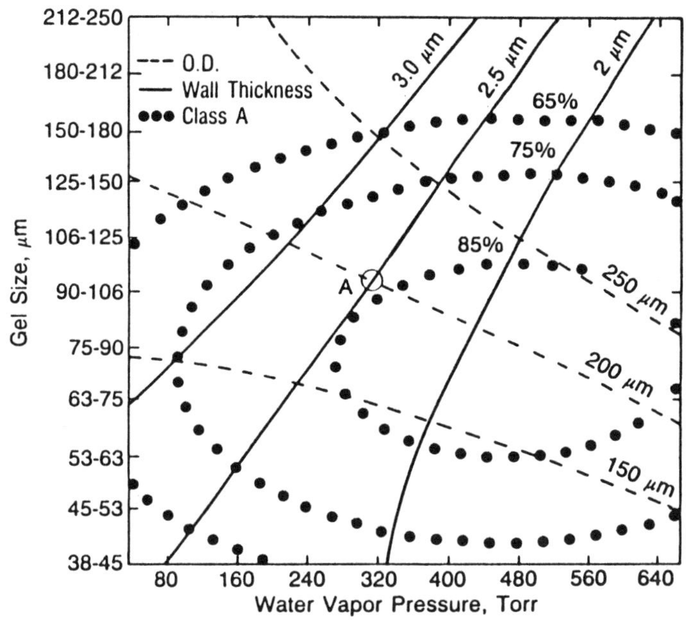

Figure 18: Constant value contours for diameter, wall thickness, and percent Class A, calculated for the batch mean from the empirical model. The point A corresponds to a mean diameter, wall thickness, and percent Class A of 200 μm, 2.5 μm, and 81% respectively, predicted from the use of 90 to 106 μm gel particles and 320 torr water vapor pressure.

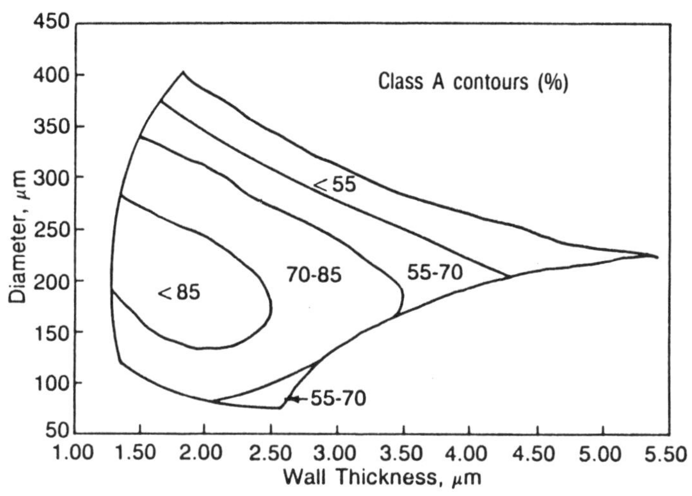

Figure 19: The range of mean values for diameter, wall thickness, and percent Class A shells calculated from an empirical model. The model predicts that any shell batch with a mean diameter and wall thickness that lies within figure boundaries can be prepared. The percentage of shells in batch with Class A uniformity is given by the contours.

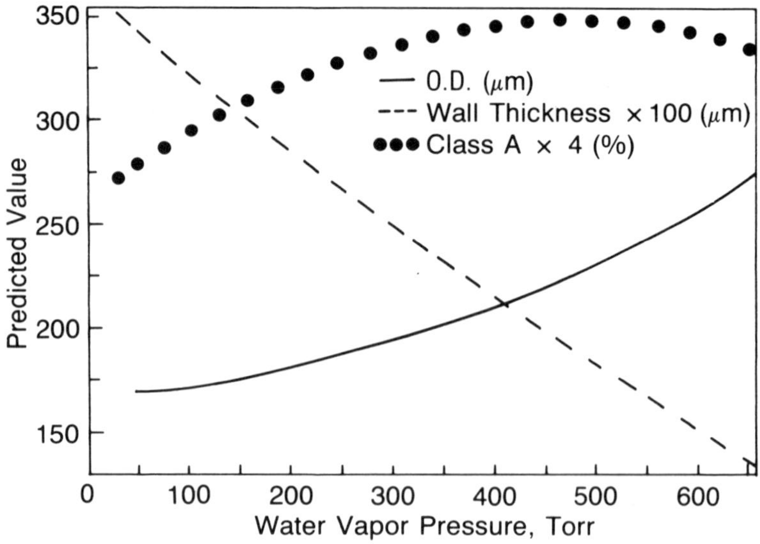

Figure 20: The effect of water vapor pressure in the furnace ambience on the resulting mean values for shell diameter, wall thickness, and uniformity.

Table 11: Model Predictions of Shell Properties Compared to Actual Results

Predicted			Actual		
			Sieved		
o.d. (μm)	Wall (μm)	Class A	o.d. (μm)	Wall (μm)	Class A
135	3.0	70%	90-125	2.9	75%-80%
150	1.3	76%	120-160	1.1	85%-90%
157	2.1	87%	125-175	2.1	>90%
161	2.8	81%	125-175	2.4	85%
164	3.3	77%	125-175	3.6	90%
171	1.4	83%	150-200	1.2	80%
180	2.9	79%	150-200	2.7	75%-80%
180	2.2	87%	175-225	2.2	85%
220	1.6	88%	200-250	1.8	>80%
214	2.7	81%	200-250	2.7	>75%
268	1.8	82%	250-300	1.7	85%-90%

cellent agreement between the predicted shell characteristics and the actual shell characteristics.

Compositional Experiments. Gel (glass) composition is important even though not included in the empirical equations. For instance response surface experiments on gels producing glasses with differing compositions have resulted in different ranges of glass shell dimensions (Figure 21). Only recently have we begun systematic experiments examining differences in gel glass compositions.

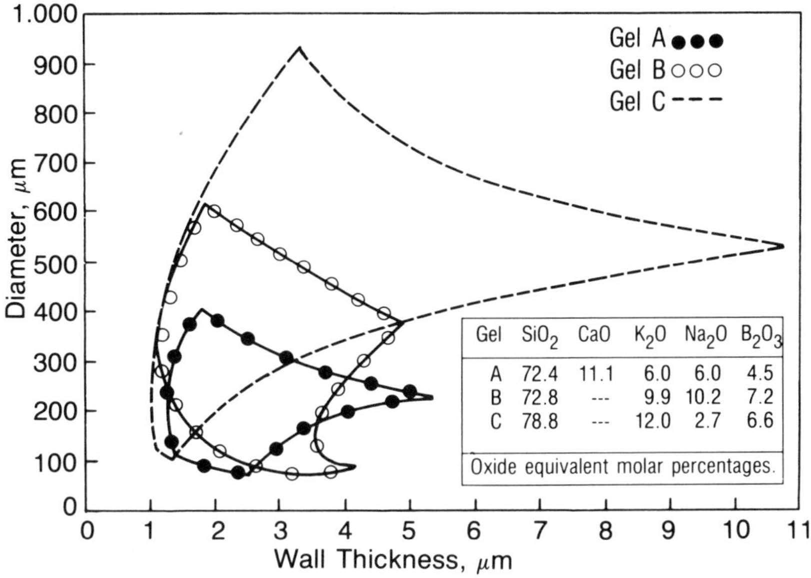

Figure 21: The outlines show the range of the computed mean values for diameter and wall thickness of glass shells which can be prepared from metal-organic-derived gels, A, B and C.

In our most recent experimental series we examined metal-organic-derived alkali-borosilicate gels with a composition of 78.5% SiO_2, 15% Alkali Oxide, and 6.5% B_2O_3 (oxide equivalent molar percentages) where the alkali oxide is divided between the oxides of sodium, potassium, and rubidium. An optimum alkali-to-alkali ratio is being sought to produce glass shells with the highest possible degree of uniformity.

Seven gels were prepared spanning the full ranges of alkali ratios (shown by the circles in Figure 22). Three more gels are being prepared to cover the gap in the ratios of the first seven gels (shown by the X's in Figure 22). All synthesis parameters were kept constant. The gels were blown into shells under a broad range of experimental parameters to ensure a wide range of shell diameters and shell wall thicknesses.

Shell batches were grouped according to the experimental conditions used. Contour plots showing gel (glass) composition and predicted shell properties were calculated. The contours were not a least squares fit since in most cases only seven shell batches were in a group and a seven parameter expression was fit to the results [Scheffe canonical form of the special cubic equation.][49] These semi-quantitative contours suffice for the qualitative comparisons required to find the best overall gel compositions.

We compared the contours over the broad range of experimental conditions to determine which gel compositions produce the most concentric shells. Figure 23 shows a contour plot. Clearly the rubidium-enriched gels are superior to the others. In future work we plan to look at other concentrations of alkali and explore the effect of cesium on shell properties.

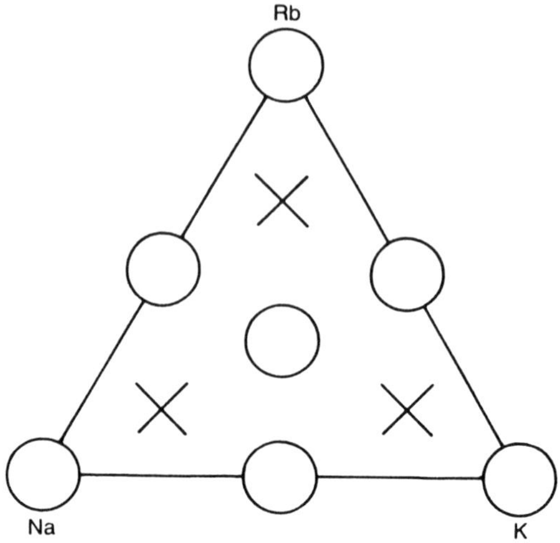

Figure 22: Compositional diagram showing alkali ratios of gels prepared with the same base composition of 78.5% SiO_2, 15% Alkali Oxide, and 6.5% B_2O_3 (oxide equivalent molar percentages). The x's represent compositions of gels planned to complete the study.

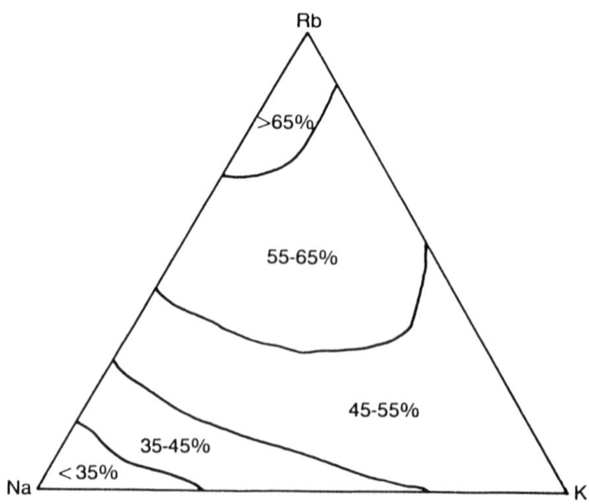

Figure 23: The contour plot has been computer generated from empirical equations relating alkali ratios in the gel to the percentage of Class A shells expected to be in the resulting shell batch.

REFERENCES

1. Ryan, R. and Tardy, B., Hollow Spherical Fillers, *Handbook of Fillers and Reinforcements for Plastics,* H.S. Katy and J.V. Milewski (Eds.), 317-330, Van Nostrand Reinhold Co. (1978).
2. Smiley, L.H., *Mech. Eng.,* 27 (Feb. 1986).
3. Wehrenberg, R.H., *Mech. Eng.,* 58 (Oct. 1978).
4. Nuckolls, J.H., Wood, L.L., Thiessen, A.R. and Zimmerman, G.B., *Nature (London), 239,* 139 (1972). See also Emmett, J.L., Nuckolls, H.H. and Wood, L.L., *Sci. Am. 230,* 24 (1974).
5. Charatis, G., Downward, J., Goforth, R., Guscott, B., Henderson, T., Hildum, S., Johnson, R., Moncur, K., Leonard, T., Mayer, F., Segall, S., Siebert, L., Solomon, D., Thomas, C., Experimental Study of Laser-Driven Compression of Spherical Glass Shells, Plasma Physics and Controlled Nuclear Fusion Research, Proc. Int. Conf., Tokyo, 1974, IAEA, Vienna (1975).
6. Souers, P.C., Tsugawa, R.T. and Stone, R.R., Fabrication of the Glass Microballoon Laser Target, Lawrence Livermore National Laboratory, Report UCRL-51609 (1974). See also by same authors, *Rev. Sci. Instrum. 46,* 682-685 (1975).
7. Draper, V.F., Production of Hollow Glass Microspheres for Laser Fusion Targets, Lawrence Livermore National Laboratory, Report UCRL-85252 (1980).
8. KMS Fusion, Inc., Annual Technical Report, KMS Fusion, Inc., Ann Arbor, MI, 1-44 to 1-58 (1980).
9. Netting, D.I.; U.S. Patents 3,794,503; Feb. 26, 1974; 3,796,777; Mar. 12, 1974; 3,888,957; June 10, 1975; assigned to the Philadelphia Quartz Company; and 4,340,642; June 20, 1982; assigned to PQ Corporation.
10. Walsh, R.J.; U.S. Patent 3,161,468; December 15, 1964; assigned to Monsanto Company.
11. Baer, K., Sperber, H., Boehre, O. and Leibner, G.; U.S. Patent 3,347,798; October 7, 1967; assigned to BASF.
12. Sowman, H.G.; U.S. Patent 4,349,456; September 14, 1982; assigned to Minnesota Mining and Manufacturing Company.
13. Torobin, L.B.; U.S. Patent 4,303,732; December 1, 1981; assigned to L.B. Torobin.
14. Coxe, E.F.; U.S. Patent 3,607,169; September 21, 1971; assigned to Esso Research and Engineering Company.
15. Johnson, W.L. and Lee, M.C., *J. Vac. Sci. Technol.,* Al(3), 1568 (1983).
16. Lee, M.C., Kendall, J.M., Bahrami, P.A. and Wang, T.G., *Aerospace America,* 72-76 (January 1986).
17. Veatch, F., Alford, H.E. and Croft, R.D.; U.S. Patent 3,030,215; April 17, 1962; assigned to The Standard Oil Company.
18. Henderson, C.; U.S. Patent 3,699,050; October 17, 1972; assigned to Emerson & Cuming, Inc.
19. Beck, W.R. and O'Brien, D.L.; U.S. Patent 3,365,315; January 23, 1968; assigned to Minnesota Mining and Manufacturing Company.
20. Veatch, F., et al.; U.S. Patent 2,797,201; June 25, 1957; assigned to The Standard Oil Company.

21. Schneider, J.M. and Hendricks, C.D., *Rev. Sci. Instrum., 35,* 1349 (1964).
22. Babil, S., Controlled Generation of Uniform Solid Aerosol Particles with Radii in the 0.5 to 20.0 micron Range, Ph.D. Thesis, Department of Electrical Engineering, University of Illinois, Urbana (1977).
23. Hendricks, C.D. and Babil, S., *J. Phys. E., 5,* 905 (1972).
24. Hendricks, C.D., Glass Spheres, *Glass Science and Technology, Vol. 2,* D.R. Ullmann and N.J. Kreidl (Eds.), 149-168, Academic Press (1984).
25. Hendricks, C.; U.S. Patent 4,133,854; January 9, 1979; assigned to The United States of America, Department of Energy.
26. Rosencwaig, A., Koo, C. and Dressler, J.; U.S. Patent 4,257,799; March 24, 1981; assigned to The United States of America, Department of Energy.
27. Hendricks, C.D. and Dressler, J.L., Production of Glass Balloons for Laser Targets, Lawrence Livermore National Laboratory, UCRL-78481 (1976).
28. Campbell, J.H., Grens, J.Z. and Poco, J.R., Preparation and Properties of Hollow Glass Microspheres for Use in Laser Fusion Experiments, Lawrence Livermore National Laboratory, UCRL-53516 (Nov. 1, 1983).
29. Budrick, R.G., King, F.T., Nolen, R.L. and Solomon, D.E.; U.S. Patent 4,021,253; May 3, 1977; assigned to KMS Fusion, Inc. See also: Budrick, R.P., King, F.T., Martin, A.J., Nolen, R.L. and Solomon, D.E.; U.S. Patent 4,017,290; April 12, 1977; assigned to KMS Fusion, Inc.
30. KMS Fusion, Inc., Annual Technical Report, KMS Fusion, Inc., Ann Arbor, MI, 1-1 to 1-24 (1977).
31. Nolen, R.L., Downs, R.L., Miller, W.J., Ebner, M.A., Doletzky, N.E. and Solomon, D.E., Fabrication of Glass Shells, Paper TuE1-1 in the Technical Digest on the Topical Meeting on Inertial Confinement Fusion, San Diego, Calif., Digest No. 78CH1310-2QEA (Optical Society of America), (Feb. 7-9, 1978).
32. Nogami, M., Moriya, Y., Hayakawa, J. and Komiyama, T., Yogyo-Kyokai-Shi, Fabrication of Hollow Glass Microspheres for Laser Fusion Targets from *Metal Alkoxides, Prep. of glass shells at Osaka University, 88,* 712 (1980).
33. Nogami, M., Hayakawa, J. and Moriya, Y., *J. Mat'l., Sci., 17,* 2845 (1982).
34. Clement, X., Coudeville, A., Eyharts, P.H., Perine, J.P. and Rouillard, R., Target Fabrication and Development in the Centre D'Etudes De Limeil, Fifth International Conference on Solid Surfaces, Madrid, Spain (Sept. 26-Oct. 1, 1983).
35. Weinstein, B.W., *J. Appl. Phys., 12:,* 5305-5306 (1975).
36. Henderson, T.M., Cielaszyk, D.E. and Simms, R.J., *Rev. Sci. Instrum., 48:,* 835-840 (1977).
37. Dislich, H., New Routes to Multicomponent Oxide Glasses, *Angew. Chem. Internat.,* Vol. 10, No. 6 (1971).
38. Levene, L. and Thomas, I.M.; U.S. Patent 3,640,093; February 8, 1973; assigned to Owens-Illinois, Inc.
39. Downs, R.L. and Miller, W.J.; U.S. Patent 4,336,338; June 22, 1982; assigned to The United States of America, Department of Energy.
40. Downs, R.L., Ebner, M.E. and Nolen, R.L., Glass Shell Manufacturing in Space, Final Report NAS8-33103, prepared for Marshall Space Flight Center (Dec. 21, 1981).

41. Downs, R.L. and Miller, W.J., Gel Precursors as Glass and Ceramic Starting Materials for Space Processing Applications Research, Final Report for JPL Contract No. 956403 under Jet Propulsion Laboratory, National Aeronautics and Space Administration Contract No. NAS 7-100 (Jan. 8, 1986).
42. Miller, W.J. and Ebner, M.A., Alumino-silicate and Calcia-aluminate Gel Preparation, KMSF Internal Report DMS-3888-WJM/MAE (Aug. 18, 1982).
43. O'Holleran, T.P., Downs, R.L. and Homyk, B.D., *J. Vac. Sci. Technol.* 18, (3), 1276 (1981).
44. Anderson, P.R. and Miller, W.J.; U.S. Patent 4,340,407; July 20, 1982; assigned to The United States of America, Department of Energy.
45. Downs, R.L., Ebner, M.A., Homyk, B.D. and Nolen, R.L., *J. Vac. Sci Technol.* 18(3), 1272-1275 (1981).
46. Belanger, R.P. and Miller, W.J., *J. Vac. Sci. Technol.* A 3 (3), 1270-1273 (1985).
47. Plackett, R.L. and Burman, J.P., *Biometrika 33,* 305 (1946).
48. Box, G.E.P. and Behnken, D.W., *Technometrics, 2,* 455 (1960).
49. Scheffe, H., *J. Royal Statistical Soc., B, 20,* 344-360 (1958).

17

Filters and Membranes by the Sol-Gel Process

Lisa C. Klein

Ceramics Department
Rutgers - The State University of New Jersey
Piscataway, New Jersey

INTRODUCTION TO FILTRATION

In the field of separations, where solids are separated from liquids, liquids from other liquids, particulates from gases or gases from other gases, the medium is selected primarily by its mechanism for separating. On the micron or greater scale, separations can be accomplished either by using a screen which separates by size or by using a centrifuge which separates by density. On the submicron scale, separations are accomplished using filters. For a review of nomenclature and mechanisms, see References 1 and 2, for example.

A filter is a porous medium. The proper filter is selected by its mechanism for separating. In many cases, the mechanism is size exclusion. The porosity of the filter is chosen because it allows some material to pass and it traps the rest. The porosity may be uniform, nonuniform, graded or assymetric, depending on the function of the filter. Microfiltration refers to size exclusion from 50 nm to 1 micron. Below nm there are reverse osmosis and dialysis filters which involve diffusivity controlled separation mechanisms. Filters in this range are often called membranes. Membrane implies some chemical reaction as well as size exclusion. In the case of oxide filters, surface reactions are important, so the use of the term membrane is appropriate.

Factors to consider, besides separation mechanism, are membrane permeability, stability and compatibility. These factors depend on size, charge and solubility of solute in the solvent or influent. In practice, a fluid is forced to flow through the filter by applying pressure on the influent or vacuum on the effluent, also known as filtrate.

The filter can consist of discrete particles such as a clay bed, fibrous material such as asbestos or porous material such as partially sintered compacts. From this list, it is evident that oxide materials are a major factor in filtration.

On the very fine scale, polymers and plastics are more common because the pore size in ceramics from conventional processing is not fine enough. For example, osmosis requires a semipermeable barrier for water. When two aqueous solutions of different concentration are separated by a semipermeable membrane flow will occur from concentrated to dilute solution to eliminate the gradient. The pressure developed is called the osmotic pressure. When a pressure is developed in the dilute solution greater than the osmotic pressure, the result is reverse osmosis. This process is effective for desalination. The pores in the reverse osmosis membrane are typically 2 nm. In the case of dialysis, the membrane is semipermeable to macromolecules. No pressure is needed for transport. When a voltage is applied across the membrane in the case of electrodialysis, electrolytes can be separated. As in the case of reverse osmosis, the membrane porosity is finer than that found in conventionally sintered ceramics.

In addition to the nature of the porosity in filters, there is the chemical nature of the surface. A surface has charge, negative or positive, known as its zeta potential. This provides electrostatic or electrokinetic effects for separation. For aqueous separation, the surface is either hydrophobic or hydrophilic. Hydrophobic membranes will trap water on the surface but not in their pores where hydrophilic membranes will condense water in their pores.

The thickness of the membrane is important. The separation may require a large surface or a thick medium depending on volume of flow or ease of cleaning. Clearly there are many areas where ceramic membranes can be used or are likely to be used such as air pollution, liquid stabilization, chemical clarification and generation of demineralized, deionized water.

FLUID FLOW EQUATIONS

There are two equations which will be described in relation to use of membranes. For a more complete treatment see Reference 3. The first equation is for the solvent flux J

$$J = V_p\, r^2\, \Delta P / 8\, n dx \qquad (1)$$

where V_p is pore volume, r is pore size, ΔP is pressure drop, n is viscosity and dx is thickness. This relates to the tortuous path that the solvent travels through interconnected porosity. The second equation is that for the bubble point pressure P

$$P = 2s \cos\theta / r \qquad (2)$$

where s is surface tension (solvent-membrane) and θ is the contact angle. A contact angle less than 40° is hydrophilic behavior and a contact angle of 90° or greater is hydrophobic behavior. The bubble point pressure is measured to determine the presence of leaks.

The factors in these equations which can be controlled are pore volume, pore size and thickness. By adjusting these factors during the synthesis of the membrane, the solvent flux, which depends on the pressure can be determined. The pressure which can be sustained by the membrane will regulate the choice of material.

PROCESSING CERAMIC MEMBRANES

Ceramic membranes offer an advantage over organic materials. That is ceramics are stable to higher temperatures. Ceramics do not decompose like plastics nor do they swell like plastics. Ceramics are harder and more resistant to abrasion. Still, the vast majority of membranes are plastics because of their ease of fabrication.[4]

There are three processes for obtaining ceramic membranes: chemical leaching, solid-state sintering and sol-gel processing. A few examples will be cited for each. The principle materials are alumina, silica and glass. The sol-gel process being the newest approach makes up the bulk of this chapter.

Chemical Leaching Approach

Chemical leaching refers to taking a glass which will phase separate and heat treating to develop interconnected phases.[5,6] One phase is removed by acid leaching, leaving behind interconnected porosity. In most cases, the glass is a sodium borosilicate where a sodium borate phase is removed leaving a silica rich phase. After leaching, the porous glass is heat treated for compaction. Two specific examples from the patent literature are described.

To prepare a reverse osmosis membrane,[7] an appropriate glass was leached. However, the leached glass was mechanically weak. The glass was treated with $TiCl_4$ in gas form. This was converted to TiO_2 by immersing the glass in water. Then the glass was partially sintered so that the pore volume was below 35% and the pore size was about 3 nm.

To prepare a hollow glass filament,[8] an appropriate glass was drawn into a 50 micron filament, phase separated and leached. The leached glass was then subjected to a plasma etching treatment containing fluorine. The plasma etching improved the uniformity of the pores. This process gave pores about 100 nm. To increase the pore size, the silich rich phase may be dissolved in strong alkaline solution, or the porosity may be altered by heat treating in a thermal gradient to create assymetric porosity.

Sintering Approach

Sintering refers to taking a compact of particles and turning the compact into a polycrystalline material. For a general review see Reference 9. The contacts between particles develop into necks joining particles. The void space in the compact is initially interconnected but may become isolated at 3 grain intersections or inside grains. Ideally, the pores remain at the grain boundaries. Occasionally, there are flaws in the compacts known as agglomerates which result in uneven consolidation.

Sintering is used to describe the progress of the compact towards full density and it involves mass transport in some form. The temperature for this process is typically 80% of the melting temperature.

The microstructure of the sintered compact depends on grain size, sintering agents and agglomerates. With respect to filters, a permeable microstructure is desired so complete densification is not the goal. To achieve porosity on the desired scale, colloidal particles are generally used. A specific example from the patent literature is described.

To prepare a catalyst support,[10] colloidal particles of alumina, silica, titania or zirconia were dispersed, coagulated, calcined and heat treated. The particles themselves were prepared by a flame hydrolysis method. The particles were spherical 4 to 50 nm and not hydrated. The particles were dispersed in water and then dried in air. Surface area was between 125 and 150 m^2/g, the mean pore diameter was 14 to 19 nm and 80% of the pores were 5 to 9 nm. During heat treatment the oxide particles were weakly consolidated.

Sol-Gel Approach

Sol-gel processing, of course, is the subject of this book. By the nature of the process, the resulting product is porous. Phase separation and leaching are not required, nor do individual particles have to be consolidated. Sol-gel processing for forming filters and membranes is relatively new. The few examples which exist will be described.

Silica. Two attempts to make silica filters by a sol-gel process have been reported. In one case, a fibrous mat was coated with a silica gel. Regions of microporous gel adhered to the substrate and modified the near-surface porosity. In another case, a gel solution was drawn through a slit to form sheets.[12] This was a step in the direction of forming unsupported membranes.

Related work has appeared on hybrid materials where a silica precursor has been mixed with a hydrolyzable silane.[13] This results in a material which shows inorganic and organic character and may be suitable for contact lenses.[14] The silica filters are noncrystalline, while alumina filters are crystalline.

Alumina. There are several reports specifically on making alumina filters by a sol-gel process. Lenaars, et al.[15,16] studied thin supported alumina layers and membranes with ultrafine pores. They prepared aluminum sec-butoxide (ASB) and H_2O (2/1 mol ASB) at 90°C and then added $HClO_4$ additions (0.05 to 0.11 mol) to form boehmite (AlOOH) gels. The solutions were boiled several hours, and then refluxed for about 16 hours. They observed no systematic change of crystallite size of calcined gels with acid addition, but pore size and porosity decreased with increasing acid. They also reported that the pore size distribution was not bimodal. A bimodal distribution had been expected for a solid structure which was aggregated. Since there was only one mode of pores present, they concluded that the aggregation of the boehmite sol, if present, broke up during drying as a result of capillary forces. Drying was accomplished by natural evaporation. These gels are termed xerogels.

They also postulated that there are plate shaped crystallites in the microstructure of the dried boehmite membranes. If they are plate like, a regular packed stacking without large pores can only be obtained in a so-called stacked card structure. They conclude that the structure is the result of gel composition due to capillary forces during H_2O extraction in the drying step. The average pore size was 2.7 nm and the porosity was 50%.

Teichner, et al.[17] prepared aerogels of alumina by hydrolysis of ASB in a

10% solution of sec-butanol in various amounts of H_2O. Aerogel refers to gels where the solvent is evacuated in an autoclave at the critical point. They reported that the best textural characteristics were achieved for a quantity of H_2O not exceeding 4.5 mols. Effect of concentration of ASB was such that up to 10% concentration, amorphous materials with high surface areas, high pore volumes and low apparent densities were observed. For concentrations greater than 10%, boehmite appeared.

They also tested aerogels and concluded that only aerogels exhibit a macropore volume of the order of 17 ml/g, while corresponding xerogels never exceed 2.4 ml/g. Vacuum treatment up to 600°C did not radically change the texture of the aerogels, whereas treatment in air at 600°C produced a large decrease in surface area.

CHARACTERIZING POROSITY

The most important properties of filter materials are their porosity and microstructure. Therefore, the subject of characterizing porosity is an important part of this chapter, especially with respect to gels.

The texture of gels is difficult to resolve with the electron microscope. Aerogels of very low bulk density can be observed,[18,19] but xerogels which are 50% dense present the experimental problem of looking through several layers at one time.[20] Generally, the texture of gels is probed with nitrogen sorption.[21] Nitrogen is used for microporosity (below 1.5 nm) and mesoporosity (below 50 nm). Mercury porosimetry is used for macroporosity (above 50 nm). The adsorption isotherm gives the surface area according to the BET equation. The desorption gives the pore size distribution from the Kelvin equation. Hysteresis in these curves is interpreted according to previously characterized materials to indicate texture. Little hysteresis is interpreted to mean uniform cross-section porosity. Presence of hysteresis is interpreted to mean narrow-neck or "ink bottle" pores.

Definitions

Specific surface area is a measure of the accessible surface area per unit mass of solid. This surface area is the sum of the internal surface area associated with pores and the external surface area of the particles. Specific porosity refers to the accessible pore void space in the particles per unit mass of solid. It includes the intergranular void space in agglomerates. Pore size distribution is the distribution of the pore volume versus the pore size. Generally, the measured pore size distribution does not give the exact volume of pores having a given size, but relates to the volume of the pores accessible through the pores of a given size. It thus depends on the shape of the pores present in the material.

In measuring surface areas and pore volumes, physical adsorption is involved. The forces between the solid (adsorbent) and the fluid (adsorbate) are van der Waals forces. The adsorbates commonly used are nitrogen, argon or krypton because they are inert and of sufficiently small size to enter the pores of the solid.

Characterization of the solid surface is obtained through determination of an isotherm. The adsorption isotherm is the function that relates, at constant temperature, the amount of substance adsorbed at equilibrium to the adsorptive pressure in the gas phase. The adsorption-desorption isotherm can be determined by a gravimetric method or, more commonly, a volumetric method.

The accuracy of results depends on the knowledge of the actual adsorption temperature. If nitrogen is the adsorbate, the adsorption isotherms are obtained at liquid nitrogen temperature (77 K). The temperature of adsorption is lower than the critical temperature of the adsorbate, so the adsorbed amount is represented as a function of the relative pressure, p/p_o, where p is the pressure of nitrogen in the gas phase, and p_o is the saturation pressure.

Calculation of surface areas and pore volumes requires use of the BET theory. The BET hypothesis is based on several assumptions: (i) adsorption is localized on well defined sites, each of which is the same energy and can only accommodate one adsorbate molecule; (ii) a multilayer adsorption occurs even at low pressure; (iii) an adsorption-desorption equilibrium is effective between molecules reaching and leaving the solid surface. Except for the first layer, the adsorbed molecules are supposed to be in the liquid state.

The limit of the BET equation is that it is normally able to describe a physical adsorption isotherm only in the relative pressure range between 0.05 and 0.35. Also, the equation itself describes only "S" shaped isotherms.

Nitrogen Sorption Analysis

Samples for nitrogen adsorption treatments were ground in an alumina mortar and pestle.[22] A sample size of 20 mg was used. Samples (-200 to +325 mesh) were weighed into the sample cell. The cell was then attached to the outgassing station of the Quantasorb™ Surface Area Analyser (Quantachrome Corp., Greenvale, NY). High purity dry nitrogen was flowed over the sample at a rate of 5 ml/min. Samples were outgassed for 12 hours at 200°C. Standard procedures recommended by the manufacturer for obtaining nitrogen adsorption-desorption isotherms were used. The volume of pores with radius less than 50 nm was determined by first adsorbing pure nitrogen then switching to a 98% N_2–2% He mixture at 77 K. As the sample was heated to room temperature, the desorption signal was integrated and calibrated. Isotherms were obtained with ten prepared mixtures of helium and nitrogen. Surface area was estimated from the isotherms.

As an alternative to the static volumetric method, a dynamic system was used.[23] The Omnisorp 360™ (Omicron Technology, Berkeley Heights, NJ) is a continuous volumetric unit. Pure N_2 is used and pressure is varied at a pressure manifold, so p/p_o could be continuously varied between about 0.05 and 1.0. Adsorption/desorption isotherms were obtained for the samples from the Omnisorp unit. Outgassing of samples was performed under vacuum at 200°C for five hours. This was better than outgassing under flowing nitrogen. Due to the nature of the equipment, sample weights used on the Omnisorp were about 0.3 g, so error due to weighing was reduced. From the adsorption isotherm and application of the BET equation, surface areas were determined. With the Omnisorp software, the linear portion of the BET equation was selected, so calculation of surface areas was more accurate than when it was assumed that the BET equation was linear over the whole range, p/p_o = 0 to 0.30, as in the case of the Quantasorb unit.

From the desorption isotherms, pore volumes, pore size distributions and average pore sizes were obtained for all samples. A typical nitrogen adsorption curve for dilute alumina gel samples is shown in Figure 1. The adsorption isotherm was used to estimate surface area and pore fraction.

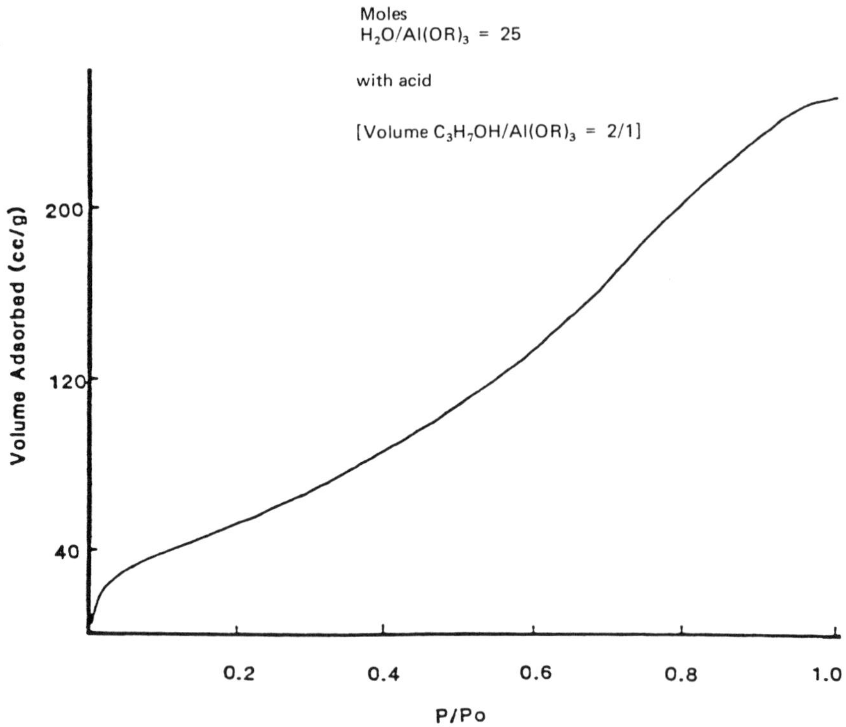

Figure 1: Typical adsorption isotherm for dilute alumina gels using Omnisorb 360™.

Texture of Microporous Silica

There are remarkable differences in appearance in newly gelled solutions starting from tetraethylorthosilicate (TEOS).[22] In acid-catalyzed solutions, the first hydrolysis of the TEOS monomer is easier than the second, so a growing polymer will have an even distribution of hydroxyls. The polymer can form an occasional crosslink by a condensation reaction, and any structural features which might scatter light remain too small during gelling or drying. Acid-catalyzed gels are transparent.

In base-catalyzed solutions, the successive hydrolysis of the TEOS monomer becomes easier, so that condensed polymers co-exist with unreacted monomer. Structural features develop in the overall polymer network that scatter light, both in solution while gelling and in the drying sediment.[24] In general, base-catalyzed solutions are cloudy. All samples show a tendency to sediment. When base-catalyzed gels are dried, the result is weakly coalesced powder.

In acid-catalyzed solutions, when the water to TEOS ratio was increased from 4 to 16 there was an increase in porosity from 10 to 40%. The trend in the surface area was an increase from 20 to 640 m^2/g with increasing water. A lack of hysteresis indicated that the pores were cylindrical. The cylinder radius was calculated from the Kelvin equation. A rough estimate for the average pore radius is 1.5 nm for 4 or more mols water.

For acid-catalyzed solutions the volume of liquid nitrogen per gram gel (V_s) increased for all nitrogen partial pressures as the water level increased. The pore volume was read directly from the isotherm at the point corresponding to 98% N_2. The single point BET value was obtained using the V_s value corresponding to 10% N_2. Each isotherm had an inflection point at intermediate values of nitrogen partial pressure which became more pronounced with increased water. The plot of the isotherm approached the y-axis at right angles at high nitrogen partial pressures for acid-catalyzed solutions but not for base-catalyzed solutions.

The isotherms for base-catalyzed gels showed increasing hysteresis at high nitrogen partial pressures with increasing water level.[25] This implies nonuniform cross-section pores. Using the desorption portion of the isotherm, the cumulative pore volume was calculated. The calculation was normalized to 50 nm. Normalizing the cumulative porosity to 50 nm is a fair assumption for acid-catalyzed gels. For base-catalyzed gels, the representation of cumulative porosity between 1 and 50 nm is not accurate. Acid-catalyzed gels are transparent, and base-catalyzed gels scatter light to produce opacity.

When comparing acid versus base, not only are the mechanisms different, the kinetics are different. For acid-catalysis, hydrolysis is complete and the number of unreacted hydroxyls per silicon decreases with decreasing acid concentration. For base-catalysis, the effect of polymerization is large.

For the moment, direct evidence of the size of features from techniques such as electron microscopy are rare. The microstructure of dried gels has to be inferred from surface area and pore volume measurements. From this, it is possible to suggest that acid-catalyzed gels have uniform interconnected porosity. Base-catalyzed gels have pores within clusters and between clusters making up the sediment. Since silicic acid is more soluble in basic solutions, the surface area decreases over time with increasing base and increasing water. With this increased solubility, silica dissolves and reattaches far more easily in basic than acidic aqueous medium.

The adsorption-desorption isotherms can be used to plot a pore size distribution. The result for acid-catalyzed gels is a narrow distribution, with the width and mode of the peak decreasing in radius and the height increasing as the water level is increased. The absence of hysteresis in the isotherm indicates cylindrical pores. The result for base-catalyzed gels is a broad distribution at low water levels which becomes bimodal at intermediate water levels. The distribution peaks are located at 1.5 and 5 nm. The 5 nm peak increases and the 1.5 nm peak decreases as the water level is increased further.

Texture of Microporous Alumina

The difference in pore volumes between catalyzed and uncatalyzed alumina gels is greater at high water levels than at low water levels. The effect of acid is greater in dilute solutions.[23] In general, the textural trends in silica gels are reversed in alumina gels.

The most useful information obtained about the texture of alumina samples comes from examining the pore size distribution. Pore size distribution is obtained by calculating the derivative of pore volume/derivative of pore radius, and plotting that versus pore radius. Figure 2 shows pore size distribution plots for samples with 25 mols H_2O/ASB, with and without acid.

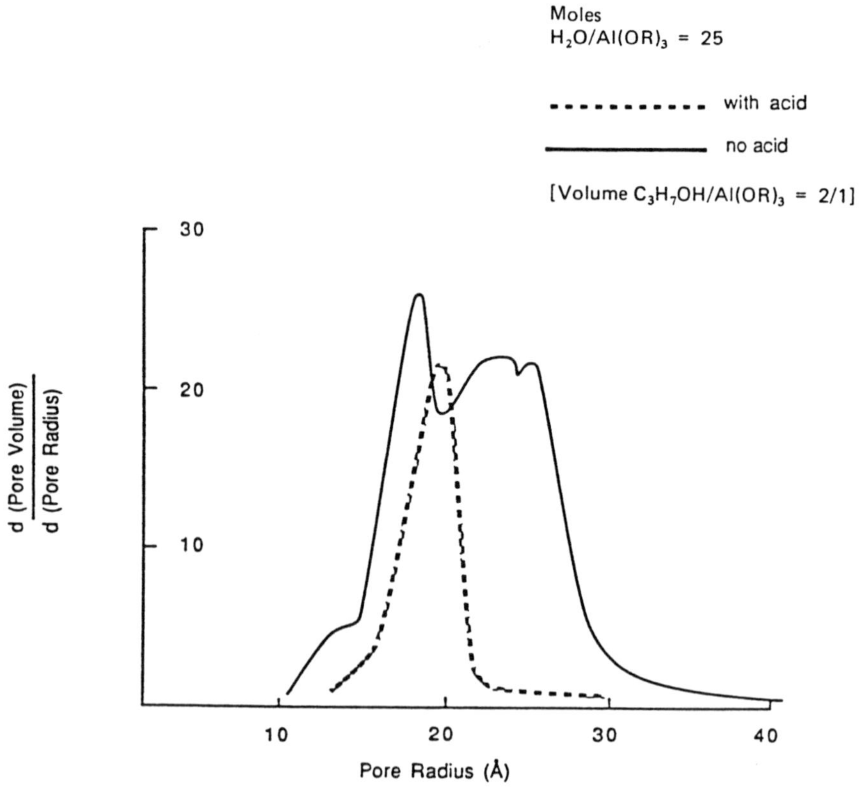

Figure 2: Pore size distribution for dilute alumina gel prepared with and without acid.

From comparing compositions with and without acids for a given molar ratio of H_2O/ASB, it is observed that acid addition leads to a narrower pore size distribution. Also, acid addition shifts the pore size distribution to smaller pore sizes. These results suggest that acid addition serves to tighten the gel structure and form smaller, more uniform pores.

It is also found that low water gels have a broader pore size distribution than high water gels. Gels with greater than 25 mols H_2O/ASB are more uniform with respect to pore structure than gels with less than 10 mols H_2O/ASB. Only for gels with greater than 25 mols H_2O/ASB are there very narrow pore size distributions with acid catalyst. Low water, acid-catalyzed gels do not have very narrow distributions.

Figure 3 is a plot of average pore radius versus composition. Average pore radius is obtained from the maximum in the pore size distribution plots. The average pore radius for compositions with greater than 25 mols H_2O/ASB is essentially constant. A maximum is observed in average pore radius at 10 mols H_2O/ASB. It should be noted also that at this composition, a broad pore size distribution and nonuniform pore structure is observed.

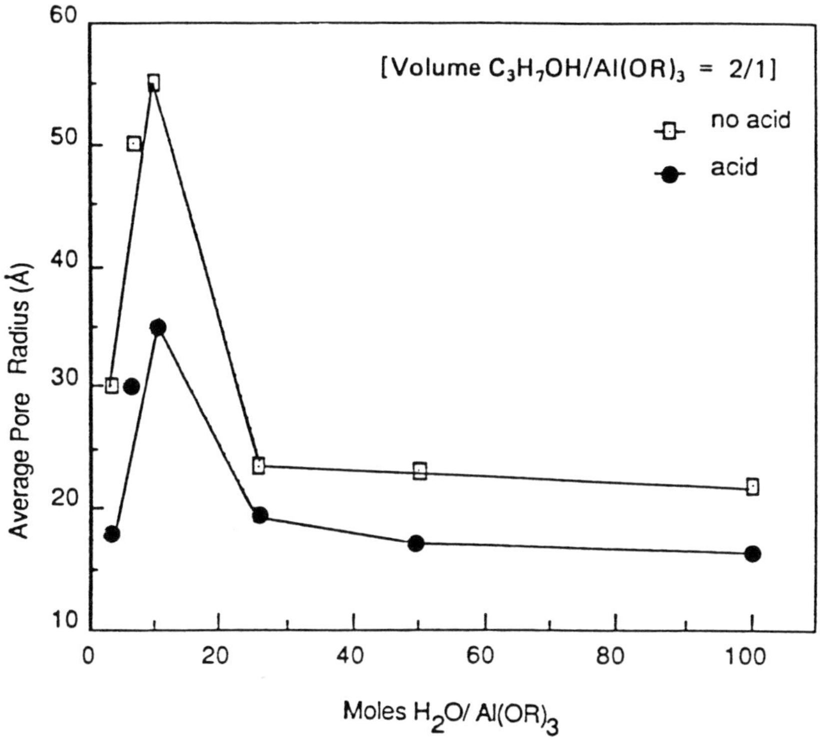

Figure 3: Average pore radius versus mol water per mol aluminum sec-butoxide prepared with and without acid.

It may appear from Figure 3 that a composition with 3 mols H_2O/ASB and acid has a similar pore structure to a composition with greater than 25 mols H_2O/ASB and acid. Although the average pore radius is similar, the pore size distribution at 3 mols H_2O/ASB is much broader and less uniform than that at greater than 25 mols. Apparently, both excess water and acid addition contribute to the formation of uniform structures with small pores.

In order to explain the textural data, a schematic of the gel structures is proposed. In high water, acid-catalyzed gels, there are presumably active end members and hydroxyls on the polymers in solution. These can then grow and crosslink to form a tight, 3-dimensional network. Pore volumes are low, average pore radius is small, and pore size distribution is very narrow. Surface areas are low due to a small pore volume.

In high water, uncatalyzed gels, crosslinking occurs, but the network is not as tight. Bulk densities are lower, pore volumes are approximately a factor of five higher, and pore radius is increased. Although pore radius is larger than in the acid-catalyzed gels, the fact that pore volume is much larger causes the surface areas to be larger.

Thus, high water gels result in crosslinked structures that dry to a relatively dense piece. Porosity is uniform and pores are small. Effect of acid is seen on the amount of crosslinking and formation of a tighter structure with much lower pore volume.

On the other hand, in the low water gels, less crosslinking occurs. There is not a lot of excess water, so the polymer ends are not as active. Polymers can link over short ranges and form clusters. There is less long range, 3-dimensional crosslinking. Thus, these gels dry to flaky powders, not consolidated pieces. Bulk density is very low. Porosity probably consists of intergranular porosity as well as intragranular porosity, whereas in high water gels the porosity is probably more uniform throughout the network structure.

Effect of acid on crosslinking and pore volume is not as pronounced in low water gels. Acid serves to tighten up the structure and form tighter clusters with smaller pores, but pore volume is relatively unaffected. The larger pores and equivalent pore volumes cause the surface areas of the low water, catalyzed gels to be lower than the uncatalyzed gels.

Thus, the role acid plays in both high water gels and low water gels is similar. It tightens the gel structure. The effect of acid is greater in high water gels, where more of a network structure is present.

In the high water regime, acid decreases pore size and pore volume, and results in lower surface areas. In the low water regime, acid decreases average pore size but has only a small effect on pore volume, so this results in higher surface areas.

The microstructure of dried alumina gels has an effect on the microstructure of fired gels, especially when bulk samples are considered. Continuous open porosity allows oxidation of residual organics, where closed porosity usually gives bloating. Bloating is the foaming of ceramics when gases become trapped.

The texture of gels is important for modeling the densification of gels. At the same time, the texture is important for the applications cited for microporous materials.

PREPARATION OF MICRO/MACROPOROUS SILICA SHEETS

Recent technological advances with microporous sol-gel derived silica point to using silica as an inorganic membrane. If silica is handled in such a way that its microstructure mimics that of cellulose acetate, then it becomes a high temperature filter.

In the sol-gel process involving acid-catalyzed tetraethylorthosilicate (TEOS)-ethanol-water solutions, the silica end product is largely determined by the water level in the solution. This means that solutions with low water levels can be spun into fibers,[26] with medium water levels can be used for coatings,[27] and high water levels are necessary for casting monolithic shapes.[28] The low water partially hydrolyzed solutions that produce fibers are also the best candidate solutions for free standing silica sheets.[29] As such, a low water solution was used to fabricate silica sheets so that their porosity and microstructure could be evaluated for a potential separation medium.

Experimental Techniques

All solutions were prepared with distilled TEOS, distilled deionized water and reagent grade ethanol. The solution compositions are listed in the table below, 1 mol TEOS: x mols water, where x is 1.5, 2.5 or 16: 0.5 mol ethanol. The ethanol and TEOS are mixed in a closed flask at room temperature. Nitric acid is added to the required water from a 1 molar solution. The acidified water is then added to the stirring solution. As hydrolysis begins, the solution becomes cloudy and the exothermic reaction heats the flask. Shortly after, the solution clears. Pressure built up in the flask is reduced by venting the flask. The flask is cool after about 3 hours.

Some of the ethanol is extracted by a so-called vacuum distillation. This permits an increase in the density of the solution without substantial change in the molecular structure. After cooling the reaction flask to about 5°C, a rush condenser leading to a cold trap and oil breather is attached. The condenser is cooled with circulating acetone-dry ice. The condenser is connected to a mechanical pump and a vacuum of about 10 mm Hg is drawn. The temperature of the flask is raised until the solution starts to boil, somewhere between 10°C and 20°C. After ethanol removal, the solutions are stored in airtight plastic bottles. They do not lose fluidity for several months.

Surface tension values for water, ethanol, TEOS and the three mixtures were measured using a Wilhelmy plate technique where a platinum plate is lifted from the surface of the liquid. Because of the large surface area exposed during measurements with a Wilhelmy plate, this technique could not be used with gelling solutions. The values reported in the table below are for solutions before vacuum distillation. From this survey, the mixture with the lowest value was used for all further sample preparation.

Sheets of gel were formed by pouring the solution onto a bath of s-tetrabromoethane (acetylene tetrabromide $CHBr_2CHBr_2$). It was chosen for low reactivity, high density (2.96 g/ml) and high surface tension 54 dynes/cm at 25°C. As the solution is poured, it spreads evenly over the bath. The solution has a density of about 1 g/ml, so it remains on the surface. A large area of solution becomes exposed to atmospheric moisture. First, a skin forms on the solution, and then moisture penetrates through this layer so that the sheet can gel. As soon as the layer has gelled, the sheet is peeled off of the bath or lifted with a wire ring. The sheet is very flexible at this point but dries quickly to a fragile sample resembling wax paper. To avoid curling, the sheets are suspended vertically in air to allow drying from both sides. Dried sheets are typically 0.1 mm thick. Dried sheets were examined with reflected light microscopy and scanning electron microscopy.

Results

In order for a solution to spread into a sheet, the solution must have a surface tension less than that of the substrate, and the interfacial tension between solution and substrate should be low. The measured surface tensions for the TEOS-water-ethanol solutions are listed in the following table. The values are all lower than that for tetrabromoethane. Nevertheless, the 2.5 and 16 mols water solutions tended to ball up. Only when the solution was in the partially hydrolyzed state with 1.5 mols water per mol TEOS was even spreading achieved.

Measured Surface Tension Using Wilhelmy Plate Technique

.....Solution at 20°C.....	Surface Tension, dynes/cm ...(10^{-3} Newtons/m)...
Ethanol	25.3
TEOS	25.4
1.5 mols water/1 mol TEOS	25.8
2.5 mols water/1 mol TEOS	26.7
16 mols water/1 mol TEOS	31.9
Water	62.1

The optical micrographs, Figures 4 and 5, show the top surface of the gel sheets. This is the surface exposed to air. Figure 4 shows the surface in a wet condition and Figure 5 shows the surface after some drying. During hydrolysis, ethanol is produced. It escapes to the top surface of the solution where the bubbles intersect the rapidly gelling skin (Figure 4). The bubbles leave a hemispherical depression. As the gel becomes rigid, the hemispherical pores are pulled into an equilibrium arrangement which approaches close packing (Figure 5).

Figure 6 shows the bottom surface of the gel sheet which became rigid in contact with the tetrabromoethane. Because gas was not free to escape through the bottom, there are a few pinholes, but no regular pattern of bubbles intersecting the surface. Tetrabromoethane has no solubility for water or ethanol.

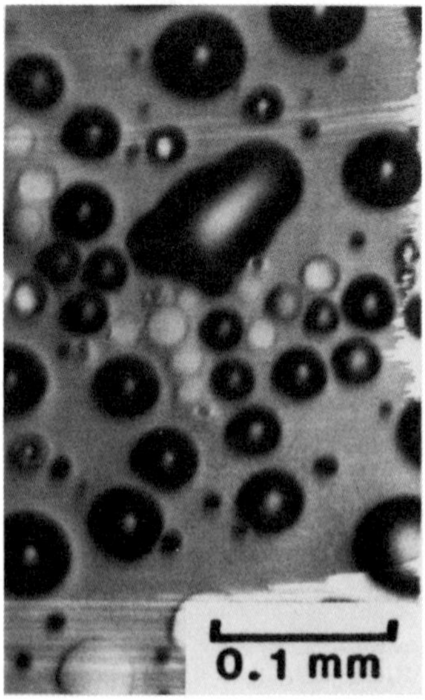

Figure 4: Top surface of wet sample immediately after removal from bath. Bubbles of various sizes intersect free surface. Solution contains 1.5 mols water per mol TEOS.

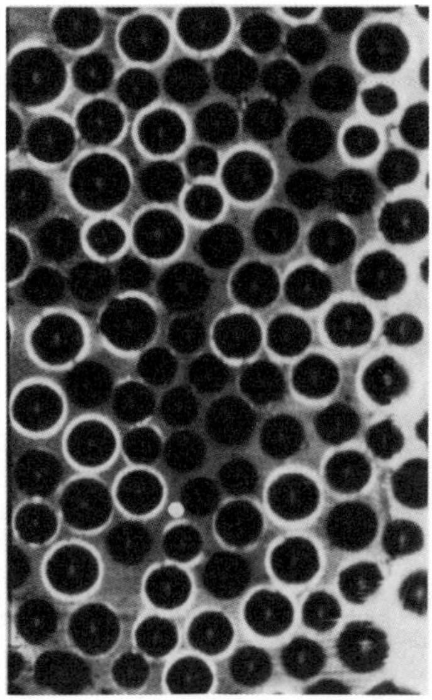

Figure 5: Top surface of sample after gel has been air dried. Bubbles of uniform size are drawn approximately into close packed arrangement. Same solution as Figure 4.

Figure 6: Bottom surface of sample after gel has been air dried. Pinholes indicate escape of gas at solution-bath interface. Same solution as Figure 4.

The large pores of the top surface are connected to the pinholes of the bottom surface by circuitous channels and the microporosity. The fine pores in the silica honeycomb between the large pores are shown in the scanning electron micrograph in Figure 7. Figure 8 shows a typical fracture surface. It shows that the bubble intersection is a diameter and the depression is a hemisphere. It also shows that smaller bubbles were trapped before they got to the exposed surface.

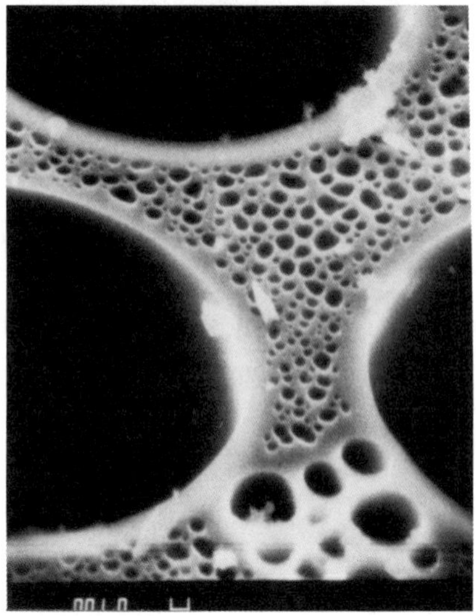

Figure 7: Scanning electron micrograph of top surface of air dried gel. Microporosity fills silica between bubbles. Same solution as Figure 4.

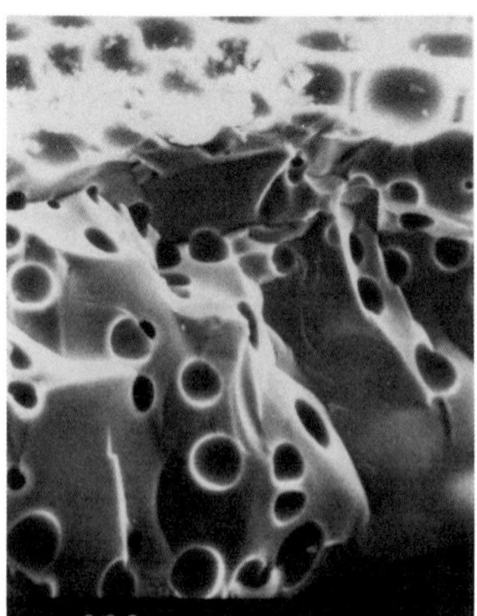

Figure 8: Scanning electron micrograph of fracture surface of air dried gel. Bubbles rising to the surface are trapped in the bulk. Same solution as Figure 4.

Discussion

There are similarities between the membrane microstructure shown in Figures 4 through 8 and the microstructure of organic membranes known as phase inversion membranes.[30] The organic membranes are solvent cast materials, and the porosity is a result of the immobilization of the gel before complete depletion of the solvent. The sequence of steps in the phase inversion process is loss of volatile solvent, gelation, contraction, capillary depletion and loss of residual solvent. In organic membranes the mechanism for separation into droplets of solvent phase in a continuous gel phase is the decreased power of the solvent as a result of evaporation. The mechanism in the silicate gels is not entirely clear, though it appears that the chemistry of the initial solution is more important than its handling during the process.

In acidified low water solutions the silicate polymers in the solution are generally assumed to be linear.[24] During further hydrolysis through the absorption of atmospheric moisture, there is no tendency for the solutions to ball up. Yet the linear polymers eventually have enough crosslinking within the layer to give cohesive sheets. The newly formed sheets are pliable before drying, but after drying the sheets are brittle and fragile, as expected for thin, glassy silicate.

The surface tension is an important factor for determining the ease of spreading. A low value for the solution and a low value for the interfacial tension allow for spreading. These conditions are only met in the low water solutions. Because the hydrolysis reaction is controlled by the permeation of moisture through the layer, in its intial stage the linear polymers are largely unhydrolyzed. The groups along the polymer chains are ethoxy groups rather than hydroxy groups, and the surface tension of the polymer is low like that of ethanol. The solutions which were intentionally hydrolyzed to a high degree with up to 16 mols of water per mol TEOS tended to ball up, indicating the conditions for spreading on an organic bath are not met.

The surface tension gradually changes with exposure to the atmosphere. The gel skin begins as a flexible layer and eventually thickens to a rigid sheet. During this maturing of the layer, the vapor pressure of ethanol establishes the size of the bubbles escaping, while the surface tension controls the arrangement of bubbles. The resulting lowest energy configuration approximates close packing of hemispherical depressions. It is this uniform configuration that brings to mind applications of these materials as filters and membranes and the likelihood of replacing organic filters in high temperature applications.

SUMMARY

Sol-gel processing is a natural fit with the requirements of filters and membranes. In this Chapter, examples have been given of ceramic membranes and a few of them involve sol-gel processing. The technique used at Rutgers is described in some detail, but it is only in a very early stage. Further work is warranted to exploit microporous and macroporous materials by the sol-gel process.

Acknowledgement

The financial support of the Center for Ceramics Research, a University-Industry-NSF Cooperative Program at Rutgers University is appreciated. In addition, a grant from Alcoa has supported the work on high surface area aluminas. Figures 1 through 3 were supplied by Tom Lombardi and Figures 4 through 8 were supplied by Dennis Gallagher. The hard work of these colleagues has contributed greatly to this Chapter. Technical assistance was provided by Omicron Technology Corporation. Claudia Kuchinow is thanked for her preparation of this manuscript and her help in coordinating this publication.

REFERENCES

1. Lonsdale, H.K., *J. Membrane Sci. 10,* 81-181 (1982).
2. Pusch, W. and Walch, A., *Angew. Chem. Int. Ed. 21,* 660-685 (1982).
3. Dullien, F.A.C., *Porous Media - Fluid Transport and Pore Structure,* Academic Press, New York (1979).
4. Lloyd, D.R. (ed.), *Materials Science of Synthetic Membranes,* Am. Chem. Soc., Washington, D.C. (1985).
5. Elmer, T.H., *J. Am. Ceram. Soc. 57,* 1051-1054 (1978).
6. Tanaka, H., et al., *J. Non-Crystal. Solids 65,* 301-309 (1984).
7. McMillan, P.W. and Maddison, R., U.S. Patent 4,473,476, Sept. 25, 1984.
8. Yamamoto, M., Sakata, J. and Doi, H., U.S. Patent 4,521,236, June 4, 1985.
9. Kingery, W.D., Bowen, H.K. and Uhlmann, D.R., *Introduction to Ceramics,* 2nd Ed., Chapter 10, Wiley-Interscience, New York (1976).
10. Day, M.A. and Reid, A., U.S. Patent 4,526,885, July 2, 1985.
11. Kaiser, A. and Schmidt, H., *J. Non-Crystal. Solids 63,* 261-271 (1984).
12. Sakka, S., Kamiya, K., Makita, K. and Yamamoto, Y., *J. Non-Crystal. Solids 63,* 223-235 (1984).
13. Kaiser, A., Schmidt, H. and Bottner, H., *J. Membrane Science 22,* 257-268 (1985).
14. Philipp, G. and Schmidt, H., *J. Non-Crystal. Solids 63,* 283-292 (1984).
15. Leenaars, A.F.M., Keizer, K. and Burggraaf, A.J., *Studies in Inorganic Chemistry,* Vol. 3, 401-404, (R. Metselaar, H.J.M. Heijligers and J. Schoonman, eds.), Elsevier, Amsterdam (1983).
16. Leenaars, A.F.M., Keizer, K. and Burggraaf, A.J., *J. Mater. Sci. 10,* 1077-1088 (1984).
17. Teichner, S.J., Nicholaon, G.A., Vicarini, M.A. and Gardes, G.E.E., *Adv. in Colloid. Interface Sci. 5,* 245-273.
18. Rubin, M. and Lampert, C.M., *Solar Energy Materials 7,* 393-400 (1983).
19. Caps, R. and Fricke, J., *Int. J. Solar Energy 3,* 13 (1984).
20. Dumas, J., et al., *J. Mat. Sci. Lett. 4,* 1089-1091 (1985).
21. Lecloux, A.J., in *Catalysis-Science and Technology,* Vol. 2, 172-230, (J.R. Anderson and M. Boudart, eds.), Springer, Berlin (1981).
22. Klein, L.C. and Garvey, G.J., in *Better Ceramics Through Chemistry,* 33-39, (C.J. Brinker, D.R. Ulrich and D.E. Clark, eds.), Elsevier, New York (1984).

23. Lombardi, T., High surface area sol-gel derived alumina, Masters thesis, Rutgers University, October, 1986.
24. Brinker, C.J., et al., *J. Non-Crystal. Solids 63,* 45-49 (1984).
25. Klein, L.C., Design of microstructures in sol-gel processed silicates, in *Design of New Materials,* (A. Clearfield and D.L. Cocke, eds.), Plenum Press, New York (1986), to appear.
26. Sakka, S. and Kamiya, K., *J. Non-Crystal. Solids 48,* 31-46 (1982).
27. Dislich, H. and Hinz, P., *J. Non-Crystal. Solids 48,* 11-16 (1982).
28. Klein, L.C., Gallo, T.A. and Garvey, G.J., *J. Non-Crystal. Solids 63,* 23-33 (1984).
29. Gallagher, D. and Klein, L.C., *J. Colloid. Interface Sci. 109,* 40-45 (1986).
30. Kesting, R.E., *Synthetic Polymer Membranes,* 116, McGraw-Hill, New York (1971).

Index